CONTROL AND
DYNAMIC SYSTEMS

Advances in Theory
and Applications

Volume 50

CONTRIBUTORS TO THIS VOLUME

BRIAN D. O. ANDERSON
ROBERT R. BITMEAD
J. S. CHEN
Y. H. CHEN
B. R. COPELAND
JAMES B. FARISON
A. M. HOLOHAN
TOSHIHIRO HONMA
ALTUĞ İFTAR
MASAKAZU IKEDA
MASAHIKO KIHARA
SRI R. KOLLA
C. T. LEONDES
TSUYOSHI OKADA
ÜMIT ÖZGÜNER
NADER SADEGH
M. G. SAFONOV
DAUCHUNG WANG
WEI-YONG YAN
RAFAEL T. YANUSHEVSKY

CONTROL AND DYNAMIC SYSTEMS

ADVANCES IN THEORY AND APPLICATIONS

Volume Editor

C. T. LEONDES

School of Engineering and Applied Science
University of California, Los Angeles
Los Angeles, California

VOLUME 50: ROBUST CONTROL SYSTEM TECHNIQUES AND APPLICATIONS
Part 1 of 2

ACADEMIC PRESS, INC.
Harcourt Brace Jovanovich, Publishers
San Diego New York Boston
London Sydney Tokyo Toronto

ACADEMIC PRESS RAPID MANUSCRIPT REPRODUCTION

This book is printed on acid-free paper. ♾

Academic Press, Inc.
1250 Sixth Avenue, San Diego, California 92101-4311

United Kingdom Edition published by
Academic Press Limited
24–28 Oval Road, London NW1 7DX

Library of Congress Catalog Number: 64-8027

International Standard Book Number: 0-12-012750-4

PRINTED IN THE UNITED STATES OF AMERICA
92 93 94 95 96 97 BB 9 8 7 6 5 4 3 2 1

CONTENTS

ROBUST CONTROL SYSTEM TECHNIQUES AND APPLICATIONS

EXTENDED CONTENTS

Volume 51

CONTRIBUTORS

Numbers in parentheses indicate the pages on which the authors' contributions begin.

Brian D. O. Anderson (1), *Department of Systems Engineering, Research School of Physical Sciences and Engineering, The Australian National University, Canberra, ACT 2601, Australia*

Robert R. Bitmead (1), *Department of Systems Engineering, Research School of Physical Sciences and Engineering, The Australian National University, Canberra, ACT 2601, Australia*

J. S. Chen (175), *The George W. Woodruff School of Mechanical Engineering, Georgia Institute of Technology, Atlanta, Georgia 30332*

Y. H. Chen (175), *The George W. Woodruff School of Mechanical Engineering, Georgia Institute of Technology, Atlanta, Georgia 30332*

B. R. Copeland (331), *Department of Electrical and Computer Engineering, Real Time Systems Group, University of the West Indies, Trinidad and Tobago, West Indies*

James B. Farison (395), *Department of Electrical Engineering, The University of Toledo, Toledo, Ohio 43606*

A. M. Holohan (297), *Department of Electrical Engineering-Systems, University of Southern California, Los Angeles, California 90089*

Toshihiro Honma (79), *Department of Aerospace Engineering, National Defense Academy, Hashirimizu, Yokosuka 239, Japan*

Altuğ İftar (255), *Department of Electrical Engineering, University of Toronto, Toronto M5S 1A4, Canada*

Masakazu Ikeda (79), *Department of Aerospace Engineering, National Defense Academy, Hashirimizu, Yokosuka 239, Japan*

Masahiko Kihara (79), *Department of Aerospace Engineering, National Defense Academy, Hashirimizu, Yokosuka 239, Japan*

Sri R. Kolla (395), *Department of General Engineering, Pennsylvania State University, Sharon, Pennsylvania 16146*

C. T. Leondes (119), *School of Engineering and Applied Science, University of California, Los Angeles, Los Angeles, California 90024*

Tsuyoshi Okada (79), *Department of Aerospace Engineering, National Defense Academy, Hashirimizu, Yokosuka 239, Japan*

Ümit Özgüner (255), *Department of Electrical Engineering, The Ohio State University, Columbus, Ohio 43210*

Nader Sadegh (223), *The George W. Woodruff School of Mechanical Engineering, Georgia Institute of Technology, Atlanta, Georgia 30332*

M. G. Safonov (297, 331), *Department of Electrical Engineering-Systems, University of Southern California, Los Angeles, California 90089*

Dauchung Wang (119), *Yuan-Ze Institute of Technology, Neih LI, Chun Li, Taoyuan Shian, Taiwan, People's Republic of China*

Wei-Yong Yan (1), *Department of Systems Engineering, Research School of Physical Sciences and Engineering, The Australian National University, Canberra, ACT 2601, Australia*

Rafael T. Yanushevsky (55), *Department of Mechanical Engineering, University of Maryland, Baltimore, Maryland 21228*

PREFACE

In the early days of modern control theory, the techniques developed were relatively simple but, nevertheless, quite effective for the relatively simple systems applications of those times. Basically, the techniques were frequency domain analysis and synthesis techniques. Then, toward the latter part of the 1950s, system state space techniques began to emerge. In parallel with these developments, computer technology was evolving.These two parallel developments (i.e., increasingly effective system analysis and synthesis techniques and increasingly powerful computer technology) have resulted in a requisite powerful capability to deal with the increasingly complex systems of today's world.

In these modern day systems of various levels of complexity, the need to deal with a wider variety of situations, including significant parameter variations, modeling large scale systems with models of lower dimension, fault tolerance, and a rather wide variety of other problems, has resulted in a need for increasingly powerful techniques, that is, system robustness techniques, for dealing with these issues. As a result, this is a particularly appropriate time to treat the issue of robust system techniques in this international series. Thus, this volume is the first volume of a two-volume sequence devoted to the most timely theme of "Robust Control Systems Techniques."

The first contribution to this volume is "Trade-offs Among Conflicting Objectives in Robust Control Design," by Brian D. O. Anderson, Wei-Yong Yan, and Robert Bitmead. This contribution presents important techniques for dealing with conflicting design objectives in control systems.

The next contribution is "Aspects of Robust Control Systems Design," by Rafael T. Yanushevsky. It provides an in-depth treatment of robustness techniques of systems described by differential-difference equations, and it also presents methods for the design of a wide class of robust nonlinear systems.

The next contribution is "System Observer Techniques in Robust Control Systems Design Synthesis," by Tsuyoski Okada, Masahiko Kihara, Masakazu Ikeda, and Toshihiro Honma. This contribution discusses techniques for dealing with the problems resulting from the use of observers in robust systems design, and it offers three distinct design techniques for treating these problems.

The next contribution is "Robust Tracking Control of Non-Linear Systems with Uncertain Dynamics," by Dauchang Wang and Cornelius T. Leondes. It presents rather effective techniques for the robust control on non-linear time varying of tracking control systems with uncertainties.

The next contribution is "Adaptive Robust Control of Uncertain Systems," by Y. H. Chen and J. S. Chen. This article sets forth techniques for incorporating adaptive control techniques into a (non-adaptive) robust control design.

The next contribution is "Robustness Techniques in Nonlinear Systems with Applications to Manipulator Adaptor Control," by Nader Sadegh. It presents techniques for achieving exponential and robust stability for a rather general class of nonlinear systems.

The next contribution is "Techniques in Modeling Uncertain Dynamics for Robust Control Systems Design," by Altuğ İftar and Ümit Özgüner. This contribution discusses a unified framework for robust control systems design for systems with both parameter uncertainties and uncertain dynamics due to the difficulties in modeling complex systems.

The next contribution is "Neoclassical Control Theory: A Functional Analysis Approach to Optimal Frequency Domain Controller Synthesis," by A. M. Holohan and M. G. Safonov. It offers techniques for the optimal synthesis (optimal in the sense defined in this contribution) of robust control systems.

The next contribution is "A Generalized Eigenproblem Solution for Singular H^2 and H^∞ Problems," by B. R. Copeland and M. G. Safonov. It presents techniques for the design of H^2 and H^∞ which apply equally well to both singular and nonsingular system cases.

The final contribution to this first volume of this two-volume sequence on the theme of "Robust Control System Techniques" is "Techniques in Stability Robustness Bounds for Linear Discrete-Time Systems," by James B. Farison and Sri R. Kollaq. This contribution provides a unified treatment of stability robustness design for discrete-time systems.

This volume is a particularly appropriate one as the first of a companion set of two volumes on robust control system analysis and synthesis techniques. The authors are all to be commended for their superb contributions, which will provide a significant reference source for workers on the international scene for years to come.

Trade-offs among Conflicting Objectives in Robust Control Design

Brian D. O. Anderson
Wei-Yong Yan
Robert R. Bitmead

Department of Systems Engineering
Research School of Physical Sciences and Engineering
The Australian National University
Canberra, ACT 2601, Australia

I. Introduction

Consider the feedback control system depicted in Figure 1.

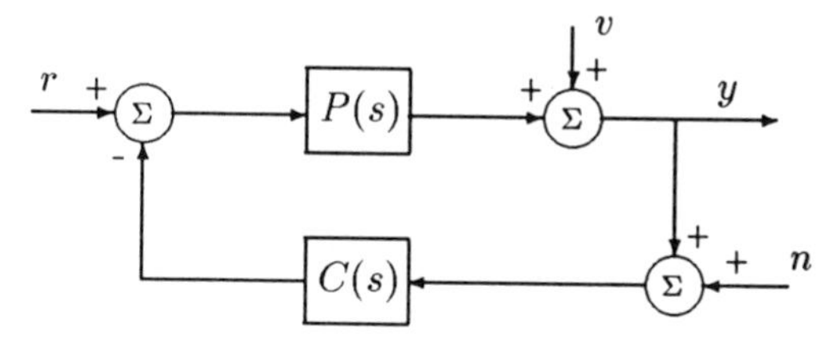

Figure 1. Feedback control system.

Here P is the plant system concerned, C is the controller and the signal r, y, u, v and n are the reference, system output, control input, output disturbance and sensor noise respectively. Classical control systems design emphasizes the securing of a number of conflicting objectives for this feedback system such as;

- securing closed loop stability from each external input to the loop signals u and y. (We refer to a controller $C(s)$ achieving this as [internally] *stabilizing* the plant $P(s)$.)

- securing the rejection of disturbance signals v and n from y. This is reflected by the properties of the closed loop sensitivity and complementary sensitivity functions.

- securing robustness to plant modeling errors and plant variations. Depending upon the nature of plant uncertainty, this objective may appear in terms of gain margins, phase margins or sensitivity functions.

A classical controller synthesis aims to achieve all these ends but usually proceeds by focusing on a single objective and then de-tunes or compromises to effect the trade-off between these sometimes conflicting desires. Alternatively, one may optimize with respect to one design issue and then select, from the class of available solutions, that controller which best meets one of the other objectives [1]. Of critical importance in such approaches is to know to what extent each of these individual design criteria are conflicting in their demands of the controller. Further, it is advantageous to know whether successive optimization of objective functions is feasible and, if so, in what order. Lastly, one may pose the question of the ability to ameliorate these conflicts through the choice of a time-varying or nonlinear controller.

In this chapter, we study the trade-offs between a number of these design objectives firstly for the class of linear, time invariant controller. We use a tool introduced in [2] for this purpose. This tool deals with the construction of functions of a complex variable which fulfill certain interpolation and analyticity conditions arising from stability requirements and design objectives. The specific problems treated are:

- the combined sensitivity-gain margin problem

- the combined sensitivity-phase margin problem

- the combined sensitivity-complementary sensitivity problem

Having made apparent some of the compromises available to the designer with time invariant linear controllers, we then move on to study the benefits achievable in reducing some of this conflict by using a periodically time varying controller. Our analysis treats the single-input/single-output case but, where extension to the multi-input/multi-output case is direct, this is noted.

II. The Optimal Sensitivity and Gain Margin Problem as Separate Problems

The material of this section is largely drawn from [2], and serves as a tutorial introduction to the main ideas of the chapter. Let $P(s)$ be a scalar linear time-invariant plant with poles $p_1, \ldots, p_n \in \text{Re}[s] \geq 0$ and zeros $z_1, \ldots, z_n$ (including possibly infinity) $\in \text{Re}[s] \geq 0$. Consider a stable closed-loop as depicted in Figure 1.

A. Optimal Sensitivity and Gain Margin

The sensitivity function $S(s)$ is defined by

$$S(s) = [1 + P(s)C(s)]^{-1} \tag{1}$$

and the *sensitivity* is defined by

$$R[C(s)] \triangleq ||S(s)||_\infty = \sup_{s \in \text{Re}[s] \geq 0} |S(s)| = \sup_\omega |S(j\omega)| \tag{2}$$

The last equality in Eq. (2) follows from the maximum modulus principle; closed-loop stability ensures that $S(s)$ is analytic in $\text{Re}[s] \geq 0$.

The sensitivity $R[C(s)]$ of course depends on $C(s)$. Its minimization through choice of $C(s)$ serves to secure a design which minimizes the maximum (over ω) of the gain from a disturbance entering at the plant output to the actual output.

A natural question is: what is

$$r_{\text{min}} = \inf_{C(s)} \{R[C(s)] : \quad C(s) \text{ stabilizes } P(s)\} \tag{3}$$

(and what is the associated $C(s)$; and how may it be found)?

With a fixed controller $C(s)$, the upper and lower gain margins b_{max} and a_{min} are defined by

$$b_{\text{max}} = \sup\{b : \quad C(s) \text{ stabilizes } kP(s)\ \forall k \in [1, b]\} \tag{4}$$

$$a_{\text{min}} = \inf\{a : \quad C(s) \text{ stabilizes } kP(s)\ \forall k \in [a, 1]\} \tag{5}$$

Of course, it is possible to have $b_{\text{max}} = \infty$ or $a_{\text{min}} = 0$ (or both), but not if the sets $\{z_i\}$ or $\{p_j\}$ are nonempty. We shall define the *gain margin* as

$$K[C(s)] \triangleq \sup\{b/a : \quad 0 < a < 1 < b \quad \text{and} \quad C(s) \text{ stabilizes } kP(s)\ \forall k \in [a, b]\} \tag{6a}$$

$$= \frac{b_{\max}}{a_{\min}} \tag{6b}$$

Evidently, K can be infinite for certain plants. Note that the gain margin K is the same for $P(s)$ and $\alpha P(s)$, for any $\alpha > 0$, so long as $C(s)$ stabilizes $\alpha P(s)$.

A natural question is: what is

$$k_{\max} = \sup\{K[C(s)] : \quad C(s) \text{ stabilizes } P(s)\}? \tag{7}$$

We shall now review how these questions can be answered. There are two key relevant ideas, one tied to interpolation properties of $S(s)$ and the other tied to mapping properties. The overall thrust is to work with $S(s)$ rather than $C(s)$; once $S(s)$ is known, $C(s)$ of course follows easily.

B. Interpolation Properties of $S(s)$

Recall that p_i is a pole in $\text{Re}[s] \geq 0$ of $P(s)$. Because $C(s)$ is stabilizing, $C(p_i) = 0$ is impossible. Hence $S(p_i) = 0$. Recall also that z_j is a zero in $\text{Re}[s] \geq 0$ of $P(s)$. Again because $C(s)$ is stabilizing, z_j cannot be a pole of $C(s)$. Hence $S(z_j) = 1$. Thus we have

$$S(p_i) = 0 \quad \forall p_i \in \text{Re}(s) \geq 0,\ p_i \text{ a pole of } P(s) \tag{8}$$

$$S(z_j) = 1 \quad \forall z_j \in \text{Re}(s) \geq 0,\ z_j \text{ a zero of } P(s) \tag{9}$$

For convenience, we shall assume poles and zeros in $\text{Re}(s) \geq 0$ of $P(s)$ are simple. The theory can be extended to cope with multiple poles and zeros, but is more complex.

C. Mapping Properties of $S(s)$

Suppose $C(s)$ achieves a sensitivity of r. Let $\bar{H}$ denote $\text{Re}[s] \geq 0$. Then, clearly

$$S(s) : \quad \bar{H} \to G_1 \triangleq \{s \in \boldsymbol{C} : \ |s| < r\} \tag{10}$$

Also, suppose $C(s)$ achieves a gain margin pair of a, b. Then for all $k \in [a, b]$, we have $\forall s \in \bar{H}$,

$$1 + kP(s)C(s) \neq 0$$

or

$$P(s)C(s) \neq -1/k, \qquad \forall s \in \bar{H}$$

or

$$S(s) = [1 + P(s)C(s)]^{-1} \neq \frac{k}{k-1}, \quad \forall s \in \bar{H}$$

or

$$S(s):\ \bar{H} \to G_2 \triangleq \mathbf{C} \setminus \left\{ \left(-\infty,\ -\frac{a}{1-a} \right] \cup \left[\frac{b}{b-1},\ \infty \right) \right\} \tag{11}$$

Evidently, if a sensitivity of r is achieved, $S(s)$ satisfies the interpolation conditions (8)-(9), and the mapping condition (10), while if a gain margin pair a, b is achieved, it again satisfies the interpolation conditions (8)-(9), but now the mapping condition (11).

Importantly, and conversely, if we can find an $S(s)$ satisfying the interpolation conditions and a mapping condition, we can then construct $C(s)$ from $S(s)$ to achieve a controller of the desired properties, i.e. one which yields a sensitivity of r or a gain margin pair of a, b. The quantities $r_{\min}$ and $k_{\max}$ are characterized by finding the infimum of the r and supremum of the ratio b/a such that $S(s)$ exists.

To examine the question of simultaneous satisfaction of mapping and interpolation conditions, we shall first look at a special case.

D. Nevanlinna-Pick Theory

The Nevanlinna-Pick theory is concerned with the existence and construction of a function $F(z)$ mapping the closed unit disk $\bar{D} = \{|z| \leq 1\}$ into the open unit disk $D = \{|z| < 1\}$. Let $\beta_1, \ldots, \beta_p$ satisfy $|\beta_i| \leq 1$ and let $\gamma_1, \ldots, \gamma_p$ satisfy $|\gamma_i| < 1$. If $\beta_i = \beta_j^*$, then $\gamma_i = \gamma_j^*$. Ask the question: does there exist

$$F:\quad \bar{D} \to D \tag{12}$$

such that

$$F(\beta_i) = \gamma_i, \qquad i = 1, \ldots, p? \tag{13}$$

Suppose first of all that all β_i are in D, so that $|\beta_i| < 1$. Then the simple answer is that F exists if and only if the following $p \times p$ matrix is positive definite:

$$\Gamma = (\Gamma_{ij})_{p \times p}, \quad \Gamma_{ij} = \frac{1 - \gamma_i \bar{\gamma}_j}{1 - \beta_i \bar{\beta}_j} \tag{14}$$

A variant on the Nevanlinna-Pick problem is to seek

$$\alpha_{\max} \triangleq \sup\{\gamma > 0 :\ \exists\, F : \bar{D} \to D \text{ for which } F(\beta_i) = \gamma\gamma_i\}$$

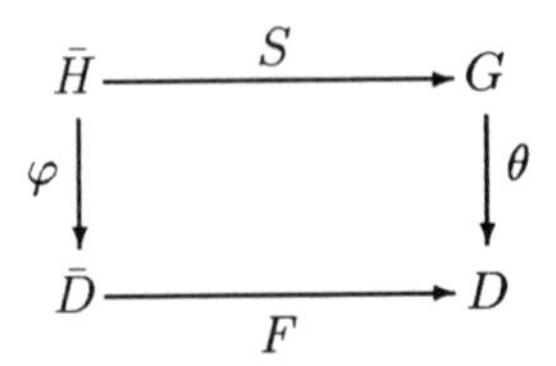

Figure 2. Commutative diagram of maps.

[One is trying to scale F upward, to make it as large as possible without violating the $F(\bar{D}) \subset D$ requirement]. It is easy to see that with

$$A = \left(\frac{1}{1 - \beta_i \bar{\beta}_j} \right) \quad \text{and} \quad B = \left(\frac{\gamma_i \bar{\gamma}_j}{1 - \beta_i \bar{\beta}_j} \right) \tag{15}$$

there holds

$$\alpha_{\max} = \sup\{\gamma > 0 : \quad A - \gamma^2 B \geq 0\} \tag{16}$$

Next, suppose that β_i, $i = 1, \ldots, p$ obey $|\beta_i| < 1$, while β_i, $i = p+1, \ldots, p+q$ obey $|\beta_i| = 1$. We are required to find F so that $F(\beta_i) = \gamma_i$, $i = 1, \ldots, p+q$ and also $F : \bar{D} \to D$. Then it turns out that F exists if the same $p \times p$ matrix Γ as before is positive definite and $|\gamma_i| < 1$ for $i = p+1, \ldots, p+q$. For the scaled problem, let

$$\bar{\gamma} = \sup\{\gamma > 0 : \quad A - \gamma^2 B \geq 0\} \tag{17}$$

Then

$$\alpha_{\max} = \min \left\{ \bar{\gamma}, \frac{1}{|\gamma_{p+1}|}, \ldots, \frac{1}{|\gamma_{p+q}|} \right\} \tag{18}$$

There exist algorithms for constructing F from the data, assuming it exists.

The Nevanlinna-Pick theory can be applied to our problem of interest by introducing additional conformal mappings that convert the region $\bar{H}$ and G_1 or G_2 to $\bar{D}$ and D.

E. Transformation of the Interpolation-Mapping Problem

Consider Figure 2, with G corresponding to either G_1 or G_2. Let φ be the map

$$\varphi(s) = \frac{s-1}{s+1} \tag{19}$$

It is clearly invertible, and maps $\bar{H}$ onto $\bar{D}$ in a one to one fashion. Further, let $\theta_i(s)$ be a conformal map from G_i to D. The existence of $\theta_i(s)$ is in general a consequence of the Riemann Mapping Theorem [3], and there exist dictionaries, e.g. [4], where such mappings can be found. For the regions G_1, G_2 we have

$$\theta_1(s) = s/r \tag{20}$$

$$\theta_2(s) = \frac{1 - \left(\frac{1-s/b'}{1+s/a'}\right)^{1/2}}{1 + \left(\frac{1-s/b'}{1+s/a'}\right)^{1/2}} \tag{21}$$

where

$$a' = \frac{a}{1-a} \qquad b' = \frac{b}{b-1} \tag{22}$$

Notice that θ_1 and θ_2 are both invertible maps, so that

$$S = \theta_i^{-1} \circ F \circ \varphi \tag{23}$$

Now because S satisfies interpolation conditions, so must F. In particular, because $S(p_i) = 0$ and $\theta_i(0) = 0$, it follows that

$$F(\varphi(p_i)) = 0$$

or

$$F\left(\frac{p_i - 1}{p_i + 1}\right) = 0 \quad i = 1, \ldots, n \tag{24}$$

Further

$$F\left(\frac{z_j - 1}{z_j + 1}\right) = F(\varphi(z_j)) = \theta_i(S(z_j)) = \theta_i(1) \quad j = 1, \ldots, m \tag{25}$$

Recall again that the maps θ_j are known even when F or S is not known. Thus interpolation conditions on F are known, even when neither F nor S is known.

Now the procedure for checking the existence of, and constructing S, is clear.

- Check that there exists $F : \bar{D} \to D$ satisfing the interpolation conditions (24)-(25).
- If so construct F, and then S via Eq. (23) and finally C. If not, the problem as posed is not solvable.

F. The Optimization Problems

We can now understand how the minimal sensitivity and maximal gain margin can be obtained. Let

$$\begin{aligned} a_i &= \varphi(p_i) \quad i = 1, \ldots, n \\ &= \varphi(z_{i-n}) \quad i = n+1, \ldots, n+m \end{aligned} \tag{26}$$

$$\alpha = \sup\{\gamma : \exists\, F : \bar{D} \to D \text{ with } F(a_i) = \begin{matrix} 0, & i = 1, \ldots, n \\ \gamma, & i = n+1, \ldots, n+m \end{matrix} \} \tag{27}$$

It is easy to characterize α: it is $\sup \gamma$ such that

$$\begin{bmatrix} \left(\frac{1}{1-a_i\bar{a}_j}\right)_{i,j=1,\ldots,n} & \left(\frac{1}{1-a_i\bar{a}_j}\right)_{\substack{i=1,\ldots,n \\ j=n+1,\ldots,n+m}} \\ \left(\frac{1}{1-a_i\bar{a}_j}\right)_{\substack{i=n+1,\ldots,n+m \\ j=1,\ldots,n}} & \left(\frac{1-\gamma^2}{1-a_i\bar{a}_j}\right)_{i,j=n+1,\ldots,n+m} \end{bmatrix} \geq 0 \tag{28}$$

Evidently, the combined interpolation-mapping problem is solvable if

$$|\theta_i(1)| < \alpha$$

For the case of the sensitivity problem, this says, using Eq. (20), that

$$\frac{1}{r} < \alpha \quad \text{or} \quad r > \alpha^{-1}$$

Evidently then,

$$r_{\min} = \alpha^{-1}$$

For the case of the gain margin problem, Eq. (21) yields

$$\left| \frac{1 - \left(\frac{1-1/b'}{1+1/a'}\right)^{1/2}}{1 + \left(\frac{1-1/b'}{1+1/a'}\right)^{1/2}} \right| < \alpha$$

or

$$\sqrt{\frac{b}{a}} < \frac{1+\alpha}{1-\alpha}$$

So evidently,

$$k_{\max} < \left(\frac{1+\alpha}{1-\alpha}\right)^2$$

Several points should be noted.

- α is determined solely by the unstable poles and zeros, p_i and z_j, of $P(s)$; nothing else matters.
- Both the least sensitivity and maximal gain margin depend (in a simple way) on α.
- In case there are no right half plane zeros, so $m = 0$, the interpolation problem can be solved for any r and k i.e. $r_{\min} = 1$ and $k_{\max} = \infty$.
- If there are no right half plane poles, so $n = 0$, then it is easily seen that $\alpha = 1$, and again $r_{\min} = 1$ and $k_{\max} = \infty$.

So the individual problems are only of interest if both unstable poles and zeros are present. This is, in a sense, unnatural. Too many issues are being swept under the rug for this conclusion to be of practical value.

G. Review

Let us recall the key points we have made:

- the presence of unstable poles and zeros forces interpolation constraints on the sensitivity function $S(s)$.
- if a certain sensitivity or gain margin is to be achieved, $S(s)$ maps $\mathrm{Re}[s] \geq 0$ into a certain region.
- the search for a controller achieving a specified sensitivity or gain margin is equivalent to a search for a sensitivity function $S(s)$ achieving certain interpolation conditions and having a mapping property.
- the search for an $S(s)$ can be converted to a standard interpolation-mapping problem, the Nevanlinna-Pick problem (see Figure 2) by introducing a bilinear transformation φ and a conformal map θ obeying $\theta(0) = 0$.

In the next section, we shall look at the *combined* sensitivity-gain margin problem.

III. The Combined Sensitivity-Gain Margin Problem

In this section, we are interested in the question of deciding whether simultaneously imposed constraints on both sensitivity and gain margin are achievable, and in studying the trade-off between the goals of low sensitivity and high gain margin. As we shall see, there is a conflict in the objectives.

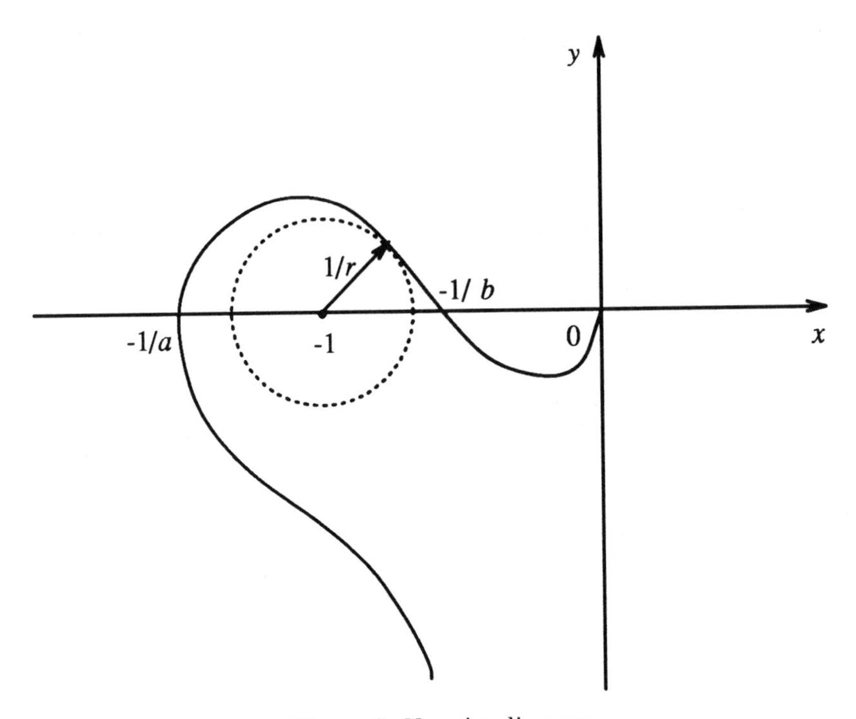

Figure 3. Nyquist diagram.

Consider Figure 3, which represents the Nyquist diagram of $P(j\omega)C(j\omega)$. The diagram depends on $C(j\omega)$, while $P(j\omega)$ is fixed. If C stabilizes P, then C stabilizes kP for all $k \in (a, b)$, i.e. the gain margin with C is b/a. The sensitivity is defined on the reciprocal of the radius of the greatest radius circle of center $-1 + j0$ which can be drawn so that it touches but does not cross the Nyquist plot.

The trade-off is one between pushing the points $-a^{-1} + j0$, $-b^{-1} + j0$ away from $-1 + j0$ (to secure better gain margin) and pushing the circle radius r^{-1} out (to secure better sensitivity), under the constraint that the curve plots $P(j\omega)C(j\omega)$, with P fixed and C adjustable. It is easy to imagine that as the points $-a^{-1} + j0$ and $-b^{-1} + j0$ move further away from $-1 + j0$ [when $C(j\omega)$ is adjusted] the radius r^{-1} gets smaller.

A loose connection between gain margin and sensitivity with $C(s)$ is easily obtained. The diagram shows clearly that

$$r^{-1} < 1 - b^{-1}, \quad a^{-1} - 1 \tag{29}$$

which can be rewritten, with Eq. (22), as

$$r > a', \quad r > b' \tag{30}$$

These inequalities say very little about the sensitivity, gain margin tradeoff, and treat only one possible topological orientation of the Nyquist diagram and $-1+j0$ point.

In the light of the material of Section II, it is evident that we can achieve a given sensitivity r and gain margin pair $(a,\ b)$ if and only if we can find $S(s)$ satisfying the interpolation conditions.

$$\begin{aligned} S(p_i) &= 0 \quad \forall p_i \in \mathrm{Re}[s] \geq 0,\ p_i \text{ a pole of } P(s) && (31)\\ S(z_i) &= 1 \quad \forall z_i \in \mathrm{Re}[s] \geq 0,\ z_i \text{ a zero of } P(s) && (32)\end{aligned}$$

and

$$\begin{aligned} S(s):\ \bar{H} &\rightarrow G_1 \cap G_2 && (33)\\ &= \{s \in \boldsymbol{C};\ |s| < r\} \cap \{\boldsymbol{C} \setminus [(-\infty,\ -a') \cup (b',\ \infty)]\} && (34)\end{aligned}$$

Our ultimate goal in this section is to understand how the existence of $S(s)$ depends on r, a and b or r and b/a. To this end, we first need to understand what the conditions are on a particular r, a, b triple for $S(s)$ to exist. In turn this requires us to introduce a conformal mapping $\theta(s)$ from $G_1 \cap G_2$ to D.

A. Conformal Mapping from $G_1 \cap G_2$ to D

The region $G \triangleq G_1 \cap G_2$ can have one of four characters, depending on the relative magnitudes of a', b' and r. These four possibilities are illustrated in Figure 4.

The conformal mapping $\theta(s)$ differs for each of the four cases and is given by

$$\theta(s) = \begin{cases} \dfrac{[b'(a'+s)(r^2+a's)]^{1/2} - [a'(b'-s)(r^2-b's)]^{1/2}}{[b'(a'+s)(r^2+a's)]^{1/2} + [a'(b'-s)(r^2-b's)]^{1/2}}, & \text{if } r > \max(a', b') \\ \dfrac{[(a'+s)(r^2+a's)]^{1/2} - [a'(r-s)^2]^{1/2}}{[(a'+s)(r^2+a's)]^{1/2} + [a'(r-s)^2]^{1/2}}, & \text{if } a' < r \leq b' \\ \dfrac{[b'(r+s)^2]^{1/2} - [(b'-s)(r^2-b's)]^{1/2}}{[b'(r+s)^2]^{1/2} + [(b'-s)(r^2-b's)]^{1/2}}, & \text{if } b' < r \leq a' \\ s/r, & \text{if } r \leq \min(a', b') \end{cases} \quad (35)$$

with

$$\theta(1) = \begin{cases} \dfrac{[b(r^2+a')]^{1/2} - [a(r^2-b')]^{1/2}}{[b(r^2+a')]^{1/2} + [a(r^2-b')]^{1/2}}, & \text{if } r > \max(a', b') \\ \dfrac{(r^2+a')^{1/2} - [a(r-1)^2]^{1/2}}{(r^2+a')^{1/2} + [a(r-1)^2]^{1/2}}, & \text{if } a' < r \leq b' \\ \dfrac{[b(r+1)]^{1/2} - (r^2-b')^{1/2}}{[b(r+1)]^{1/2} + (r^2-b')^{1/2}}, & \text{if } b' < r \leq a' \\ 1/r, & \text{if } r \leq \min(a', b') \end{cases} \quad (36)$$

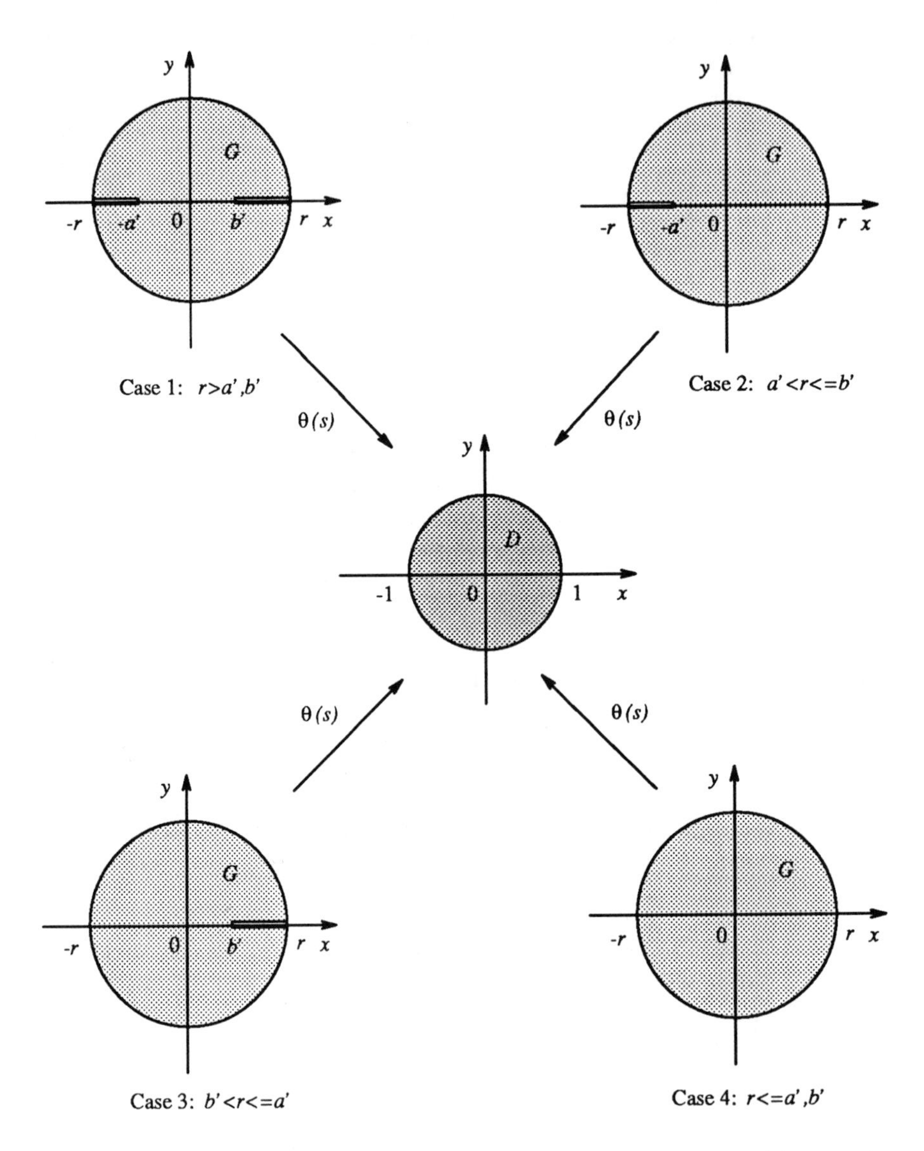

Figure 4: Conformal mapping associated with combined sensitivity-gain margin problem.

The constructions are available in dictionaries, such as [4]. They are also fairly easy to derive from first principles. For example, in the case $r > \max(a',\ b')$, θ can be viewed as the composition of four elementary maps

$$\theta = \theta_4 \circ \theta_3 \circ \theta_2 \circ \theta_1,$$

with

$$\theta_1(s) = \left(\frac{r+s}{r-s}\right)^2 :$$

$$G \rightarrow \boldsymbol{C} \setminus \left\{\left(-\infty,\ \left(\frac{r-a'}{r+a'}\right)^2\right] \cup \left[\left(\frac{r+b'}{r-b'}\right)^2,\ \infty\right)\right\} \tag{37}$$

$$\theta_2(s) = \frac{(\frac{r-a'}{r+a'})^2 - s}{s - (\frac{r+b'}{r-b'})^2} :$$

$$\boldsymbol{C} \setminus \left\{\left(-\infty,\ \left(\frac{r-a'}{r+a'}\right)^2\right] \cup \left[\left(\frac{r+b'}{r-b'}\right)^2,\ \infty\right)\right\} \rightarrow \boldsymbol{C} \setminus (-\infty,\ 0] \tag{38}$$

$$\theta_3(s) = \frac{1-\sqrt{s}}{1+\sqrt{s}} : \quad \boldsymbol{C} \setminus (-\infty,\ 0] \rightarrow D \tag{39}$$

$$\theta_4(s) = \frac{s - s_o}{s_o s - 1} : \quad D \rightarrow D \tag{40}$$

where

$$s_0 = \frac{(r+a')\sqrt{b'} - (r-b')\sqrt{a'}}{(r+a')\sqrt{b'} + (r-b')\sqrt{a'}}$$

[The function of $\theta_4(s)$ is solely to ensure that $\theta(0) = 0$.]

Notice that

$$\overline{\theta(\bar{s})} = \theta(s) \tag{41}$$

This ensures that the complex-conjugate paired interpolation constraints on $S(s)$ map into complex-conjugate paired constraints for $F(s)$, that $F(s)$ (if it exists) can be taken to be real rational in s, and then that $S(s)$ (though not rational), when computed from $F(s)$ has a real inverse Laplace transform. Such a property is essential, since at the end of the day we want a physically realizable controller, i.e. one with a real impulse response.

We can now apply the theory presented in the last section; the sole difference is that the particular $\theta(1)$ which we have differs from that used before. The quantity α is obtained as before, *viz.* $\alpha = \sup \gamma$ such that Eq. (28) holds. The combined problem has a solution if and only if any of the following mutually exclusive condition holds:

$$r > \max(a',\ b') \quad \text{and} \quad \frac{b(r^2+a')}{a(r^2-b')} < \left(\frac{1+\alpha}{1-\alpha}\right)^2 \tag{42a}$$

$$a' < r \leq b' \quad \text{and} \quad \frac{r^2+a'}{a(r-1)^2} < \left(\frac{1+\alpha}{1-\alpha}\right)^2 \tag{42b}$$

$$b' < r \leq a' \quad \text{and} \quad \frac{b(r+1)^2}{r^2-b'} < \left(\frac{1+\alpha}{1-\alpha}\right)^2 \tag{42c}$$

$$\frac{1}{\alpha} < r \leq \min(a',\ b') \tag{42d}$$

Of course if $r \leq 1/\alpha$, no matter what a and b are, one cannot obtain a stabilizing controller, since α^{-1} is the (unconstrained) sensitivity index. Notice also that if $r = \infty$, the only active condition is the first, and it yields

$$b/a < \left(\frac{1+\alpha}{1-\alpha}\right)^2 \tag{43}$$

which is the already known gain margin condition.

Using Eq. (42), it now proves possible to define the least sensitivity consistent with a given gain margin pair, or the maximal upper gain margin b and minimal lower gain margin a consistent with a prescribed sensitivity r. Let us define

$$\beta \triangleq \left(\frac{1+\alpha}{1-\alpha}\right)^2 \tag{44}$$

Observe first that, via simple calculation, the sets in $\{r > 0\}$ of values of r defined by Eq. (42) are, in order

$$\begin{aligned} I_1 &= \{r > 0 : \quad r \text{ satisfies Eq. (42a)}\} \\ &= (\max\left\{a',\ b',\ \sqrt{a'b'\left(\frac{\beta-1}{a\beta-b}-1\right)}\right\},\ \infty) \text{ if } b/a < \beta \\ &= \emptyset \quad \text{otherwise} \end{aligned} \tag{45}$$

$$\begin{aligned} I_2 &= \{r > 0 : \quad r \text{ satisfies Eq. (42b)}\} \\ &= (a',\ b'] \quad \text{if } a' \geq 1/\alpha \\ &= \left(\frac{a\beta+\sqrt{a'(\beta-1)}}{a\beta-1},\ b'\right] \text{ if } a' < 1/\alpha \text{ and } \beta > \frac{a+b^2-2ab}{a-a^2} \\ &= \emptyset \quad \text{otherwise} \end{aligned} \tag{46}$$

$$\begin{aligned} I_3 &= \{r > 0: \quad r \text{ satisfies Eq. (42c)}\} \\ &= (b',\ a'] \quad \text{if } b' \geq 1/\alpha \\ &= \left(\frac{b+\sqrt{b'\beta(\beta-1)}}{\beta-b},\ a'\right] \quad \text{if } b' < 1/\alpha \text{ and } \beta > \frac{b^2-b}{2ab-a^2-b} \\ &= \emptyset \quad \text{otherwise} \end{aligned} \tag{47}$$

$$\begin{aligned} I_4 &= \{r > 0: \quad r \text{ satisfies Eq. (42d)}\} \\ &= (\alpha^{-1},\ \min(a',\ b')], \qquad \text{if } \min(a',\ b') > \alpha^{-1} \\ &= \emptyset \quad \text{otherwise} \end{aligned} \tag{48}$$

Now for given $a,\ b$ the task of defining the sensitivity index, or infimum of the achievable sensitivities is the task of finding which of $I_1,\ I_2,\ I_3$ and I_4 is nonempty, what the leftmost part of the closures of the nonempty I_j is, and what is the least value of such points. It is evident from the definition of I_j that at most one of I_2 and I_3 is nonempty, and that

$$\inf I_4 < \inf I_2 < \inf I_1 \text{ or } \inf I_4 < \inf I_3 < \inf I_1 \tag{49}$$

assuming that the intervals are each nonempty. So the task of finding

$$\mathcal{R}(a,b) = \inf_r\{(r,\ a,\ b) \text{ satisfies Eq. (42)}\} \tag{50}$$

is one of deciding which I_j are nonempty, and what the left most points in the closure are. The result is

Theorem 3.1 *With quantities as defined above,*

$$\mathcal{R}(a,b) = \alpha^{-1} \textit{ if } \min(a',\ b') \geq \alpha^{-1} \tag{51}$$

$$\mathcal{R}(a,b) = \frac{a\beta+\sqrt{a'(\beta-1)}}{a\beta-1} \textit{ if } a' < \alpha^{-1} \textit{ and } \beta > \frac{a+b^2-2ab}{a-a^2} \tag{52}$$

$$\mathcal{R}(a,b) = \frac{b+\sqrt{b'\beta(\beta-1)}}{\beta-b} \textit{ if } b' < \alpha^{-1} \textit{ and } \beta > \frac{b^2-b}{2ab-a^2-b} \tag{53}$$

$$\mathcal{R}(a,b) = \left[\frac{a'b'(\beta-1)}{a\beta-b}-1\right] \textit{ if } b/a < \beta \text{ and } \textit{none of first three alternatives holds} \tag{54}$$

If $b/a \geq \beta$, there is no stabilizing controller.

Of particular interest are the circumstances under which the unconstrained minimal sensitivity α^{-1} can be achieved also with a gain margin constraint, i.e. when we have

$$\mathcal{R}(a,b) = \alpha^{-1} \tag{55}$$

A sufficient condition is certainly

$$\min(a', b') \geq \alpha^{-1} \tag{56}$$

It turns out that this is necessary, i.e. the other three possibilities for $\mathcal{R}(a,b)$ in the theorem all yield $\mathcal{R}(a,b) > \alpha^{-1}$. Notice that $\min(a',b') \geq \alpha^{-1}$ implies that $b/a < \frac{1+\alpha}{1-\alpha} = \sqrt{\beta}$. So it is only possible to have the optimal constrained sensitivity equal to the unconstrained optimal sensitivity if the gain margin constraint is no greater than the square root of the maximal gin margin. We shall later exhibit a converse of this statement.

Let us now fix b/a and denote it by k; then $\mathcal{R}(a,b)$ depends on only one parameter a or b. We can then choose a or b so that $R(a,b)$ is minimized. This tells us the best sensitivity consistent with a given gain margin $k = b/a$, as well as how to achieve it by correct choice of one of a or b.

Theorem 3.2 *With quantities as defined above, and with $b/a = k$ fixed:*

(a) *If $k \leq \sqrt{\beta}$, $\mathcal{R}(a,b) = 1/\alpha$ for all $a \in \left[\frac{1}{2}(1 + 1/\sqrt{\beta}),\ \frac{1}{2k}(1+\sqrt{\beta})\right]$ and $b = ak$.*

(b) *If $\sqrt{\beta} < k < \beta$, $\mathcal{R}(a,b)$ attains its unique minimum*

$$(1+\sqrt{\beta})\sqrt{\frac{k}{(k-1)(\beta-k)}}$$

at $a = (1+\sqrt{\beta})/(k+\sqrt{\beta})$ and $b = k(1+\sqrt{\beta})/(k+\sqrt{\beta})$.

(c) *with $r_k = \inf\{\mathcal{R}(a,b) :\ 0 < a < 1 < b \ \text{ and } \ b/a = k\}$, there holds*

$$r_k = 1/\alpha \qquad 1 < k \leq \sqrt{\beta} \tag{57a}$$

$$= (1+\sqrt{\beta})\sqrt{\frac{k}{(k-1)(\beta-k)}} \qquad \sqrt{\beta} < k < \beta. \tag{57b}$$

The proof of this theorem uses routine calculation based on the formulas of Theorem 3.1.

The general shape of the curve relating k (gain margin constraint) and r_k (least sensitivity consistent with this constraint) is shown in Figure 5. Notice that

- any k, r pair above the curve defines a pair achievable by a stabilizing controller; equivalently, there exist a, b with $k = b/a$ such that r, a, b satisfy one of Eq. (42)

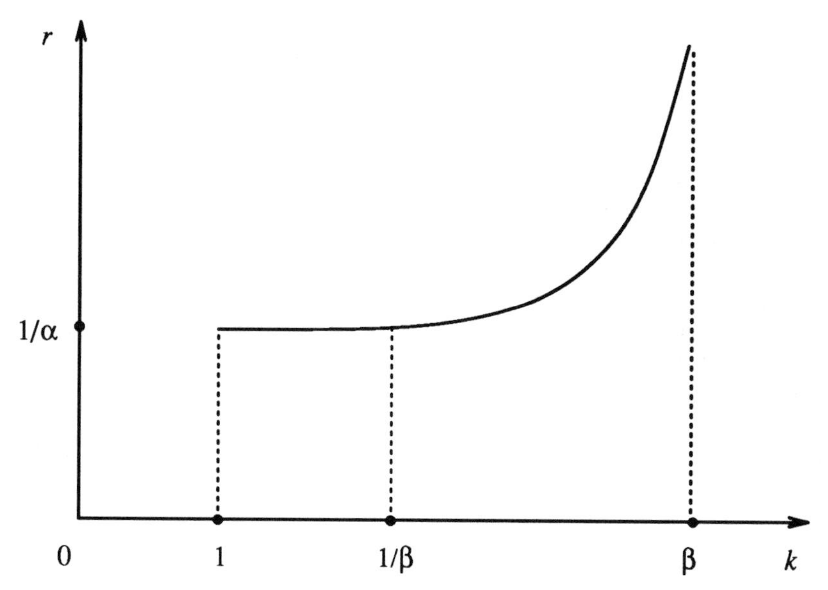

Figure 5. General curve relating r_k to k.

- the maximal unconstrained gain margin can only be approached with an infinitely sensitive design, while the minimal unconstrained sensitivity can be approached with a design of gain margin equal to the square root of the maximum
- the curve also obviously relates a sensitivity constraint with the maximal gain margin achievable for that constraint; in obvious notation, there holds

$$k_r = \left[\frac{\sqrt{r^2 - 1} + \sqrt{\alpha^2 r^2 - 1}}{r(1-\alpha)} \right]^2 \quad \text{for } r > 1/\alpha$$

This expression is simply the inverse of Eq. (57b).

IV. The Combined Sensitivity-Phase Margin Problem

As the title of the section implies, we study in this section a problem in which there are simultaneous constraints on sensitivity and phase margin. We need first to consider the question of phase margin by itself.

A. Optimal Phase Margin

Let $P(s)$ be a scalar plant and $C(s)$ a stablizing controller. Suppose that

$$1 + e^{j\phi}P(s)C(s) \neq 0 \quad \forall\phi \in [-\phi_1,\ \phi_2],\ \mathrm{Re}[s] \geq 0 \tag{58}$$

Let $\bar{\phi}_1$, $\bar{\phi}_2$ be the supremal values for which Eq. (58) holds. Then they define respectively the positive and negative phase margin. In applications, $\bar{\phi}_1$ is normally the quantity of interest. It is the additional phase lag needed just to lose stability.

We can study the question of achieving prescribed phase margin and characterizing the maximal phase margin by the methods introduced in Section II. Let $P(s)$ have poles p_i, $i = 1, \ldots, n$ in $\mathrm{Re}[s] \geq 0$ and zeros z_i, $i = 1, \ldots, m$ in $\mathrm{Re}[s] \geq 0$. We know that

$$S(p_i) = 0 \quad i = 1, \ldots, n \quad \text{and} \quad S(z_i) = 1 \quad i = 1, \ldots, m \tag{59}$$

Also, Eq. (58) implies that

$$P(s)C(s) \neq -e^{-j\phi} \quad \text{for } \phi \in [-\phi_1,\ \phi_2],\ \mathrm{Re}[s] \geq 0$$

or that

$$S(s) \neq (1 - e^{-j\phi})^{-1} \quad \text{for } \phi \in [-\phi_1,\ \phi_2],\ \mathrm{Re}[s] \geq 0$$

or equivalently

$$S(s) \neq \frac{1}{2}[1 - j\cot\phi/2] \quad \text{for } \phi \in [-\phi_1,\ \phi_2],\ \mathrm{Re}[s] \geq 0$$

Thus $S(s)$ in addition to satisfying the interpolation constraints Eq. (59) must satisfy

$$S(s):\ \bar{H} \to G_3 \triangleq \{s \in \mathbb{C}:\ s \neq \frac{1}{2}[1 - j\cot(\phi/2)],\ \phi \in [-\phi_1,\ \phi_2]\} \tag{60}$$

This region is depicted in Figure 6.

Using a conformal mapping dictionary [4] it is straightforward to obtain a conformal mapping $\theta(\cdot)$ from G_3 to D with $\theta(0) = 0$ (to appear in the commutative diagram of Figure 2). It is

$$\theta(s) = j\frac{\sqrt{(s - s_1)/(s_2 - s)} - \sqrt{-s_1/s_2}}{\sqrt{(s - s_1)/(s_2 - s)} + \sqrt{-\bar{s}_1/\bar{s}_2}} \times \frac{1 + \sqrt{-\bar{s}_1/\bar{s}_2}}{1 + \sqrt{-s_1/s_2}} \tag{61}$$

where $s_1 = \frac{1}{2}[1 + i\cot(\phi_1/2)]$, $s_2 = \frac{1}{2}[1 - i\cot(\phi_2/2)]$. Now this mapping does not generally have the property that $\overline{\theta(\bar{s})} = \theta(s)$ and thus its inverse will not have the property. It follows that $S(s)$ is unlikely to obey $\overline{S(\bar{s})} = S(s)$, which is unacceptable, since the inverse Laplace transform of $S(s)$

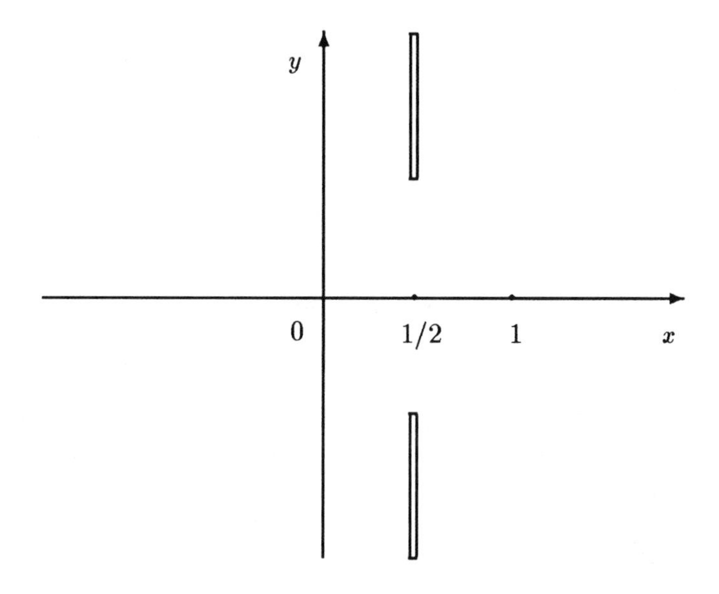

Figure 6. The region G_3 associated with phase margin problem.

must be a real time function. If however $\phi_1 = \phi_2$, so that $s_1 = \bar{s}_2$, then the desired property is obtained. So we shall henceforth assume

$$\phi_1 = \phi_2 \tag{62}$$

This is an unfortunate restriction from the practical point of view, as there is normally more interest in securing a large ϕ_1 than ϕ_2. With this assumption, it follows after some calculation that

$$\theta(s) = \frac{|s_1| - \sqrt{(s - s_1)(s - \bar{s}_1)}}{s - 1} \tag{63}$$

and

$$|\theta(1)| = \sin(\phi_1/2) \tag{64}$$

By the main result of Section II, it follows that with α as defined earlier (using a Nevanlinna-Pick problem), the phase margin problem is solvable if

$$\sin(\phi_1/2) < \alpha$$

or

$$\phi_1 < 2\sin^{-1}\alpha \tag{65}$$

and the maximal phase margin is

$$\bar{\phi} = 2\sin^{-1}\alpha \tag{66}$$

In the case of a minimum phase plant, $\alpha = 1$ and $\bar{\phi} = \pi$. In the case of a stable plant, use of the zero controller establishes again that $\bar{\phi} = \pi$.

Summing up

Theorem 4.1 *With the notation as previously defined, and with phase margin bounds being defined by $\phi_1 = \phi_2$ the maximal phase margin is $\bar{\phi} = 2\sin^{-1}\alpha$. For a minimum phase or a stable plant, the maximal phase margin is π.*

B. Combined Phase Margin and Sensitivity Problem

The formulation of this problem should be clear, as it is a simple combination of a number of the previous ideas. Suppose the unstable poles and zeros of $P(s)$ are, respectively $p_1, \ldots, p_n$ and $z_1, \ldots, z_m$, and suppose we are interested in securing a phase margin defined by $\phi_1 = \phi_2$ and a sensitivity index r.

The sensitivity function $S(s)$ satisfies the usual interpolation constraints, and also must map $\bar{H}$ into the region

$$\begin{aligned} G_4 &= G_1 \cap G_3 \\ &= \{|s| < r\} \cap \{s \in C : s \neq \frac{1}{2}[1 - j\cot(\phi/2)],\ \phi \in [-\phi_1,\ \phi_1]\} \end{aligned}$$

To study the existence of $S(s)$, we must construct a conformal mapping $\theta(s)$ of G_4 onto D, with $\theta(0) = 0$. This can be expressed, see Figure 7, as a composition of three mappings, i.e. $\theta = \theta_3 \circ \theta_2 \circ \theta_1$, where

$$\begin{aligned} \theta_1(s) &= -\sqrt{\frac{r}{4r-1}}\ \frac{2s-1-j\sqrt{4r-1}}{2s-1+j\sqrt{4r-1}} \\ \theta_3(s) &= \frac{s - \theta_2(\theta_1(0))}{s - \overline{\theta_2(\theta_1(0))}} \end{aligned}$$

and $\theta_2(s)$ is the inverse of the following Schwarz-Christoffel conformal transformation.

$$\zeta(s) = \int_0^s \frac{(v-v_1)(v-v_2)}{(v-v_3)}(v-v_4)^{1+(\eta/\pi)}dv + c$$

with $\eta = \tan^{-1}(\sqrt{4r-1})$ and with the parameters $c,\ v_i,\ i = 1, \ldots, 4$ determined by

$$\zeta(v_1) = \sqrt{\frac{r}{4r-1}\frac{\sqrt{4r-1}-\cot(\theta/2)}{\sqrt{4r-1}+\cot(\theta/2)}}$$

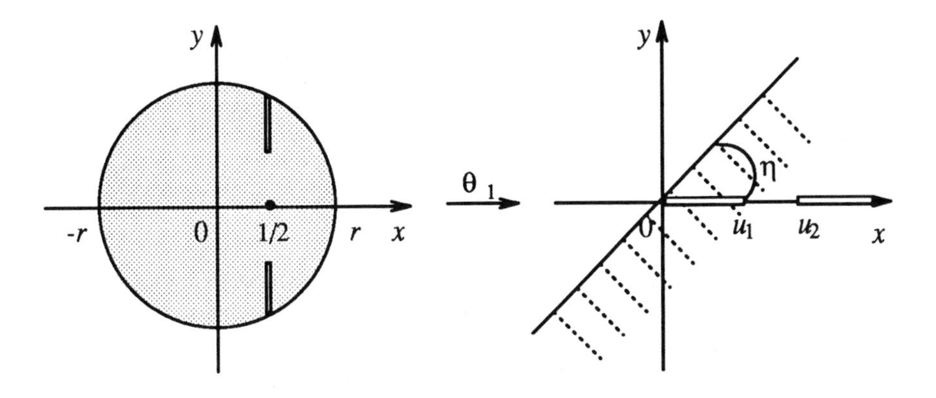

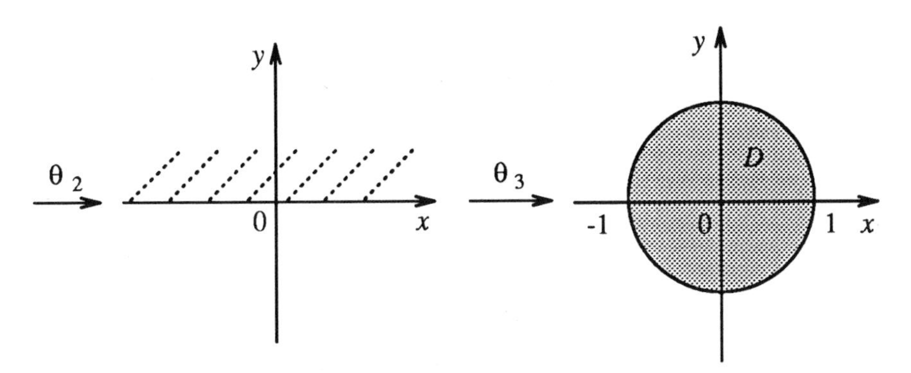

Figure 7. Construction of conformal mapping from G_4 to D.

$$
\begin{aligned}
\zeta(v_2) &= \sqrt{\frac{r}{4r-1}\frac{\sqrt{4r-1}+\cot(\theta/2)}{\sqrt{4r-1}-\cot(\theta/2)}} \\
\zeta(v_3) &= 0 \\
\zeta(v_4) &= \infty \\
\zeta(\infty) &= \infty
\end{aligned}
$$

We remark that this construction depends on having $r > 1$ and $\phi < \pi$. The analytic or explicit determination of the v_i and c seems impossible, as does the explicit determination of θ_2. Not surprisingly, the explicit evaluation of $|\theta(1)|$ then appears out of the question. Hence tidy results for the combined sensitivity phase margin problem are not available. Nevertheless, there are some valuable practical results.

First, for a minimum phase plant, we require simply that $|\theta(1)| < 1$, since $\alpha = 1$, and this condition is guaranteed by the existence of θ (which maps G into D, with $1 \in G$). Existence of θ in turn simply requires $r > 1$

and $\phi_1 < \pi$.

Of course, for a stable plant, the problem in effect evanesces when one uses $C(s) = 0$, which results in a sensitivity index of 1 and phase margin of π.

Next, suppose that ϕ is such that

$$\phi \leq 2\sin^{-1}\left(\frac{1}{2r}\right) \tag{67}$$

This means that in Figure 7, the vertical lines shown as penetrating the circle of radius r do not actually penetrate the circle and $G_4 = \{s \in \mathbf{C} : |s| < r\}$. This means that the combined problem reduces to the ordinary sensitivity problem when $\phi \leq 2\sin^{-1}(1/2r)$. Thus we have

Theorem 4.2 *Consider a nonminimum phase unstable plant $P(s)$ with α defined in the usual way. Then the combined sensitivity phase margin problem is solvable whenever $r > \alpha^{-1}$ and $\phi_1 < 2\sin^{-1}(1/2r)$.*

This result is of course reminiscent of a result for the combined sensitivity gain-margin problem where we showed the combined problem is solvable whenever $r > \alpha^{-1}$ and

$$k < \left[\frac{\sqrt{r^2 - 1} + \sqrt{\alpha^2 r^2 - 1}}{r(1 - \alpha)}\right]^2 \tag{68}$$

and in particular,

$$k < \frac{1 + \alpha}{1 - \alpha} = \sqrt{\beta} \tag{69}$$

To make further progress, we need to modify the problem formulation slightly. This we do by introducing some additional sensitivity indices.

C. Sensitivity, Complementary and Generalized Sensitivity

The sensitivity function $S(s)$ has earlier been defined as $[1 + P(s)C(s)]^{-1}$. We can also define the *complementary sensitivity* $T(s)$ by

$$T(s) = 1 - S(s) = \frac{P(s)C(s)}{1 + P(s)C(s)} \tag{70}$$

[Small $|T(j\omega)|$ is associated with securing low amplification of sensor noise, and with securing robustness to unstructured multiplicative perturbation in the plant]. Let us also define a generalized sensitivity by

$$GS(s) = [1 + P(s)C(s)]^{-1} - x \tag{71}$$

The generalized sensitivity index is

$$\|GS\|_\infty = \sup_{s \in \bar{H}} |[1 + P(s)C(s)]^{-1} - x| \tag{72}$$

With $x = 0$ and $x = 1$, the sensitivity and complementary sensitivity indices are obtained, while with $x = 1/2$, minimization aims to secure both the sensitivity and complementary sensitivity as close to $1/2$ as possible. Evidently, a generalized sensitivity index of r is attained if and only if $S(s)$ has the mapping property

$$S(s) : \bar{H} \rightarrow G_5 \triangleq \{S \in \boldsymbol{C} : |s - x| < r\} \tag{73}$$

Let us assume $x \in [0,\ 1]$.

The mapping θ from G_5 to D is a composition of

$$\theta_1(s) = \frac{s - x}{r} \quad (\text{ mapping } G_4 \text{ to } D)$$

and

$$\theta_2(s) = \frac{s + \frac{x}{r}}{-\frac{x}{r}s - 1} \quad (D \text{ to } D)$$

so that the composition obeys $\theta(0) = 0$. Evidently,

$$\theta(s) = \frac{rs}{-xs + x^2 - r^2} \quad \text{and} \quad \theta(1) = \frac{r}{x^2 - x - r^2}$$

Then we have $|\theta(1)| < \alpha$ provided

$$r > \frac{1/\alpha + \sqrt{(2x-1)^2 - 1 + (1/\alpha^2)}}{2} \tag{74}$$

The conditions for solvability of the generalized sensitivity problem are now easily determinable:

a) Suppose the plant has no unstable zeros. It is trivial then to satisfy the interpolation constraints on F [they are of the form $F(a_i) = 0$]. We require that $S(p_i) = 0 \in G_5$ so that $r > |x|$.

b) Suppose the plant has no unstable poles. Then to satisfy the interpolation constraints on F we require that $|\theta(1)| < 1$ or $r > x$ and to ensure that $S(z_i) = 1 \in G_5$, we require $r > |x - 1|$.

c) Suppose the plant has unstable poles and zeros. Then we require Eq. (74), which implies that $r > |x|$, $r > |x - 1|$, and of course S satisfies the interpolation constraints.

We shall now return to the combined phase-margin sensitivity problem, but with a generalized sensitivity where $x = 1/2$. In this case Eq. (74) becomes

$$r > \frac{1 + \sqrt{1 - \alpha^2}}{2\alpha} \tag{75}$$

D. Combined Phase Margin-Generalized Sensitivity Problem

In order to perform more explicit calculation, we shall consider the problem formed by combining phase margin and generalized sensitivity with $x = 1/2$.

Figure 8 now displays the region G_5 intersected with G_3. Call this combined region G_6:

$$G_6 = \{|s - 1/2| < r\} \cap \{s \neq \frac{1}{2}[1 - j\cot(\phi/2)],\ \phi \in [-\phi_1,\ \phi_1]\} \tag{76}$$

In case the segments $\frac{1}{2}[1 - j\cot\phi/2]$, $\phi \in [-\phi_1,\ \phi_1]$ do not intersect G_5, then $G_6 = G_5$, and the combined problem is very easily solved, as per the previous section. This corresponds to the condition

$$\cot\frac{\phi_1}{2} > 2r \quad \text{or} \quad \phi_1 < 2\cot^{-1} 2r \tag{77}$$

Put another way, achieving a generalized sensitivity of r automatically yields a phase margin of $2\cot^{-1} 2r$.

The combined problem is also very easily solved for a stable plant or a minimum phase plant. In the case of a stable plant, use of the zero controller ensures that $||GS|| = 1/2$ and we can tolerate simultaneously a phase margin of π. In the case of a minimum phase plant, we simply require the existence of a conformal mapping (which is guaranteed) and $S(p_i) = 0 \in G_6$, i.e. $r > 1/2$. The interesting case is (as before) thus an unstable, nonminimum phase plant. Note that, by virtue of results for the generalized sensitivity problem with no phase margin constraint, we must have $r > 1/2$.

The main result is as follows.

Theorem 4.3 *Suppose that $P(s)$ is a nonminimum phase, unstable plant. Let $r > 1/2$ and $0 \leq \phi_1 < \pi$ with r and ϕ_1 the generalized sensitivity index ($x = 1/2$) and phase margin respectively. Define $x_\phi = \frac{1}{2}\cot(\phi_1/2)$. Then with α taking its usual meaning and defined in terms of the unstable poles and zeros of the plant, the combined generalized sensitivity ($x = 1/2$) phase margin problem is solvable if and only if*

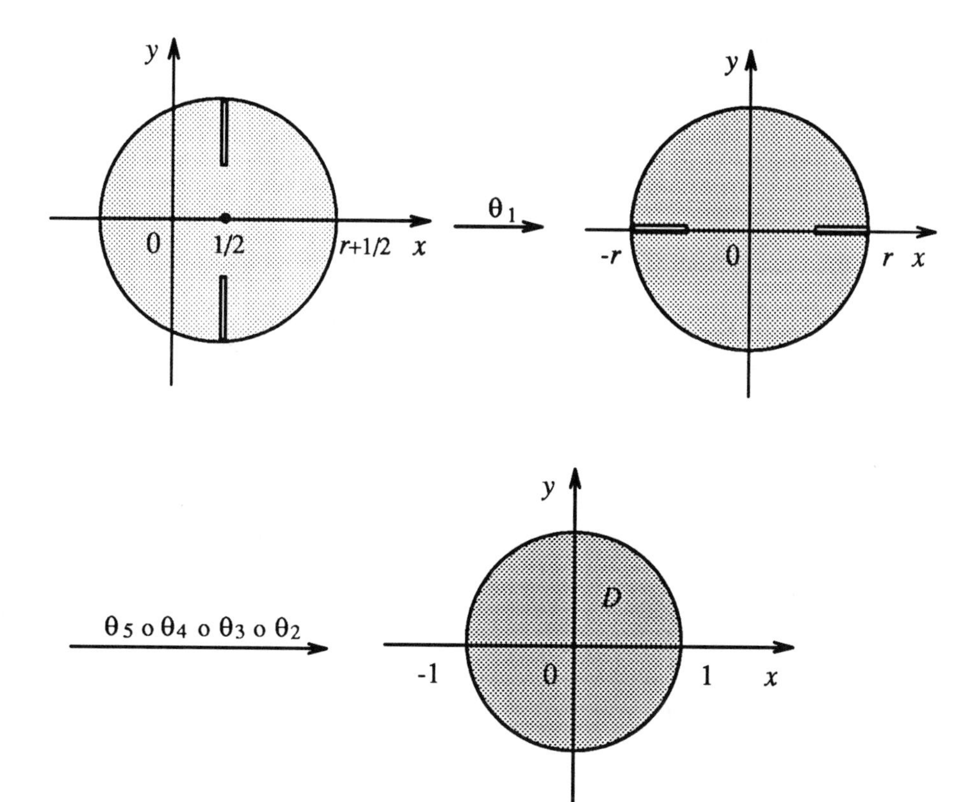

Figure 8. Construction of conformal mapping from G_6 to D.

$$\sin(\phi_1/2) < \alpha \tag{78}$$

and either

$$\frac{1+\sqrt{1-\alpha^2}}{2\alpha} < r \leq x_\phi \tag{79a}$$

or

$$r > \max\left(x_\phi, \sqrt{f(x_\phi)}\right) \tag{79b}$$

where

$$f(x) \triangleq \frac{x(2\sqrt{1-\alpha^2}\,x+\alpha)}{2(2\alpha x-\sqrt{1-\alpha^2})} \tag{80}$$

Proof: Equation (78) is certainly necessary by Theorem 4.1. Assume it holds. Then

$$x_\phi > \frac{\sqrt{1-\alpha^2}}{2\alpha} \tag{81}$$

Now suppose that $r \le x_\phi$. Then G_6 is identical with G_5, and, as established in the previous section, a necessary and sufficient condition for solvability is that $r > \frac{1+\sqrt{1-\alpha^2}}{2\alpha}$. The other case of $r > x_\phi$ is more difficult. The conformal mapping from G_6 to D with $\theta(0) = 0$ turns out to be

$$\theta(s) = j\,\frac{\lambda(s)\sqrt{(x_\phi - j/2)(r^2 - jx_\phi/2)} \; - \; \overline{\lambda(\bar{s})}\sqrt{(x_\phi + j/2)(r^2 + jx_\phi/2)}}{\lambda(s)\sqrt{(x_\phi + j/2)(r^2 + jx_\phi/2)} \; + \; \overline{\lambda(\bar{s})}\sqrt{(x_\phi - j/2)(r^2 - jx_\phi/2)}}$$

where

$$\lambda(s) = \sqrt{[x_\phi - j(s-1/2)][r^2 - jx_\phi(s-1/2)]}$$

The mapping $\theta(s)$ is a composition of maps (see Figure 8) separately defined as follows:

$$\theta_1(s) = j(s-1/2): \qquad G_6 \to D_r \setminus \{(-\infty,\, -x_\phi] \cup [x_\phi,\, \infty)\} \tag{82a}$$

$$\theta_2(s) = \left(\frac{r+s}{r-s}\right)^2: \quad D_r \setminus \{(-\infty,\, -x_\phi] \cup [x_\phi,\, \infty)\}$$
$$\to C \setminus \left\{\left(-\infty,\, \left(\frac{r-x_\phi}{r+x_\phi}\right)^2\right] \cup \left[\left(\frac{r+x_\phi}{r-x_\phi}\right)^2,\, \infty\right)\right\} \tag{82b}$$

$$\theta_3(s) = \frac{\left(\frac{r-x_\phi}{r+x_\phi}\right)^2 - s}{s - \left(\frac{r+x_\phi}{r-x_\phi}\right)^2}:$$
$$C \setminus \left\{\left(-\infty,\, \left(\frac{r-x_\phi}{r+x_\phi}\right)^2\right] \cup \left[\left(\frac{r+x_\phi}{r-x_\phi}\right)^2,\, \infty\right)\right\} \to C \setminus (-\infty,\, 0] \tag{82c}$$

$$\theta_4(s) = \frac{1-\sqrt{s}}{1+\sqrt{s}}: \qquad C \setminus (-\infty,\, 0] \to D \tag{82d}$$

$$\theta_5(s) = \rho\frac{s - x_0}{1 - \bar{x}_0 s}: \qquad D \to D \tag{82e}$$

with

$$x_o = \frac{1-\lambda}{1+\lambda} \qquad \rho = i\frac{1+\lambda}{1+\bar{\lambda}}\sqrt{\frac{(x_\phi + j/2)(r^2 + jx_\phi/2)}{(x_\phi - j/2)(r^2 - jx_\phi/2)}}$$

$$\lambda = \sqrt{\frac{(r - x_\phi)^2(x_\phi - i/2)(r^2 - jx_\phi/2)}{[(r + x_\phi)^2(x_\phi + i/2)(r^2 + jx_\phi/2)]}} \tag{83}$$

The composition $\theta_5 \circ \theta_4 \circ \theta_3 \circ \theta_2$ accounts for the last mapping in Figure 8, and is found in just the same way as the map we used in discussing the combined sensitivity gain margin problem except that θ_5 must be chosen to ensure that $\theta(0) = 0$ rather than $\theta_5 \circ \theta_4 \circ \theta_3 \circ \theta_2(0) = 0$.

With the definition of $\theta(s)$, some work will show that $\overline{\theta(\bar{s})} = \theta(s)$. A simple calculation yields

$$\theta(1) = \frac{r^2 + x_\phi^2}{\sqrt{(1 + 4x_\phi^2)(r^4 + x_\phi^2/4)}}$$

Solvability is equivalent to $\theta(1) < \alpha$, and this can be rewritten as $r^2 > f(x_\phi)$. This completes the proof. □

The above theorem also allows us to identify the infimum of the sensitivities consistent with a given phase margin. Call this infimum r_ϕ. Then clearly, if

$$x_\phi \geq \frac{1 + \sqrt{1 - \alpha^2}}{2\alpha}$$

there holds, by Eq. (79)

$$r_\phi = \frac{1 + \sqrt{1 - \alpha^2}}{2\alpha}$$

The condition $x_\phi = \frac{1}{2} \cot \frac{\phi_1}{2} \geq \frac{1+\sqrt{1-\alpha^2}}{2\alpha}$ is equivalent to

$$\phi_1 \leq \sin^{-1} \alpha$$

Next, if $2 \sin^{-1} \alpha > \phi_1 > \sin^{-1} \alpha$, we have

$$r_\phi = \max\left(x_\phi, \sqrt{f(x_\phi)}\right)$$

It is not hard to check that $f(x_\phi) - x_\phi^2 > 0$, and so

$$r_\phi = \sqrt{f(x_\phi)}$$

The graph of r_ϕ vs ϕ_1 is shown in Figure 9, and is represented by the middle curve of that figure.

Theorem 4.3 also gives us easily separate necessary and sufficient conditions for the conventional combined sensitivity phase margin problem. Notice that

$$\|GS\|_\infty < r - 1/2 \implies \|S\|_\infty < r$$
$$\|S\|_\infty < r \implies \|GS\|_\infty < r + 1/2$$

Now consider the three curves depicted in Figure 9. We claim:

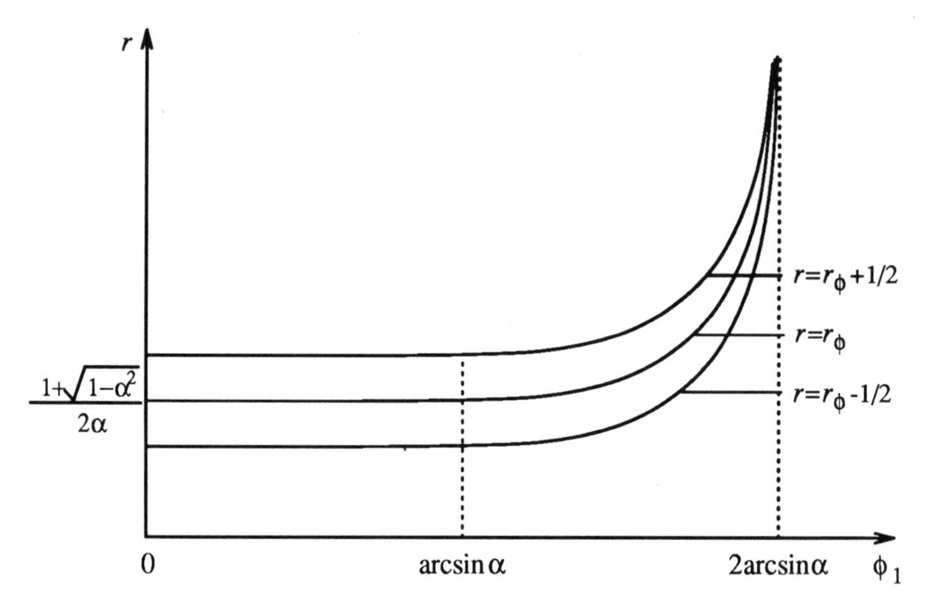

Figure 9. Curve of r_ϕ depending on ϕ_1.

(a) in order that $||S||_\infty < r$ and ϕ be attainable, it is necessary that the point (r, ϕ) lie above the lower curve. [For if $||S||_\infty < r$, $||GS||_\infty < r + 1/2$. The point $(r + 1/2, \phi)$ must be above the middle curve, and so (r, ϕ) must lie above the lower curve]

(b) in order that $||S||_\infty < r$ and ϕ be attainable, it is sufficient that the point (r, ϕ) lies above the upper curve. [For if the point (r, ϕ) lies above the upper curve, $(r - 1/2, \phi)$ lies above the middle curve, $||GS||_\infty < r - 1/2$ and ϕ are attainable and if $||GS||_\infty < r - 1/2$, $||S||_\infty < r$,]

Where the actual boundary is of the allowed and disallowed regions for the (ordinary) sensitivity-phase margin problem is unknown, save that it clearly must be between the two curves.

V. Combined Sensitivity-Complementary Sensitivity Problem

In this section, we will characterize all pairs of sensitivity and complementary sensitivity indices which can be simultaneously achieved in a closed-loop system. Meanwhile, we will also establish a connection between the two separate design problems for sensitivity and for complementary sensitivity.

A. Problem Formulation

As before, we consider a proper scalar LTI continuous-time plant $P(s)$. The assumption that $P(s)$ only has simple poles in $\bar{H}$ and simple zeros in $\bar{H}$ will remain enforced in this section. Associated with a stabilizing controller $C(s)$, the sensitivity function $S(s)$ and the complementary sensitivity function $T(s)$ have been defined by Eq. (1) and Eq. (70), respectively.

Then specifically, the combined sensitivity and complementary sensitivity (CSCS) problem is that of determining the existence of and then finding a proper stabilizing compensator $C(s)$ such that the following constraints are simultaneously satisfied

$$||S(s)||_\infty < r_1 \quad \text{and} \quad ||T(s)||_\infty < r_2 \tag{84}$$

where r_1, r_2 are two given positive constants. Note that the pure sensitivity or complementary sensitivity problem is in fact a particular form of the CSCS problem corresponding to $r_2 = \infty$ or $r_1 = \infty$. Both the problems have been considered as special cases of the generalized sensitivity problem in Section III.D. At this point, it may be worthwhile to recall simple relevant facts concerning the optimal unconstrained sensitivity r_{min} and optimal unconstrained complementary sensitivity t_{min}.

1. If $P(s)$ has unstable poles and zeros, then $r_{\text{min}} = t_{\text{min}} = 1/\alpha$, where α is defined as in Eq. (27).
2. If $P(s)$ has no unstable poles nor unstable zeros, then $r_{\text{min}} = t_{\text{min}} = 0$.
3. If $P(s)$ has unstable poles but no unstable zeros, then $r_{\text{min}} = 0$ and $t_{\text{min}} = 1$.
4. If $P(s)$ has unstable zeros but no unstable poles, then $r_{\text{min}} = 1$ and $t_{\text{min}} = 0$.

The above obviously gives necessary conditions for the CSCS problem to be solvable.

B. Solvability

First of all, it is a simple matter to observe that the CSCS problem is always solvable for the point (r_1, r_2) in the domain $E_2 \cup E_3 \cup E_4$ and is unsolvable for (r_1, r_2) outside the domain $\cup_{i=1}^{4} E_i$, where E_i, $i = 1, \ldots, 4$ are defined as

$$E_1 \triangleq \{(x, y): \quad r_{\min} < x < 1 + r_{\min} \text{ and } t_{\min} < y < 1 + t_{\min}\} \quad (85)$$

$$E_2 \triangleq \{(x, y): \quad r_{\min} < x < 1 + r_{\min} \quad \text{and} \quad y \geq 1 + r_{\min}\} \quad (86)$$

$$E_3 \triangleq \{(x, y): \quad x \geq 1 + r_{\min} \quad \text{and} \quad y \geq 1 + t_{\min}\} \quad (87)$$

$$E_4 \triangleq \{(x, y): \quad x \geq 1 + t_{\min} \quad \text{and} \quad t_{\min} < y < 1 + t_{\min}\} \quad (88)$$

Thus, the only unclear domain for the solvability of the CSCS problem is E_1, in which (r_1, r_2) is obviously desired to be. On the other hand, similar to the combined sensitivity-gain margin problem, the CSCS problem can be shown to be equivalent to that of finding a real rational analytic function

$$S(s): \quad \bar{H} \to G(r_1, r_2) \triangleq \{s \in \boldsymbol{C}: \quad |s| < r_1 \text{ and } |s - 1| < r_2\} \quad (89)$$

satisfying the interpolation conditions (8)-(9).

Further, observe that the function $S(s) \equiv 0$ is a mapping from $\bar{H}$ to $G(r_1, r_2)$ and satisfies the interpolation conditions (8) for all pairs (r_1, r_2) with $r_1 > 0$, $r_2 > 1$ and the function $S(s) \equiv 1$ is a mapping from $\bar{H}$ to $G(r_1, r_2)$ and satisfies Eq. (9) for all pairs (r_1, r_2) with $r_1 > 1$, $r_2 > 0$. Also, there always exists a real positive number a such that $S(s) \equiv a$ is a mapping from $\bar{H}$ to $G(r_1, r_2)$ for all pairs (r_1, r_2) with $r_1 + r_2 > 1$. Combining these observations with the facts listed in the preceding subsection yields necessary and sufficient conditions for solvability of the CSCS problem in the trivial cases, as stated in the following result.

Theorem 5.1 *There hold*

(i) *for the plant $P(s)$ without unstable zeros but with unstable poles, the CSCS problem is solvable for (r_1, r_2) if and only if $r_1 > 0$ and $r_2 > 1$;*

(ii) *for the plant $P(s)$ without unstable poles but with unstable zeros, the CSCS problem is solvable for (r_1, r_2) if and only if $r_1 > 1$ and $r_2 > 0$;*

(iii) *for the plant $P(s)$ with neither unstable poles nor unstable zeros, the CSCS problem is solvable for (r_1, r_2) if and only if $r_1 + r_2 > 1$.*

In view of the above discussion, we need only to concentrate on the remaining nontrivial case where the plant $P(s)$ has both unstable zeros and u stable poles, and (r_1, r_2) is in

$$E_1 = \{(x, y): \quad 1/\alpha < x < 1 + 1/\alpha \quad \text{and} \quad 1/\alpha < y < 1 + 1/\alpha\} \quad (90)$$

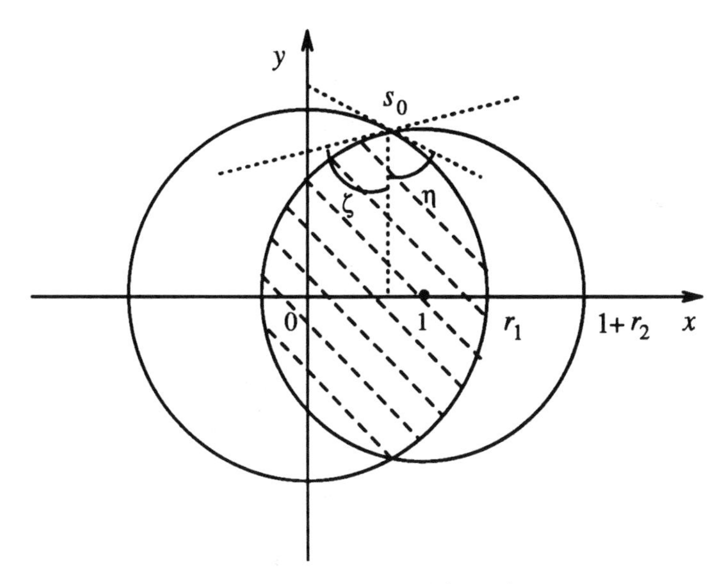

Figure 10. The region $G(r_1, r_2)$.

Since $\alpha \leq 1$, $G(r_1, r_2)$ in Eq. (89) is a region containing 0 and 1, and is depicted in Figure 10 where

$$\begin{aligned} s_0 &= x_0 + jy_0 \\ &= \frac{1}{2}\left\{r_1^2 - r_2^2 + 1 + j\sqrt{[(r_1+1)^2 - r_2^2][r_2^2 - (r_1-1)^2]}\right\} \\ \zeta &= tan^{-1}\left(\frac{y_0}{1-x_0}\right), \quad \eta = tan^{-1}\left(\frac{y_0}{x_0}\right), \\ \zeta + \eta &= tan^{-1}\left(\frac{y_0}{x_0 - r_1^2}\right) \; (\frac{\pi}{2} < \zeta + \eta < \pi) \end{aligned}$$

It turns out that the conformal mapping θ from $G(r_1, r_2)$ to D is the composition of the following three conformal mappings.

$$\begin{aligned} \theta_1(s) &= \rho\frac{s_0 - s}{s - \bar{s}_0} : \quad G(r_1, r_2) \to \{s : 0 < arg(s) < \zeta + \eta\} \\ \theta_2(u) &= u^{\pi/(\zeta+\eta)} : \quad \{u : 0 < arg(u) < \zeta + \eta\} \to \{v : 0 < arg(v) < \pi\} \\ \theta_3(v) &= k\frac{v - v_0}{v - \bar{v}_0} : \quad \{v : 0 < arg(v) < \pi\} \to D \end{aligned}$$

where

$$\rho = \sqrt{\frac{r_1^3}{2y_0^2(r_1 - x_0)}}\, e^{j\zeta}, \quad v_0 = \theta_2 \circ \theta_1(0)), \quad \text{and} \quad |k| = 1$$

With the equality

$$\left(\rho\frac{s_0 - s}{s - \bar{s}_0}\right)^{\pi/(\zeta+\eta)} = \frac{[\rho(s_0 - s)]^{\pi/(\zeta+\eta)}}{(s - \bar{s}_0)^{\pi/(\zeta+\eta)}}, \quad \forall s \in G(r_1, r_2)$$

The composition can be simplified as

$$\theta(s) = \theta_3 \circ \theta_2 \circ \theta_1(s) = k\frac{\left(\rho\frac{s_0 - s}{s - \bar{s}_0}\right)^{\pi/(\zeta+\eta)} - \left(\rho\frac{s_0}{-\bar{s}_0}\right)^{\pi/(\zeta+\eta)}}{\left(\rho\frac{s_0 - s}{s - \bar{s}_0}\right)^{\pi/(\zeta+\eta)} - \overline{\left[\left(\rho\frac{s_0}{-\bar{s}_0}\right)^{\pi/(\zeta+\eta)}\right]}} \tag{91}$$

$$= k\rho^{\frac{\pi}{\zeta+\eta}}\frac{(-s_0)^{\frac{\pi}{\zeta+\eta}}}{(-\bar{s}_0)^{\frac{\pi}{\zeta+\eta}}}\frac{[\bar{s}_0(s - s_0)]^{\frac{\pi}{\zeta+\eta}} - [s_0(s - \bar{s}_0)]^{\frac{\pi}{\zeta+\eta}}}{[\rho s_0(s - s_0)]^{\frac{\pi}{\zeta+\eta}} e^{j\frac{2\pi^2}{\zeta+\eta}} - [\bar{\rho}\bar{s}_0(s - \bar{s}_0)]^{\frac{\pi}{\zeta+\eta}}} \tag{92}$$

Apparently, $\theta(0) = 0$. To require that $\overline{\theta(\bar{s})} = \theta(s)$ be satisfied, the constant k needs to be chosen such that $|k| = 1$ and $\theta'(0)$ is real and positive, which can always be done. In addition, it is not hard to calculate that

$$|\theta(1)| = |\rho|^{\frac{\pi}{\zeta+\eta}}\left|\frac{[\bar{s}_0(1 - s_0)]^{\frac{\pi}{\zeta+\eta}} - [s_0(1 - \bar{s}_0)]^{\frac{\pi}{\zeta+\eta}}}{[\rho s_0(1 - s_0)]^{\frac{\pi}{\zeta+\eta}} e^{j\frac{2\pi^2}{\zeta+\eta}} - [\bar{\rho}\bar{s}_0(1 - \bar{s}_0)]^{\frac{\pi}{\zeta+\eta}}}\right|$$

$$= |\rho|^{\frac{\pi}{\zeta+\eta}}\left|\frac{Im\left\{[\bar{s}_0(1 - s_0)]^{\frac{\pi}{\zeta+\eta}} e^{-j\frac{\pi^2}{\zeta+\eta}}\right\}}{Im\left\{[\rho s_0(1 - s_0)]^{\frac{\pi}{\zeta+\eta}}\right\}}\right| = \left|\frac{\sin[\pi^2/(\zeta+\eta)]}{\sin[\eta\pi/(\zeta+\eta)]}\right|.$$

Now a direct application of the theory presented in Section II yields

Theorem 5.2 *Consider the plant $P(s)$ with unstable zeros and poles. Let α be defined as in Eq. (27). Then given $(r_1, r_2) \in E_1$, the CSCS problem is solvable if and only if*

$$\left|\frac{\sin(\pi^2/\mu)}{\sin(\nu\pi/\mu)}\right| < \alpha \tag{93}$$

where

$$\mu = \tan^{-1}\left\{-\frac{\sqrt{[(r_1+1)^2 - r_2^2][r_2^2 - (r_1-1)^2]}}{r_1^2 + r_2^2 - 1}\right\}$$

$$\nu = \tan^{-1}\left\{\frac{\sqrt{[(r_1+1)^2 - r_2^2][r_2^2 - (r_1-1)^2]}}{r_1^2 - r_2^2 + 1}\right\}$$

In the special case where $r_1 = r_2$, the solvability condition Eq. (93) has a much simpler form to be given below.

Corollary 5.1 *With the same assumption as in Theorem 5.2, the CSCS problem is solvable for $r_1 = r_2 = r$ with $r > 1/\alpha$ if and only if*

$$r > \frac{1}{2\cos[\frac{\pi^2}{2(\pi+\sin^{-1}\alpha)}]} \tag{94}$$

Proof: With $r_1 = r_2 = r$, the following hold

$$\mu = 2\nu \quad \text{and} \quad \nu = \cos^{-1}(1/2r) \tag{95}$$

Since $r > 1/\alpha \geq 1$, one can see that

$$\pi < \frac{\pi^2}{2\nu} < \frac{3\pi}{2} \tag{96}$$

In view of Eqs. (95)-(96), the condition (93) becomes

$$\sin\left(\frac{\pi^2}{2\nu} - \pi\right) < \alpha$$

which is obviously equivalent to Eq. (94) because of $0 < \frac{\pi^2}{2\nu} - \pi < \frac{\pi}{2}$. □

Finally, it is worth pointing out that for a given plant with open right half plane zeros and poles, it is impossible to achieve simultaneously the sensitivity and the complementary sensitivity both arbitrarily close to their common minimum $1/\alpha$. The reason is that the right-hand expression of Eq. (94) as a function of α is always greater than $1/\alpha$.

C. Connection between Sensitivity Problem and Complementary Sensitivity Problem for Multivariable Plants

We now turn to establish a connection between the sensitivity problem and the complementary sensitivity problem for a multivariable plant. This connection will allow us to solve either of the two problems by way of tackling the other one. Recall that the sensitivity (respectively complementary sensitivity) problem is to find a stabilizing compensator $C(s)$ such that its associated sensitivity (respectively complementary sensitivity) function is less than a given $r > 0$ in H^∞-norm.

Let RH_+^∞ denote the set of proper rational matrices which have no poles in the closed RHP. Suppose the transfer matrix $P(s)$ is given and has a stable doubly-coprime factorization in RH_+^∞:

$$P(s) = ND^{-1}(s) = \tilde{D}^{-1}\tilde{N}(s)$$

with the Bezout identities

$$\begin{bmatrix} X & Y \\ \tilde{D} & -\tilde{N} \end{bmatrix} \begin{bmatrix} N & \tilde{Y} \\ D & -\tilde{X} \end{bmatrix} = \begin{bmatrix} I & 0 \\ 0 & I \end{bmatrix} \tag{97}$$

or equivalently,

$$\begin{bmatrix} N & \tilde{Y} \\ D & -\tilde{X} \end{bmatrix} \begin{bmatrix} X & Y \\ \tilde{D} & -\tilde{N} \end{bmatrix} = \begin{bmatrix} I & 0 \\ 0 & I \end{bmatrix} \tag{98}$$

Then it is a standard result [5] that the set of all internally stabilizing proper compensators for $P(s)$ can be parametrized by

$$\{(Y - Z\tilde{N})^{-1}(X + Z\tilde{D}) : \quad Z \in RH_{+}^{\infty} \quad \text{and} \quad |Y - Z\tilde{N}| \neq 0\} \tag{99}$$

or

$$\{(\tilde{X} + DZ)(\tilde{Y} - NZ)^{-1} : \quad Z \in RH_{+}^{\infty} \quad \text{and} \quad |\tilde{Y} - NZ| \neq 0\} \tag{100}$$

It follows that with $C(s) = (\tilde{X} + DZ)(\tilde{Y} - NZ)^{-1}$, there holds

$$S(s) = (\tilde{Y} - NZ)\tilde{D}(s) \tag{101}$$

Let us now state the main result of this subsection.

Theorem 5.3 *Let $P(s)$ denote the $p \times m$ transfer matrix of a multivariable LTI plant, and $r > 1$ be given. Then, $C(s)$ solves the sensitivity problem for r if and only if $\frac{r^2}{r^2-1}C(s)$ solves the complementary sensitivity problem for r.*

Proof: Assume that $C_0(s)$ solves the sensitivity problem for r with $S_0(s)$ its corresponding sensitivity function. Then there exists $Z_0 \in RH_{+}^{\infty}$ such that

$$C_0(s) = (\tilde{X} + DZ_0)(\tilde{Y} - NZ_0)^{-1} \tag{102}$$

Let us show that $C_1(s) \triangleq \frac{r^2}{r^2-1}C_0(s)$ is a stabilizing controller. In other words, we need to find a $Z \in RH_{+}^{\infty}$ with $|Y - Z\tilde{N}| \neq 0$ such that

$$\frac{r^2}{r^2 - 1}(\tilde{X} + DZ_0)(\tilde{Y} - NZ_0)^{-1} = (Y - Z\tilde{N})^{-1}(X + Z\tilde{D}) \tag{103}$$

Using the Bezout identities Eq. (97), we can verify that under the condition that $|Y - Z\tilde{N}| \neq 0$, Eq. (103) is equivalent to

$$r^2 Z_0 + X(\tilde{Y} - NZ_0) = Z[r^2 I - \tilde{D}(\tilde{Y} - NZ_0)] \tag{104}$$

Since

$$||S_0(s)||_\infty = ||(\tilde{Y} - NZ_0)\tilde{D}(s)||_\infty < r,$$

$r^2I-(\tilde{Y}-NZ_0)\tilde{D}$ is unimodular in RH_+^∞; hence, so is $r^2I-\tilde{D}(\tilde{Y}-NZ_0)$. In this way, it follows from Eq. (104) that

$$Z_1 \triangleq [r^2 Z_0 + X(\tilde{Y}-NZ_0)][r^2I-\tilde{D}(\tilde{Y}-NZ_0)]^{-1} \in RH_+^\infty$$

Further, we have

$$\begin{aligned} Y - Z_1\tilde{N} &= Y-[r^2Z_0+X(\tilde{Y}-NZ_0)][r^2I-\tilde{D}(\tilde{Y}-NZ_0)]^{-1}\tilde{N} \\ &= Y-[r^2Z_0+X(\tilde{Y}-NZ_0)][(r^2-1)I+\tilde{N}(\tilde{X}+DZ_0)]^{-1}\tilde{N} \\ &= Y-[r^2Z_0+X(\tilde{Y}-NZ_0)]\tilde{N}[(r^2-1)I+(\tilde{X}+DZ_0)\tilde{N}]^{-1} \\ &= (r^2-1)(Y-Z_0\tilde{N})[(r^2-1)I+(\tilde{X}+DZ_0)\tilde{N}]^{-1} \end{aligned}$$

which implies that $|Y-Z_1\tilde{N}| \neq 0$ amounts to $|Y-Z_0\tilde{N}| \neq 0$ because of

$$\det[(r^2-1)I+(\tilde{X}+DZ_0)\tilde{N}] = \det[r^2I-\tilde{D}(\tilde{Y}-NZ_0)]$$

But

$$\begin{aligned} |(Y-Z_0\tilde{N})D| &= |I-(X+Z_0\tilde{D})N| \\ &= |I-N(X+Z_0\tilde{D})| = |(\tilde{Y}-NZ_0)\tilde{D}| \neq 0 \end{aligned}$$

so that $|Y-Z_1\tilde{N}| \neq 0$. Therefore, $Z = Z_1$ satisfies Eq. (103). Having shown that $C_1(s)$ stabilizes $P(s)$, we now verify that for this $C_1(s)$ the complementary sensitivity is less than r, i.e.

$$||I-(I+PC_1)^{-1}||_\infty < r$$

To this end, note that

$$I-(I+PC_1)^{-1} = r^2(S_0-I)(r^2I-S_0)^{-1}$$

Hence it suffices to check that

$$||r(S_0-I)(r^2I-S_0)^{-1}||_\infty < 1$$

By definition, the above inequality is equivalent to requiring that $\forall \omega \in \mathbb{R}$,

$$r^2[r^2I-S_0'(-j\omega)]^{-1}[S_0'(-j\omega)-I][S_0(j\omega)-I][r^2I-S_0(j\omega)]^{-1}-I<0 \quad (105)$$

Since the left-hand term of the above equals

$$(r^2-1)[r^2I-S_0'(-j\omega)]^{-1}[S_0'(-j\omega)S_0(j\omega)-r^2I][r^2I-S_0(j\omega)]^{-1},$$

$||S_0||_\infty < r$ implies Eq. (105). Thus, the "only if" part of the theorem is proved.

The "if" part can be proved by showing with a similar argument as above that if $C(s)$ solves the complementary sensitivity problem for r, then $\frac{r^2-1}{r^2}C(s)$ solves the sensitivity problem for r. □

As a by-product of the above theorem, we infer that for a multivariable plant, solvability of the sensitivity problem for $r > 1$ is equivalent to solvability of the complementary sensitivity problem for the same r. This obviously implies that if the minimal achievable sensitivity or the minimal achievable complementary sensitivity of the plant is greater than 1, they must be equal.

VI. Reduction of Conflict between Gain/Phase Margin and Sensitivity Using Periodic Digital Control

Periodic controllers have been shown to be superior to LTI controllers in many aspects such as robustness, pole assignment, deadbeat control, removing decentralized fixed modes, and simultaneous stabilization, see e.g. [6, 7, 8, 9, 10, 11]. One remarkable example [9] is that an arbitrarily large gain margin can be achieved for a discrete-time LTI nonminimum phase plant with periodic control. This result has been extended to the multivariable case in [8]. Obviously, the main reason for superiority of periodic control to constant control in a number of ways is that the former offers substantially more design freedom than the latter. Of course, this extra freedom does not always enable periodic controllers to overcome every design limitation associated with LTI control. For instance, it is known [9, 12] that periodic compensation cannot achieve better sensitivity than time-invariant compensation for a discrete-time LTI plant.

Recall from Sections III and IV that there always exists conflict between gain/phase margin and sensitivity for a nonminimum phase unstable LTI plant in the context of LTI compensation. This conflict is twofold. On the one hand, sensitivity minimization will cause the gain/phase margin to be less than the square root of the maximal gain margin. On the other hand, gain margin maximization will lead to an arbitrarily large sensitivity. Apparently, the second kind of conflict is much more serious than the first kind in practice.

In this section, we consider a particular kind of periodic dynamic controllers which consist of a sampler, an LTI discrete-time compensator, and a generalized sampled-data hold function (GSHF) gain. A controller of such a kind will be called a GSHF dynamic compensator. Our main purpose here is to display advantages of GSHF dynamic compensators over LTI compensators in reduction of the above mentioned conflict associated with

LTI compensation, as well as in gain margin improvement for multivariable continuous-time LTI systems.

A. Stabilization Results

Consider a continuous-time LTI plant $P(s)$ with a minimal state-space model:

$$\dot{x}(t) = Ax(t) + Bu(t) \tag{106}$$
$$y(t) = Cx(t) + Du(t) \tag{107}$$

where $x(t) \in I\!R^n$ is the state, $u(t) \in I\!R^m$ is the control, $y(t) \in I\!R^r$ is the output, and A, B, C, D are real matrices of appropriate dimensions.

The GSHF dynamic compensator is composed of an LTI compensator and a GSHF control law as follows:

$$z_{k+1} = A_c z_k + B_c y(kT) \tag{108}$$
$$v_k = C_c z_k + D_c y(kT) \tag{109}$$
$$u(t) = F(t)v_k, \quad \text{for } t \in [kT,\ (k+1)T), \quad k = 0, 1, 2, \cdots \tag{110}$$

where $T > 0$ is the sampling period, A_c, B_c, C_c, and D_c are constant matrices of compatible dimensions, the dimension of v_k is less than or equal to n, and $F(t)$ is a T-periodic integrable and bounded hold function matrix of a compatible dimension.

The following equation

$$\int_0^T \exp[A(T-t)]BF(t)dt = G \tag{111}$$

for the unknown $F(t)$ with G a given constant matrix, plays an important role in the design of the GSHF control law Eq. (110). The properties with respect to this equation are summarized in the following result.

Lemma 6.1 *Let $(A,\ B)$ be controllable, G be given and*

$$W(A,B,T) = \int_0^T \exp[A(T-t)]BB^\tau \exp[A^\tau(T-t)]dt. \tag{112}$$

Then

(i)

$$F_0(t) = B^\tau \exp[A^\tau(T-t)]W^{-1}(A,B,T)G \tag{113}$$

is the uniquely optimal one among all the solutions of Eq. (111) in the sense of minimizing $tr\int_0^T F^\tau(t)F(t)dt$;

(ii) *for almost all $T > 0$, there exists a piecewise constant solution of Eq. (111) taking at most n different values in the interval $[0, T]$;*

(iii) *for almost all $T > 0$, the optimal solution $F_0(t)$ can be approximated by a sequence of piecewise constant solutions $F_k(t)$ of Eq. (111) uniformly in the interval $[0, T]$ in a usual matrix norm.*

Proof: The proof is omitted here. □

Applying the GSHF control law Eq. (110) to the system Eqs. (106)-(107) and sampling the continuous-time state and output, we obtain the corresponding discrete-time system from v_k to $y(kT)$:

$$x((k+1)T) = \exp(AT)x(kT) + Gv_k \tag{114}$$

$$y(kT) = Cx(k) + DG_0 v_k \tag{115}$$

with $G_0 = F(0)$ and G is given as in Eq. (111). The appearance of $exp(AT)$ is no surprise. On the other hand, the fact that G can be arbitrary is perhaps a surprise; the zeros of Eqs. (114)-(115) baer no relation to those of Eqs. (106)-(107), or the discrete time system obtained with a conventional zero order hold. Now the necessary and sufficient condition for the original continuous-time system to be stabilized by a GSHF dynamic compensator can be stated as follows.

Theorem 6.1 *There exists a GSHF dynamic compensator (108)-(110) stabilizing the continuous-time plant (106)-(107) if and only if there exist constant matrices G_0 and G such that the induced discrete-time system (114)-(115) is stabilized by the LTI controller (108)-(109).*

Proof: The proof is omitted here since it is standard upon noting that for a given matrix pair (G, G_0) and any fixed $t_1 \in [0, T)$, there are an infinite number of T-periodic integrable and bounded function matrices $F(t)$ satisfying

$$F(t) = G_0, \quad t \in [0, t_1] \tag{116}$$

$$F(t+T) = F(t), \quad t \geq 0 \tag{117}$$

$$\int_0^T \exp[A(T-t)]BF(t)dt = G \tag{118}$$

□

Quite obviously, for almost all $T > 0$ and almost all constant matrices G_0 and G the induced discrete-time system (114)-(115) is controllable and observable.

B. Gain/Phase Margin Improvement

Let us first fix some notation. Given the multivariable plant $P(s)$ with the minimal realization (106)-(107), $\mathbf{Z} \triangleq \{z_1, \ldots, z_{N_1}\}$ denotes the set of all its zeros in $\bar{H}$, $\mathbf{P} \triangleq \{p_1, \ldots, p_{N_2}\}$ the set of its poles in $\bar{H}$, For simplicity, all the p_i's and z_j's will be assumed to be simple. For a fixed sampling time $T > 0$ the maximal achievable gain margin $\mathbf{K}_T$ of $P(s)$ with respect to GSHF dynamic compensation is defined as

$$\begin{aligned}\mathbf{K}_T \quad \triangleq \quad & \sup\{b/a \,:\, 0 < a < 1 < b \quad \text{and} \quad \exists \text{ a controller } (108) - (110) \\ & \text{stabilizing } kP(s) \text{ for all } k \in [a,\, b]\}.\end{aligned}$$

We are in a position to derive the explicit expression for $\mathbf{K}_T$ in the case where $P(s)$ is a SISO strictly proper plant with $\mathbf{P} \cap H \neq \emptyset$ and $\mathbf{Z} \cap H \neq \emptyset$. To this end, it is not hard to note from Theorem 6.1 that $\mathbf{K}_T$ is equal to the supremum among all the maximal gain margins of the discrete-time plant (114)-(115) with respect to LTI compensation which are corresponding to all pairs of constant matrices $(G,\ G_0)$. Since $D = 0$, the maximal gain margin of the plant (114)-(115) with respect to LTI compensation is independent of G_0 and denoted by $\mathbf{K}(G)$. Further, observe that the unstable poles of (114)-(115) are $\exp(p_i T)$ $(i = 1,\ \ldots,\ N_2)$ independent of G and that (114)-(115) always has a zero at infinity irrespective of choice of G. Since $\mathbf{K}(G)$ is solely determined by the unstable zeros and poles, and is nonincreasing as the set of unstable zeros expands, it follows that $\mathbf{K}(G)$ reaches its maximum if G can be found for which (114)-(115) has no unstable finite zeros and the zero at infinity is simple (i.e. $CG \neq 0$). Such a G evidently exists since finite zeros of (114)-(115) can be arbitrarily placed by suitable choice of a vector G whenever $(\exp(AT),\ C)$ is observable. Therefore for the sampling time T for which $(\exp(AT),\ C)$ is observable and $\exp(p_i T) \neq \exp(p_j T)$ if $i \neq j$, $\mathbf{K}_T = \max_G \mathbf{K}(G)$ can be expressed as

$$\mathbf{K}_T = \left(\frac{1 + \alpha_T}{1 - \alpha_T}\right)^2 \tag{119}$$

where

$$\begin{aligned}\alpha_T \quad = \quad & \max\{\gamma > 0; \quad \exists f \text{ analytic: } \bar{D} \to D \text{ such that} \\ & f(0) = \gamma \quad \text{and} \quad f(\exp(-p_i T)) = 0, \quad 1 \leq i \leq N_2\} \end{aligned} \tag{120}$$

By the Nevanlinna-Pick interpolation theory, the formula for α_T is given by

$$\alpha_T = \max\{\gamma > 0; \quad \begin{bmatrix} 1 - \gamma^2 & e \\ e^T & L_T \end{bmatrix} \geq 0\} \quad \textit{and} \quad e = [1 \ \cdots \ 1]$$

which turns out to be

$$\alpha_T = \sqrt{1 - [1 \cdots 1] L_T^{-1} [1 \cdots 1]^\tau} \tag{121}$$

with

$$L_T \triangleq \left[\frac{1}{1 - \exp[-(p_i + \bar{p}_j)T]} \right]_{N_2 \times N_2}$$

The bounds for α_T are given by

$$\left[\frac{N_2}{1 + \sum_{i=1}^{N_2} \frac{1}{1 - \exp(-2Re(p_i)T)}} \right]^{N_2/2} \prod_{1 \le i < j \le N_2} |\exp(-p_j T) - \exp(-p_i T)| \times$$

$$\prod_{1 \le i \le N_2} \exp(-p_i T) \le \alpha_T \le \sqrt{1 - N_2^2 \left(\sum_{1 \le i,j \le N_2} \frac{1}{1 - \exp[-(p_i + \bar{p}_j)T]} \right)^{-1}} \tag{122}$$

The main properties of $\mathbf{K}_T$ with respect to T are stated in the following result.

Theorem 6.2 *Let* $\mathbf{K}_T$ *be the maximal gain margin of a scalar plant with respect to GSHF dynamic compensation. Then*

(i) $\mathbf{K}_T$ *is strictly decreasing with respect to* T

(ii) $\lim_{T \to 0} \mathbf{K}_T = \infty$ and $\lim_{T \to \infty} \mathbf{K}_T = 1$

Proof: (i) Recall that the original expression for α_T is given in Eq. (120). Now we claim that $\alpha_{T_1} \ge \alpha_{T_2}$ for $T_2 > T_1 > 0$. In fact, for any fixed $\gamma < \alpha_{T_2}$, there exists an analytic function $f_{T_2}(s) : \bar{D} \to D$ such that

$$f_{T_2}(0) = \gamma \quad \text{and} \quad f_{T_2}(\exp(-p_i T_2)) = 0, \quad i = 1, 2, \ldots, N_2$$

Construct

$$f_{T_1}(s) = \left\{ \prod_{i=1}^{N_2} f_{T_2}(\exp[(T_1 - T_2)p_i]s) \right\}^{1/N_2}$$

Then it is routine to verify that $f_{T_1}(s)$ is an analytic function from $\bar{D}$ to D such that

$$f_{T_1}(0) = \gamma \quad \text{and} \quad f_{T_1}(\exp(-p_i T_1)) = 0, \quad i = 1, 2, \ldots, N_2$$

As a consequence, $\alpha_{T_1} \ge \alpha_{T_2}$. From Eq. (121), it is easily seen that $(\alpha_T)^2$ can be regarded as an analytic function of T in the complex plane. Thus,

$(\alpha_T)^2$ cannot be identically constant on any real interval of T otherwise it is constant in the whole plane. But, $(\alpha_T)^2$ is nonincreasing; hence, it must be strictly decreasing, which is evidently equivalent to the condition that $\mathbf{K}_T$ is strictly decreasing with respect to T.

(ii) Since

$$0 \leq \alpha_T \leq \sqrt{1 - N_2^2 \left(\sum_{1 \leq i,j \leq N_2} \frac{1}{1 - \exp[-(p_i + \bar{p}_j)T]} \right)^{-1}}$$

and the right-hand expression approaches zero as T goes to infinity, it follows that

$$\lim_{T \to \infty} \alpha_T = 0$$

leading to

$$\lim_{T \to \infty} \mathbf{K}_T = 1$$

To see $\lim_{T \to 0} \mathbf{K}_T = \infty$, let us now prove a stronger result. In fact,

$$\lim_{T \to 0} TL = \lim_{T \to 0} \left[\frac{T}{1 - \exp[-(p_i + \bar{p}_j)T]} \right]_{N_2 \times N_2} = \left[\frac{1}{p_i + \bar{p}_j} \right]_{N_2 \times N_2} > 0$$

Consequently,

$$\lim_{T \to 0} L^{-1} = \lim_{T \to 0} T(TL)^{-1} = 0$$

which implies that $\lim_{T \to 0} \mathbf{K}_T = \infty$. □

Theorem 6.2 shows that GSHF dynamic compensators can achieve an arbitrarily large gain margin provided that the sampling time is chosen sufficiently small. In addition, an important relation between the maximal gain margins associated with LTI compensation and GSHF dynamic compensation, respectively, can readily be deduced as follows.

Corollary 6.1 *For a scalar continuous-time plant, there exists a sampling time $T_0 > 0$ such that*

$$\bar{\mathbf{K}} - \mathbf{K}_T \begin{cases} < 0, & T < T_0 \\ = 0, & T = T_0 \\ > 0, & T > T_0 \end{cases} \tag{123}$$

where $\bar{\mathbf{K}}$ denotes the maximal gain margin of the plant $P(s)$ with respect to LTI compensation.

In the special case where the scalar and strictly proper plant $P(s)$ only has one open RHP pole p and one open RHP zero z, T_0 is easily derived as

$$\frac{1}{2\text{Re}p} \ln \left| \frac{p + \bar{z}}{p - z} \right|$$

From this, it can be seen that the closer the unstable pole is to the unstable zero, the more flexible the choice of the period T can be in order for GSHF dynamic compensation to be superior to LTI compensation in respect of gain margin; of course, LTI compensation is bad when the pole and zero are close.

Next, we turn to a discussion of the multivariable case as well as the scalar and nonstrictly proper case. But first, let us note a result on the gain margin of an MIMO discrete-time LTI plant with an LTI compensator.

Lemma 6.2 *Let $P(z)$ be an MIMO LTI discrete-time plant. Assume that $P(z)$ has a minimal realization (A, B, C, D) with $A \in I\!R^{n\times n}$ and all its poles in D^c are denoted by $p_1, \ldots, p_N$. If*

$$\operatorname{rank} D \geq n - \min_i \operatorname{rank}(A - p_i I) \tag{124}$$

$$\operatorname{rank}\begin{bmatrix} A - zI & B \\ C & D \end{bmatrix} \geq 2n - \min_i \operatorname{rank}(A - p_i I), \quad \forall z \in D^c \setminus \{p_1, \ldots, p_N\} \tag{125}$$

then for any given pair a, b with $0 < a < 1 < b$ there exists an LTI controller stabilizing $kP(z)$ for all $k \in (a, b)$.

Proof: Let $\mathbf{R}$ denote the ring of proper rational functions which are stable in the discrete-time sense. Then $P(z)$ has a Smith-McMillan form over $\mathbf{R}$ as follows:

$$\begin{bmatrix} \operatorname{diag}(n_1/d_1, n_2/d_2, \ldots, n_l/d_l) & 0 \\ 0 & 0 \end{bmatrix},$$

where $l = \operatorname{rank} P(z)$, 0 represents the zero matrix of appropriate size, n_i divides n_{i+1}, d_{i+1} divides d_i, and (n_i, d_i) are coprime over $\mathbf{R}$. It is clear that the maximal gain margin of $P(z)$ with respect to LTI compensation is greater than or equal to $\min(k_1, \ldots, k_l)$, where k_i represents the maximal gain margin of n_i/d_i with respect to LTI compensation. Thus, to prove Lemma 6.2, it suffices to show that n_i/d_i $(i = 1, \ldots, l)$ has either no zeros in D^c or no poles in D^c since this implies that k_i is unbounded . To this end, in order to secure a contradiction suppose that there exist $1 \leq j \leq l$, $1 \leq \gamma \leq N$, and $z_0 \in D^c$, $z_o \neq p_\gamma$ such that

$$n_j(z_0) = 0 \quad \text{and} \quad d_j(p_\gamma) = 0\,.$$

Obviously, the former implies that $\operatorname{rank} P(z_0) \leq j - 1$, which is equivalent to

$$\operatorname{rank} D \leq j - 1, \quad \text{if } z_0 \text{ is infinite} \tag{126}$$

$$\operatorname{rank}\begin{bmatrix} A - z_o I & B \\ C & D \end{bmatrix} \leq n + j - 1, \quad \text{if } z_0 \text{ is finite.} \tag{127}$$

On the other hand, since

$$P(z) = [(C - DK)(zI - A + BK)^{-1}B + D][I - K(zI - A + BK)^{-1}B]^{-1}$$

is a coprime factorization over **R** provided that K is an arbitrary matrix such that all the eigenvalues of $A - BK$ lie in D, the nonunit invariant factors (over **R**) of $[I - K(zI - A + BK)^{-1}B]$ and $\text{diag}(d_1, d_2, \ldots, d_l)$ are the same, see [5]. As a consequence,

$$\text{rank}[I - K(p_\gamma I - A + BK)^{-1}B] \leq n - j$$

or equivalently,

$$\text{rank}\begin{bmatrix} I & K \\ B & p_\gamma I - A + BK \end{bmatrix} \leq m + n - j$$

from which it follows that

$$\text{rank}(A - p_\gamma I) \leq n - j$$

This together with Eq. (126) or Eq. (127) readily leads to a contradiction to the lemma condition (124) or (125). Therefore, n_i/d_i has either no zeros in D^c or no poles in D^c for all $i = 1, \ldots, l$. □

Theorem 6.3 *Consider the MIMO LTI continuous-time plant $P(s)$. Let $\mathbf{K}_T$ be defined as in Section VI.B. Assume that $P(s)$ has a minimal realization (106)-(107) with all its poles in $\bar{H}$ denoted by $p_1, \ldots, p_{N_2}$. If*

$$\text{rank}D \geq n - \min_i \text{rank}(A - p_i I) \tag{128}$$

then

$$\mathbf{K}_T = \infty, \quad \forall T > 0 \tag{129}$$

Proof: Let T be any fixed sampling period with $T \neq \frac{2k\pi}{Im(p_i - p_j)}$ for all integers k whenever $\text{Re}(p_i - p_j) = 0$. Then, $(\exp(AT), C)$ is observable. It will now be shown that $\mathbf{K}_T = \infty$. But in view of Lemma 6.2, it obviously suffices to show that there exist two constant matrices G and G_0 of appropriate dimensions such that

$$\text{rank}DG_0 \geq n - \min_i \text{rank}[\exp(AT) - \exp(p_i T)I] \tag{130}$$

$$\text{rank}\begin{bmatrix} \exp(AT) - zI & G \\ C & DG_0 \end{bmatrix} \geq 2n - \min_i \text{rank}[\exp(AT) - \exp(p_i T)I], \quad \forall z \in \mathbf{C} \tag{131}$$

To do this, choose $m' \geq n - \min_i \operatorname{rank}(A - p_i I) + 1$. Since

$$\operatorname{rank}[\exp(AT) - \exp(p_i T)I] = \operatorname{rank}(A - p_i I)$$

Eq. (128) implies that Eq. (130) holds for almost all $m \times m'$ constant matrices G_0. In addition, from

$$\operatorname{rank} C \geq n - \min_i \operatorname{rank}(A - p_i I)$$

it is easy to see that for an arbitrary constant matrix G_0 and $\forall z \in \mathbf{C}$

$$\begin{aligned} \operatorname{rank}\begin{bmatrix} \exp(AT) - zI \\ C \end{bmatrix} + m' &\geq 2n - \min_i \operatorname{rank}(A - p_i I) + 1 \\ \operatorname{rank}\begin{bmatrix} \exp(AT) - zI & 0 & I \\ C & DG_0 & 0 \end{bmatrix} &\geq 2n - \min_i \operatorname{rank}(A - p_i I) \end{aligned}$$

Thus, our appeal to Lemma 7 in [13] shows that Eq. (131) generically holds with respect to $(G, G_0) \in \mathbb{R}^{n \times m'} \times \mathbb{R}^{m \times m'}$, from which the theorem follows. □

Several observations can be made.

- Eq. (128) clearly holds if $P(s)$ is scalar and bicausal; (recall that the p_i are all simple, by assumption).
- G can still be chosen to remove the possible effect of finite unstable zeros on the gain margin even if the condition (128) fails.
- If $p_1, p_2, \ldots, p_{N_2}$ are all simple, then Eq. (128) implies that $P(s)$ does not have a blocking zero at infinity.
- in the design of a GSHF compensator (108)-(110), the dimension of v_k can be chosen to be any integer greater than $n - \min_i \operatorname{rank}(A - p_i I)$ and the GSHF gain $F(t)$ can be chosen to have at most $n+1$ different values.

If the plant is bicausal, to avoid a delay-free loop, the controller should be strictly proper. Moreover, as has been shown in [14], stabilization by a nonstrictly proper controller is never robust against singular perturbations at the plant, whereas stabilization by a strictly proper controller is robust against singular perturbations. The following result shows the existence of a strictly proper GSHF dynamic compensator for achieving a prescribed gain margin in case the sampling time is allowed to be small.

Corollary 6.2 *Adopt the same hypothesis as in Theorem 6.3. Further assume that all the unstable poles $p_1, \ldots, p_{N_2}$ are simple and $D \neq 0$. Given k_1 and k_2 with $0 < k_1 < 1 < k_2$, there always exists a strictly proper GSHF compensator Eqs. (108)-(110) stabilizing $kP(s)$ for all $k \in [k_1, k_2]$.*

Proof: From the proofs of Theorem 6.3 and Lemma 6.2, we can find (G, G_o) such that $\bar{P}(z) = DG_0 + C(zI - \exp(AT))^{-1}G$ has a Smith-McMillan form over **R** as follows:

$$\begin{bmatrix} \mathrm{diag}(n_1/d_1,\ n_2/d_2,\ \ldots,\ n_l/d_l) & 0 \\ 0 & 0 \end{bmatrix}$$

Moreover, since all poles are simple and d_{i+1} divides d_i, we see that n_1/d_1 has unstable poles $\exp(p_i T)$ ($i = 1, \ldots, N_2$) but no unstable zeros while $n_2/d_2, \ldots, n_l/d_l$ have no unstable poles. Thus, there is no problem with n_i/d_i ($i = 1, \ldots, l$) for the construction of a strictly proper controller achieving any prescribed gain margin provided that T is allowed to be sufficiently small. This evidently implies that there exists a strictly proper LTI controller stabilizing $kP(z)$ for all $k \in [k_1,\ k_2]$. □

In the case where the condition (128) fails, we have

Theorem 6.4 *Let $P(s)$ be an MIMO LTI continuous-time plant with a minimal realization (106)-(107) with $D = 0$. Suppose that the plant unstable poles $p_1, \ldots, p_{N_2}$ are simple and lie in H. Then for almost all sampling periods T*

$$\mathbf{K}_T \geq \left(\frac{1+\alpha_T}{1-\alpha_T}\right)^2 \quad \text{and} \quad \lim_{T\to 0} \mathbf{K}_T = \infty \tag{132}$$

where α_T is defined as in Eq. (120).

Proof: Take arbitrary $T > 0$ such that $T \neq \frac{2k\pi}{Im(p_i - p_j)}$ for all integers k whenever $Re(p_i - p_j) = 0$. Then,

$$\min_i \mathrm{rank}[\exp(AT) - \exp(p_i T)I] = \min_i \mathrm{rank}(A - p_i I) = n - 1$$

and $(\exp(AT),\ C)$ is observable. Now from the proof of Theorem 6.3, we can generically choose $n \times m'$ constant matrix G with $m' > 1$ such that

$$\begin{aligned} \mathrm{rank} CG &\geq 1 \\ \mathrm{rank}\begin{bmatrix} \exp(AT) - zI & G \\ C & 0 \end{bmatrix} &\geq n+1, \quad \forall z \in \mathbf{C} \end{aligned}$$

For such a chosen G, the above inequalities yield that the first diagonal element of the Smith-McMillan form in **R** of $\bar{P}(z) = C(zI - \exp(AT))^{-1}G$ has a unique unstable zero at infinity of multiplicity 1 while all the other nonzero diagonal elements have no unstable poles and thus have the maximal gain margin of ∞. Consequently,

$$\mathbf{K}_T \geq \mathbf{K}(G,\ G_0) \geq \left(\frac{1+\alpha_T}{1-\alpha_T}\right)^2$$

which implies $\lim_{T\to 0} \mathbf{K}_T = \infty$, as was shown in the proof of (ii) of Theorem 6.2. □

To conclude this subsection, we mention that most of the results for gain margin in this subsection are valid for phase margin after obvious modifications. For example, under the condition of Theorem 6.3, the maximal phase margin of $P(z)$ is equal to π with respect to GSHF dynamic compensation.

C. Reduction of Conflict between Gain/Phase Margin and Sensitivity

In this subsection, we will show that periodic compensation can reduce the conflict between gain/phase margin and sensitivity for a MIMO LTI discrete-time plant.

Let $P(z)$ denote a $p \times m$ discrete-time LTI plant and $0 < a < 1 < b$. Define the minimal sensitivity $\mathcal{R}(a,b)$ subject to a gain margin interval constraint in the context of LTI compensation as in Eq. (50) and its counterpart $\tilde{\mathcal{R}}(a,b)$ in the context of periodic compensation as follows

$$\begin{aligned}\tilde{\mathcal{R}}(a,b) \quad &\triangleq \quad \inf\{\|(I+PC)^{-1}\|; \quad C \text{ is a linear periodic controller} \\ &\qquad \text{stabilizing } kP(z) \text{ for each } k \in [a,\, b]\}\end{aligned}$$

where $\|(I+PC)^{-1}\| \triangleq \sup\{\|(I+PC)^{-1}u\|_2 : u \text{ in } (h^2)^m,\ \|u\|_2 = 1\}$.

It is known from [9] that by a procedure of "lifting" or "blocking" involving rearrangement of inputs and outputs, any periodic system such as C or P can be uniquely represented by a transfer matrix $\tilde{C}(z)$ or $\tilde{P}(z)$ with $\tilde{C}(\infty)$ or $\tilde{P}(\infty)$ lower block triangular, and vice versa (of course, a time-invariant P is *a fortiori* periodic). Moreover:

1. $\|(I+PC)^{-1}\| = \|(I+\tilde{P}\tilde{C})^{-1}\|$
2. C stabilizes kP if and only if $\tilde{C}(z)$ stabilizes $k\tilde{P}(z)$

where $\tilde{P}(z)$ and $\tilde{C}(z)$ are the transfer matrix representations for the periodic plant P and controller C, and k is any constant. Thus it can be seen that

$$\begin{aligned}\tilde{\mathcal{R}}(a,b) \quad &= \quad \inf\{\|(I+\tilde{P}(z)\tilde{C}(z))^{-1}\|_\infty : \tilde{C}(\infty) \text{ is lower block} \\ &\qquad \text{triangular and } \tilde{C}(z) \text{ stabilizes } k\tilde{P}(z)\ \forall\, k \in [a,\, b]\}.\end{aligned}$$

Recall from Sections III and IV that if $P(z)$ is a scalar nonminimum phase unstable plant, the maximal gain or phase margin associated with LTI controllers is finite and seeking to achieve this maximum will inevitably lead to an infinite sensitivity. In contrast, with periodic compensation the sensitivity can be guaranteed to be uniformly bounded at the same time as any prescribed gain margin is achieved, as is shown below.

Theorem 6.5 *Let $P(z)$ be a $p \times m$ LTI discrete-time plant with distinct unstable poles $p_1, \ldots, p_N$. Assume that $P(z)$ has a minimal realization (A, B, C, D) with $A \in \mathbb{R}^{n \times n}$. If*

$$\text{rank} D \geq n - \min_i \text{rank}(A - p_i I) \tag{133}$$

then there exists some constant M dependent only upon $P(z)$ such that for any interval $[a, b]$ with $0 < a < 1 < b$,

$$\tilde{\mathcal{R}}(a, b) \leq M \tag{134}$$

Proof: First, choose an integer $\nu \geq n$ such that $p_i^\nu \neq p_j^\nu, \quad i \neq j$. As a consequence,

$$\text{rank}(A^\nu - p_i^\nu I) = \text{rank}(A - p_i I) \tag{135}$$

Viewing $P(z)$ as a ν-periodic plant, we can write down its LTI transfer matrix representation as follows

$$\tilde{P}(z) = \begin{bmatrix} D & 0 & \cdots & 0 \\ CB & D & \cdots & 0 \\ \vdots & \vdots & \ddots & \vdots \\ CA^{\nu-2}B & CA^{\nu-3}B & \cdots & D \end{bmatrix} + \begin{bmatrix} C^T & A^T C^T & \cdots & (A^T)^{\nu-1} C^T \end{bmatrix}^T (zI - A^\nu)^{-1} \begin{bmatrix} A^{\nu-1}B & \cdots & AB & B \end{bmatrix}$$

Let $U(z) = \text{block diag}(z^{-1} I_m \cdots z^{-1} I_m \; I_m)$. Then $\tilde{P}(z)U(z)$ has a stabilizable and detectable realization $(\tilde{A}, \tilde{B}, \tilde{C}, \tilde{D})$ with

$$\tilde{A} = \begin{bmatrix} A^\nu & A^{\nu-1}B & \cdots & AB & 0 \\ 0_{\nu m \times n} & & 0_{\nu m \times \nu m} & & \end{bmatrix}, \quad \tilde{B} = \begin{bmatrix} 0 & \cdots & 0 & B \\ I_m & \cdots & 0 & 0 \\ \vdots & \ddots & \vdots & \vdots \\ 0 & \cdots & I_m & 0 \\ 0 & \cdots & 0 & I_m \end{bmatrix}$$

$$\tilde{C} = \begin{bmatrix} C & D & \cdots & 0 & 0 \\ CA & CB & \cdots & 0 & 0 \\ \vdots & \vdots & \ddots & \vdots & 0 \\ CA^{\nu-1} & CA^{\nu-2}B & \cdots & CB & 0 \end{bmatrix}, \quad \tilde{D} = \begin{bmatrix} 0 & \cdots & 0 & 0 \\ \vdots & \ddots & \vdots & \vdots \\ 0 & \cdots & 0 & 0 \\ 0 & \cdots & 0 & D \end{bmatrix}$$

Also, note that all poles of $\tilde{P}(z)U(z)$ in D^c are given by $p_1^\nu, \ldots, p_N^\nu$. Using the relations

$$\text{rank}[A^{\nu-1}B \cdots AB \; B] = n \quad \text{and} \quad \text{rank}[C^T \; A^T C^T \cdots (A^T)^{\nu-1} C^T] = n$$

together with Eqs. (133)-(135), it is trivial to check that $(\tilde{A}, \tilde{B}, \tilde{C}, \tilde{D})$ fulfills the conditions (124)-(125). In this way, it follows from the proof of

Lemma 6.2 that if $\tilde{P}(z)U(z)$ has the following Smith-McMillan form over **R**

$$V_1(z)\tilde{P}(z)U(z)V_2(z) = \begin{bmatrix} \text{diag}(p_1(z), \ldots, p_l(z)) & 0 \\ 0 & 0 \end{bmatrix}$$

where $V_1(z)$ and $V_2(z)$ are unimodular matrices over **R**, then $p_i(z)$ has either no unstable zeros or no unstable poles for $i = 1, \ldots, l$. Hence for any given $\epsilon > 0$ and interval $[a, b]$ with $0 < a < 1 < b$ there exists $c_i(z)$ stabilizing $kp_i(z)$ for each $k \in [a, b]$ such that $\|(1 + p_i c_i)^{-1}\| < 1 + \epsilon$. Setting

$$\tilde{C}(z) = U(z)V_2(z) \begin{bmatrix} \text{diag}(c_1(z), \ldots, c_l(z)) & 0 \\ 0 & 0 \end{bmatrix} V_1(z)$$

we can see that $\tilde{C}(z)$ stabilizes $k\tilde{P}(z)$ for each $k \in [a, b]$ and $\tilde{C}(\infty)$ is lower block triangular. In addition,

$$\begin{aligned} \|(I + \tilde{P}\tilde{C})^{-1}\| &= \|V_1^{-1}(z)(I + V_1(z)\tilde{P}(z)\tilde{C}(z)V_1^{-1}(z))^{-1}V_1(z)\| \\ &\leq \|V_1^{-1}(z)\|\|V_1(z)\| \max_i \|(1 + p_i(z)c_i(z))^{-1}\| \\ &< (1 + \epsilon)\|V_1^{-1}(z)\|\|V_1(z)\| \end{aligned}$$

which implies that $\tilde{\mathcal{R}}(a, b) < (1 + \epsilon)M$, where $M \triangleq \|V_1^{-1}(z)\|\|V_1(z)\|$ only depends on $P(z)$. Since ϵ is arbitrary, Eq. (134) is concluded. □

In the same way, it can be shown that with periodic compensation any prescribed phase margin can be achieved without causing an arbitrarily large sensitivity.

D. An Example

In this subsection, we present an example to illustrate how a GSHF dynamic compensator can be designed to achieve any prescribed gain margin.

Consider a SISO LTI continuous-time plant described by

$$\begin{aligned} \dot{x}(t) &= Ax(t) + bu(t) \\ y(t) &= cx(t) + du(t) \end{aligned}$$

with

$$A = \begin{bmatrix} 0 & 1 \\ -12 & 7 \end{bmatrix}, \; b = \begin{bmatrix} 0 \\ 1 \end{bmatrix}, \; c = [-10 \;\; 4], \; d = 1$$

and the transfer function

$$P(s) = \frac{(s-1)(s-2)}{(s-3)(s-4)}$$

It has been indicated in [2] that this plant has a maximal gain margin of 1.114 with respect to LTI compensators. But by Theorem 6.3, the plant has a maximal gain margin of ∞ with respect to GSHF compensators. To construct a GSHF dynamic compensator which stabilizes $kP(s)$ for all $k \in [k_1,\ k_2]$ with $k_1,\ k_2$ given and $0 < k_1 < 1 < k_2$, first choose

$$T = 1/4,\ G_0 = [1\ \ 0],\ G = \begin{bmatrix} 0 & 0 \\ 0 & 1 \end{bmatrix}$$

Then

$$\bar{P}(z) = dG_0 + c[zI - \exp(AT)]^{-1}G = \begin{bmatrix} 1 & \frac{4z-7.265}{(z-2.117)(z-2.718)} \end{bmatrix}$$

which has the following Smith-McMillan form in $\mathbf{R}$:

$$\bar{P}(z)U(z) = \begin{bmatrix} \frac{(z+0.5)^2}{(z-2.117)(z-2.718)} & 0 \end{bmatrix}$$

with

$$U(z) = \begin{bmatrix} \frac{z+43.837}{z+0.5} & \frac{-4z+7.265}{(z+0.5)^2} \\ \frac{-9.376z+34.709}{z+0.5} & \frac{z^2-4.835z+5.755}{(z+0.5)^2} \end{bmatrix}$$

It is easy to see that

$$p_1(z) = \frac{(z+0.5)^2}{(z-2.117)(z-2.718)}$$

has a maximal gain margin of ∞ with respect to LTI compensators since it has no unstable zeros. Thus, there exists an LTI compensator which stabilizes $kp_1(z)$ for all $k \in [k_1,\ k_2]$. In [2], it is shown that all solutions to the gain margin problem for an LTI plant with LTI compensation can be found by parameterizing all the corresponding sensitivity functions, which are simply solutions to a certain interpolation problem. Using this idea, one of the required sensitivity functions in our case can be constructed as follows

$$s_1(z) = [4k_1k_2(z-2.117)(z-2.718)(2.117z-1)(2.718z-1)]/A(z)$$

and the corresponding LTI compensator is

$$c_1(z) = \frac{1-s_1(z)}{s_1(z)p_1(z)} = B(z)/[4k_1k_2(1-2.117z)(1-2.718z)(z+0.5)^2]$$

where

$$\begin{aligned} A(z) &= (23.016k_1k_2 + 22.603k_2 - 45.619k_1)(z^4-1) - \\ &\quad 130.62(k_1k_2 - k_1)(z^3 - z) + \\ &\quad (229.944k_1k_2 - 45.201k_2 - 185.743k_1)z^2 \\ B(z) &= (22.603k_2 - 45.619k_1)(z^4-1) + 130.62k_1(z^3 - z) - \\ &\quad (45.201k_2 + 185.743k_1)z^2 \end{aligned}$$

Define

$$\begin{aligned} C(z) &\triangleq U(z)\begin{bmatrix} c_1(z) \\ 0 \end{bmatrix} = \begin{bmatrix} \frac{z+43.837}{z+0.5} \\ \frac{-9.376z+34.709}{z+0.5} \end{bmatrix} c_1(z) \\ &= C_0 + \frac{C_1 z^4 + C_2 z^3 + C_3 z^2 + C_4 z + C_5}{z^5 + 0.66z^4 - 0.337z^3 - 0.245z^2 + 0.025z + 0.022} \end{aligned}$$

with

$$\begin{aligned} C_0 &= \begin{bmatrix} \frac{0.982}{k_1} - \frac{1.982}{k_2} \\ -\frac{9.199}{k_1} + \frac{18.567}{k_1} \end{bmatrix}, \quad C_1 = \begin{bmatrix} \frac{42.402}{k_1} - \frac{79.904}{k_2} \\ \frac{40.156}{k_1} - \frac{134.209}{k_2} \end{bmatrix} \\ C_2 &= \begin{bmatrix} -\frac{1.633}{k_1} + \frac{240.05}{k_2} \\ \frac{15.297}{k_1} + \frac{278.83}{k_2} \end{bmatrix}, \quad C_3 = \begin{bmatrix} -\frac{85.85}{k_1} - \frac{348.58}{k_2} \\ -\frac{70.42}{k_1} + \frac{231.49}{k_1} \end{bmatrix} \\ C_4 &= \begin{bmatrix} -\frac{1.01}{k_1} + \frac{246.85}{k_2} \\ \frac{9.429}{k_1} - \frac{216.01}{k_1} \end{bmatrix}, \quad C_5 = \begin{bmatrix} -\frac{43.07}{k_1} + \frac{86.933}{k_2} \\ -\frac{33.88}{k_1} + \frac{68.385}{k_1} \end{bmatrix}. \end{aligned}$$

Then $C(z)$ stabilizes $k\bar{P}(z)$ for each $k \in [k_1,\ k_2]$. On the other hand, one of the GSHF gains associated with G_0 and G by Eqs. (116)-(118) can be found as follows:

$$F(t) = \begin{cases} [1\ \ 0], & 0 \le t < 1/8 \\ [-7.276\ \ -5.102], & 1/8 \le t < 3/16 \\ [5.602\ \ 20.408], & 3/16 \le t < 1/4 \end{cases}$$

Thus, the GSHF compensator stabilizing $kP(s)$ for each $k \in [k_1,\ k_2]$ can be constructed as

$$\begin{aligned} z_{k+1} &= \begin{bmatrix} -0.66 & 0.337 & 0.245 & -0.025 & -0.022 \\ 1 & 0 & 0 & 0 & 0 \\ 0 & 1 & 0 & 0 & 0 \\ 0 & 0 & 1 & 0 & 0 \\ 0 & 0 & 0 & 1 & 0 \end{bmatrix} z_k + \begin{bmatrix} 1 \\ 0 \\ 0 \\ 0 \\ 0 \end{bmatrix} y(kT) \\ v_k &= [C_1 \quad C_2 \quad C_3 \quad C_4 \quad C_5]\, z_k + C_0 y(kT) \\ u(t) &= F(t) v_k, \quad \text{for } t \in [kT,\ (k+1)T), \quad k = 0, 1, 2, \cdots \end{aligned}$$

It is interesting to note that the order of the GSHF compensator does not increase as the required closed-loop gain margin (i.e. k_2/k_1) increases, and that the GSHF compensator becomes strictly proper when $k_2/k_1 = 45.619/22.603$.

VII. Conclusions

We have derived and presented results detailing the fundamental trade-offs involved between individual objectives in several combined objective designs. The key tool here has been the interpolation methods of [2]. With this tool we have reposed these combined problems as conformal mapping problems which are solvable from first principles or using a conformal map dictionary. Our aim has been to make explicit what is achievable and upon what properties of the plant this depends. From this latter viewpoint we then were able to manipulate some effective plant properties by using periodically time-varying feedback strategies, thereby permitting subsequent amelioration of certain design conflicts.

There are several obvious directions which appear for further research:

- How do these results extend to multi-input/multi-output systems? Are there natural counterparts to the design objectives? Are there counterparts to the trade-off conclusions?
- To what extent is frequency weighting able to be in corporated in the sensitivity and complementary sensitivity functions? How does this alter the nature of the compromises?
- What are the fundamental limitations to control performance with a particular plant and with the controller constrained only to be stabilizing and causal?

This latter question arises here because, while it appears from Section VI that several design constraints can be relaxed by using GSHF periodic control, the fact is that the intersample behavior of the plant is often sacrificed to achieve these zero positioning strategies [15].

These questions of achievable robustness and performance become even less well resolved in the case of restricted complexity controller. Conformal mapping theorems as presented here frequently produce irrational controllers, which need reduction before they cab be realized. There are clearly many areas of investigation available whose resolution is necessary before these multi-objective designs can be finalized. Our aim here has been to explore methods and results in in the compromises existing between particular objectives.

References

1. M. C. Tsai, D.-W. Gu, and I. Postlethwaite, "A State Space Approach to Superoptimal H^∞ Control Problems," *IEEE Trans. Autom. Control* AC-33, pp. 833–843 (1988).

2. P. P. Khargonekar and A. Tannenbaum, "Non-Euclidian Metrics and the Robust Stabilization of Systems with Parameter Uncertainty," *IEEE Trans. Autom. Control* AC-30, pp. 1005–1013 (1985).

3. J. H. Curtiss, *Introduction to Functions of a Complex Variable*, Marcel Dekker, New York and Basel, 1978.

4. H. Kober, *Dictionary of Conformal Representations*, Dover Publications, New York, 1952.

5. M. Vidyasagar, *Control Systems Synthesis: A Factorization Approach*, MIT Press, Cambridge, Massachusetts, 1985.

6. B. D. O. anderson and J. B. Moore, "Time-Varying Feedback Laws for Decentralized Control," *IEEE Trans. Autom. Control* AC-26, pp. 1133–1139 (1981).

7. A. B. Chammas and C. T. Leondes, "On the Design of Linear Time Invariant Systems by Periodic Output Feedback," *Int. J. Control* 27, pp. 885–903 (1978).

8. B. A. Francis and T. T. Georgiou, "Stability Theory for Linear Time-Invariant Plants with Periodic Digital Controllers," *IEEE Trans. Autom. Control* AC-33, pp. 820–832 (1988).

9. P. Khargonekar, K. Poolla, and A. Tannenbaum, "Robust Control of Linear Time-Invariant Plants Using Periodic Compensation," *IEEE Trans. Autom. Control* AC-30, pp. 1088–1096 (1985).

10. P. T. Kabamba, "Control of Linear Systems Using Generalized Sampled-Data Hold Functions," *IEEE Trans. Autom. Control* AC-32, pp. 772–783 (1987).

11. S. Lee, S. M. Meerkov, and T. Runolfsson, "Vibrational Feedback Control: Zeros Placement Capabilities," *IEEE Trans. Autom. Control* AC-32, pp. 604–611 (1987).

12. A. Feintuch and B. A. Francis, "Uniformly Optimal Control of Linear Feedback Systems," *Automatica* 21, pp. 563–574 (1985).

13. W.-Y. Yan and R. R. Bitmead, "Decentralized Control of Multi-Channel Systems with Direct Control Feedthrough," *Int. J. Control* 49, pp. 2057–2075 (1989).

14. M. Vidyasagar, "Robust Stabilization of Singularly Perturbed Systems," *Systems and Control Letters* 5, pp. 413–418 (1985).

15. C. Zhang and R. J. Evans, "Discrete Time Periodic Controller for Multivariable Linear Systems," *Proceedings of the 28th IEEE Conference on Decision and Control*, Flrioda, pp. 2319–2324 (1989).

16. P. L. Duren, *Theory of H_p Space*, Academic Press, New York, 1970.

17. A. Tannenbaum, *Invariance and System Theory: Algebraic and Geometric Aspects*, Springer-Verlag, Berlin and New York, 1981.

18. A. Tannenbaum, "Feedback Stabilization of Plants with Uncertainty in the Gain Factor," *Int. J. Control* 32, pp. 1–16 (1980).

19. J. Walsh, *Interpolation and Approximation by Rational Functions in the Complex Domain*, AMS Colloquium Publications 20, Fourth Edition, 1965.

20. J. S. Freudenberg and D. P. Looze, "Right Half Plane Poles and Zeros and Design Tradeoffs in Feedback Systems," *IEEE Trans. Autom. Control* AC-30, pp. 555–565 (1985).

21. T. Kailath, *Linear Systems*, Prentice-Hall, Englewood Cliffs, New Jersey, 1980.

22. G. Zames and B. A. Francis, "Feedback, Minimax Sensitivity, and Optimal Robustness," *IEEE Trans. Autom. Control* AC-28, pp. 585–601 (1983).

23. I. Horowitz and M. Sidi, "Optimum Synthesis of Nonminimum-Phase Feedback Systems with Parameter Uncertainty," *Int. J. Control* 27, pp. 361–386 (1978).

24. S. D. O'Young and B. A. Francis, "Sensitivity Tradeoffs for Multivariable Plants," *IEEE Trans. Autom. Control* AC-30, pp. 625–632 (1985).

Aspects of Robust Control Systems Design

Rafael T. Yanushevsky

University of Maryland
Mechanical Engineering Department
Baltimore, Maryland 21228

I. INTRODUCTION

Robustness of control systems refers to the ability of the system to maintain stability and/or performance characteristics in the presence of parameter uncertainties.

Usually, we consider systems with the so-called nominal parameters. It is natural to examine the behavior of the system under consideration when its parameters defer from their nominal values. The changes can be assumed small. In this case, the problem of robustness reduces to the investigation of the influence of small parameters on the system dynamics. Small parameters can arise also when a more precise mathematical description is considered. The analysis of systems with small parameters is the first important step in examining the robustness problem. A more rigorous analysis may be required to investigate the system dynamics assuming that some parameters of the system belong to certain domains. Two types of the problems mentioned are connected with robustness "in small" and robustness "at large".

The robustness "in small" problem was formulated and examined in the 1940's by Andronov [1]. A system was considered as unrobust if small changes of its parameters made the system unstable. The implementation of the Andronov approach to analyze robustness of a class of linear optimal control systems was given in [2]. The robustness problem was reduced to the analysis of differential equations with small parameters which may change the dimension of the state space vector representing the dynamic behavior of the system under consideration.

The definition of robustness, which is used in the up-to-date literature, differs from the definition proposed by Andronov. The new definition is of a more general nature because it assumes perturbed parameters to belong to a definite domain.

The relationship between the definitions mentioned is similar to the relationship between stability "in small" and "at large".

Of course, a system which is unrobust in the Andronov sense will be unrobust in the up-to-date sense. However, in practice, the robustness approach based on the Andronov definition can be applied to a specific limited number of control problems, e.g., problems of control systems design with a large gain, some problems with a simplified mathematical description. In the problems mentioned, the influence of small parameters on system dynamics should be examined.

The robustness "at large" needs different mathematical tools. The main problem consists in difficulty of applying a unified criterion not for one but for a family of dynamical systems. The most interesting results have been obtained for special classes of linear systems and for special types of domains the system uncertainties belong to [3-22].

Intensive research in the area of robust control was inspired by the Kharitonov theorem related to the asymptotic stability of a family of systems described by linear differential equations [3]. Different generalizations of the Kharitonov theorem for continuous and discrete linear systems in the time and frequency domains have been discussed in [4-12].

Based on criterion similar to the well-known ones in the classical automatic control theory, the problem of robust stability in linear time-invariant closed-loop systems with parametric uncertainty is formulated mostly, in essence, as the analysis problem. Using classical techniques, the stability of the system under consideration can be examined and the controller parameters, which guarantee the asymptotic stability of the closed-loop system, can be determined. Usually, it is assumed that a controller has been constructed for the so-called nominal plant. By using the Kharitonov-type theorem, one can only check if the system with perturbed parameters remains stable. There exists no effective design procedure of changing the controller parameters to make the real (not nominal) system asymptotically stable.

Parallel with the approach mentioned above, the Lyapunov-Bellman method is used to design controllers for systems with uncertain dynamics (see, e.g., [13-18]). In [13], a measure of stability robustness was presented in terms of the solution of a Lyapunov matrix equation. A modified Riccati equation is used in [12-18] to design the controller, which guarantees robust stability. The robustness problem is combined with the known H_2 and H_∞ problems.

Unlike the literature devoted to robustness of systems described by ordinary differential equations, we can point out only some papers related to the robustness problem for systems with delays [19-21] which generalize results obtained in [3,4,7].

The linear differential-difference systems represent the most general class of linear systems. Robustness will be considered for such a class of dynamical systems. The corresponding results for systems described by ordinary differential equations can be considered as a particular case of results discussed below.

The synthesis of robust control systems with uncertain parameters is based on the consideration of an optimal control problem with a specified performance index. The introduced estimate of the location of the eigenvalues of the plant with uncertain parameters allows us to formulate the optimal control problem, the solution of which guarantees the robustness of the control system. The main advantage of the proposed procedure is its simplicity. It is similar to the well-known procedure of the analytical controller design based on the solution of the linear quadratic optimal control problem with the integral quadratic performance index.

Uncertainties of plants represent only one part of the robustness problem. This part was paid significant attention in the up-to-date literature. However, these uncertainties are not the only source of uncertainties in dynamical systems. Parameters of controller are the second part of uncertainties which in some cases should be taken into account. Usually desired control laws are implemented approximately in practice, and in some cases such an approximation may change significantly the closed-loop system dynamics. As a result, the real system will have a higher order than the system with an "ideal" controller and may be unrobust. We will examine such a problem as a robustness problem in the Andronov sense.

II. ROBUSTNESS OF DIFFERENTIAL-DIFFERENCE SYSTEMS WITH STABLE D-OPERATOR

Let us consider the linear uniformly controllable plant described by linear differential-difference equations of the form

$$\sum_{i=0}^{l} C_i \dot{x}(t-\tau_i) = \sum_{i=0}^{l} A_i x(t-\tau_i) + B u(t) \tag{1}$$

$$x(t) = \varphi_x(t), \quad \dot{x}(t) = \varphi_{\dot{x}}(t), \quad -\tau \le t \le 0$$

where $x(t)$ is an m-dimensional vector; $u(t)$ is an n-dimensional control vector ($n \le m$); $0 = \tau_0 < \tau_1 ... < \tau_\ell = \tau$ are time delays associated with the system coordinates; $\varphi_x(t)$ and $\varphi_{\dot{x}}(t)$ are the initial functions, respectively $x(t)$ and $\dot{x}(t)$; C_i, $C_0 = I$, A_i and B are constant matrices of order compatible with $x(t)$ and $u(t)$.

By choosing a free basis $\{h_1,...,h_K\}$ for the Z-module $Z\tau_1 + ... + Z\tau_\ell$ and then defining delay operators

$$z_j x(t) = x(t-h_j) \quad j = 1,...,k \tag{2}$$

the system (1) may be written as a generalized linear system over the polynomial ring $R[z] = R[z_1,...,z_k]$

$$D(z)\dot{x}(t) = A(z)x(t) + Bu(t) \tag{3}$$

with appropriate definitions of $D(z)$, $A(z)$, and with the multi-index notation $z = (z_1,...,z_k)$ (see [23]).

It is assumed that the system (1) has a stable D-operator [24], and elements of its state matrices belong to some domains, i.e.,

$$A_{1i} \le A_i \le A_{2i}, \quad C_{1i} \le C_i \le C_{2i} \tag{4}$$

where

$$A_{1i} = [a_{ikj}^{(1)}],\ A_{2i} = [a_{ikj}^{(2)}],\ C_{1i} = [c_{ikj}^{(1)}],\ \text{and}\ C_{2i} = [c_{ikj}^{(2)}]$$

characterize the upper and lower bounds of A_i and C_i, respectively (the inequalities (4) are componentwise).

The robust control problem consists of finding controller equations which make the closed-loop system asymptotically stable for all matrices of the form (4).

The well-known procedure of analytical controller design for systems of the form (1) is based on minimization of the functional

$$J_o = \frac{1}{2}\int_o^\infty (x^T(t)Qx(t) + cu^T(t)u(t))dt \tag{5}$$

where $Q = [q_{ij}]$ is a non-negative definite symmetric matrix, c is a positive constant (see, e.g., [25,26]).

The optimal control law has the form [25,26]

$$u(t) = -\frac{1}{c}B^T[Wx(t) + \int_{-\tau}^{o}(B_2(\xi) + P_4^T(\xi,0))x(t+\xi)d\xi + (B_3(\xi)+P_6(0,\xi))\dot{x}(t+\xi))d\xi] \tag{6}$$

where the matrices $B_2(\xi)$, $B_3(\xi)$ and W, together with $P_i(\xi,\sigma)$ (i = 2,4,6) satisfy the Riccati equations

$$Q + A_o^T W + WA_o - \frac{1}{c}WBB^T W + B_2(0) + B_2^T(0) = 0 \tag{7}$$

$$\frac{dB_2(\xi)}{d\xi} - A_o^T(B_2(\xi) + P_4^T(\xi,0)) - P_2(0,\xi) + \frac{1}{c}WBB^T(B_2(\xi) + P_4^T(\xi,0)) = 0 \qquad -\tau \le \xi \le 0 \tag{8}$$

$$\frac{dB_3(\xi)}{d\xi} - A_o^T(B_3(\xi) + P_5(0,\xi)) - P_4^T(\xi,0) + \frac{1}{c}WBB^T(B_3(\xi) + P_5(0,\xi) = 0 \qquad -\tau \le \xi \le 0 \tag{9}$$

$$\frac{\partial P_2(\xi,\sigma)}{\partial \xi} + \frac{\partial P_2(\xi,\sigma)}{\partial \sigma} + \frac{1}{c}(B_2^T(\xi) + P_4(\xi,0)) \times BB^T(B_2(\sigma) + P_4^T(\sigma,0)) = 0 \qquad (10)$$
$$-\tau \le \xi \le 0,\ -\tau \le \sigma \le 0$$

$$\frac{\partial P_4(\xi,\sigma)}{\partial \xi} + \frac{\partial P_4(\xi,\sigma)}{\partial \sigma} + \frac{1}{c}(B_2^T(\xi) + P_4(\xi,0)) \times BB^T(B_3(\sigma) + P_6(0,\sigma)) = 0 \qquad (11)$$

$$\frac{\partial P_5(\xi,\sigma)}{\partial \xi} + \frac{\partial P_5(\xi,\sigma)}{\partial \sigma} + \frac{1}{c}(B_3^T(\xi) + P_6(\xi,0)) \times BB^T(B_3(\sigma) + P_6(0,\sigma)) = 0 \qquad (12)$$

$$\left.\begin{array}{c} (W + B_3(0))A_i - B_2(-\tau_i^+) + B_2(-\tau_i^-) = 0 \\ (i = 1, \ldots, l-1) \\ (W + B_3(0))\ A_l - B_2(-\tau) = 0 \end{array}\right\} \qquad (13)$$

$$\left.\begin{array}{c} (W+B_3(0))C_i - B_3(-\tau_i^+) + B_3(-\tau_i^-) = 0 \\ (W+B_3(0))C_l - B_3(-\tau) = 0 \end{array}\right\} \qquad (14)$$

$$\left.\begin{array}{c} A_i^T(B_2(\xi) + P_4^T(\xi,0)) - P_2(-\tau_i^+,\xi) + P_2(-\tau_i^-,\xi) = 0 \\ (i = 1, \ldots, l-1) \\ A_l^T(B_2(\xi) + P_4(\xi,0) - P_2(-\tau,\xi) = 0 \\ -\tau \le \xi \le 0 \end{array}\right\} \qquad (15)$$

$$\left.\begin{array}{c} (B_2^T(\xi) + P_4(\xi,0))\ C_i + P_4(\xi,-\tau_i^+) - P_4(\xi,-\tau_i^-) = 0 \\ A_i^T(B_3(\sigma) + P_6(0,\sigma)) - P_4(-\tau_i^+,\sigma) + P_4(-\tau_i^-,\sigma) = 0 \\ (i = 1, \ldots, l-1) \\ (B_2^T(\xi) + P_4(\xi,0))\ C_l + P_4(\xi,-\tau) = 0 \\ A_l^T\ (B_3(\sigma) + P_6\ (0,\sigma)) - P_4(-\tau,\sigma) = 0 \\ -\tau \le \xi(\sigma) \le 0 \end{array}\right\} \qquad (16)$$

$$\left.\begin{array}{c} C_i^T(B_3(\sigma) + P_5(0,\sigma)) + P_5(-\tau_i^+,\sigma) - P_6(-\tau_i^-,\sigma) = 0 \\ (i = 1, \ldots, l-1) \\ C_l^T(B_3(\sigma) + P_5(0,\sigma)) + P_5(-\tau,\sigma) = 0 \end{array}\right\} \qquad (17)$$

(The above equations are given in [25]; the upper indices "+" and "-" denote the left $(\tau_i - 0)$ and the right $(\tau_i + 0)$ limits at the points of discontinuity).

The minimal value of the functional (5) $J(x(\xi),\ \dot{x}(\xi))$ follows from the expression

$$\begin{aligned} 2J(x(t+\xi),\dot{x}(t+\xi)) &= x^T(t)Wx(t) + \\ &\int_{-\tau}^{0}(x^T(t)B_2(\xi)x(t+\xi) + x^T(t+\xi)B_2^T(\xi)x(t)\,d\xi \\ &+ \int_{-\tau}^{0}(x^T(t)B_3(\xi)\dot{x}(t+\xi) + \dot{x}^T(t+\xi)B_3^T(\xi)x(t))\,d\xi \\ &+ \int_{-\tau}^{0}\int_{-\tau}^{0}(x^T(t+\xi)P_2(\xi,\sigma)x(t+\sigma) + \dot{x}^T(t+\xi)P_5(\xi,\sigma)\dot{x}(t+\sigma))\,d\xi d\sigma \\ &+ \int_{-\tau}^{0}\int_{-\tau}^{0}(x^T(t+\xi)P_4(\xi,\sigma)\dot{x}(t+\sigma) + \dot{x}^T(t+\sigma)P_4^T(\xi,\sigma)x(t+\xi))\,d\sigma d\xi \end{aligned} \qquad (18)$$

It is known [26] (see also Appendix 1) that if the control law described by Eq. (6) minimizes the functional (5) subject to the system described by Eq. (1), then the control law

$$\begin{aligned} u(t) = -\frac{1}{c}B^T[Wx(t) &+ \int_{-\tau}^{0}(B_2(\xi) + P_4^T(\xi,0))e^{\gamma\xi}x(t+\xi)\,d\xi \\ &+ \int_{-\tau}^{0}(B_3(\xi) + P_5(0,\xi))\,\gamma e^{\gamma\xi}\dot{x}(t+\xi)\,d\xi] \end{aligned} \qquad (19)$$

minimizes the functional

$$J_1 = \frac{1}{2}\int_{0}^{\infty} e^{2\gamma t}\,(x^T(t)\,Qx(t) + cu^T(t)\,u(t))\,dt \qquad (20)$$

subject to the system described by Eq. (1).

We will demonstrate how this property can be used to solve the problem of robust control of the system (1).

Definition 1. The state matrices A_{i0} and C_{i0} (i = 0, 1, ..., ℓ) of the family (4) of the system (1) are called the strong dominant state matrices if the corresponding minimal value of $J_0\ (x(\zeta),\ \dot{x}(\zeta)) > J(x(\zeta),\ \dot{x}(\zeta))$, where $J(x(\zeta),\ \dot{x}(\zeta))$ is the minimal value of (5) for arbitrary A_i and C_i (i = 0,1,...,ℓ) of family (4).

The concept of the system with the strong dominant matrices is connected with the case of the most unfavorable displacement of the characteristic roots of the system (1) with state matrices (4).

In contrast to linear systems described by linear differential equations, for systems with delays there exist no criteria of positive definiteness of functionals of the form (18). Hence, there is no criteria to establish the positive definiteness of $J_0(x(\zeta), \dot{x}(\zeta)) - J(x(\zeta), \dot{x}(\zeta))$. In practice, the positive definiteness of (18) can be established by considering its approximation. Moreover, the solution of the Riccati equation (7) - (17) is obtained approximately on the basis of a discrete approximation. In the future, we will use the finite dimensional approximation of the system (1)

$$\dot{x}_0(t) = A_0 x_0(t) + \sum_{i=1}^{l} A_i z_{\tau_i}(t) - \sum_{i=1}^{l} C_i z^o_{\tau_i}(t) + Bu(t) \quad (21)$$

$$\frac{\tau}{N}\dot{z}_i(t) + z_i(t) = z_{i-1}(t) \quad (i = 1, \ldots, N)$$

$$\frac{\tau}{N}\dot{z}^o_i(t) + z^o_i(t) = z^o_{i-1}(t) \quad (i = 1, \ldots, N)$$

$$z_{\tau_i}(t) = z_{N\frac{\tau_i}{\tau_l}}(t),\ z^o_{\tau_i}(t) = z^o_{N\frac{\tau_i}{\tau_l}}(t),\ z_0(t) = x_0(t),\ z^o_0(t) = \dot{x}_0(t)$$

and the difference approximation of Eqs. (7) - (17) (we do not give here the boundary and discontinuity conditions; see [25]).

$$Q + A_0^T W + W A_0 - \frac{1}{c} WBB^T W + B_2[0] + B_2^T[0] = 0 \quad (22)$$

$$\frac{N}{\tau}(B_2[1-i] - B_2[-i]) - A_0^T(B_2[1-i] + P_4^T[1-i,0])$$
$$-P_2[0,1-i] + \frac{1}{c} WBB^T(B_2[1-i] + P_4^T[1-i,0]) = 0 \quad (23)$$

$$\frac{N}{\tau}(B_3[1-i] - B_3[-i]) - A_0^T(B_3[1-i] + P_6[0,1-i])$$
$$- P_4^T[1-i,0] + \frac{1}{c} WBB^T(B_3[1-i] + P_6[0,1-i]) = 0 \quad (24)$$

$$\frac{N}{\tau}(P_2[1-i,1-j] - P_2[-i,1-j] + P_2[1-i,1-j] - P_2[1-i,-j])$$
$$+ \frac{1}{c}(B_2^T[1-i] + P_4[1-i,0])\, BB^T(B_2[1-j] + P_4^T[1-j,0]) = 0$$
$$(25)$$

$$\frac{N}{\tau}(P_4[1-i,1-j] - P_4[-i,1-j] + P_4[1-i,1-j] - P_4[1-i,-j])$$
$$+ \frac{1}{c}(B_2^T[1-i] + P_4[1-i,0])\, BB^T(B_3[1-j] + P_5[0,1-j]) = 0$$
$$(26)$$

$$\frac{N}{\tau}(P_6[1-i,1-j] - P_6[-i,1-j] + P_5[1-i,1-j] - P_6[1-i,-j]) + \frac{1}{c}(B_3[1-i] + P_5[1-i,0])\, BB^T(B_3[1-j] + P_6[0,1-j]) = 0 \tag{27}$$

where $x_o(t)$, $z(t)$ and $z^o(t)$ are m-dimensional vectors; N is an integer.

By introducing the state and input matrices

$$B_N = \begin{bmatrix} B \\ \cdots \\ 0 \\ \cdots \\ \frac{N}{\tau}B \\ \cdots \\ 0 \\ \cdots \\ 0 \end{bmatrix} \tag{28}$$

$$A_N = \begin{bmatrix} A_o & \cdots & 0 & A_i & \cdots & 0 & A_l & \cdots & -C_i & \cdots & -C_l \\ \frac{N}{\tau}I & \cdots & 0 & 0 & \cdots & 0 & 0 & \cdots & 0 & \cdots & 0 \\ \cdots & \cdots & \cdots & \cdots & \cdots & \cdots & \cdots & \cdots & \cdots & \cdots & \cdots \\ 0 & \cdots & \frac{N}{\tau}I & -\frac{N}{\tau}I & \cdots & 0 & 0 & \cdots & 0 & \cdots & 0 \\ \cdots & \cdots & \cdots & \cdots & \cdots & \cdots & \cdots & \cdots & \cdots & \cdots & \cdots \\ 0 & \cdots & 0 & 0 & \cdots & \frac{N}{\tau}I & -\frac{N}{\tau}I & \cdots & 0 & \cdots & 0 \\ \frac{N}{\tau}A_o & \cdots & 0 & \frac{N}{\tau}A_i & \cdots & 0 & \frac{N}{\tau}A_l & \cdots & -\frac{N}{\tau}C_i & \cdots & -\frac{N}{\tau}C_l \\ \cdots & \cdots & \cdots & \cdots & \cdots & \cdots & \cdots & \cdots & \cdots & \cdots & \cdots \\ 0 & \cdots & 0 & 0 & \cdots & 0 & 0 & \cdots & 0 & \cdots & -\frac{N}{\tau}I \end{bmatrix} \tag{29}$$

and the positive definite matrix

$$W_N = \begin{bmatrix} W_{11} & W_{12} & \cdots\cdots & W_{1,N+1} \\ W_{21} & W_{22} & \cdots\cdots & W_{2,N+1} \\ \cdots & \cdots & \cdots\cdots & \cdots \\ W_{N+1,1} & W_{N+1,2} & \cdots\cdots & W_{N+1,N+1} \end{bmatrix} \tag{30}$$

with the elements

$$W_{11} = W, \quad \frac{N}{\tau}W_{1,i+1} = B_2[1-i], \quad \frac{N}{\tau}W_{1,i+N+1} = B_3[1-i] \text{ and}$$

$$(\frac{N}{\tau})^2 W_{i+1,j+1} = P_2[1-i,1-j], \quad (\frac{N}{\tau})^2 W_{i+1,j+N+1} = P_4[1-i,1-j],$$

$$(\frac{N}{\tau})^2 W_{i+N+1,j+N+1} = P_6[1-i,1-j]$$

the approximate system (14) with the state vector $x_N(t) = \{x_o(t), z_1(t), ..., z_N(t), z_1{}^o(t), ..., z_N{}^o(t)\}$, the system of difference equations (22) - (27), the optimal control $u_N[x_N(t)]$ in this system with respect to the functional (5) and the approximate optimal value J_N of the performance index (18) can be presented in the form [25,26]:

$$\dot{x}_N(t) = A_N x_N(t) + B_N u_N(t) \tag{31}$$

$$u_N(t) = -\frac{1}{c} B_N W_N x_N(t) \tag{32}$$

$$Q_N + A_N^T W_N + W_N A_N - \frac{1}{c} W_N B_N B_N^T W_N = 0 \tag{33}$$

$$J_N(x_o, z_i, z_i^o) = \frac{1}{2} x_N^T W_N x_N \tag{34}$$

where $u_N(t)$ denotes the control of the approximate system and

$$Q_N = \begin{bmatrix} Q & 0 \\ 0 & 0 \end{bmatrix} \tag{35}$$

The next consideration is based on the following theorems [25,26].

Theorem 1. Let $u[x(t)]$, $x[u(t)]$, $J(x(\xi),\dot{x}(\xi))$ and $u_N[x_N(t)]$, $x_N[u_N(t)]$, J_N be the optimal solutions of the original (1), (6), (18) and approximate (31), (32), (34) quadratic cost optimal problems. Then, for any arbitrary small $\delta > 0$, a number N_o can be found such that for all $N \geq N_o$

$$\int_0^\infty |x(t) - x_o(t)|^2 \, dt < \delta \tag{36}$$

$$| J(x(\xi), \dot{x}(\xi)) - J_N | \leq 0.5(1 + q)\delta \tag{37}$$

(The symbol $\| \cdot \|^2$ denotes the square of the euclidean norm of the corresponding vector; $q = max_{ij}\, q_{ij}$).

This theorem justifies the approximation of the type (21) - (34) for the study of the given class of delayed systems. It allows us to give the equivalent definition of the strong dominant state matrices of the original system (1).

Definition 2. The state matrices A_{io} and C_{io} (i = 0, 1, ..., ℓ) of the family (2) of the system (1) are called the strong dominant matrices if, for the corresponding family A_N of approximate systems (29) and (4), a number N_o can be found such that for all $N \geq N_o$ the minimal value J_{ON} of the performance index (5) of the approximate optimal problem is more than the minimal values J_N for arbitrary matrices of this family, i.e., $J_{ON} > J_N$ or $W_{ON} > W_N$, where W_{ON} is the positive definite solution of the Riccati equation (22) - (27) which is the approximation of the solution of Eqs. (7) - (17).

The condition $W_{ON} > W_N$ or $W_{ON} - W_N > 0$ has an advantage with respect to $J_{ON} > J_N$ because it is not connected with initial conditions and can be easily checked in practice on the basis of the Sylvester criterion [26] - [27].

If the system (1) with the state matrices of family (4) has no strong dominant A_{oi} and C_{oi}, we can build such matrices by extending the domain (4), i.e., by introducing, for example, $A_{1i}^o \leq A_{1i}$, $C_{1i}^o \leq C_{i1}$ and $A_{2i}^o \geq A_{2i}$, $C_{2i}^o \geq C_{2i}$ $(i = 0, 1, ..., \ell)$. Even if the system has the strong dominant A_{oi} and C_{oi}, sometimes it is difficult to find them in practice. That is why the way of extending the domain (4) to form the strong dominant matrices in this domain looks more attractive than the direct way to determine A_{oi} and C_{oi} of (4).

Definition 3. The matrices A_{oi} and C_{oi} (i = 0, 1, ..., ℓ) are called the dominant state matrices of the family (4) of the system (1) if they are the strong dominant state matrices of an extended family of the state matrices.

Below, we consider one of the simplest approaches to build A_{oi} and C_{oi}, which is based on shifting the eigenvalues of (1) (see Appendix 1). By choosing certain A_{oi} and C_{oi} (i = 0, 1, ..., ℓ) of family (4) and by including the matrices $(A_{oi} + \gamma C_{oi}) e^{\gamma \tau_i}$ and $C_{oi}\, e^{\gamma \tau_i}$, $\gamma \geq 0$ $(i = 0, 1, ..., \ell)$ in the extended domain (4), i.e.,

$$\left.\begin{aligned} A_{1i} \leq A_i \leq A_{2i} \quad & or \quad A_i - (A_{0i} + \gamma C_{0i})\, e^{\gamma\tau_i} \\ C_{1i} \leq C_i \leq C_{2i} \quad & or \quad C_i - C_{0i} e^{\gamma\tau_i} \} \end{aligned}\right\} \tag{38}$$

we can expect that the minimal value of the performance index (5), which is determined for a linear delayed system with $(A_{oi} + \gamma C_{oi}) e^{\gamma \tau_i}$ and $C_{oi} e^{\gamma \tau_i}$ $(i = 0, 1, ..., \ell)$, and the input matrix B, will be more than the minimum of (5) determined for the whole family (4) of the systems (1), i.e., these state matrices are the dominant state matrices of the system (1). This follows from the relationship between the optimal problems with the cost functionals (5) and (20) (see Appendix 1). The existence of γ, for which the matrices $(A_{oi} + \gamma C_{oi}) e^{\gamma \tau_i}$ and $C_{oi} e^{\gamma \tau_i}$ $(i = 0, 1, ..., \ell)$ are the dominant state matrices of (1) follows from the fact that the minimum value of (20) increases with an increase of γ.

The above consideration shows the importance of the optimal problem (1) and (20) in building robust time-lag systems.

The problem of robust control of the system described by Eq. (1) can be solved in two steps (see also [22]). First, an estimate of the appropriate γ should be established. Then, the optimal control problem with the performance index (20) for the system (1) with fixed constant parameters is considered.

The next theorem establishes the condition for the dominant state matrices to satisfy.

Let $W_{ON} = W_N + \Delta W_N$, $A_{oi} = A_i + \Delta A_i$, $C_{oi} = C_i + \Delta C_i$, $A_{ON} = A_N + \Delta A_N$ and

$$\Delta A_N = \begin{bmatrix} 0 & \Delta A_0 \ldots 0 & \Delta A_1 \ldots 0 & \Delta A_l \ldots 0 & \Delta C_l \\ 0 & 0 \ldots 0 & 0 \ldots 0 & 0 \ldots 0 & 0 \\ \ldots & \ldots & \ldots & \ldots & \ldots \\ 0 & 0 \ldots 0 & 0 \ldots 0 & 0 \ldots 0 & 0 \end{bmatrix} \tag{39}$$

where ΔA_i, ΔC_i, ΔW_N, and ΔA_N are some matrices, ΔA_N being formed from A_{ON} (see (29)) corresponding to A_{oi} and C_{oi} ($i = 0, 1, \ldots, \ell$).

Theorem 2. Suppose W_{0N} and W_N correspond to the optimal solutions (33) for the approximate systems (21) ((31)) with the state submatrices $(A_{oi} + \gamma C_{oi}) e^{\gamma\tau_i}$ and $C_{oi}e^{\gamma\tau_i}$ and with the state submatrices $A_i = A_{oi} - \Delta A_i$, $C_i = C_{oi} - \Delta C_i$ $(i = 0, 1, \ldots, \ell)$ of family (4). Then the matrices $(A_{oi} + \gamma C_{oi}) e^{\gamma\tau_i}$ and $C_{oi} e^{\gamma\tau_i}$ $(i = 0, 1, \ldots, \ell)$ are the dominant matrices of this family if there exists such N_0 that for all $N \geq N_0$

$$Q_N(W_{ON}) = 2\gamma W_{ON} + \Delta A_N^T W_{ON} + W_{ON}\Delta A_N \tag{40}$$

is a positive definite matrix for all ΔA_i and ΔC_i corresponding to A_i and C_i of family (4).

Proof: The optimal values W_{ON} and W_N of the functional (5) corresponding to the dominant matrices $(A_{oi} + \gamma C_{oi}) e^{\gamma\tau_i}$ and $C_{oi} e^{\gamma\tau_i}$ and arbitrary matrices of family (4), respectively, satisfy the Riccati equations

$$Q_N + 2\gamma W_{ON} + A_{ON}^T W_{ON} + W_{ON}A_{ON} - \frac{1}{c} W_{ON}B_N B_N^T W_{ON} = 0 \tag{41}$$

$$Q_N + A_N^T W_N + W_N A_N - \frac{1}{c} W_N B_N B_N^T W_N = 0 \tag{42}$$

Subtracting (42) from Eq. (41), we have

$$2\gamma W_{ON} + \Delta A_N^T W_{ON} + W_{ON}\Delta A_N + A_N^T \Delta W_N + \Delta W_N A_N$$
$$- \frac{1}{c}\Delta W_N B_N B_N^T W_N - \frac{1}{c} W_N B_N B_N^T \Delta W_N - \frac{1}{c}\Delta W_N B_N B_N^T \Delta W_N = 0 \tag{43}$$

or, according to (40),

$$Q_N(W_{ON}) + (A_N - \frac{1}{C}B_N B_N^T W_N)^T \Delta W_N$$

$$+ \Delta W_N (A_N - \frac{1}{C}B_N B_N^T W_N) - \frac{1}{C}\Delta W_N B_N B_N^T \Delta W_N = 0 \qquad (44)$$

The Riccati equation (44) corresponds to an asymptotically stable system with the state and input matrices $A_N - \frac{1}{C}B_N B_N^T W_N$ and B_N, respectively. It has a positive definite solution $\Delta W_N > 0$ if $Q_N(W_{ON}) > 0$ [28]. The positive definiteness of $Q_N(W_{ON})$ for all $N \geq N_o$ guarantees the existence of the dominant matrices for the approximate systems and, hence, for the original system (1).□

Remark: When $\gamma = 0$, then the condition (40) is a sufficient condition for the existence of the strong dominant state matrices.

In the future, only dominant matrices of the form $(A_o + \gamma C_{oi}) e^{\gamma\tau_i}$ and $C_{oi}e^{\gamma\tau_i}$ $(i = 1, ..., \ell)$ will be considered. We also assume that for systems with dominant and strong dominant matrices, the condition (40) is satisfied.

Theorem 3. Suppose the system described by Eq. (1) with the elements of the state matrices A_i and C_i (i = 0, 1, ..., ℓ) satisfying (4) has strong dominant matrices A_{oi} and C_{oi} (i = 0, 1, ..., ℓ). Then, the optimal control law (6) which minimizes the cost functional (5) subject to the system (1) with these matrices makes the system (1) and (6) asymptotically stable for all matrices A_i and C_i (i = 0, 1, ..., ℓ) satisfying (4).

Proof: For arbitrary A_i and C_i (i = 0, 1, ..., ℓ) of family (4) there exists the optimal control (6) and the positive definite solution W, $B_2(\xi)$, $B_3(\xi)$, and $P_i(\xi,\sigma)$ (i = 2, 4, 6) of the Riccati equation (7) - (17) (the quantities corresponding to the strong dominant matrices are denoted by the lower index "0"). Consider the system with arbitrary state matrices A_i, C_i and the control (6) determined for the strong dominant matrices

$$\sum_{i=0}^{l} C_i \dot{x}(t-\tau_i) = \sum_{i=0}^{l} A_i x(t-\tau_i) + B u_o(t) \qquad (45)$$

$$u_o(t) = -\frac{1}{C}[B^T(W_o + B_{30}^T(0))x(t)$$

$$+ \int_{-\tau}^{0}(B_{20}(\zeta) + P_{40}^T(\zeta,0))x(t+\zeta)d\zeta + \int_{-\tau}^{0}(B_{30}(\zeta) + P_{60}(0,\zeta))\dot{x}(t+\zeta)d\zeta] \qquad (46)$$

Its finite approximation has the form

$$\dot{x}_N(t) = A_N x_N(t) - \frac{1}{C}B_N B_N^T W_{ON} x_N(t) \qquad (47)$$

The derivative of the positive definite form $J_{ON} = \frac{1}{2} x_N^T(t)\, W_{ON}\, x_N(t)$ along the equation (47) is

$$2\frac{dJ_{ON}}{dt} = x_N^T (A_N^T W_{ON} + W_{ON} A_N - \frac{2}{c} W_{ON} B_N B_N^T W_{ON})\, x_N$$

$$= x_N^T (-Q_N - \frac{1}{c} W_{ON} B_N B_N^T W_{ON} - (\Delta A_N^T W_{ON} + W_{ON} \Delta A_N))\, x_N < 0 \quad (48)$$

Its negative definiteness follows immediately from the nonnegative definiteness of $\Delta A_N^T W_{ON} + W_{ON}\,\Delta A_N$ (see the condition (40)).

Hence, the system (47) is asymptotically stable for all state matrices A_i and C_i (i = 0, 1, ..., ℓ) of family (4). Theorem 1 establishes the proximity of behavior of the approximate and original closed-loop optimal systems (1), (6), and (31), (32). Analogous to the proof of Theorem 1, the proximity of the solutions of (45), (46), and (47) can be established. Therefore, the system (45) and (46) is asymptotically stable for all matrices A_i and C_i (i = 0, 1, ..., ℓ) of (4). □

Theorem 4. Suppose the system (1) with uncertain parameters (4) has dominant state matrices $(A_{oi} + \gamma C_{oi})e^{\gamma\tau_i}$ and $C_{oi}e^{\gamma\tau_i}$, where A_{oi} and C_{oi} $(i = 0, 1, ..., \ell)$ are fixed matrices of family (4). Then, the optimal control law (19) which minimizes the cost functional (20) subject to the system (1) with the fixed A_{oi} and C_{oi} (i = 0, 1, ..., ℓ) makes the system (1) and (19) asymptotically stable for all state matrices satisfying (4).

Proof: The minimization of the functional (20) subject to Eq. (1) is equivalent to the minimization of (5) subject to the system with state matrices $(A_{oi} + \gamma C_{oi})e^{\gamma\tau_i}$ and $C_{oi}e^{\gamma\tau_i}$ (see Appendix 1). The existence of dominant state matrices is equivalent to the existence of the strong dominant matrices of type $(A_{oi} + \gamma C_{oi})e^{\gamma\tau}$ and $C_{oi}e^{\gamma\tau_i}$ for the system described by Eq. (1) with uncertain parameters belonging to the domain (42). The needed result follows immediately from Theorem 3. □

As mentioned, the problem of robust control of the system (1) is solved in two steps. First, an estimate of the appropriate γ should be established. Then, the optimal control problem (20) for the system (1) with fixed constant parameters is considered. The chosen state dominant matrices should satisfy the condition (40).

As an estimate of γ, the upper bound to the real parts of the characteristic roots of Eq. (1) of family (4) can be used, i.e., such a number that for any matrices of A_i and C_i $(i = 0, 1, ..., \ell)$ satisfying (4) the zeros of the characteristic determinant

$$\det\,(s\sum_{i=0}^{l} C_i e^{-\tau_i s} - \sum_{i=0}^{l} A_i e^{-\tau_i s})$$

are in the half plane $\{s\colon Re\, s \leq \gamma\}$. For a class of differential-difference systems with $C_i = 0$, $i \neq 0$, the following estimates of γ can be used (see Appendix 2):

$$\gamma = \max_{\substack{r,\, a_{ikj} \\ Re\, s\geq 0}} [1, (m+\varepsilon) \max | M_r(s) |] \tag{49a}$$

$$\gamma = \min [\max_{k, a_{ikj}} \sum_{j=1}^{m} \sum_{i=0}^{l} | a_{ikj} |, \max_{j, a_{ikj}} \sum_{j=1}^{m} \sum_{i=0}^{l} | a_{ikj} |] \tag{49b}$$

$$\gamma = \frac{1}{2} m \max_{\substack{k,j \\ a_{ikj}}} \sum_{i=0}^{l} | a_{ikj} + a_{ijk} | \tag{49c}$$

where $M_r(s)$ denotes the sum of all principle minors of order r, $1 \leq r \leq m$; ϵ is a small positive number.

For ordinary linear differential systems, the width of the s- plane strip containing the eigenvalues of family (4) (the spread of real parts of the eigenvalues) can be used as an estimate of γ (see [27]).

$$\gamma = \max_{a_{ij}} (2 \| A_o \|^2 - \frac{2}{m} | trA_o |^2)^{1/2} \tag{49d}$$

$$\gamma = \max_{a_{ij}} \sqrt{2} \| A_o \| \tag{49e}$$

$$\gamma = \max_{a_{ij}} (2(1-\frac{1}{m}) (trA_o)^2 - 4\, E_2(A_o))^{1/2} \tag{49f}$$

where $E_2(A_o)$ is a sum of all principal minors of A_o of order 2, $trA_o = \sum_{i=1}^{m} a_{ii}$ is a trace of A_o; $\| A_o \|^2 = tr(A_o^T A_o)$.

Now we assume that the elements of the input matrix B belong to a domain, which is described by

$$B_1 \leq B \leq B_2 \tag{50}$$

where, unlike (4), the upper and lower bounds of b_{kj} is assumed to have the same sign.

Such a condition corresponds to the realistic case when an investigator knows qualitatively (but not quantitatively) how the controls influence the state variables.

Theorem 5. Let the control (19) be determined for the functional (20), arbitrary state matrices of the form (4), which correspond to the dominant state matrices $(A_{oi} + \gamma C_{oi})e^{\gamma\tau_i}$ and $C_{oi}e^{\gamma\tau_i}$, and the input matrix equal to B_1. Then, the closed-loop system described by Eqs. (1) and (19) is asymptotically stable for all the matrices A_i and C_i ($i = 0, 1, ..., \ell$) and B of the form (4) and (50), respectively.

This theorem can be considered as a part of the more general theorem which will be given below.

Let us consider the nonlinear controllable plant of the type

$$\sum_{i=0}^{l} C_i \dot{x}(t-\tau_i) = \sum_{i=0}^{l} A_i x(t-\tau_i) + Bg(u(t)) \tag{51}$$

where A_i, C_i, and B are matrices of the form (4) and (50); $g(u) = (g(u_1), ..., g(u_n))$ is a vector function whose elements satisfy the conditions

$$h_{ii}^{o}\, u_i^2 \leq u_i g_i(u_i), \quad h_{ii}^{o} > 0 \quad (i = 1, \ldots, n) \tag{52}$$

(It is assumed that $g(u)$ is smooth enough to guarantee the existence of the solution of (51) for any given initial conditions and controls $u(t)$).

Theorem 6. Let there exist the optimal solution (19) of the problem (1) and (20) for A_i of the form (4) which correspond to the dominant state matrices $(A_{gi} + \gamma C_{oi})e^{\gamma\tau_i}$, $C_{oi}e^{\gamma\tau_i}$, and $B = B_1 H$, $H^o - H \geq 0$, $H = [h_{ii}] > 0$ and $H^o = [h_{ii}^o] > 0$. Then the nonlinear system (19) and (51) is absolutely stable for all A_i, C_i, and B of the form (4) and (50), respectively.

Proof: For any fixed A_i, C_i, and B, the control (6) of the system (1) with the input matrix BH, which is optimal with respect to the performance index (5), or the control (19), which is optimal with respect to the performance index (20), makes the nonlinear system described by Eq. (51) asymptotically stable (see the corresponding theorem in [29] and Appendix 3). In the case of B belonging to the domain (50) we can describe the term $Bg(u)$ by introducing a set of nonlinear functions $\bar{g}(u)$ satisfying the inequality (52) and the condition $Bg(u) = B_1\bar{g}(u)$. This is due to the fact that only the lower bound is used in (52).

Hence, instead of the system described by Eq. (51) with the family of input matrices B and the given $g(u)$, we can examine the nonlinear system with the constant matrix B_1

$$\sum_{i=0}^{l} C_i \dot{x}(t-\tau_i) = \sum_{i=0}^{l} A_i x(t-\tau_i) + B_1 \bar{g}(u) \tag{53}$$

where the vector function $\bar{g}(u)$ satisfies the conditions (52).

According to the theorem mentioned above, the system described by Eqs. (51) and (19) under the control determined for $B = B_1 H$ is asymptotically stable. Finally, according to Theorem 4, the optimal control (19) determined for the system (1) with fixed A_i, C_i, and $B = B_1 H$ makes the system described by Eqs. (1) and (19) asymptotically stable for the entire family of A_i, C_i, and B ($i = 0, 1, ..., \ell$). □

In conclusion, we discuss how to check the positive definiteness of the matrix (40). This is a problem of the positive definiteness of a matrix with some elements belonging to domains of type (4) and it is enough to establish the lower bounds of positive definiteness with respect to the state matrices of family (4).

According to the Sylvester criterion of positive definiteness, all principal minors of (40) should be positive definite. If the minimal values of these minors with respect to A_i and C_i ($i = 0, 1, ..., \ell$) of family (4) are positive, then they will be positive for all A_i and C_i of family (4).

III. UNROBUSTNESS OF DIFFERENTIAL-DIFFERENCES SYSTEMS WITH UNSTABLE D-OPERATOR

The problem of feedback stabilization of differential-difference systems with unstable D-operator was considered in [30]. This work has shown that such systems cannot be stabilized by state feedback alone, but stability can be achieved by using derivative feedback. The synthesis procedure for the stabilization of differential-difference systems with unstable D-operator is discussed in [23,30]. A stabilizing control law has the form

$$u(t) = K(z)\dot{x}(t) + v(t) \tag{54}$$

where $K(z)$ is a polynomial in z, such that the resulting system has a stable D-operator, and v is a stabilizing control using state feedback (see, e.g., Eq. (6) and [23,25]).

We will call the solution (54) the "ideal" control law in contrast to the actual control law $u_a(t)$ because of the impossibility to reproduce exactly the derivative of the system coordinates in accordance with the control law (54).

We consider digital and analog variants of the approximate implementation of derivative feedback.

The simplest difference approximation of a derivative has the following form (more precise expressions have no significant difference)

$$\dot{x}(t) = \frac{1}{\varepsilon}(x(t) - x(t-\varepsilon)) \tag{55}$$

where ϵ is a small positive parameter.

Using this expression in the control law (54), we obtain the closed-loop system of the type

$$D(z)\dot{x}(t) = A(z)x(t) + \frac{1}{\varepsilon}BK(z)(x(t) - x(t-\varepsilon)) + Bv(t) \tag{56}$$

This system of differential-difference equations contains the same unstable D-operator as the original (1) and, therefore, is unstable.

For a law $u_a(t)$ approximating the "ideal" control law by using analog computations, we examine the following equation

$$\varepsilon \dot{u}_a(t) + u_a(t) = K(z)\dot{x}(t) + v(t) \qquad (57)$$

For $\epsilon \to 0$ we have $u_a(t) = u(t)$. The corresponding transfer matrix function $W_a(s)$ of the controller satisfies the condition $\lim_{\omega \to \infty} | W_a(i\omega) | < \infty$. This control law can be easily realized by operational amplifiers (different realizations give expressions similar to (57)).

The closed-loop system described by Eqs. (1) and (57) can be written in the form

$$D_1(z)\dot{y}(t) = A_1(z)y(t) + B_0 v(t) \qquad (58)$$

where $y(t) = \{x(t), u_a(t)\}$,

$$D_1(z) = \begin{bmatrix} D(z) & 0 \\ -K(z) & \varepsilon \end{bmatrix} \qquad (59)$$

$$A_1(z) = \begin{bmatrix} A(z) & B \\ 0 & -I \end{bmatrix} \qquad (60)$$

$$B_0 = \begin{bmatrix} 0 \\ I \end{bmatrix} \qquad (61)$$

Since the D-operator of the closed-loop system is unstable, the plant described by Eq. (1) cannot be stabilized by the control law (57).

The above consideration shows that "ideal" stabilizing controls cannot guarantee in practice the stability of differential-difference systems with unstable D-operator. Hence, the differential-difference systems with unstable D-operator are structurally unrobust [31].

IV. EXAMPLES

First, we consider the problem of robust control for the system

$$\dot{x}_1 = a_{11} x_1 + u$$
$$\dot{x}_2 = a_{22} x_2 + u$$

where

$$-1 \le a_{11} \le 0, \quad a_{22} = 1.$$

According to (49), we can choose $\gamma = 2$. Let

$$A_{00} = \begin{bmatrix} 0 & 0 \\ 0 & 1 \end{bmatrix}.$$

To prove that $A_{00} + 2I$ is the dominant matrix, we will solve the optimal problem with $q_{ii} = 1$, $q_{ij} = 0$, $i \neq j$, and $c = 1000$ (see Eq. (20)) and then use the condition (40) of Theorem 2. The corresponding solution W_0 of the algebraic Riccati equation of type (41) is

$$W_0 = \begin{bmatrix} 1 & -1.2 \\ -1.2 & 1.501 \end{bmatrix} \cdot 10^5$$

The condition (40) has the following form:

$$4.10^5 \cdot \begin{bmatrix} 1+0.5\Delta a_{11} & . & -1.2(1+0.25\Delta a_{11}) \\ -1.2(1+0.25\Delta a_{11}) & . & 1.501 \end{bmatrix} > 0$$

where $0 \leq \Delta a_{11} \leq 1$.

It is easy to check the positive definiteness of this matrix. The robust control law follows from the expression (19)

$$u = -20x_1 + 30x_2$$

Let the first equation of the system examined contains an additional term with delay. We consider the problem of the analytical design of a controller for the system

$$\dot{x}_1 = a_{11}x_1 + 0.05\, x_1(t-0.01) + u$$

$$\dot{x}_2 = x_{22} + u$$

where $-1 \leq a_{11} \leq 0$.

The approximate solutions of optimal control problem (20) for

$$\gamma = 2, \quad A_{01} = \begin{bmatrix} 0.05 & 0 \\ 0 & 0 \end{bmatrix}, \quad A_{00} = \begin{bmatrix} 0 & 0 \\ 0 & 1 \end{bmatrix}$$

showed little change between the second-order and the first-order solutions (i.e., $N = 1$ and $N = 2$) and the computations were terminated at $N = 2$ (this is stipulated by the smallness of the delayed term). We give here the approximate solution which corresponds to $N = 1$:

$$A_N = \begin{bmatrix} 2 & 0.05e^{0.02} & 0 \\ 100 & -98 & 0 \\ 0 & 0 & 3 \end{bmatrix},$$

$$W_N = \begin{bmatrix} 1.17 & 6.10^{-4} & -1.38 \\ 6.10^{-4} & 10^{-5} & -7.10^{-4} \\ -1.38 & -7.10^{-4} & 1.7 \end{bmatrix} \cdot 10^5$$

As done before, we can see that the condition (40) is satisfied and determine the robust control law

$$u(t) = -21.8x_1 - 0.01x_1(t-0.01) + 31.9x_2.$$

V. CONCLUSIONS

The proposed approach of designing a class of robust control systems is based on the consideration of the special optimal control problem for the system with the specified constant parameters. The given procedure allows us to build robust linear systems, as well as a wide class of robust nonlinear systems.

Robustness of linear differential-difference systems with unstable difference operator was considered. The robustness analysis is stipulated by the impossibility to implement precisely derivative feedback. As shown, differential-difference systems with unstable D-operator belong to a class of unrobust structures.

APPENDIX 1

Let the control law (6) minimize the functional (5) subject to the system (1). Consider the problem of minimizing the functional (20) subject to the system (1). By introducing

$$x_\gamma(t) = x(t)e^{\gamma t}, \quad u_\gamma(t) = u(t)e^{\gamma t}$$

the functional (20) becomes of the form (5) with respect to $x_\gamma(t)$ and $u_\gamma(t)$, and the system (1) is transformed into

$$\sum_{i=0}^{l} C_i e^{\gamma\tau_i} \dot{x}_\gamma(t-\tau_i) = \sum_{i=0}^{l} (A_i + \gamma C_i) e^{\gamma\tau_i} x_\gamma(t-\tau_i) + Bu_\gamma(t)$$

Hence, the problem of minimizing the functional (20) subject to the system (1) is equivalent to the usual problem of minimizing the functional of the form (5) subject to the system given above, and the optimal control law $u_\gamma(t)$ can be written as

$$u_\gamma(t) = u(t)\,e^{\gamma t} = -\frac{1}{c}\,B^T[Wx(t)\,e^{\gamma t} + \int_{-\tau}^{0}(B_2(\xi) + P_4^T(\xi,0))\,e^{\gamma t}e^{\gamma\xi}x(t+\xi)\,d\xi + \int_{-\tau}^{0}(B_3(\tau) + P_6(0,\xi))\,\gamma e^{\gamma t}e^{\gamma\xi}\,x(t+\xi)\,d\xi]$$

i.e., $u(t)$ has the form (19).

APPENDIX 2

The characteristic determinant $f(s)$ of the system described by Eq. (1) with $C_i = 0, i \neq 1$ is

$$f(s) = det(sI - \sum_{i=0}^{l} A_i e^{-\tau_i s}) = s^m + P_1(s)\,s^{m-1} + \ldots + P_m(s)$$

where $P_r(s) = (-1)^r M_r(s)$ and $M_r(s)$ is the sum of all the principal minors of order r of the matrix $\sum_{i=0}^{l} A_i e^{\tau_i s}$. The coefficients of $P_r(s)$ belongs to some domains which are defined by the elements of A_i $(i = 0, 1, \ldots, \ell)$ in (1). The location of the eigenvalues of (1) depends on these domains.

We will determine a domain of the right half of the complex plane, $Re\, s \geq 0$, which does not contain eigenvalues of the system (1) for all possible a_{iKj} satisfying (4).

From the above expression of $f(s)$, we have

$$|f(s)| \geq |s|^m\,[1 - \sum_{r=1}^{m} |P_r(s)|\,|s|^{-r}] \geq |s|^m\,[1 - \sum_{r=1}^{m} K_{max}\,|s|^{-r}]$$

where

$$K_{max} = \max_{\substack{r,\,a_{iKj}\\ Re s \geq 0}} |M_r(s)|$$

Assuming $|s| \geq 1$, we have $|s_o| \leq \max \sum_{k=1}^{m} \sum_{i=0}^{l} |a_{iKj}|$.

The inequality (49b) follows immediately from the given expressions.

Finally, the inequality (49c) can be obtained by using the procedure proposed by Hirsh [27]. Let x be an unit eigenvector, which corresponds to an eigenvalue s_o, i.e., $A(s_o)x = s_o x$ and the scalar product $(x, x) = 1$. Then,

$$(A(s_o)x, x) = s_o\,, \quad (A^*(\bar{s}_o)x, x) = (x, A(s_o)x) = \bar{s}_o$$

where the symbol "*" denotes the transpose complex-conjugated matrix and the bar "-" denotes a complex conjugated value.

Hence

$$Re\ s_o = \frac{1}{2}[(A(s_o)\ x, x) + (A^*(\bar{s}_o)\ x,\ x)]$$
$$= \frac{(A(s_o) + A^*(\bar{s}_o)}{2}\ x, x) = (B(s_o)\ x, x)$$

and in the right complex halfplane

$$Re\ s_o \leq \sum_{i,j=1}^{n} |b_{ij}\ (s_o)|\ |x_i||x_j| \leq \gamma$$

where γ is described by inequality (49c).

APPENDIX 3

We will use the functional (18) to examine the stability of the nonlinear control system, which is formed by the plant described by Eq. (51) and the controller, described by Eq. (6), determined by solving the auxiliary optimal problem (1) and (5). Its derivative $\frac{dJ_n}{dt}$ along the equations of the nonlinear system has the form [26,29]

$$\frac{dJ_n}{dt} = \frac{dJ_e}{dt} + cu^T(t)\ u(t) + (x^T(t)\ (W - B_3(0))$$
$$+ \int_{-\tau}^{0} (x^T(t+\xi)\ (B_2^T(\xi) + P_4(\xi, 0)) + \dot{x}^T(t+\xi)\ (B_3^T(\xi) + P_5^T(0, \xi))\ d\xi$$

where $\frac{dJ_e}{dt}$ denotes the derivative of J along the equation of the linear system (1) with the input matrix BH and the corresponding control (6). According to (6) and (52), we have

$$cu^T(t)\ u(t) + (x^T(t)\ (W + B_3(0))$$
$$+ \int_{-\tau}^{0} (x^T(t+\xi)\ (B_2^T(\xi) + P_4(\xi, 0)) + \dot{x}^T(t+\xi)\ (B_3^T(\xi) + P_5^T(0, \xi))\ d\xi B g(u)$$

$$= cu^T(t)\ u(t) - cu^T(t) H^o H^{-1} u(t)$$

$$\leq cu^T(t)\ u(t) - cu^T(t) H^o H^{-1} u(t)$$

If $H^o - H \geq 0$, $H > 0$, then the right part of the inequality is not positive so that $\frac{dJ_n}{dt} \leq \frac{dJ_e}{dt}$ and, hence, the nonlinear system is absolutely stable.

The case of the control (19) which is determined from the condition of the minimum of the functional (20) has no principal differences from the case considered above.

VI. REFERENCES

1. A. Andronov, A. Vitt, and S. Khaikin, Theory of Oscillations, Pergamon Press, Oxford, New York (1966).

2. R. Yanushevsky, "The Coarseness of Solutions of the Problem of Analytic Construction of Controls", *Automation and Remote Control*, 27(3), pp. 356-363 (1966).

3. V. Kharitonov, "Asymptotic Stability of a Family of Linear Differential Equations", *Differential'nye Uravneniya*, 14(11), pp. 2086-2088 (1978).

4. B. Barmish, "Invariance of Strict Hurwitz Property for Polynomial with Perturbed Coefficients", *IEEE Transactions on Automatic Control*, AC-29, pp. 935-936 (1984).

5. C. Soh, C. Berger, and K. Dabre, "On the Stability Properties of Polynomials with Perturbed Coefficients", *IEEE Transactions on Automatic Control*, AC-30, pp. 1033-1036 (1985).

6. R. Biernacki, H. Hwang, and H. Bhattacharyya, "Robust Stability with Structural Real Parameter Perturbation", *IEEE Transactions on Automatic Control*, AC-32, pp. 495-506 (1987).

7. B. Barmish, "A Generalization of Kharitonov's Four Polynomial Concept for Robust Stability Problems with Linearly Dependent Coefficient Perturbations", *Proceedings of the 1988 Automatic Control Conference*, Atlanta, GA, pp. 1869-1875 (1988).

8. A. Bartlett, C. Hollot, and L. Huang, "Root Locations for an Entire Polytope of Polynomials: It Suffices to Check the Edges", *Mathematics of Control Signals and Systems*, 1, pp. 61-71 (1987).

9. D. Siljak, "Polytopes of Nonnegative Polynomials", *Proceedings of the 1989 American Control Conference*, Pittsburgh, PA, pp. 193-199 (1989).

10. B. Anderson and E. Jury, "On Robust Hurwitz Polynomials", *IEEE Transactions on Automatic Control*, AC-32, pp. 1001-1008 (1987).

11. J. Cieslki, "On Possibilities of the Extension of Kharitonov's Stability Test for Interval Polynomials to the Discrete-time Case", *IEEE Transactions on Automatic Control*, AC-32, pp. 237-238 (1987).

12. N. Argoun, "Frequency Domain Conditions for Stability of Perturbed Polynomials", *IEEE Transactions on Automatic Control*, AC-32, pp. 913-916 (1987).

13. R. Patel, M. Toda, and B. Shridar, "Robustness of Linear Quadratic State Feedback Designs in the Presence of System Uncertainty", *IEEE Transactions on Automatic Control*, AC-22, pp. 945-949 (1977).

14. A.A. Abdul-Wahab, "Robustness Measure Bounds for Optimal Model Matching Control Design", *IEEE Transactions on Automatic Control*, AC-33, pp. 1178-1180 (1988).

15. I. Peterson and C. Hollot, "A Riccati Equation Approach to the Stabilization of Uncertain Systems", *Automatica*, 22, pp. 397-411 (1986).

16. K. Zhou and P. Khargonekar, "Stability Robustness Bounds for Linear State Space Models with Structured Uncertainty", *IEEE Transactions on Automatic Control*, AC-32, pp. 621-623 (1987).

17. D. Bernstein and W. Haddad, "The Optimal Projection Equations with Peterson-Hollot Bounds: Robust Stability and Performance via Fixed-order Dynamic Compensation for Systems with Structured Real-Valued Parameter Uncertainty", *IEEE Transactions on Automatic Control*, AC-33, pp. 578-582 (1988).

18. K. Zhou and P. Khargonekar, "Robust Stabilization of Linear Systems with Non-Bounded Time-Varying Uncertainty", *Systems & Control Letters*, 10, pp. 17-20 (1988).

19. M. Fu, A.W. Olbrot, and M. Polis, "Robust Stability for Time-Delay Systems: The Edge Theorem and Grafical Tests", *IEEE Transactions on Automatic Control*, AC-34, pp. 813-821 (1989).

20. T. Mori and H. Kakame, "An Extension of Kharitonov's Theorem and its Application", *Proceedings of the 1987 American Control Conference*, Minneapolis, MN, pp. 892-896 (1987).

21. B. Barmish and Z. Shi, "Robust Stability of Perturbed Systems with Time Delays", *Automatica*, 25 (3), pp. 371-387 (1989).

22. R. Yanushevsky, "An Approach to Robust Control Systems Design", *AIAA Journal of Guidance and Control*, 14(1), pp. 218-220 (1991).

23. C. Byrnes, M. Spong, and T.-J. Tarn, "A Several Complex Variable Approach to Feedback Stabilization of Linear Neutral Delay-Difference Systems", *Mathematical Systems Theory*, 17, pp. 97-133 (1987).

24. M. Cruz and J. Hale, "Stability of Functional Differential Equation of Neutral Type", *Journal of Differential Equations*, 7, pp. 334-355 (1970).

25. R. Yanushevsky, "Optimal Control of Linear Differential-Difference Systems of Neutral Type", *International Journal of Control*, 49(6), pp. 1835-1850 (1989).

26. R. Yanushevsky, Control of Plants with Time-Lag, Nauka, Moscow (1978).

27. M. Marcus and H. Minc, A Survey of Matrix Theory and Matrix Inequalities, Allyn and Bacon, Inc., Boston (1964).

28. J. Willems, "Least Squares Stationary Optimal Control and the Algebraic Riccati Equation", *IEEE Transactions on Automatic Control*, AC-16, pp. 621-634 (1971).

29. R. Yanushevsky, "A Class of Nonlinear Differential-Difference Systems of Neutral Type", *Proceedings of the 1989 American Control Conference*, Pittsburg, PA, pp. 409-410 (1989).

30. D. O'Connor and T.-J. Tarn, "On Stabilization by State Feedback for Neutral Differential-Difference Equations", *IEEE Transactions on Automatic Control*, AC-28, pp. 615-618 (1983).

31. R. Yanushevsky, "On Robust Stabilizability of Linear Differential-Difference Systems with Unstable D-Operator", *IEEE Transactions on Automatic Control*, AC-36 (1991).

System Observer Techniques in Robust Control Systems Design Synthesis

Tsuyoshi Okada
Masahiko Kihara
Masakazu Ikeda
Toshihiro Honma

Department of Aerospace Engineering
National Defense Academy
Hashirimizu, Yokosuka, Japan

I. INTRODUCTION

The cases are often encountered where the state variables in a given plant are not available for measurement. When the optimal feedback control is intended in such a case, an observer will be needed to estimate the immeasurable variables. If an observer is used, however, the influence of the parameter variation of the controlled plant on the control characteristics becomes great.

In aircraft and missiles for example, the plant parameters vary with their altitude and/or velocity variation. It is desirable however that the stability and response characteristics of the control system remain essentially unchanged despite the parameter variation. A control system with this property is said to be a robust control system. Therefore when a system including an observer is constructed, the robustness should be considered.

To synthesize a robust control system, methods using singular values have been recently developed for multi–input/multi–output (MIMO) systems. In this method, a measure of the robustness is the minimum singular value of the return difference matrix at the input or the output of the plant. The robustness of systems has been analyzed with the condition for robust stability and reduced sensitivity expressed in terms of singular values [1,2], and the concept of classical stability margin has been extended to MIMO systems [3,4]. Numerical optimization techniques have also been studied to enhance the robustness [5,6].

The regulator with optimal feedback is known to have the excellent robust stability and reduced sensitivity irrespective of the selected weighting matrices in the performance index [7–9]. Some of the present authors have demonstrated that the use of optimal feedback and the double perfect model following will further improve robustness [10]. Expanding this method, they have also developed a robust model following system which has perfect robustness about the steady–state characteristics [11]. This robustness of the optimal regulator is limited to the ideal case, though, where all the state variables are detectable and available for feedback. As mentioned above, the optimal feedback using the state variables estimated by an observer or Kalman filter results in considerable degradation in the robustness compared with the ideal case. As the solution to this problem, the linear quadratic Gaussian/loop transfer recovery (LQG/LTR) method of Doyle [12,13] and design method using perfect regulation and perfect observation of Kimura [14] have been proposed. But since these methods are asymptotic, the gains for some signals become large, and saturation is probable in a practical system. A method of recovering robustness by selecting poles of observer has also been proposed [15]. On the other hand, some of the authors have proposed a method of designing robust control systems which have an additional output feedback loop besides the observer loop so that the return difference matrix satisfies the circle condition [16]. In this method the closed-loop part is designed to recover robustness so that the return difference matrix coincides with that of the optimal regulator, and a precompensator part is used to equalize the response of the designed system to the desired one.

This article reviews the methods of recovering the robustness in the control system containing observer. The first method is a method where the system is formed based on the aforementioned idea of output feedback and designed using the singular values [17]. This method is readily applicable to MIMO systems. In particular, instead of requiring robustness as large as the optimal regulator, we recover the robustness to the extent that the designer requires by a numerical optimization technique of Newsom et al.[5]. In this method, the design requirement of robust stability and reduced sensitivity is designated as a specification on the minimum singular value, which is transformed to objective functions. The robust controller gain is then determined by a numerical calculation to minimize the feedback gain as well as these objective functions. As a numerical example, a two-input/two-output control system for the lateral motion of an aircraft is designed by this method. The resulting control system is simulated for a step input to show the expected performance.

It is a drawback of the first method to increase the dimension of the control system. To evade this, the second method is introduced which uses an output estimating error feedback loop. Because this loop does not change the response characteristics of the closed-loop system, the resulting system does not require the precompensator, and is called model matching system. It is well-known in this system that the use of transfer function element in the added loop permits the complete recovery of the robustness. Since the use of transfer function element does not remove the drawback of the increased dimension, however, the second method uses only a fixed gain in the added loop. A design by this method is made for the same example as before, and their simulated responses are compared with those for the previous method.

The third method shown here uses the same configuration of the control system as the output feedback system, where the robustness is recovered by using the pole placement technique in the observer channel while the response characteristics of the optimal regulator is recovered in the output feedback channel. Since the precompensator or model following system is not needed in this method, the dimension of the system does not increase. This method takes robustness at both the input and the output into consideration because the design is made using mean singular value curves. Besides, it is considered in this method that the difference between the maximum and the minimum singular values becomes small, and elements of each gains does not have large values. A design calculation of this method is shown for the previous example, and the simulated results is compared.

II. ROBUSTNESS

The sufficient conditions for a system to have robust stability and reduced sensitivity are expressed using singular values as follows.

A. ROBUST STABILITY

Assume that the nominal closed-loop system is stable. Then a sufficient condition for the system to remain robustly stable after the open-loop transfer function $G(s)$ is perturbed at the output to $L(s)G(s)$ is

$$\underline{\sigma}[I+G(j\omega)] \equiv \underline{\sigma}[T(j\omega)] > \bar{\sigma}[L^{-1}(j\omega)-I], \quad (0<\omega<\infty) \tag{1}$$

where $\bar{\sigma}$ and $\underline{\sigma}$ represent the maximum and the minimum singular values, respectively [1].

Consider the following diagonal perturbation as L in Eq.(1)

$$L=\mathrm{diag}[k_n \exp(j\phi_n)], \quad n=1,2,\cdots,m \tag{2}$$

where m is the number of outputs. L reduces to the identity matrix for the nominal system. Substitution of Eq.(2) into L in the right-hand side of Eq.(1) yields

$$\underline{\sigma}[I+G(j\omega)] = \underline{\sigma}[T(j\omega)] > \max_{n}\{(1-\frac{1}{k_n})^2+\frac{2}{k_n}(1-\cos\phi_n)\}^{\frac{1}{2}}, \tag{3}$$

which corresponds to the case where the gains and phases vary simultaneously in all loops. From Eq.(3) the diagram for multiloop phase and gain margin evaluation in Fig.1 is obtained [3,5]. From this diagram the minimum value (σ_M) of $\underline{\sigma}[I+G(j\omega)]$ corresponding to the required gain

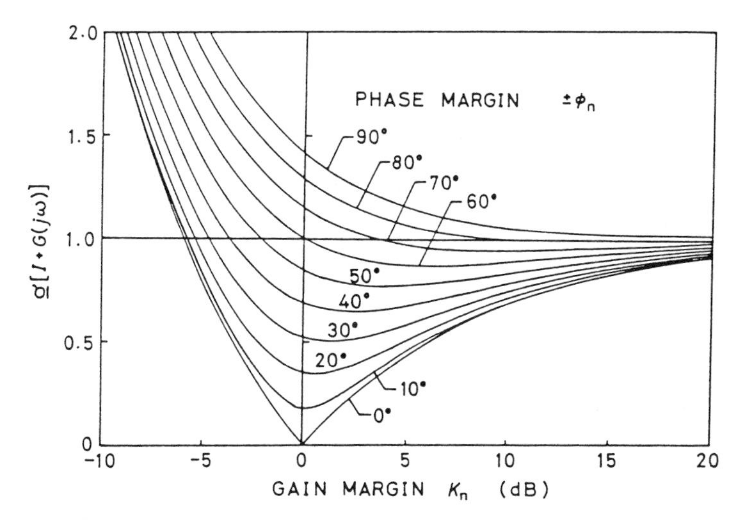

Fig. 1. Diagram for multiloop phase and gain margin evaluation.

margin (GM) or phase margin (PM) is obtained. For example, when both the gain and phase vary, the stability margin of the system for $\underline{\sigma}_M = 0.6$ is -1.5 dB < GM < $+5.3$ dB and PM = ±30 deg, whereas when either the gain or the phase varies ($\phi_n = 0$ or $K_n = 1$), the stability margin of the system is -4.2 dB < GM < $+8$ dB or PM = ±35 deg.

B. SENSITIVITY REDUCTION

Sensitivity reduction here means that the effect of plant parameter variation appears smaller in the closed–loop system than in the nominally equivalent open–loop system. When the parameter variation is small and continuous, a sufficient condition for this is [1]

$$\underline{\sigma}[I+G(j\omega)]=\underline{\sigma}[T(j\omega)]\geq 1. \quad (\omega\leq\omega_0) \tag{4}$$

Though this condition is satisfied for the optimal regulator over the entire frequency range, it is considered sufficient for other regulators that this condition is satisfied within the frequency range up to the system band–width ω_0.

III. SYNTHESIS OF ROBUST CONTROL SYSTEM USING OUTPUT FEEDBACK FORM [17]

A. CONFIGURATION OF THE SYSTEM

Consider a controlled system described by the following state equa–tions:

$$\dot{x}=Ax+Bu \tag{5}$$

$$y=Cx \tag{6}$$

where $x\in R^n$, $u\in R^r$, $y\in R^m$ are vectors, A, B, C are constant matrices with appropriate dimensions, and the system is assumed to be controllable and observable. Since the output is y as shown in Fig.2, we use an observer whose characteristics are described by

$$\dot{\hat{x}}=(A-FC)\hat{x}+Fy+Bu \tag{7}$$

where $\hat{x}$ is an estimate of x, and F is the optimal gain of the observer. The feedback gain K is taken as that of the optimal regulator.

The optimal regulator is generally used for the design of controllers for two reasons. The first is to optimize the response of the nominal plant, and the second is to realize the guaranteed robustness for reduction of sensitivity to plant parameter variation. When an observer is included, however, the return difference matrix at the output of the plant is different from that when the observer is not included. This explains that robustness is not guaranteed in this case even if the optimal feedback gain is used.

To recover the robustness and the optimal response in this case, we add an output feedback $\boldsymbol{H}$ shown by the broken line in Fig.2. The closed-loop is primarily used to recover robustness and the precompensator is used to obtain a desired response.

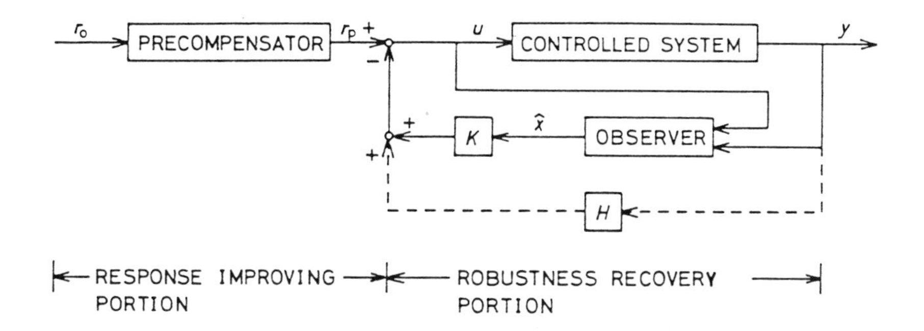

Fig. 2. Configuration of output feedback form system.

Then the open-loop transfer function at the output of this system is

$$G(s)=[C(sI-A)^{-1}B][K(sI-A+BK+FC)^{-1}(F-BH)+H]. \quad (8)$$

B. ROBUSTNESS RECOVERY DESIGN

For robustness recovery, we determine $\boldsymbol{H}$ in Eq.(8) so that the return difference matrix at the output satisfies the robustness condition shown by Eqs.(3) and (4). It is also required that the synthesized closed-loop system is stable.

The characteristic equation of this system is described by the following two equations:

$$\det(sI - A + BK + BHC) = 0 \tag{9}$$

$$\det(sI - A + FC) = 0. \tag{10}$$

Since Eq.(10) is the characteristic equation of the Kalman filter, it is evidently stable. Therefore the closed-loop system is stable if Eq.(9) is stable. The conditions that Eq.(9) remains stable no matter how large H becomes are given as follows [19].

Condition 1:

$$\text{rank } CB = m \quad \text{and} \quad m \le r \tag{11}$$

Condition 2:

$$Q(s) = \begin{bmatrix} sI - A + BK & B \\ C & 0 \end{bmatrix}$$

satisfies

$$\text{rank } Q(s) = n + m \tag{12}$$

for any $s \in C^+$ where C^+ is the closed right-half complex plane divided at and including the imaginary axis.

If these conditions are satisfied, then H may be chosen as

$$H = gP, \tag{13}$$

where P is a matrix and g is a scalar. Since the dimension of P is constrained by Eq.(11), we treat a case of the maximum dimension of output where $m = r$ and then P becomes square. In the case of $m < r$, one may apply our method for an m-dimensional subset of control loops with the remaining $(r-m)$ loops closed. The scalar g is selected as the minimum positive number, which reduces the following cumulative measures $J_1(g)$ and $J_2(g)$ to zero simultaneously to satisfy the robustness conditions of Eqs.(3) and (4).

Robust stability:

$$J_1(g) = \sum_i(\max\{0, [\underline{\sigma}_D - \underline{\sigma}(j\omega_i, g)]\})^2 \quad (0<\omega<\infty) \tag{14}$$

Sensitivity reduction:

$$J_2(g) = \sum_i(\max\{0, [1-\underline{\sigma}(j\omega_i, g)]\})^2 \quad (0<\omega<\omega_0) \tag{15}$$

where Σ means that the summation is calculated for many frequency points ω_i, which is chosen suitably within the specified frequency range. The cumulative measure of Eq.(14) aims to increase the minimum value $\underline{\sigma}_M$ of $\underline{\sigma}$ up to the desired value $\underline{\sigma}_D$ over the entire frequency range. The $\underline{\sigma}_D$ is selected as the minimum value of $\underline{\sigma}$ that gives the desired stability margin from the diagram for phase and gain margin evaluation in Fig.1. On the other hand, Eq.(15) aims to make $\underline{\sigma}$ larger than unity within the specified frequency range where the bandwidth ω_0 is selected from the characteristics of the given system.

Now P is assumed to be a nonsingular diagonal matrix for simplicity in calculation. Though nondiagonal P can be used, it brings computational complexity. When only either input or output return difference is considered, the diagonal P suffices. A g is obtained, which makes Eq.(14) and Eq.(15) zero simultaneously for the given values of $\underline{\sigma}_D$ and ω_0, and then H is determined. Large H is not preferable because it amplifies the effect of measurement noise since H is a direct feedback element of the system. Therefore minimization of H is desired. Now let us define the following norm:

$$\|H\| = \left[\sum_{i=1}^{m} |h_{ii}|^2\right]^{\frac{1}{2}} = \left[\sum_{i=1}^{m} g^2 |p_{ii}|^2\right]^{\frac{1}{2}}. \tag{16}$$

The flow of the computation is shown in Fig.3. First, an initial value for P is given so that $(-PCB)$ is stable. An initial value of increment ΔP of P is also given previously. Then computation is started and an H, which makes Eqs.(14) and (15) zero, is obtained and $\|H\|$ for this H is computed

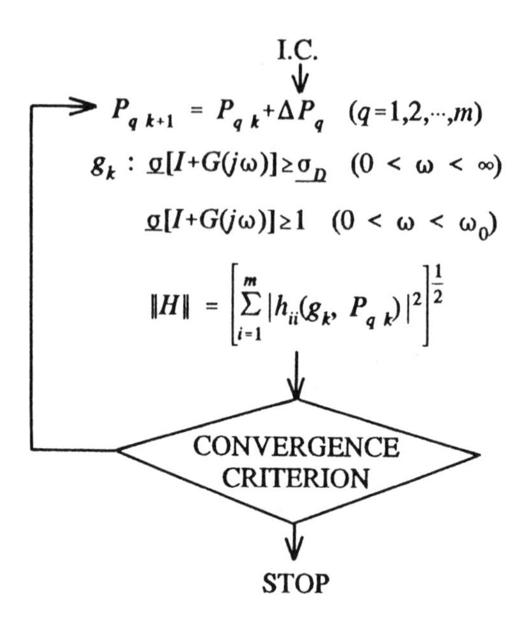

Fig. 3. Flow of computation for output feedback method.

from Eq.(16). Second, P is increased by ΔP and a new value of $\|H\|$ is computed by the same procedure for a new P. The new value of $\|H\|$ is compared with the old one, and the smaller value is kept. The same computation is repeated, altering ΔP automatically through a computer program based on the simplex method. When the variation of $\|H\|$, according to the variation of P, becomes smaller than a specified value, then the computation is terminated and H is adopted as the designed value.

C. DESIGN TO IMPROVE THE RESPONSE

The closed-loop transfer function of the optimal regulator

$$G_c(s) = C(sI - A + BK)^{-1}B \tag{17}$$

is different from that of the system having the additional loop of H,

$$G_c'(s) = C(sI - A + BK + BHC)^{-1}B. \tag{18}$$

The resulting response characteristics of Eq.(18) are not necessarily desirable. Therefore a precompensator based on the model following system is added to obtain a desirable response. A precompensator is designed so that the system behaves like that of the optimal regulator of Eq.(17). Let the transfer function of the precompensator be $G_f(s)$, then the overall transfer function from the input r_0 to the system output y is written as

$$\begin{aligned} C(sI-A+BK+BHC)^{-1}BG_f(s) &= C(sI-A+BK)^{-1}B \\ &\quad \times\{I+HC(sI-A+BK)^{-1}B\}^{-1}G_f(s). \end{aligned} \tag{19}$$

In order to equalize Eq.(19) with Eq.(17), the transfer function of the precompensator is chosen as

$$G_f(s) = I+HC(sI-A+BK)^{-1}B. \tag{20}$$

If other arbitrary response characteristics are desired, one may use prefilter [20] or extended perfect model following [21] methods that are applicable to a nonsquare case.

A block diagram of the designed system is shown in Fig.4.

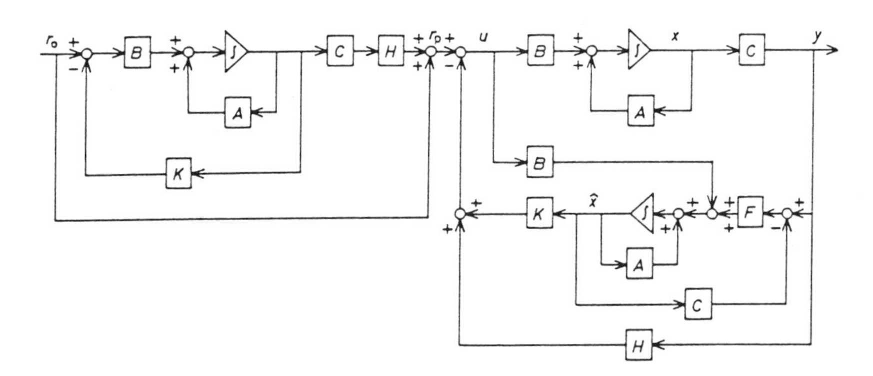

Fig. 4. Block diagram of the designed system.

So far we treated the robustness at output. If the robustness at input is needed, the same procedure is applicable for a return difference matrix based on the previous open–loop transfer function at input. Methods of evaluating the robustnesses at both input and output are discussed in several papers [22,23], and also treated in the later section of this article.

D. EXAMPLE OF APPLICATION TO A FLIGHT CONTROL SYSTEM

The state equation for the lateral motion of an aircraft flying at a flight condition of 40,000 feet altitude and Mach=0.8 is written as [24]

$$\dot{x} = \begin{bmatrix} -43.2 & 0.0416 & 0 & -1 \\ 0 & 0 & 1 & 0 \\ -3.05 & 0 & -0.465 & 0.388 \\ 0.598 & 0 & -0.0318 & -0.15 \end{bmatrix} x + \begin{bmatrix} 0 & 0.00729 \\ 0 & 0 \\ 0.143 & 0.153 \\ 0.00775 & -0.475 \end{bmatrix} u \tag{21}$$

where

$$x = [\beta \ \phi \ p \ r]^T, \qquad u = [\delta_a \ \delta_r]^T$$

and

β = sideslip angle

p = roll rate

δ_a = aileron angle

ϕ = roll angle

r = yaw rate

δ_r = rudder angle.

Assuming that p and r are measurable, we obtain the following output equation

$$y = \begin{bmatrix} 0 & 0 & 1 & 0 \\ 0 & 0 & 0 & 1 \end{bmatrix} x. \tag{22}$$

We assume that an optimal regulator has the desired response characteristics when the weighting matrices in the performance index are chosen as Q = diag[100, 100] and R = diag[1, 1]. For these weighting matrices, the optimal gain is obtained as follows:

$$K = \begin{bmatrix} -0.463 & 0 & 7.25 & 2.37 \\ -0.134 & 0.00300 & 0.418 & -10.1 \end{bmatrix} \tag{23}$$

On the other hand, we use a Kalman filter–type observer and choose the covariance matrices for fictitious noise at input and output as the same values of Q and R. Then the observer gain F is obtained as

$$F = \begin{bmatrix} 0.0171 & 0.00689 & 1.34 & -0.787 \\ -0.102 & -0.117 & -0.787 & 4.56 \end{bmatrix}^T . \tag{24}$$

Substituting the preceding system parameters into Eqs.(11) and (12), we see that the conditions of Eqs.(11) and (12) are satisfied, and therefore a stable H exists and the proposed robust control system can be designed.

Now we designate the following performance about the robustness of the designed system.

Robust stability:

$$\underline{\sigma_D} = 0.90$$

This corresponds to

$$-5.6 \text{ dB} < \text{GM} < +20.0 \text{ dB}$$

$$\text{PM} = \pm 54 \text{ deg}$$

as obtained from Fig.1, and these stability margins are adequate for servo systems.

Sensitivity reduction:

$$\omega_0 = 10 \text{ rad/sec}$$

This value is considered to be adequate for the bandwidth of the lateral motion of aircraft.

The numerical calculation is performed according to the flow in Fig.3. The simplex method was used as the minimization procedure. Using the initial vertex value

$$p(1, 1) = 1, \quad p(2, 2) = -1,$$

we obtain the final values of P and g as

$$P = \begin{bmatrix} 1.244 & 0 \\ 0 & -0.857 \end{bmatrix}$$

$$g = 11.31$$

from which $\boldsymbol{H}$ for the minimum norm becomes

$$\boldsymbol{H} = g\boldsymbol{P} = \begin{bmatrix} 14.07 & 0 \\ 0 & -9.47 \end{bmatrix}. \tag{25}$$

To confirm that the designed system with this $\boldsymbol{H}$ improves robustness, the minimum singular value vs. frequency plot is shown by a one dotted broken line in Fig.5.

Those for the optimal regulator (solid line) and for the optimal regulator with observer (broken line) are also shown in the same figure. The optimal regulator has the guaranteed stability margins

$$-6 \text{ dB} < \text{GM} < +\infty \text{ dB}$$

$$\text{PM} = \pm 60 \text{ deg.}$$

The stability margins for the optimal regulator with observer are

$$-5.4 \text{ dB} < \text{GM} < +17.7 \text{ dB}$$

$$\text{PM} = \pm 52 \text{ deg.}$$

since the minimum value of $\underline{\sigma}$ is 0.87. As for the sensitivity of the optimal regulator with observer, $\underline{\sigma}$ is less than unity within $\omega > 0.9$ rad/sec, which does not satisfy the design specification. On the other hand, the designed system constructed with the above $\boldsymbol{H}$ has $\underline{\sigma} \geq 1$ up to the frequency range of $\omega_0 = 10$ rad/sec, satisfying the design specification on sensitivity. It has also the minimum value of 0.98, which fully satisfies the design specification on stability.

The precompensator $G_f(s)$ that results in the same response as the optimal regulator in Eq.(23) is designed using Eq.(20), and the system is constructed as shown in Fig.4.

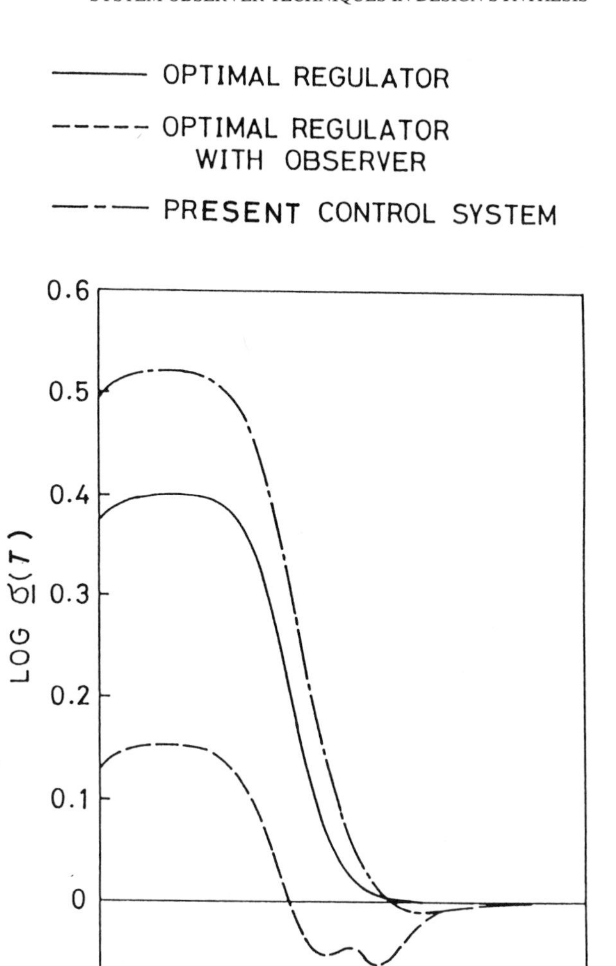

Fig. 5. Singular value curves for output feedback method.

E. SIMULATION

Simulated responses for the following systems are shown in Figs.6 and

7 as nominal: 1) an optimal regulator for the plant of Eq.(21) with $\boldsymbol{K}$ of Eq.(23); 2) an optimal regulator with observer for the plant of Eqs.(21), (22), and (9) with $\boldsymbol{K}$ of Eq.(23) and $\boldsymbol{F}$ of Eq.(24). (This system is from $\boldsymbol{r}_p$ to y without $\boldsymbol{H}$ in Fig.2); and 3) the control system in Fig.4 with $\boldsymbol{K}$, $\boldsymbol{F}$, and $\boldsymbol{H}$ in Eqs.(23)–(25).

A step input

$$\boldsymbol{r}_p = \begin{bmatrix} 0.1 \\ 0.1 \end{bmatrix}$$

was applied to systems 1 and 2 and a step input

$$\boldsymbol{r}_0 = \begin{bmatrix} 0.1 \\ 0.1 \end{bmatrix}$$

was applied to system 3. All initial conditions were zero. The output p (roll rate) and r (yaw rate) of nominal systems are shown by solid lines in Fig.6 and Fig.7, respectively. Since systems 1–3 are nominally equivalent to each other, their responses perfectly coincide as seen in these figures.

To confirm the robustness, a parameter variation is considered. When the flight condition of the aircraft is varied to 20,000 feet altitude and Mach=0.80, the parameters are varied as follows:

$$\boldsymbol{A} = \begin{bmatrix} -99.4 & 0.0388 & 0 & -1 \\ 0 & 0 & 1 & 0 \\ -4.12 & 0 & -0.974 & 0.292 \\ 1.62 & 0 & -0.0157 & -0.232 \end{bmatrix} \quad (26)$$

$$\boldsymbol{B} = \begin{bmatrix} 0 & 0.0124 \\ 0 & 0 \\ 0.310 & 0.183 \\ 0.0127 & -0.922 \end{bmatrix} \quad (27)$$

Simulated responses of the system with a perturbed plant for the previous inputs are shown in Figs.6 and 7 by broken lines for system 1, by one dotted broken line for system 2, and by two dotted broken lines for system 3. The responses of the optimal regulator with the observer (system 2)

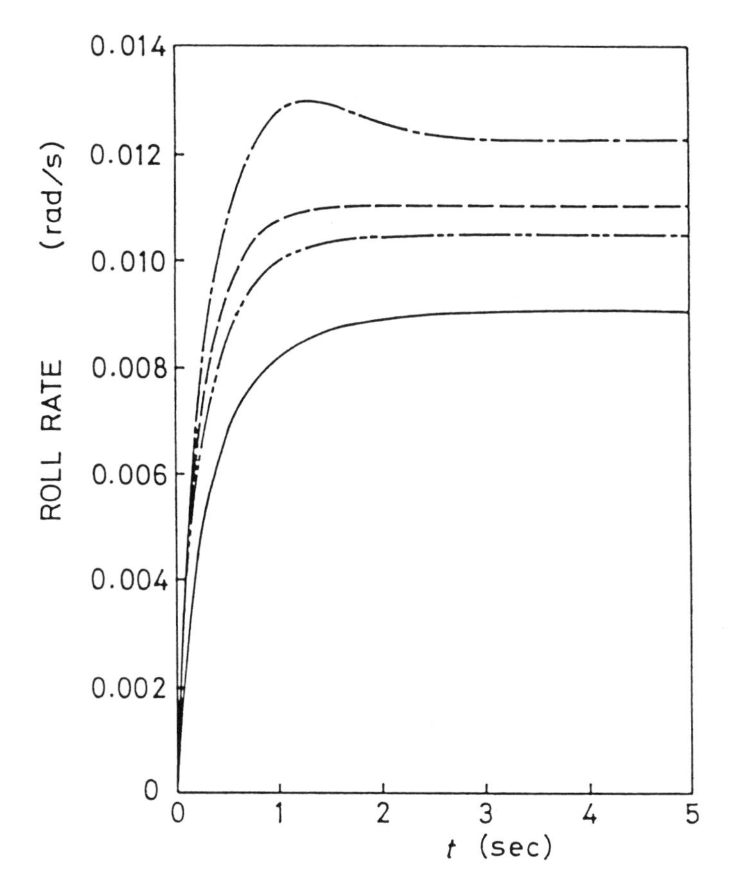

Fig. 6. Roll-rate step response.

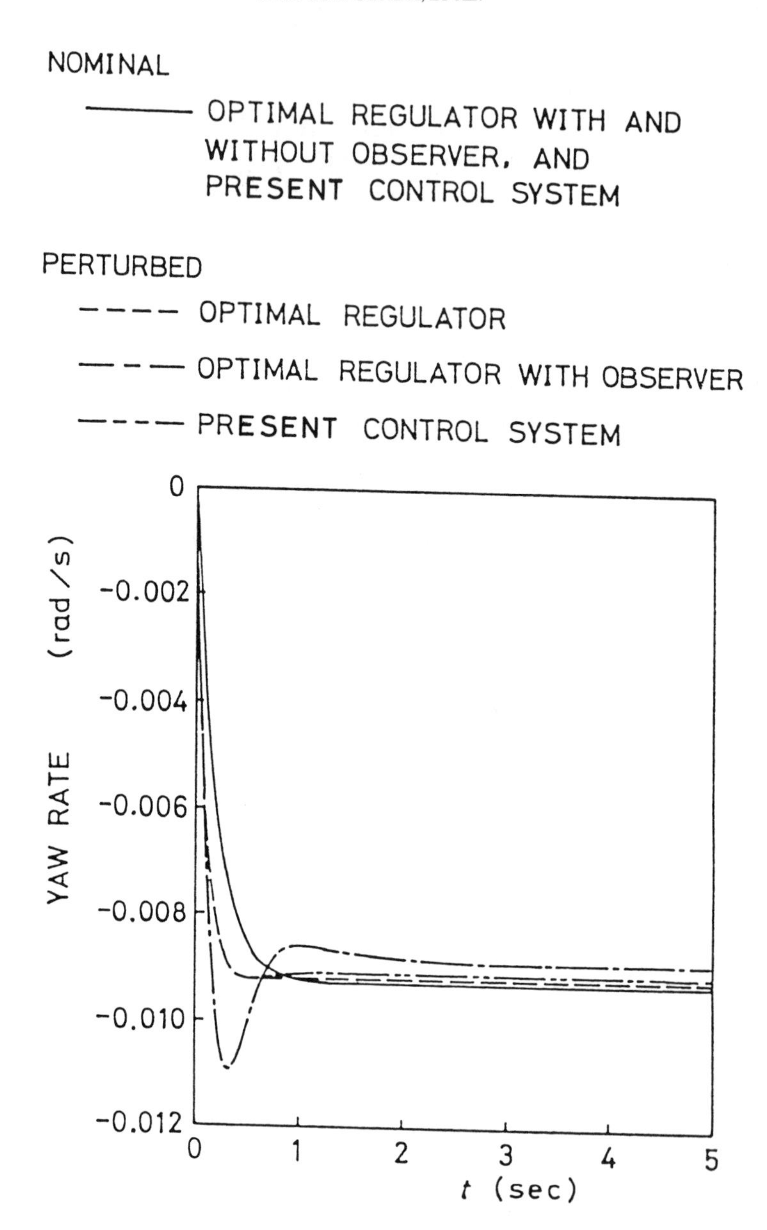

Fig. 7. Yaw–rate step response.

show large overshoot and are deviated from the nominal responses (solid lines) considerably. The responses of the present designed system (system 3) are close to those of the optimal regulator without the observer (system 1). Particularly, the response of the roll rate is rather close to the nominal response (solid line).

As shown in this example, compared with the optimal regulator without an observer, the system designed by the present method has more or equally reduced effect for parameter variation and has a sufficient stability margin, and therefore its robustness is considered to be recovered.

IV. SYNTHESIS OF ROBUST CONTROL SYSTEM USING MODEL MATCHING FORM

A. CONFIGURATION OF THE SYSTEM

The design method of robust control system with output feedback presented in the previous chapter has a drawback to increase the dimension of the designed system because it synthesizes the robustness recovering part and response characteristics improving part separately.

The robust control system using model matching method presented in this chapter has a configuration shown in Fig.8 where feedback of the difference of the plant output and the observer output is made to the plant input through the element $\boldsymbol{M}$. Though this system has the additional loop $\boldsymbol{M}$, the response characteristic of the closed-loop from input $\boldsymbol{r}$ to output y is described by the following transfer function, which includes no term of $\boldsymbol{M}$.

$$y(s) = C(sI-A+BK)^{-1}B\ r(s) \tag{28}$$

Therefore any precompensator or model following system as in the previous chapter is not needed. (From this reason, the word model matching resulted.)

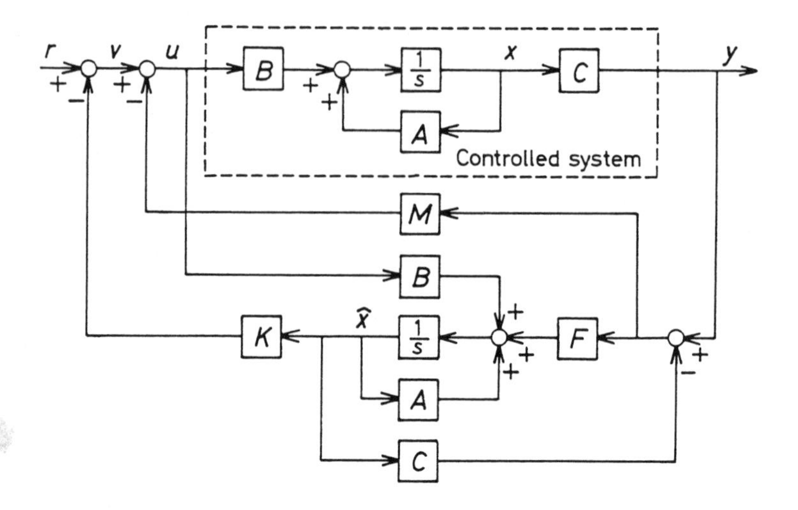

Fig. 8. Configuration of model matching form system.

On the other hand, the return difference matrices at the input and at the output include terms of $\boldsymbol{M}$ as follows:

$$T_i(s) = \{I+\underline{(K-M_iC)(sI-A+FC)^{-1}B}\}^{-1}\{I+K(sI-A)^{-1}B\} \tag{29}$$

$$T_o(s) = \{I+C(sI-A)^{-1}F\}\{I+\underline{C(sI-A+BK)^{-1}(F-BM_o)}\}^{-1} \tag{30}$$

where the subscripts i and o imply input and output respectively. Therefore if value of $\boldsymbol{M}$ is selected so that the underlined part of Eq.(29) or Eq.(30) becomes zero or very small, the return difference matrix of this system becomes equal or nearly equal to that of the optimal regulator and then recovers robustness.

If

$$M_i = [(s+a)\{K(sI-A+FC)^{-1}B\}][(s+a)\{C(sI-A+FC)^{-1}B\}]^{-1} \tag{31}$$

and

$$M_o = [\{C(sI-A+BK)^{-1}B\}(s+a)]^{-1}[\{C(sI-A+BK)^{-1}F\}(s+a)] \tag{32}$$

are taken as $\boldsymbol{M}$ in Eqs.(29) and (30) respectively, the underlined parts become zero [18]. In the above equations, a is a positive real number and terms $(s+a)$ are included to facilitate implementation. If the given plant is non–minimum phase system, $M's$ as Eqs.(31) and (32) have some unstable poles and therefore $M's$ need to be modified to remove the unstable poles and hence the underlined parts in Eqs.(29) and (30) can not become zero. In this case, the robustness is partially recovered.

In any case, the use of M in Eqs.(31) and (32) is not consistent with the objective to reduce the dimension of the system. If fixed gain element such as $K = M_iC$ or $F = BM_o$ could be chosen, the underlined part in Eq.(29) or Eq.(30) would become zero. But since such M is generally difficult to find, a method of finding a fixed gain matrix M is presented in this chapter that satisfies the robustness condition described in the previous chapter.

B. ROBUSTNESS RECOVERY DESIGN

In order to recover the robustness, the return difference matrix $T_i(s)$ (or $T_o(s)$) in Eq.(29) (or Eq.(30)) is applied to the condition for robustness in Eqs.(3) and (4) to find M_i (or M_o) satisfying them.

For this method, the same algorithm as the previous chapter can be applied. Now M is written as

$$M = gP, \tag{33}$$

where a minimum positive number g is determined to reduce the cumulative measures in Eqs.(14) and (15) to zero simultaneously. The elements of M are also desirable to be small because the large M brings large effect of noise. Therefore the following norm is defined and P is coordinated to minimize this norm.

$$\|M\| = \left[\sum_{i=1}^{m} \sum_{j=1}^{m} g^2 |P_{ij}|^2\right]^{\frac{1}{2}} \tag{34}$$

The poles related to this system are $\lambda[A-BK]$, $\lambda[A-FC]$, $\lambda[A-BMC]$, and $\lambda[A-BK-FC+BMC]$, where the former two are stable when K and F are taken as the optimal regulator gain and the optimal observer gain respectively. Therefore it is needed to take care for the latter two to become stable. In particular, $\lambda[A-BMC]$ is the consequence of the feedback $u = -My$ applied to the controlled system, which is different from the system in the previous section. The flow of the computation is shown in Fig.9.

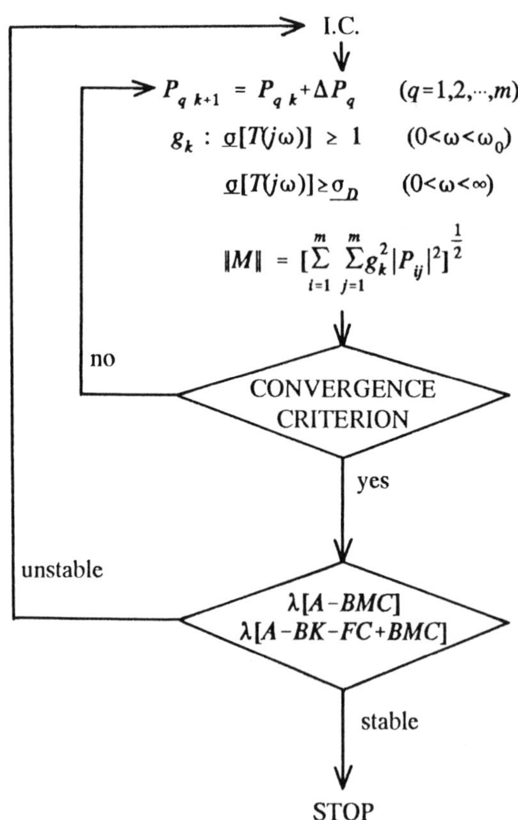

Fig. 9. Flow of computation for model matching method.

C. EXAMPLE OF APPLICATION TO A FLIGHT CONTROL SYSTEM AND ITS SIMULATION

The present design method is applied to the example of aircraft whose dynamics is described by Eqs.(21) and (22) in the previous chapter. K and F in Eqs.(23) and (24) are again used. Required performance as to robustness is specified as $\underline{\sigma_D} = 0.9$ and $\omega_0 = 10$ rad/sec as the previous chapter. Then M is calculated according to the algorithm without limiting P to diagonal matrix, and obtained as

$$M = \begin{bmatrix} 8.436 & 3.612 \\ 0.992 & -11.404 \end{bmatrix}. \tag{35}$$

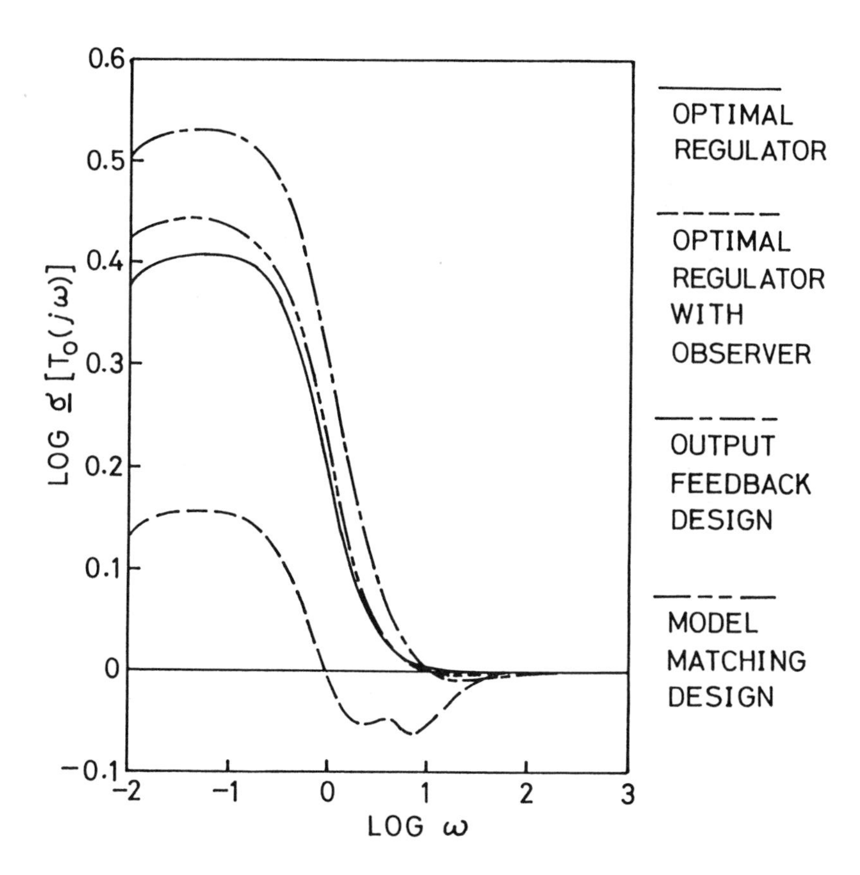

Fig. 10. Singular value curves for model matching method.

Since $\lambda[A-BMC] = \{-43.19, -1.66, -5.74, -0.002\}$ and $\lambda[A-BK-FC+BMC] = \{-43.19, -4.27, -1.33, -0.002\}$ with this M, the designed system is stable.

The singular value curve of the output return difference matrix of the control system in Fig.8 with this M is shown in Fig.10 by two dotted broken line. The required performance is satisfied since $\underline{\sigma} \geq 1$ up to $\omega_0 = 10$ rad/sec, and the minimum value of $\underline{\sigma}$ is 0.99. Compared with the

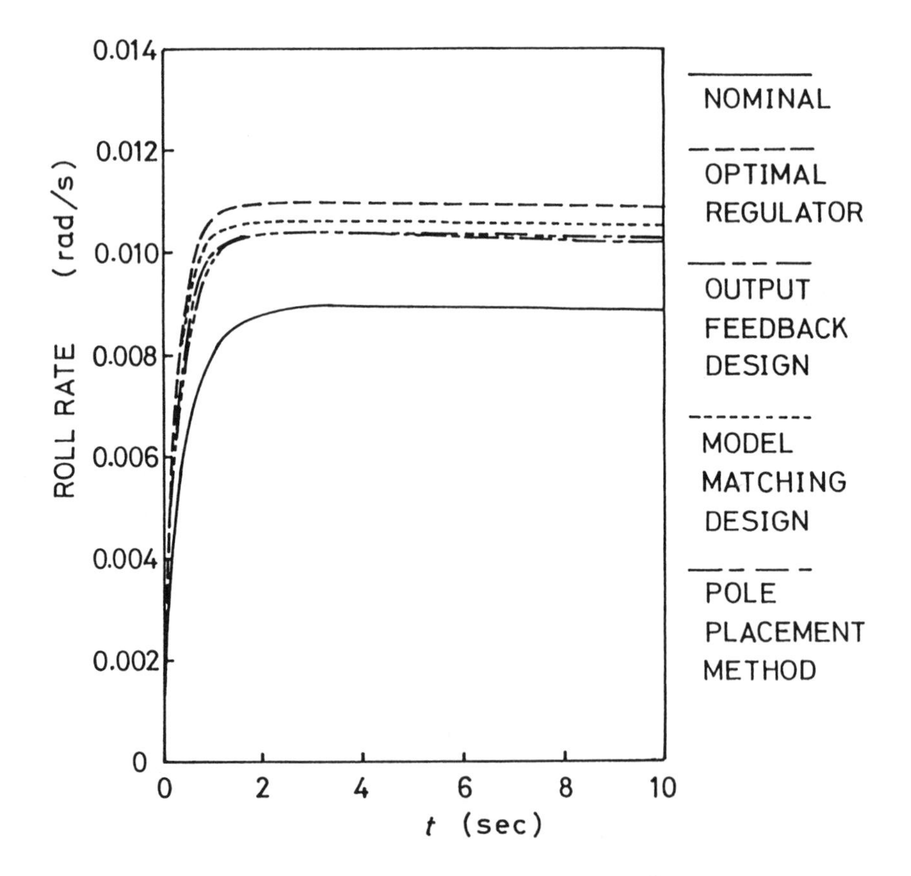

Fig. 11. Simulated step responses of roll rate.

result of the output feedback system in the previous chapter, which is shown by one dotted broken line in Fig.10, the present method brings a little higher minimum value of $\underline{\sigma}$ and a little lower maximum value of $\underline{\sigma}$. But latter value is higher than that of the optimal regulator shown by solid line. The norm of $\boldsymbol{H}$ in Eq.(25) and the norm of $\boldsymbol{M}$ in eq.(35) are considered to be substantially same value though the latter is slightly smaller.

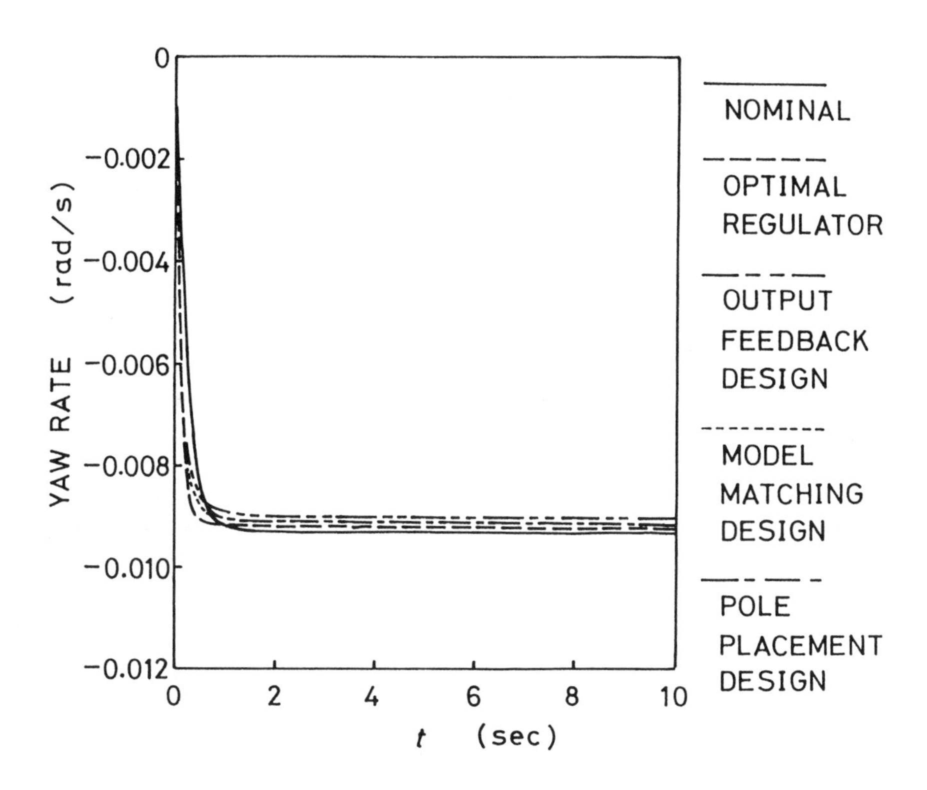

Fig. 12. Simulated step responses of yaw rate.

Simulated step responses of the designed system are shown in Fig.11 and 12 for the perturbed flight condition given by Eqs.(26) and (27) as in the previous chapter. No precompensator is needed in this simulation unlike the previous case. Although the response characteristics of the present system shown by dotted line is somewhat inferior to those of the system in the previous chapter shown by two dotted broken line, the response of the present system is nearer to that of the nominal system shown by solid line than that of the perturbed optimal regulator system shown by broken line.

It is concluded from these results that compared with the method of the previous chapter, the present method is somewhat inferior in achieving robustness while it is superior in realizing low dimension.

V. SYNTHESIS OF A ROBUST CONTROL SYSTEM USING POLE PLACEMENT METHOD

A. CONFIGURATION OF THE SYSTEM

A block diagram of a system used in this method is the same as that in Fig.2 except that the precompensator is deleted, but a feedback element K from observer is used to place poles to improve robustness and a output feedback element H is determined as an optimal regulator gain for the system including the plant and observer. Since the entire system have the optimal control structure, it does not need to have a precompensator as in chapter III and therefore its dimension can be decreased.

The return difference matrices of this system at input and at output are

$$T_i(s) = [I+K(sI-A+FC)^{-1}B]^{-1}[I+(K+HC)(sI-A)^{-1}B] \tag{36}$$

and

$$T_o(s) = [I+C(sI-A)^{-1}F][I+C(sI-A+BK+BHC)^{-1}(F-BH)]^{-1} \tag{37}$$

respectively, and the response characteristics of the closed-loop system from input r to output y are described using the transfer function matrix as follows:

$$y(s) = C(sI-A+BK+BHC)^{-1}B\ r(s). \tag{38}$$

B. ROBUSTNESS RECOVERY DESIGN

The robustness dealt with in chapter III and IV laid emphasis on that at the output. Although the robustness at the input can be treated by the same method, simultaneous recovery of robustness at both input and output has not been considered. In this chapter, the robustness is considered at both input and output simultaneously.

The determinants of the return difference matrices of Eqs.(36) and (37) are

$$\det T_i(s) = \det T_o(s) = \frac{\det(sI-A+FC)\det(sI-A+BK+BHC)}{\det(sI-A)\det(sI-A+FC+BK)}. \tag{39}$$

It is seen from this expression that both the return difference matrices are related to the eigenvalues of $\lambda[A]$, $\lambda[A-FC]$, $\lambda[A-FC-BK]$ and $\lambda[A-BK-BHC]$.

In the first step of the procedure, the observer gain F is selected. This gain is usually determined as a optimal gain for some weighting matrices. Then the eigenvalues of $\lambda[A-FC]$ are determined. In the second step, K is selected so as for the system to recover the robustness. K affects the eigenvalues of $\lambda[A-FC-BK]$ and $\lambda[A-BK-BHC]$. This K has pole positional freedom and directional freedom of $(A-FC-BK)$.

First, the pole positional freedom of $(A-FC-BK)$ is adjusted to satisfy the following inequality while its directional freedom is fixed.

$$\sigma_{ave}[T_i(j\omega)] \geq 1.0, \qquad (\omega \leq \omega_0) \tag{40}$$

where σ_{ave} represents average gain defined by [25]

$$\begin{aligned} \log\sigma_{ave}[T_i(j\omega)] &\equiv r^{-1}\{\log\sigma_1[T_i(j\omega)]+\cdots+\log\sigma_n[T_i(j\omega)]\} \\ &= r^{-1}\log|\det T_i(j\omega)|, \end{aligned} \tag{41}$$

where $\sigma_i[\cdot]$ represents i–th singular value. The average gain is also defined for the output as follows:

$$\log\sigma_{ave}[T_o(j\omega)] \equiv m^{-1}\log|\det T_o(j\omega)|. \tag{42}$$

Since a relation

$$\log\sigma_{ave}[T_i(j\omega)] = r^{-1}m \log\sigma_{ave}[T_o(j\omega)] \tag{43}$$

is derived from Eq.(39), both input and output robustnesses are taken into consideration by using one of these logarithmic averages of singular values (=average gains).

H needs to be calculated to examine the condition of Eq.(40). H is calculated as an optimal output feedback gain for a system

$$\dot{x} = (A-BK)x+Bu \equiv \bar{A}x+Bu \tag{44}$$

$$y = Cx \tag{45}$$

which has output feedback

$$u = -Hy \tag{46}$$

with respect to a performance index

$$J = \int_0^\infty (y^TQy+u^TRu)dt, \qquad Q>0, \quad R>0. \tag{47}$$

This gain is determined by solving the following simultaneous equations [26].

$$H = R^{-1}B^TPLC^T(CLC^T)^{-1} \tag{48}$$

$$0 = (\bar{A}-BHC)L+L(\bar{A}-BHC)^T+I \tag{49}$$

$$0 = (\bar{A}-BHC)^TP+P(\bar{A}-BHC)+Q+C^TH^TRHC \tag{50}$$

where L and P are non-negative symmetric matrices.

In this calculation, Q and R are coordinated to satisfy the following conditions.

$$\max \log\{\sigma_{ave}[T_i(j\omega)]\} \geq \log\sigma_D \tag{51}$$

$$\max |H_{ij}| \leq \zeta \tag{52}$$

where σ_D is a design parameter to specify the stability margin and ζ is one to specify maximal value in the elements of output feedback H. In practical calculation, Q and R are first taken as identity matrices and increased gradually (multiplying scalars for example) until Eq.(51) is satisfied. But the increase should be limited to the range where Eq.(52) holds because the left hand side of Eq.(52) is increased along with the increase of Q and R. Q

and R determined after this coordination are denoted by Q_D and R_D. K determined in this step is tentative.

In the third step of this design procedure, the directions of the eigenvectors of $(A-FC-BK)$ are coordinated to minimize the condition number of the return difference matrix as follows:

$$\underset{k}{\text{minimize}}\ \Sigma[\text{cond}\ T_i(j\omega_k)] = \text{minimize} \sum_k \frac{\bar{\sigma}[T_i(j\omega_k)]}{\underline{\sigma}[T_i(j\omega_k)]}, \tag{53}$$

where ω_k are several frequency points chosen by designer within the effective frequency range. By this coordination, the maximum and the minimum singular values are forced to approach the average singular value. In this coordination, K is adjusted with keeping the pole positions of $\lambda[A-FC-BK]$ fixed, and Q_D and R_D at the values determined in the second step. Then when K is changed, $\bar{A}$ changes, and H is also changed according to Eqs.(44)–(50). However, since the average singular value of the return difference matrix is almost unchanged through the coordination of K and H in this step, K and H obtained in this step are adopted as the final values without returning to the second step. A flow of the computation is shown in Fig.13.

Since the obtained F and H are the optimal observer gain and the optimal regulator gain respectively, $\lambda[A-BK-BHC]$ and $\lambda[A-FC]$ are certainly stable. The coordination in each step can be visualized using singular value curves as demonstrated in Appendix.

C. EXAMPLE OF APPLICATION TO A FLIGHT CONTROL SYSTEM AND ITS SIMULATION

The present method is applied to the aircraft described by Eqs.(21) and (22) in chapter III. F of Eq.(24) is adopted as an observer gain to be determined in the first step.

In the second step, $\lambda[A-FC-BK]$ is assigned to {−0.003, −1.48, −4.88,

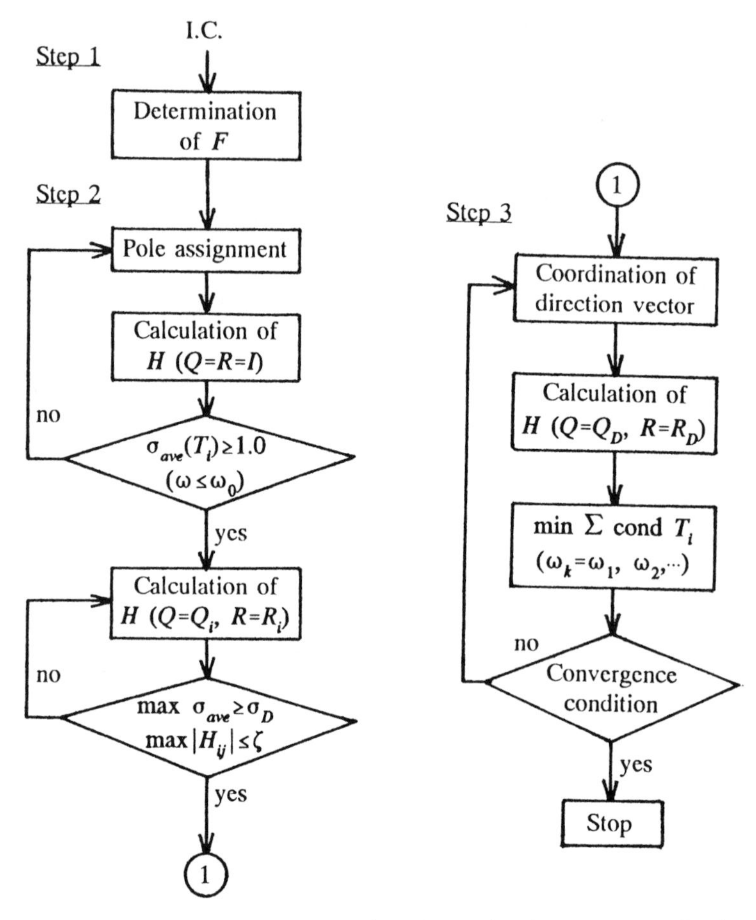

Fig. 13. Flow of computation for pole placement method.

−43.2} to satisfy the condition of Eq.(40) within bandwidth of ω_0 = 10 rad/sec. (Refer to appendix as to procedure of this assignment.) With requirement for design as σ_D = 0.8 in Eq.(51) and ζ = 15 in Eq.(52), Q_D = diag[250, 250] and R_D = diag[1, 1] are obtained through the coordination of Q and R.

In the third step, the directional vectors of $(A-FC-BK)$ are coordinated using $\omega_k = \omega_1 = 0.1$ rad/sec in Eq.(53), and

$$K = \begin{bmatrix} 0.6542 & 0.0118 & -0.1523 & -0.0038 \\ -3.653 & 0.0159 & 0.0962 & 0.1858 \end{bmatrix} \tag{54}$$

and

$$H = \begin{bmatrix} 9.6145 & 2.8233 \\ 1.5734 & -11.129 \end{bmatrix} \tag{55}$$

are obtained. The magnitude of elements of these gains are of the same order as that of the previous design examples. The poles of the closed-loop system are shown as $\lambda[A-BK-BHC] = \{-0.00247, -1.743, -5.671, -43.2\}$ and stable.

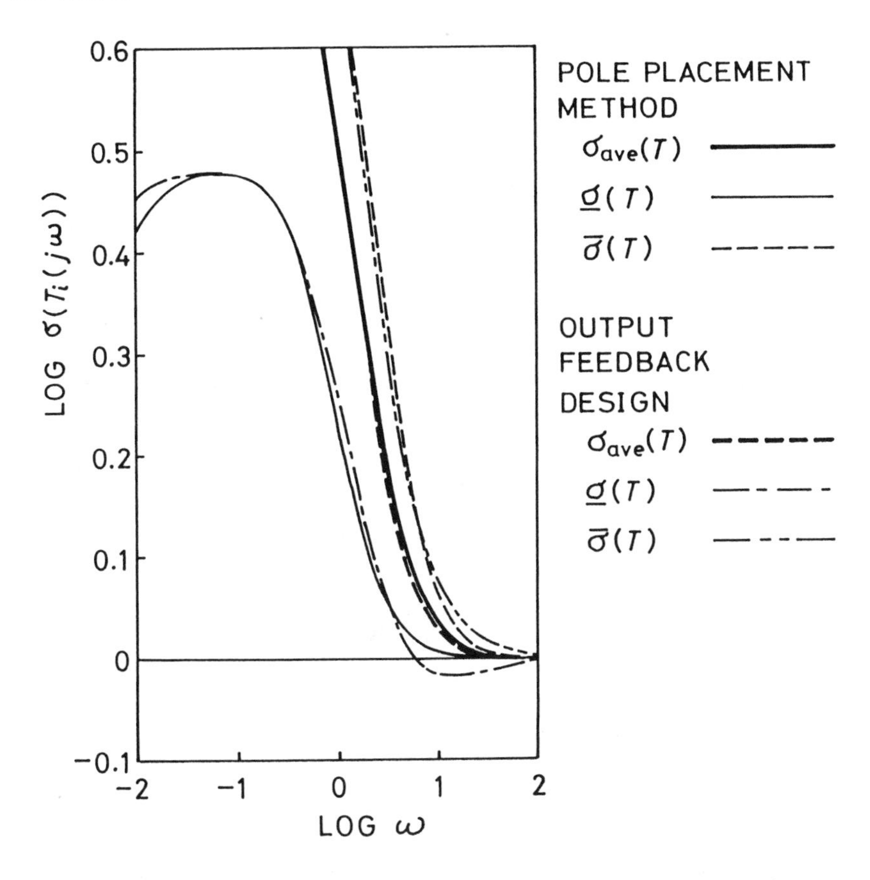

Fig. 14. Singular value curves for pole placement method at input.

Fig.14 and Fig.15 show the singular value curves (maximum, minimum, and average singular value curves) of the return difference matrices at input and at output of the present designed system respectively. These figures also include for comparison's sake those for the design using the output feedback method. It is seen that the curves for both the systems almost coincide, and satisfy the design specification at the output. But at the input, the former system (present designed system) fully satisfies the specification, while the latter system has frequency range below $\omega = 10$ rad/sec where the minimum singular value is less than unity because its design has not taken the specification into consideration.

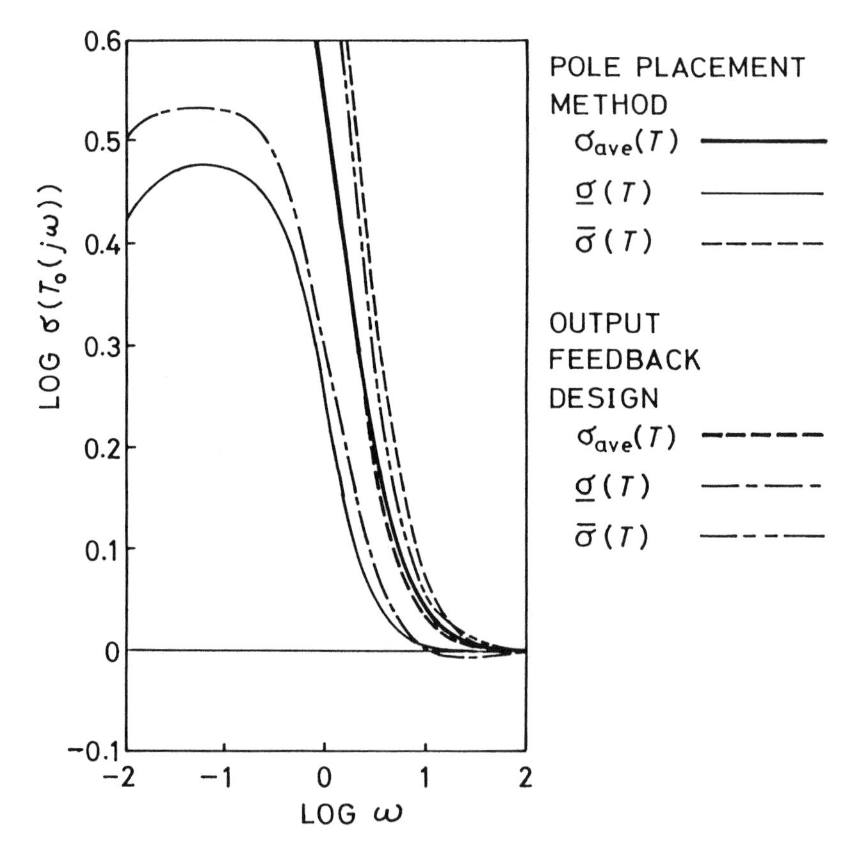

Fig. 15. Singular value curves for pole placement method at output.

Step responses for perturbed flight condition of Eqs.(26) and (27) with input $r = [0.1, 0.1]^T$ are shown in Figs.11 and 12 again. To enable comparison with the previous example, a precompensator to equalize both responses at nominal parameter value was added in this simulation. (This precompensator was used only for comparison purpose. In practical application, it is not needed and therefore no increase of the dimension results.)

As seen from these figures where the responses of the previous design examples are also shown, the responses of the present design shown by one dotted broken line is nearer to the nominal response shown by solid line than those of the model matching design shown by dotted line. Therefore the present design method is considered to have the same degree of robustness recovery as the output feedback method.

VI. SUMMARY

When all variables in control system can not be measured an observer (or Kalman filter) is needed, but if an observer is used the parameter variation of the controlled system have a great influence on the response characteristics. Therefore one should pay attention to the robustness in constructing and designing the control system, when an observer is used.

Explanation has been made for three such design methods as (1) output feedback method, (2) model matching method, and (3) pole placement method. In the method of (1), the robustness is recovered by selecting the adequate output feedback gain while the desired response characteristics is obtained by using model following system. In this method, the gain margin and the phase margin is considered through singular value of the return difference matrix to recover the robustness while the gain is coordinated not to become too large. This method can obtain good characteristics with relative ease though it has a drawback that the system dimension increases. In the method of (2), a loop is used which makes feedback of the difference between the output of the plant and the output of the observer, and the robustness is recovered by the selection of the element included there. In this method, the model following system is not needed because the response characteristics of the closed–loop system is not changed by the added loop, and therefore the dimension of the system is not increased. If dynamics are introduced in the feedback element to recover the complete robustness, however, the increase in the dimension will occur. In order to design the feedback element under constraint of fixed gain to avoid the dimensional increase, the method similar to (1) is used where the robustness is recovered considering the gain margin and the phase margin using the singular value of the return difference, while the increase in gain is moderated. This method has a merit that no increase in the dimension results while it has a drawback that the resulting robustness is slightly less than that of the method (1). In the method of (3), the same output feedback loop as in the method of (1) is used, but the robustness is recovered by the placement of the poles in the observer channel, and the response characteristics of the closed–loop system between input and output are improved by the output feedback channel. In this method, the robustnesses at input and output are equally recovered by using the average singular value of the return differences while moderating the increase in the gain. This method has the advantage that the robustness recovery as much as that in the method of (1) is obtained without increase in the dimension though the procedure for the placement of poles are somewhat complicated.

All of the three methods have their merits and demerits, but they are all effective in recovering the robustness of the system including observer.

APPENDIX. PROCEDURE FOR COORDINATING POLES

Specify first the frequency range ω_0 in Eq.(40) where the sensitivity reduction is desired. Take the initial value of the poles of $\lambda[A-FC-BK]$ at vicinity of those of an optimal feedback system, and move them gradually so that the average singular value curve is reformed from its initial shape like Fig.16(a) to its final shape like Fig.16(b).

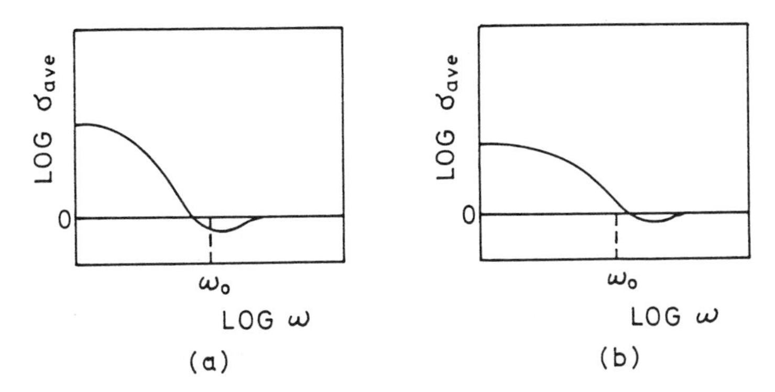

Fig. 16. Coordination of frequency range for robustness of average singular value.

Specify second σ_D in Eq.(51) as high as possible and increase Q and R gradually from their initial values of identity matrices so that the average singular value curve is reformed from its initial shape like Fig.17(a) to its final shape like Fig.17(b). But in this coordination, the condition of Eq.(52) must be also taken into consideration.

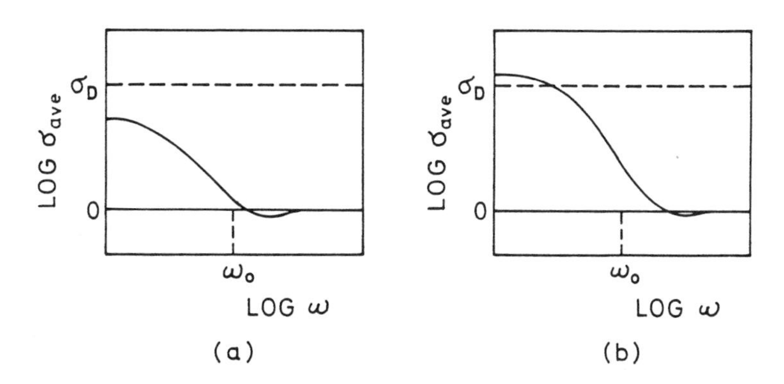

Fig. 17. Coordination of magnitude of average singular value.

As shown in Fig.18, even if the maximum values of the average singular values are identical, the width of their maximum values may differ. If its width is narrow as case 2 in Fig.18, coordinate the positions of poles so that it becomes as wide as possible, as case 1 in Fig.18.

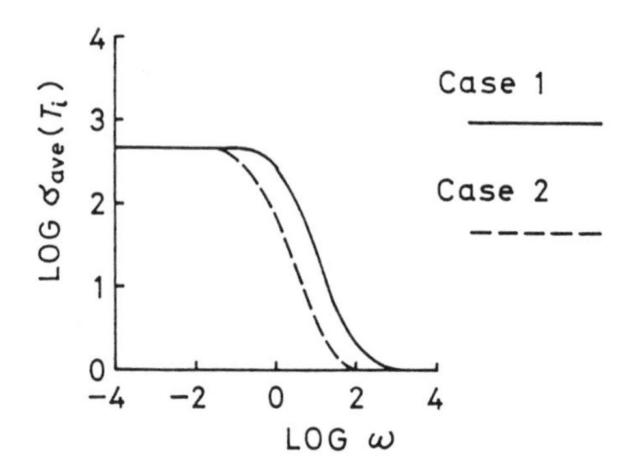

Fig. 18. Coordination of frequency bandwidth of large average singular value part.

Then select several points of ω_k smaller than ω_0 and coordinate the directional vectors so that the minimization of the condition number as in Eq.(53) is achieved. By this coordination, the maximum and the minimum singular value curves, which are far separated as in Fig.19(a), approach the average singular value curve as in Fig.19(b).

The design can be performed by above procedure.

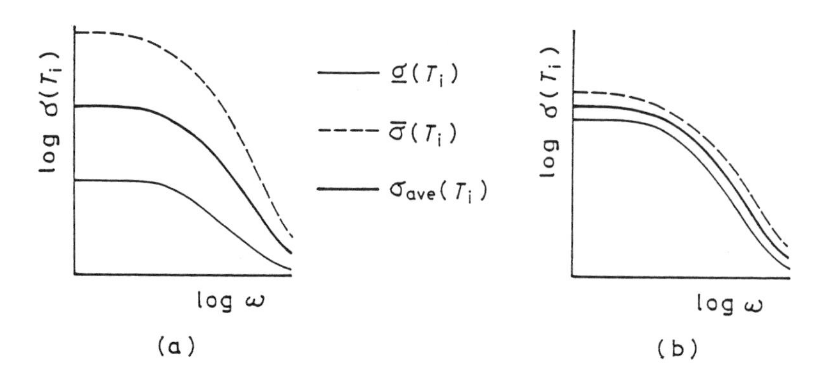

Fig. 19. Coordination of condition number.

If the magnitude of the parameter variation is known in advance, write the transfer function of the perturbed plant $\tilde{G}(s)$ using the additive perturbation as follows:

$$\tilde{G}(s) = G(s) + \Delta G(s) \tag{56}$$

where it is assumed that $\tilde{G}(s)$ is strictly proper and has no pole on the imaginary axis, and $\tilde{G}(s)$ and $G(s)$ have the same number of unstable poles. Then when the corresponding variation of the return difference matrix at input or output is denoted by $\Delta T(s)$, a sufficient condition for the closed-loop system after parameter variation to be asymptotically stable is [25],

$$\frac{\|\Delta T(j\omega)\|}{\underline{\sigma}(T(j\omega))} < 1. \tag{57}$$

Fig.20 shows examples which satisfy this condition.

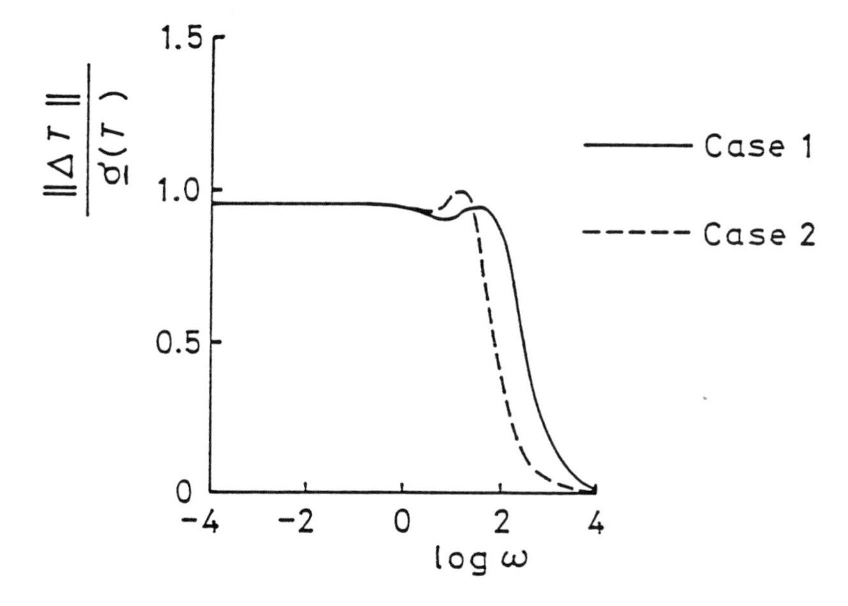

Fig. 20. Stability consideration.

When a closed-loop transfer function $G_{CL}(s)$ is varied by the parameter variation as

$$\tilde{G}_{CL}(s) = G_{CL}(s) + \Delta G_{CL}(s), \tag{58}$$

the following relation is satisfied at the input

$$\|\Delta G_{CL}(j\omega)\| \leq S_i \frac{\|\Delta G(j\omega)\|}{\|T_i(j\omega)\|}. \tag{59}$$

where

$$S_i = \frac{\text{cond } T_o(j\omega)}{\alpha \underline{\sigma}[T_i(j\omega)]}, \tag{60}$$

where α is a constant between 0 and 1. (At the output, the relation (59) and (60) where subscripts i and o are interchanged is satisfied.) It is noticed from Eq.(59) that $\|\Delta G(j\omega)\| / \|T_i(j\omega)\|$ should be small for the variation of the closed-loop transfer function to be small. Logarithm of inverse of this quantity is depicted in Fig.21. In this figure, since it is desirable for the range of large value to be wider and for the range of small value to be narrower, the case of solid line is considered to be more robust than the case of broken line. The pole placement using these figures is one of the way.

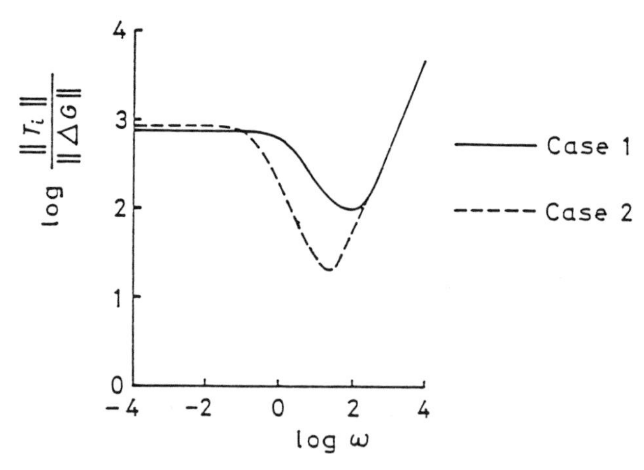

Fig. 21. Another approach to pole placement.

ACKNOWLEDGMENT

This article is partly based on earlier paper [17]. The authors would like to thank to the former student Mr.Seiichi Sekizuka for his cooparation in this work.

REFERENCES

1. J. B. Cruz, J. S. Freudenberg, and D. P. Looze, "A Relationship between Sensitivity and Stability of Multivariable Feedback System",IEEE Transactions on Automatic Control AC–26, pp. 66–74 (1981).
2. V. Mukhopadhyay and J. R .Newsom, "A Multiloop System Stability Margin Study Using Matrix Singular Values", Journal of Guidance, Control, and Dynamics 7, pp. 582–587 (1984).
3. U. Ly, "Robustness Analysis of a Multiloop Flight Control System", AIAA Paper, 83–2189 (1983).
4. A. Herrera–Vaillard, J. Paduano, and D. Downing, "Sensitivity Analysis of Automatic Flight Control System Using Singular Value Concepts", AIAA Paper, 85–1899 (1985).
5. J. R. Newsom and V. Mukhopadhyay, "A Multiloop Robust Controller Design Study Using Singular Value Gradients", Journal of Guidance, Control, and Dynamics 8, pp. 514–519 (1985).
6. V. C. Gordon and D. J. Collins, "Multi–Input Multi–Output Automatic Design Systhesis for Performance and Robustness", Journal of Guidance, Control, and Dynamics 9, pp. 281–287 (1986).
7. B. D. O. Anderson and J. B. Moore, Linear Optimal Control, Prentice–Hall, Engelwood Cliffs, NJ (1971).
8. M. G. Safonov and M. Athans, "Gain and Phase Margin for Multiloop LQG Regulators", IEEE Transactions on Automatic Control AC–22, pp.173–179 (1977).
9. H. Kwakernaak and R. Sivan, Linear Optimal Control Systems, Wiley, New York (1972).
10. T. Okada, M. Kihara, and K. Motoyama, "Sensitivity Reduction by Double Perfect Model Following", IEEE Transactions on Aerospace and Electronic Systems AES–18, pp. 29–38 (1982).
11. T. Okada, M. Kihara, and T. Takei, "Robust Model Following Systems", International Journal of Control 36, pp. 905–923 (1982).
12. J. C. Doyle and G. Stein, "Robustness with Observers", IEEE Transactions on Automatic Control AC–24, pp. 607–611 (1979).
13. J. C. Doyle and G. Stein, "Multivariable Feedback Design : Concepts for a Classsical/ Modern Synthesis", IEEE Transactions on Automatic Control AC–26, pp. 4–16 (1981).

14. H. Kimura and O. Sugiyama, "Design of Robust Controller Using Perfect Regulation and Perfect Observation", Transactions of the Society of Instrument and Control Engineers 18, pp. 955–960 (1982).

15. C. Tsui, "New Approach to Robust Observer Design", International Journal of Control 47, pp. 745–751 (1988).

16. T. Okada, M. Kihara, and H. Furihata, "Robust Control System with Observer", International Journal of Control 41, pp. 1207–1219 (1985).

17. T. Okada, M. Kihara, and M. Ikeda, "Robust Control System Design Synthesis with Observers", Journal of Guidance, Control, and Dynamics 13, pp. 337–342 (1990).

18. J. B. Moore and T. T. Tay, "Loop Recovery via H^{∞}/H^{2} Sensitivity Recovery", International Journal of Control 49, pp. 1249–1271 (1989).

19. Y. Ando and M. Suzuki, "A Design of High–gain Feedback Regulators", Transactions of the Society of Instrument and Control Engineers 19, pp. 374–380 (1983).

20. N. A. Lehotmaki, G. Stein, and J. E. Wall Jr., "Multivariable Prefilter Design for Command Shaping", AIAA Paper No. 84–1829, AIAA Guidance and Control Conference, Seattle, Washington, pp. 38–43 (1984).

21. T. Okada and M. Kihara, "Extended Perfect Model Following",Bulletin of the JSME 125, pp. 1985–1993 (1982).

22. V. Mukhopadhyay, "Stability Robustness Improvement Using Constrained Optimization Techniques", Journal of Guidance, Control, and Dynamics 10, pp. 172–177 (1987).

23. V. Mukhopadhyay, "Digital Robust Control Law Synthesis Using Constrained Optimization", Journal of Guidance, Control, and Dynamics 12, pp. 175–181 (1989).

24. R. K. Heffley and W. F. Jewell, "Aircraft Handling Qualities Data : Section IX", NASA CR–2144 (1972).

25. H. Kajiwara, Computer–Aided Control System Design, Korona, Tokyo (1988).

26. T. Sönderström, "On Some Algorithms for Design of Optimal Constrained Regulators", IEEE Transactions on Automatic Control AC–23, pp. 1100–1101 (1978).

Robust Tracking Control of Non-linear Systems with Uncertain Dynamics

DAUCHUNG WANG
Yuan-Ze Institute of Technology
135 Yuan-Tung Road, Neih LI, Chun Li,
Taoyuan Shian, Taiwan, R.O.C.
CORNELIUS T. LEONDES
Electrical Engineering Department, FT-10
University of Washington,
Seattle, Washington 98195, U.S.A.

Part I Theory

1. Introduction

A robust control scheme for treating the tracking problem of non-linear time-varying systems with uncertainties is investigated. A class of physical system composed of several interconnected subsystems is then considered as an example for the application of this robust tracking control scheme. Also, we assume that each subsystem has its own control input.

Several concepts are realized by this robust control scheme.

(a) Linearization of a non-linear system by the control scheme, i.e. the non-linear system acts like a linear system after the feedback compensation (feedback linearization).

(b) Decentralization of a coupled system to a set of nearly independent subsystems by the control algorithm.

(c) The time-varying system dynamics,unmodelled dynamics,uncertainties and disturbances are all treated as "general disturbances" to the system. Thus, a nominal model can be used for the control purpose.

(d) Controls, based on two unequal criteria, are then designed to overcome the effect of general disturbances on the system performance. Robust control is thus realized.

(e) Owing to these two unequal criteria, the set of nearly independent subsystems can be treated as a set of independent subsystems by taking the coupled terms as disturbances of the subsystems. Thus, the statement about decentralization of a coupled system (b) can be treated as fact.

Generally speaking, a precise mathematical model and exact understanding of the system are required for the control purposes of the system. System identification techniques are usually needed in the design of a control law for the time-varying system. This is also based on a full understanding of the system. Unfortunately, a full investigation of the physical system is very costly; in reality, it is sometimes almost impossible, due to the existence of uncertain dynamics, unmodelled dynamics, disturbances, and the complexity of the real system. Thus, difficulties in deriving an exact mathematical model for a physical system are always there - i.e. we usually have an approximate model on hand only when we are dealing with the control of physical systems. The non-linearity of the system certainly causes further difficulties in solving the problem.

A robust tracking control algorithm that is based mainly on the theory of variable structure systems (VSS) and Lyapunov's direct method is developed here, to treat the tracking control problem of non-linear time-varying systems with uncertain dynamics. An approximate model, instead of an exact one, is used for this robust tracking control algorithm. This algorithm greatly eases the difficulties in deriving the precise model for the control purposes, and in solving the tracking control problem for systems with uncertainties. This robust tracking control scheme also shows superior ability in handling non-linear time-varying systems.

For a system composed of several subsystems, the skeleton of the control scheme is listed step by step, as in the following.

Step 1. Decouple the interactions between the subsystems using a control scheme such that the system can be represented by a set of decentralized subsystems.

Step 2. Linearize each non-linear subsystem by applying a control to it, such that the compensated non-linear subsystem acts like a linear one.

Step 3. Steps 1 and 2 should be carried out together by a single control law designed via the theory of VSS.

Step 4. Depending on the decentralized linear subsystem, Lyapunov's direct method can then be easily applied to designing the control law for the non-linear subsystem.

Step 5. Tracking control of each subsystem is reached by applying two consecutive controls to that subsystem. The first one, based on VSS, is used to linearize and decentralize the non-linear subsystem, and the second one, based on Lyapunov's direct method, is used to smooth the control chattering of the first control and to complete the tracking.

Step 6. For realization, this control scheme is usually needed more than once for the switching between the two controls.

Step 7. The goal of the non-linear system is reached if the goals of the subsystems are all reached.

This combines the ideals of VSS, feedback linearization, decentralized control, Lyapunov's direct method, and the perturbation method. The linearized subsystems are chosen via the perturbation method. None-linear subsystems are driven separately to their decentralized linear correspondence, by the method of VSS. Then, the Lyapunov direct method eliminates the control chattering caused by the method of VSS. Tracking is also realized by this algorithm. For a system without subsystems, the control scheme is almost the same, except without Steps 1 and 7, since no decentralization action is needed in this case.

The method of Richter et al. (1982) combines the ideals of VSS with the methods of eigenvalue placement, in a decentralized context. Two restrictions that the subsystem model has to be in a special canonical form and that the interaction between the subsystems has to be linear are posed in this method. There is a great deal of literature containing research on VSS (Itkis 1976, Utkin 1971,1972,1977, Morgan and Ozguner 1985, Bartolini and Zolezzi 1985, Slotine and Sastry 1983, Slotine 1984, 1985, Drazenovic 1969). This approach drives a non-linear system through a fast switching control law, which forces the original non-linear system to behave as a stable linear time-invariant system. Control chattering is the main disadvantage of this approach. Several model-following control systems are discussed by

Young (1978), Nicosia and Tomei (1984), and Singh (1985). A simpler way to deal with the path tracking control problem exists if the desired path equation can be used directly, instead of using a reference model. A technique of system linearization that is suitable for VSS is presented by Guo and Koivo (1984). The stability of uncertain dynamical systems has also been studied (Gutman 1979, Corless and Leitmann 1981, Gutman and Palmer 1982, Barmish *et al.* 1983 a, b, Balestrino *et al.* 1984). A robust model tracking control for non-linear plants is discussed by Ambrosino *et al.* (1985). It is based on Lyapunov's method, with a rather restrictive assumption on the non-linear plants. Thus, the application of this method is somewhat limited. The ideal of robust tracking control developed in this context is mainly based on the research cited here. Some improvemcnts have been made, such as relieving the restrictive assumption on systems (Ambrosino et al. 1985), giving a desired path instead of a reference model for the tracking problem (Young 1978, Nicosia and Tomei 1984, Singh 1985), using a decentralized control scheme via VSS, and eliminating the control chattering caused by the method of VSS by using Lyapunov's method.

2. Variable structure systems

2.1. Introduction

Variable structure systems, as suggested by their name, differ from other control systems mainly in that their structure is not constant, but is varied during the control process. The non-continuous control laws developed in the theory of VSS usually provide for changes in the structure of the system, whenever the representing point crosses certain hypersurfaces in the state space of the system. Essentially, the theory of VSS is the theory of selecting rational switching surfaces and structures in the state space regions they define. A more clear idea of the theory of VSS can be given by the following example.

Example 1

Consider the case of a simple pendulum with small motion. The equation of motion of the system is

$$\ddot{\theta} + \alpha\,\theta = 0$$

The state-space representation of the equation can be represented as

$$\dot{x}_1 = x_2$$
$$\dot{x}_2 = -a\,x_1$$

where $a > 0$, $x_1 = \theta$,and $x_2 = \dot{\theta}$

There are two possible phase portraits for the system (see Fig. 1), and they are both conservative systems, as shown by the phase portrait. A control u is introduced into the system to form a VSS:

$$\dot{x}_1 = x_2$$
$$\dot{x}_2 = -a\,x_1 + u$$

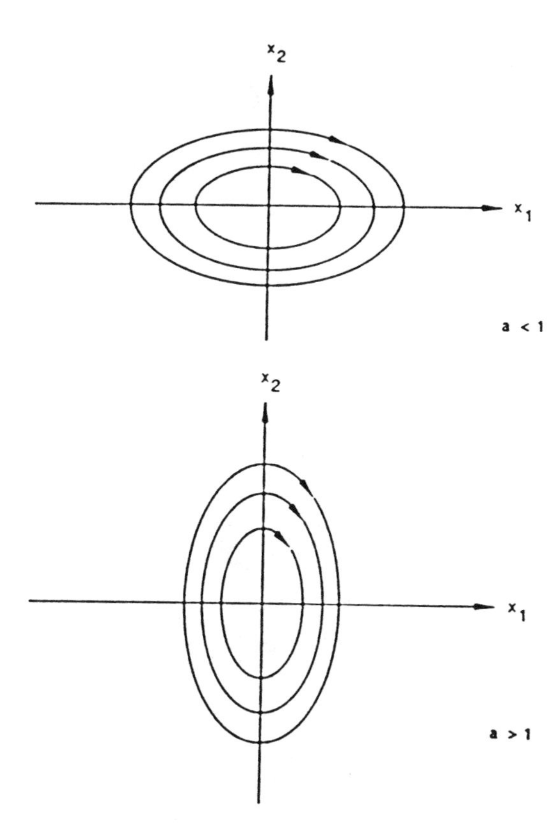

Figure 1. Phase portraits for $a < 1$ and $a > 1$.

The control law u is defined as

$$u = \begin{cases} -Kx_1, & \text{if } x_1 x_2 \geqslant 0 \\ 0, & \text{if } x_1 x_2 < 0 \end{cases}$$

where $0 < a < 1$, $K > 0$, and $a + K > 1$. The switching planes are chosen as $x_1=0$ and $x_2=0$. The control u switches to $-Kx_1$ when the representing point crosses the switching plane $x_1=0$, and u is zero when the representing point crosses the plane $x_2=0$. The phase portrait of the VSS is therefore representing an asymptotically stable system (see Fig. 2). Thus, a conservative system has been switched to an asymptotically stable system by the variable structure control algorithm (VSCA).

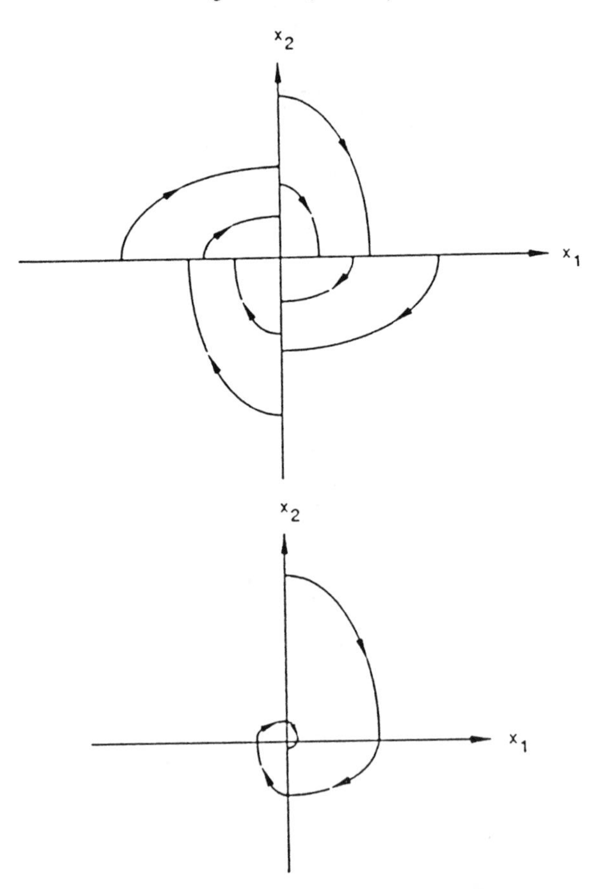

Figure 2. Phase portrait of VSS.

The physical meaning of the analysis of the above system can be explained as follows (see Fig. 3).

The control torque u =$-K\theta$ and the angular displacement θ are both defined as positive in the counter-clockwise direction. The control u switches back and forth between $-K\theta$ and zero at $\theta = 0$, $\theta = -\theta_{max}$, and $\theta = \theta_{max}$. The control u is always against the motion of the pendulum when the pendulum swings away from the equilibrium position $\theta = 0$. The system is free from control when the pendulum swings

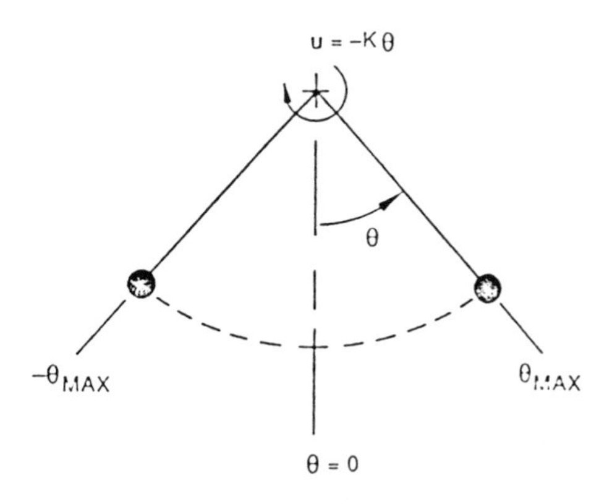

Figure 3. VSS of simple pendulum.

back towards the equilibrium position $\theta = 0$. Thus, the control u finally causes the pendulum to stay at the stable position $\theta = 0$ and to cnsurc the asymptotic stability of the VSS.

2.2. Sliding regime

Now consider VSS in the more general case. Suppose S(X,t)=0 is a chosen switching hyperplane in the state space of a non-linear system with uncertainties. The non-linear uncertain system can be driven toward .S (X, t) = 0 by a suitable choicc of control law, such that the compensated system meets the "sliding condition". The sliding condition is defined as

$$\frac{1}{2}\frac{dS^2}{dt} < 0 \tag{1}$$

This is illustrated by Fig. 4.

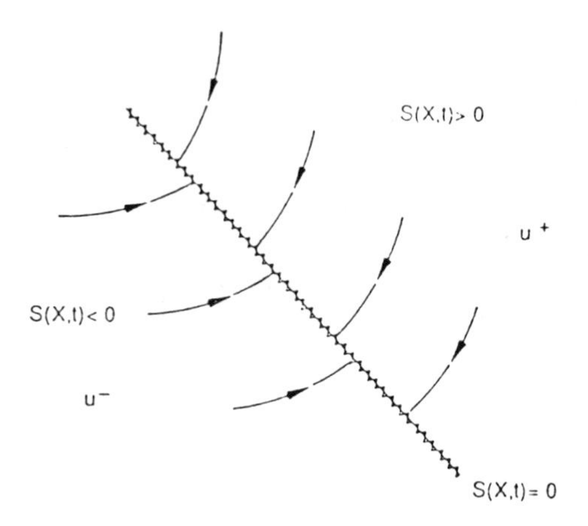

Figure 4. Sliding mode.

The state space is separated into two parts by the switching hyperplane S(X, t) = 0. The states of the system should be either in the region of S(X, t) > 0 or in the region of S(X, t) < 0. Suppose u^+ is the control that is chosen to ensure the existence of the sliding condition (1), such that the system at the same side as S(X, t) > 0 is driven towards the switching hyperplane S(X, t) = 0. Also, u^- is the control that is chosen to ensure the existence of sliding condition (1), such that the system at the same side as S(X, t) < 0 is also driven towards the switching hyperplane S(X, t) = 0. The motion of the VSS takes place in a special regime called a "sliding regime" or "sliding mode", defined as follows: the structure switchcs back and forth at infinite frequency, and the representing point performs infinitesimal oscillations about the switching hyperplane S(X, t) = 0. Therefore, once the system reaches the sliding regime, the fast switching control keeps the system trajcctory near the switching hyperplane S(X, t) = 0.

Since the switching hyperplane S(X, t) = 0 can be chosen at will, the systems can be driven to the switching hyperplane S(X, t) = 0 by satisfying the sliding condition (1), the desired system performance can be achieved by driving the system to, and keeping the system on, the switching hyperplane S(X, t) = 0. Thus, the motion of the system is described approximately by the equation of the switching hyperplane S(X, t) = 0.

2.3. Comments

(a) Sliding regimes have an important property, in that the

corresponding motion of the system, described approximately by the switching hyperplane, is independent of changes in the plant parameters and of external disturbances. Therefore, exact modelling (for example in the load forecasting of a robot manipulator) is not required in general. A much simplified model can be used for control purposes.

(b) VSS can be used as a linearization technique. This is carried out by choosing the switching hyperplane $S(X, t) = 0$ as a linear system and driving the nonlinear system to the switching hyperplane $S(X, t) = 0$. Actually, we can say that it is a state-feedback linearization technique.

(c) For a non-linear system composed of several interconnected subsystems, the switching hyperplanes of the non-linear system, each assigned to a subsystem, can be chosen as a set of decentralized linear equations. Each non-linear subsystem can then be driven to its corresponding switching hyperplane via the VSS method. The non-linear system is thus represented by a set of decentralized linear subsystems. Thus, VSS can also be used as a decentralization technique.

(d) The main drawback of VSS is the existence of a discontinuous control law across the switching hyperplane $S(X, t) = 0$. This leads to control chattering about the switching hyperplane, which is undesired in practice, since it involves high control activity and is hard to realize. It may also excite the high frequency unmodelled dynamics.

3. Robust tracking control

3.1. Control scheme

The basic idea of our robust tracking control algorithm is to linearize a nonlinear system, via the theory of VSS, to a linearized error equation that is derived from the chosen switching hyperplane. Then, a second control law is designed, based on the linearized system, to eliminate the control chattering and to complete the tracking of the non-linear system via Lyapunov's direct method. Four difficulties should be overcome when realizing the above control concept:

(a) the choice of switching hyperplane leads to a linearized error equation;

(b) the control chattering problem at the switching hyperplane is caused by the VSS method;

(c) system performance is degraded by the existence of system uncertainties; and

(d) non-lincar and time-varying systems are being dealt with.

The way to overcome these problems is described in the process of the control law design of a non-linear time-varying system.

A wide variety of physical systems can be described generally by a nth-order ordinary differential equation of the type

$$\begin{aligned} H(q^{(n)}, q^{(n-1)}, \ldots, \dot{q}, q, u) &= h_1(q^{(n-1)}, \ldots, \dot{q}, q)q^{(n)} \\ &\quad + h_2(q^{(n-1)}, \ldots, \dot{q}, q) - u + \Delta h \\ &= 0 \end{aligned} \tag{2}$$

where q is the generalized coordinate, $q(j) = d^jq(t)/dt$, and Δh is the known maximum system error due to unmodelled dynamics and disturbances.

The type of system to be considered here is a system with uncertainties. Three kinds of uncertainties are generally involved in the system. These are: uncertain dynamics in $h_1(\bullet)$ and $h_2(\bullet)$; unmodelled dynamics; and disturbances in the h term. Before we go through the design process for the control laws, several assumptions are posed on the system.

Assumption 1

All the states of the system are measurable.

Assumption 2

We know the range of variation of the system uncertain dynamics, such that $|h_i(\cdot)|_{max}$ and $|h_i(\cdot)|_{min}$ are completely defined, i= 1, 2.

Assumption 3

The maximum system error caused by the unmodelled dynamics and disturbances is known.

Based on these assumptions, the design of the robust tracking control laws for the proposed non-linear system with uncertainties (2) is outlined as follows.

3.1.1. *Deriving Variational equation.* We obtain the variational equation of the nonlinear system using the perturbation method. For example, if we define the nonlinear system as

$$H(\ddot{q}, \dot{q}, q, u) = 0$$

and the desired system as

$$H(\ddot{q}_d, \dot{q}_d, q_d, u_d) = 0$$

then the variational equation of the non-linear system is defined as

$$\left(\frac{\partial H}{\partial \ddot{q}}\right)_d \delta\ddot{q} + \left(\frac{\partial H}{\partial \dot{q}}\right)_d \delta\dot{q} + \left(\frac{\partial H}{\partial q}\right)_d \delta q + \left(\frac{\partial H}{\partial u}\right)_d \delta u = 0$$

where

$$\partial\ddot{q} = \ddot{q} - \ddot{q}_d, \quad \delta\dot{q} = \dot{q} - \dot{q}_d$$

$$\delta q = q - q_d, \quad \delta u = u - u_d$$

3.1.2. *Choosing switching hyperplane.* The switching hyperplane $S(\dot{q},q)=0$, which is one order less than the variational equation, is chosen according to the derived variational equation, such that a linearized error equation of the non-linear system can be deduced from it. The linearized error equation is a modified version of the variational equation of the non-linear system.

3.1.3. *Variable-structure control.* Driving the non-linear system, $H(\ddot{q}, \dot{q}, q, u) = 0$, approaches the switching hyperplane $S(\dot{q}, q) = 0$ by choosing a control law u such that it guarantees the existence of the sliding condition

$$\frac{1}{2}\frac{dS^2}{dt} < 0$$

The design of u is based on the method of VSS.

3.1.4. *System linearization.* The non-linear system is then represented approximately by the switching hyperplane, when the non-linear system is kept in the vicinity of the switching hyperplane.

3.1.5. *Smoothing control chattering.* The variable structure control in §3.1.3 results in control chattering at the switching hyperplane. The control chattering can be eliminated by switching the control law from the method of VSS to Lyapunov's direct method when the system trajectory enters a thin boundary layer of the switching hyperplane $S(\dot{q}, q) = 0$. The thin boundary layer is defined as $B=\{(\dot{q}, q): | S(\dot{q}, q) | \leq \varepsilon \}$, where ε is a very small constant. This situation is shown in Fig. 5.

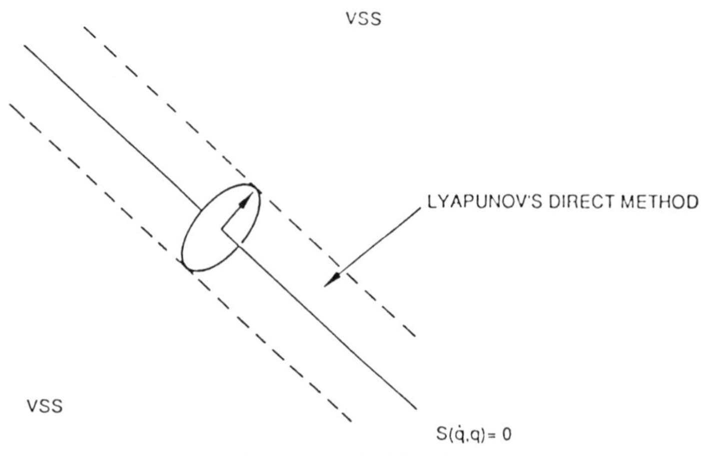

Figure 5. Control switching boundary.

3.1.6. *Lyapunov's direct method.* In order to apply Lyapunov's direct method, the linearized error equation is deduced from the switching hyperplane to represent approximately the non-linear system. According to Lyapunov's direct method, the linearized error equation (modified variational equation) that is derived from the switching hyperplane is driven asymptotically to the origin by a suitable choice of the control law u . It thus follows the tracking control of the non-linear system.

3.1.7. *Realization of tracking control.* Depending on the situation - whether the representing point of the non-linear system is staying in the thin boundary layer B or not - Lyapunov's direct method or the VSS method is applied to the system accordingly. This combined control scheme guarantees approximation of the nonlinear system by the linearized error equation, elimination of the control chattering caused by applying the VSS method alone, and tracking. A switch of the control is usually needed more than once.

3.1.8. *Robustness.* The robustness of the tracking control is assured by choosing controls u and u such that the sliding condition

$$\frac{1}{2}\frac{dS^2}{dt} < 0$$

exists and the time rate of Lyapunov's function is negative definite (V < 0). Since all the system uncertainties are on the left-hand side of these two criteria, then, if we know the range of variation in the system uncertainties, the controls can be chosen to guarantee the negative definite property of these two criteria and to overcome the effect of system uncertainties. Thus, the effect of uncertainties on the system stability is eliminated. The robust property of the compensated system is obtained by using this kind of control technique.

3.2. Robust control Laws

Depending on the control scheme given in the previous section, the robust tracking control laws for the physical system described by (2) are derived as follows.

3.2.1. *Deriving the variational equation.* $(q_d^{(n)}, q_d^{(n-1)}, \ldots, \dot{q}_d, q_d, u_d)$ is the desired path we are going to track. Thus, the variational equation of the physical system is

$$\sum_{j=0}^{n}\left(\frac{\partial H}{\partial q^{(j)}}\right)_d \delta q^{(j)} - \delta u = [h_1(q^{(n-1)}, \ldots, \dot{q}, q)]_d \delta q^{(n)} + \sum_{j=0}^{n-1}\left(\frac{\partial H}{\partial q^{(j)}}\right)_d \delta q^{(j)} - \delta u$$

$$= 0$$

where

$$\sum_{j=0}^{n-1}\left(\frac{\partial H}{\partial q^{(j)}}\right)_d \delta q^{(j)} = \sum_{j=0}^{n-1}\left[\frac{\partial h_1(q^{(n-1)}, \ldots, \dot{q}, q)}{\partial q^{(j)}} q^{(n)} + \frac{\partial h_2(q^{(n-1)}, \ldots, \dot{q}, q)}{\partial q^{(j)}}\right]_d \delta q^{(j)}$$

3.2.2. *Switching hyperplane.* According to the variational equation, the switching hyperplane S(•) = 0 is chosen as follows:

$$h_1(q^{(n-1)}, \ldots, \dot{q}, q)\delta q^{(n-1)} + [h_1(q^{(n-1)}, \ldots, \dot{q}, q)]_{md} \delta q^{(n-1)}$$
$$+ \sum_{j=1}^{n-1}\left(\frac{\partial H}{\partial q^{(j)}}\right)_{md} \delta q^{(j-1)} = S(q^{(n-1)}, \ldots, \dot{q}, q) = 0 \tag{3}$$

where

$$[h_1(\cdot)]_{md} \quad \text{and} \quad \left(\frac{\partial H}{\partial q^{(j)}}\right)_{md}$$

correspond to

$$|h_1(\cdot)|_{max} \quad \text{and} \quad \left|\frac{\partial H}{\partial q^{(j)}}\right|_{max}$$

with the states replaced by the known desired states. It is a linear time-varying system with time-varying coefficients, completely defined except for the first term. The first term

$$h_1(q^{(n-1)}, \ldots, \dot{q}, q)\delta q^{(n-1)}$$

is used to involve the physical system in the switching hyperplane.

3.2.3. *Variable structure control.* The control u that drives the physical system towards the switching hyperplane S(•) =0 is now determined. According to the chosen switching hyperplane (3), the time rate of the switching hyperplane is

$$\begin{aligned} h_1(q^{(n-1)}, \ldots, \dot{q}, q)\delta q^{(n)} &+ \frac{d}{dt}[h_1(q^{(n-1)}, \ldots, \dot{q}, q)]\delta q^{(n-1)} \\ &+ \frac{d}{dt}\{[h_1(q^{(n-1)}, \ldots, \dot{q}, q)]_{md}\}\delta q^{(n-1)} \\ &+ [h_1(q^{(n-1)}, \ldots, \dot{q}, q)]_{md}\delta q^{(n)} \\ &+ \sum_{j=1}^{n-1}\left(\frac{\partial H}{\partial q^{(j)}}\right)_{md}\delta q^{(j)} + \sum_{j=1}^{n-1}\frac{d}{dt}\left[\left(\frac{\partial H}{\partial q^{(j)}}\right)_{md}\right]\delta q^{(j-1)} \\ = \dot{S}(q^{(n-1)}, q^{(n-2)}, \ldots, \dot{q}, q) = 0 \end{aligned} \tag{4}$$

The physical system is related to the switching hyperplane by substituting (2) into (4). After substitution and rearrangement, we have

$$\begin{aligned} [h_1(q^{(n-1)}, \ldots, \dot{q}, q)]_{md}\delta q^{(n)} &+ \left\{\frac{d}{dt}[(h_1(q^{(n-1)}, \ldots, \dot{q}, q))_{md}] + \left(\frac{\partial H}{\partial q^{(n-1)}}\right)_{md}\right\}\delta q^{(n-1)} \\ &+ \sum_{j=2}^{n-1}\left\{\left(\frac{\partial H}{\partial q^{(j-1)}}\right)_{md} + \frac{d}{dt}\left[\left(\frac{\partial H}{\partial q^{(j)}}\right)_{md}\right]\right\}\delta q^{(j-1)} \end{aligned}$$

$$+\frac{d}{dt}\left[\left(\frac{\partial H}{\partial \dot{q}}\right)_{md}\right]\delta q - h_1(q^{(n-1)}, \ldots, \dot{q}, q)q_d^{(n)}$$

$$+\frac{d}{dt}[h_1(q^{(n-1)}, \ldots, \dot{q}, q)]\delta q^{(n-1)}$$

$$-h_2(q^{(n-1)}, \ldots, \dot{q}, q) + u - \Delta h$$

$$= \dot{S}(q^{n-1)}, \ldots, \dot{q}, q) = 0$$

From (3) and (5), we have

$$\frac{1}{2}\frac{dS^2}{dt} = S\dot{S}\left\{[h_1(q^{(n-1)}, \ldots, \dot{q}, q)]_{md}\delta q^{(n)}\right.$$

$$+\left\{\frac{d}{dt}[(h_1(q^{(n-1)}, \ldots, \dot{q}, q))_{md}] + \left(\frac{\partial H}{\partial q^{(n-1)}}\right)_{md}\right\}\delta q^{(n-1)}$$

$$+\sum_{j=2}^{n-1}\left\{\left(\frac{\partial H}{\partial q^{(j-1)}}\right)_{md} + \frac{d}{dt}\left[\left(\frac{\partial H}{\partial q^{(j)}}\right)_{md}\right]\right\}\delta q^{(j-1)}$$

$$+\frac{d}{dt}\left[\left(\frac{\partial H}{\partial \dot{q}}\right)_{md}\right]\delta q + \frac{d}{dt}[h_1(q^{(n-1)}, \ldots, \dot{q}, q)]\delta q^{(n-1)}$$

$$-h_1(q^{(n-1)}, \ldots, \dot{q}, q)q_d^{(n)}$$

$$\left.-h_2(q^{(n-1)}, \ldots, \dot{q}, q) - \Delta h + u\right\}$$

The control u on the above equation is chosen to guarantee the existence of the sliding condition.If we choose the control law u as

$$u = [-h_1(q^{(n-1)}, \ldots, \dot{q}, q)]_{md}\delta q^{(n)}$$

$$-\left\{\frac{d}{dt}[(h_1(q^{(n-1)}, \ldots, \dot{q}, q))_{md}] + \left(\frac{\partial H}{\partial q^{(n-1)}}\right)_{md}\right\}\delta q^{(n-1)}$$

$$-\sum_{j=2}^{n-1}\left\{\left(\frac{\partial H}{\partial q^{(j-1)}}\right)_{md} + \frac{d}{dt}\left[\left(\frac{\partial H}{\partial q^{(j)}}\right)_{md}\right]\right\}\delta q^{(j-1)}$$

$$\left.-\frac{d}{dt}\left[\left(\frac{\partial H}{\partial \dot{q}}\right)_{md}\right]\delta q + \Delta h - \gamma \operatorname{sgn} S\right] \tag{6}$$

We then have

$$\frac{1}{2}\frac{dS^2}{dt} = S\dot{S}$$

$$= S\left\{-h_1(q^{(n-1)}, \ldots, \dot{q}, q)q_d^{(n)} + \frac{d}{dt}[h_1(q^{(n-1)}, \ldots, \dot{q}, q)]\delta q^{(n-1)}\right.$$

$$- h_2(q^{(n-1)}, \ldots, \dot{q}, q) \Big\} - \gamma |S|$$

The sliding condition

$$\frac{1}{2} \frac{dS^2}{dt} < 0$$

is guaranteed if the constant γ is chosen as

$$\gamma > \Big| - h_1(q^{(n-1)}, \ldots, \dot{q}, q) q_d^{(n)} + \frac{d}{dt} [h_1(q^{(n-1)}, \ldots, \dot{q}, q)] \delta q^{(n-1)} - h_2(q^{(n-1)}, \ldots, \dot{q}, q) \Big|_{max} \tag{7}$$

Also, the physical system approaches the switching hyperplane S(•) =0. The attractiveness of the switching hyperplane is then guaranteed by applying the control u (6) to the physical system, with γ defined as in (7).

3.2.4. *Linearization of the physical system*. The control u designed in §3.2.3 finally drives the physical system to the vicinity of the switching hyperplane. If the switching hyperplane is chosen such that a linearized error equation can be deduced from it then the linearized error equation can be used to represent approximately the physical system, when the system is in the vicinity of the switching hyperplane. Thus, the physical system is linearized by the method of VSS. However, this method results in control chattering around the switching hyperplane.

3.2.5. *Smoothing the control chattering*. The control chattering resulting from VSS can be eliminated by switching the control law to one that is designed by Lyapunov's direct method. Control switching occurs when the system trajectory enters a thin boundary layer of the switching hyperplane, defined as

$$B = \{(q^{(n-1)}, \ldots, \dot{q}, q) : |S(q^{(n-1)}, \ldots, \dot{q}, q)| \leqslant \varepsilon\}$$

The physical system can be represented approximately by the linearized error equation whenever it remains inside the boundary B. Thus, a linearized error equation is used to replace the physical system for the design of control law u_ε when we switch the control method from VSS to Lyapunov's direct method. Also, u_ε is the control law used to eliminate the control chattering and to complete the tracking when the physical system is in the boundary B. The linearized error equation derived from the switching hyperplane is shown below.

The physical system is represented approximately by

$$\begin{aligned}
&[h_1(q^{(n-1)}, \ldots, \dot{q}, q)]_{md}\delta q^{(n)} + \left\{\frac{d}{dt}[(h_1(q^{(n-1)}, \ldots, \dot{q}, q))_{md}] + \left(\frac{\partial H}{\partial q^{(n-1)}}\right)_{md}\right\}\delta q^{(n-1)} \\
&\quad + \sum_{j=2}^{n-1}\left\{\left(\frac{\partial H}{\partial q^{(j-1)}}\right)_{md} + \frac{d}{dt}\left[\left(\frac{\partial H}{\partial q^{(j)}}\right)_{md}\right]\right\}\delta q^{(j-1)} \\
&\quad + \frac{d}{dt}\left[\left(\frac{\partial H}{\partial \dot{q}}\right)_{md}\right]\delta q \\
&\quad - h_1(q^{(n-1)}, \ldots, \dot{q}, q)q_d^{(n)} + \frac{d}{dt}[h_1(q^{(n-1)}, \ldots, \dot{q}, q)]\delta q^{(n-1)} \\
&\quad - h_2(q^{(n-1)}, \ldots, \dot{q}, q) - \Delta h + O(\varepsilon) + u_\varepsilon \\
&= S(q^{(n-1)}, \ldots, \dot{q}, q) = 0
\end{aligned} \tag{8}$$

whenever the system trajectory is within the boundary layer B. The system (8) differs from the system (5) in the control term u_ε and the error correction term $O(\varepsilon)$. u_ε represents the control law designed using Lyapunov's direct method, and it is applied to the system when the system is in the boundary layer B. The error correction term $O(\varepsilon)$ is used to compensate the error due to the approximation of the physical system by a linearized system when the physical system is in B.

For the practical consideration of the realization of control laws, the linearized error equation should be guaranteed to be in the form of (8). If there are missing terms, the control law u_ε is chosen to supply the missing terms and to form the linearized error equation in the form of (8). This can be represented in a more convenient form as follows:

$$\begin{aligned}
&[h_1(q^{(n-1)}, \ldots, \dot{q}, q)]_{md}\delta q^{(n)} + \left\{\frac{d}{dt}[(h_1(q^{(n-1)}, \ldots, \dot{q}, q))_{md}] + \left(\frac{\partial H}{\partial q^{(n-1)}}\right)_{md}\right\}\delta q^{(n-1)} \\
&\quad + \sum_{j=2}^{n-1}\left\{\left(\frac{\partial H}{\partial q^{(j-1)}}\right)_{md} + \frac{d}{dt}\left[\left(\frac{\partial H}{\partial q^{(j)}}\right)_{md}\right]\right\}\delta q^{(j-1)}
\end{aligned}$$

$$+\frac{d}{dt}\left[\left(\frac{\partial H}{\partial \dot{q}}\right)_{md}\right]\delta q - D + u_\varepsilon$$
$$= \dot{S}(q^{(n-1)}, \ldots, \dot{q}, q) = 0 \qquad (9)$$

where

$$D = h_1(q^{(n-1)}, \ldots, \dot{q}, q)q_d^{(n)} - \frac{d}{dt}[h_1(q^{(n-1)}, \ldots, \dot{q}, q)]\delta q^{(n-1)}$$
$$+ h_2(q^{(n-1)}, \ldots, \dot{q}, q) + \Delta h - O(\varepsilon)$$

It is obvious that (9) is a linearized system if D is treated as a general disturbance to the system. Since the physical system has been linearized to a linear system, Lyapunov's direct method can easily be applied to design the control law u_ε . The control law u_ε eliminates the control chattering and completes the tracking control. In order to represent (9) in a state-space form, the system states are defined as follows:

$$\begin{array}{ll} x_1 = \delta q, & \dot{x}_1 = x_2 \\ x_2 = \delta \dot{q}, & \dot{x}_2 = x_3 \\ \vdots & \vdots \\ x_{n-1} = \delta q^{(n-2)}, & \dot{x}_{n-1} = x_n \\ x_n = \delta q^{(n-1)}, & \dot{x}_n = \delta q^{(n)} \end{array}$$

$$X = \begin{bmatrix} x_1 \\ \vdots \\ x_n \end{bmatrix}$$

The state-space rorm is then represented as follows:

$$\dot{X} = AX - Bu_\varepsilon + BD \qquad (10)$$

with

$$A = \begin{bmatrix} 0 & 1 & 0 & \ldots & 0 \\ & 0 & 1 & & \\ & & 0 & & \\ \vdots & \vdots & \vdots & & 0 \\ 0 & 0 & 0 & & 1 \\ a_{n1} & a_{n2} & & \ldots & a_{nn} \end{bmatrix}$$

$$B = \begin{bmatrix} 0 \\ \vdots \\ 0 \\ \dfrac{1}{[h_1(\cdot)]_{md}} \end{bmatrix}$$

where

$$a_{n1} = -\frac{1}{[h_1(\cdot)]_{md}} \frac{d}{dt}\left[\left(\frac{\partial H}{\partial \dot{q}}\right)_{md}\right]$$

$$a_{nj} = -\frac{1}{[h_1(\cdot)]_{md}} \left\{\left(\frac{\partial H}{\partial q^{(j-1)}}\right)_{md} + \frac{d}{dt}\left[\left(\frac{\partial H}{\partial q^{(j)}}\right)_{md}\right]\right\}, \quad j = 2, \ldots, n-1$$

$$a_{nn} = -\frac{1}{[h_1(\cdot)]_{md}} \left\{\frac{d}{dt}[(h_1(\cdot))_{md}] + \left(\frac{\partial H}{\partial q^{(n-1)}}\right)_{md}\right\}$$

The last term, D, is treated as the general disturbance to the system (10). It includes part of the system dynamics, modelling error, error correction term due to system approximation, and disturbances to the system. As a feature of the robust control, we see that the general disturbance term D does not affect the asymptotic stability of the system. So, the general disturbance term does nothing to the stability of the system, i.e. D has no effect on the performance of the system tracking.

3.2.6. *Lyapunov's direct method.* By Lyapunov's direct method, if a Lyapunov function of the system V can be found such that $V > 0$ and $\dot{V} < 0$ then the system approaches the origin asymptotically. Thus, if we define the Lyapunov function V of (10) as

$$V = X^T P X$$

and we find a symmetric positive definite matrix

$$P = [\, a_{ij} \,]_{n*n} > 0$$

where

$$a_{ij} = a_{ji}$$

such that

$$\dot{V} = \dot{X}^T P X + X^T P \dot{X} < 0 \qquad (11)$$

then (10) approaches the origin asymptotically. In other words, we have

$$\delta q^{(j)} \to 0, \quad \text{i.e. } q^{(j)} \to q_d^{(j)}$$

for

$$j = 0,1,2,....,n-1$$

It therefore fellows the tracking automatically.

Now, the problem boils down to the choice of control law u_ε such that the existence of $\dot{V}<0$ is guaranteed. From (10) and (11), we have

$$\dot{V} = \dot{X}^T PX + X^T P\dot{X}$$

$$= X^T (A^T P + PA)X - 2X^T PBu_\varepsilon + 2X^T PBD$$

If the contral u_ε is chosen as

$$u_\varepsilon > \frac{\text{sgn}(X^T PB)}{|2X^T PB| + \eta} |X^T (A^T P + PA)X + 2X^T PBD|_{\max}$$

when

$$X^T PB > 0$$

and

$$u_\varepsilon < \frac{\text{sgn}(X^T PB)}{|2X^T PB| + \eta} |X^T (A^T P + PA)X + 2X^T PBD|_{\max}.$$

when

$$X^T PB < 0 \tag{12}$$

where is a small positive constant such that $| 2X^T PB | + \eta \simeq | 2X^T PB |$ and sgn(A) is the sign of A, then $\dot{V} < 0$ is guaranteed and therefore does the tracking.

4. Application

Let us consider the control of a class of non-linear second-order systems composed of n interconnected subsystems, under the assumption that each subsystem has its own control input. The non-linear system is represented generally by

$$[a_{ij}(q)]\ddot{q} + [b_{ij}(q, \dot{q})]\dot{q} + [c_i(q)] + [\Delta h_i] = [u_i] \tag{13}$$

where $q = [q_1, q_2, ..., q_n]^T$, i,j= 1, 2, ..., n, q_i are the generalized coordinates, $[a_{ij}(q)]$, $[b_{ij}(\dot{q}, q)]$, and $[c_i(q)]$ are the system matrices with uncertainties, and Δh_i is the known maximum system error of the ith subsystem due to modelling errors and disturbances. It can be represented in a more general form as follows:

$$[\,H_i\,(\ddot{q},\dot{q},q,u_i)\,] = [0], \quad i = 1,\ldots,n$$

Depending on the design process used (listed in the previous section), the robust tracking control laws for the system described by (13) are derived as follows.

4.1. Deriving n variational equations

From (13), the ith non-linear subsystem is represented by

$$H_i(\ddot{q},\dot{q},q,u_i) = \sum_{j=1}^{n} a_{ij}(q)\ddot{q}_j + \sum_{j=1}^{n} b_{ij}(\dot{q},q)\dot{q}_j + c_i(q) + \Delta h_i - u_i = 0$$

and the desired motion of the ith non-linear subsystem is described by

$$H_i\,(\ddot{q}_d,\dot{q}_d,q_d,u_{id}) = 0$$

Thus, the variational equation of the ith subsystem is determined accordingly. It is given by

$$a_{ii}(q_d)\delta\ddot{q}_i + \left[\sum_{j=1}^{n}\left(\frac{\partial}{\partial\dot{q}_i}b_{ij}(\dot{q},q)\right)_d \dot{q}_{jd} + b_{ii}(\dot{q}_d,q_d)\right]\delta\dot{q}_i$$
$$+\left[\sum_{j=1}^{n}\left(\frac{\partial}{\partial q_i}a_{ij}(q)\right)_d \ddot{q}_{jd} + \sum_{j=1}^{n}\left(\frac{\partial}{\partial q_i}b_{ij}(\dot{q},q)\right)_d \dot{q}_{jd}\right.$$
$$\left.+\left(\frac{\partial}{\partial q_i}c_i(q)\right)_d\right]\delta q_i - \delta u_i = 0$$

The n variational equations of the non-linear system can be represented in matrix form as

$$[e_{ii}^{\times}]\delta\ddot{q} + [f_{ii}^{\times}]\delta\dot{q} + [g_{ii}^{\times}]\delta q = [\delta u_i] \tag{14}$$

where $i = 1, 2, \ldots, n$, $e_{ii}^{\times} = a_{ii}(q_d)$,

$$f_{ii}^{\times} = \sum_{j=1}^{n}\left(\frac{\partial}{\partial\dot{q}_i}b_{ij}(\dot{q},q)\right)_d \dot{g}_{jd} + b_{ii}(\dot{q}_d,q_d)$$

$$g_{ii}^{\times} = \sum_{j=1}^{n}\left(\frac{\partial}{\partial q_i}a_{ij}(q)\right)_d \ddot{q}_{jd} + \sum_{j=1}^{n}\left(\frac{\partial}{\partial q_i}b_{ij}(\dot{q},q)\right)_d \dot{q}_{jd} + \left(\frac{\partial}{\partial q_i}c_i(q)\right)_d$$

and $[e_{ii}^{\times}]$, $[f_{ii}^{\times}]$, and $[g_{ii}^{\times}]$ are $n \times n$ diagonal matrices.

4.2. N switching hyperplanes

According to the n variational equations, n switching hyperplanes of the system are chosen as follows:

$$[a_{ij}(q)]\delta\dot{q} + [e_{ii}]\delta\dot{q} + [f_{ii}]\delta q = [S_i(\dot{q}, q)] = [0] \tag{15}$$

where e_{ii} and f_{ii} correspond to $|e_{ii}^{\times}|_{max}$ and $|f_{ii}^{\times}|_{max}$, with the uncertainties replaced by their extreme values. For the choice of (15), two points should be emphasized here. Firstly, the diagonal matrices $[e_{ii}]$ and $[f_{ii}]$ are time-varying, but are completely defined, and secondly, $[a_{ij}(q)]$ is used to involve the physical system described by (13) in (15).

4.3. Variable structure control

The control u_i that drives the ith non-linear subsystem to the ith switching hyperplane S_i is determined here. According to the chosen switching hyperplanes (15), the time rate of the switching hyperplanes is

$$[a_{ij}(q)]\delta\ddot{q} + [\dot{a}_{ij}(q)]\delta\dot{q} + [e_{ii}]\delta\ddot{q} + [\dot{e}_{ii}]\delta\dot{q} + [f_{ii}]\delta\dot{q} + [\dot{f}_{ii}]\delta q$$
$$= [\dot{S}_i(\dot{q}, q)] = [0] \tag{16}$$

The dynamical system is related to these switching hyperplanes by substituting (13) into (16), as

$$[e_{ii}[\delta\ddot{q} + [f_{ii} + \dot{e}_{ii}]\delta\dot{q} + [\dot{f}_{ii}]\delta q - [a_{ij}(q)]\ddot{q}_d + [a_{ij}(q)]\delta\dot{q} - [b_{ij}(\dot{q}, q)]\dot{q}$$
$$- [c_i(q)] - [\Delta h_i] + [u_i] = [\dot{S}_i(\dot{q}, q)] = [0] \tag{17}$$

From (15) and (17) ,we have

$$\frac{1}{2}\frac{dS_i^2}{dt} = S_i\dot{S}_i$$
$$= S_i\Bigg[e_{ii}\delta\ddot{q}_i + (f_{ii} + \dot{e}_{ii})\delta\dot{q}_i + \dot{f}_{ii}\delta q_i - \sum_{j=1}^{n} a_{ij}(q)\ddot{q}_{jd} + \sum_{j=1}^{n} \dot{a}_{ij}(q)\delta\dot{q}_j$$
$$- \sum_{j=1}^{n} b_{ij}(\dot{q}, q)\dot{q}_j - c_i(q) - \Delta h_i + u_i\Bigg]$$

for the ith switching hyperplane. If the control u_i is chosen as

$$u_i = -e_{ii}\delta\ddot{q}_i - (f_{ii} + \dot{e}_{ii})\delta\dot{q}_i - \dot{f}_{ii}\delta q_i + \Delta h_i - \gamma_i \operatorname{sgn} S_i \tag{18}$$

then we have

$$\frac{1}{2}\frac{dS_i^2}{dt} = S_i\dot{S}_i$$

$$= S_i\left[-\sum_{j=1}^{n} a_{ij}(q)\ddot{q}_{jd} + \sum_{j=1}^{n} \dot{a}_{ij}(q)\delta\dot{q}_j - \sum_{j=1}^{n} b_{ij}(\dot{q}, q)\dot{q}_i - c_i(q)\right] - \gamma_i|S_i|$$

The sliding condition

$$\frac{1}{2}\frac{dS_i^2}{dt} < 0$$

is guaranteed if the constant γ_i is chosen as

$$\gamma_i > \left|-\sum_{j=1}^{n} a_{ij}(q)\ddot{q}_{jd} + \sum_{j=1}^{n} \dot{a}_{ij}(q)\delta\dot{q}_j - \sum_{j=1}^{n} b_{ij}(\dot{q}, q)\dot{q}_j - c_i(q)\right|_{max} \tag{19}$$

whence the ith subsystem approaches the ith switching hyperplane S_i. The attractiveness of the ith switching hyperplane is then guaranteed by applying the control u_i, defined as in (18), to the ith subsystem, with γ_i defined as in (19). The same process is carried out for each subsystem, and n control laws are chosen to attract n subsystems to n switching hyperplanes.

4.4. Linearization and decentralization

The control u_i, designed for the ith non-linear subsystem, drives the ith subsystem to the ith switching hyperplane. A linearized error equation that is deduced from the ith switching hyperplane can then be used to represent approximately the ith nonlinear subsystem when the subsystem is kept in the vicinity of the ith switching hyperplane. Thus, n non-linear subsystems are linearized to n linear error equations by the method of VSS. Since n linearized error equations are chosen independently to each other, the non-linear system is also decentralized into n decoupled subsystems by the variable structure control method. The control of the non-linear system is then broken down into the control of n decentralized subsystems. The advantages gained in the simplification of the control law design for coupled non-linear systems are obvious when variable structure control is used as the control method. However, control chattering around the switching hyperplane is the major setback of VSS.

4.5. Smoothing the control chattering

The control chattering resulting from VSS can be eliminated by switching the control law to one that is designed by Lyapunov's direct method. Control switching occurs when the ith subsystem trajectory enters a thin boundary layer of the ith switching hyperplane, defined as $B_i = \{ (\dot{q}, q): |S_i(\dot{q}, q) \leqslant \varepsilon \}$. The ith non-linear subsystem is represented approximately by the ith linearized error equation whenever it remains inside the boundary of B_i. Thus, a linearized error equation is used to replace the ith non-linear subsystem for the design of the control law $u_{i\varepsilon}$ when we switch the control method from VSS to Lyapunov's direct method. The n linearized error equations used as an approximation of the non-linear system are shown in matrix form as follows:

$$[e_{ii}]\delta\ddot{q} + [f_{ii} + \dot{e}_{ii}]\delta\dot{q} + [\dot{f}_{ii}]\delta q - [a_{ij}(q)]\ddot{q}_d + [\dot{a}_{ij}(q)]\delta\dot{q} - [b_{ij}(\dot{q}, q)]\dot{q}$$

$$- [c_i(q)] - [\Delta h_i] + [u_{i\varepsilon}] + [O_i(\varepsilon)] = [\dot{S}_i(\dot{q}, q)] = [0] \tag{20}$$

The system (20) differs from the system (17) in the control term $u_{i\varepsilon}$ and the error correction term $O_i(\varepsilon)$. $u_{i\varepsilon}$ represents the control law designed using Lyapunov's direct method and applied to the ith subsystem when the subsystem is in the boundary layer B_i. The error correction term $[O_i(\varepsilon)]$ occurs due to approximation of the nonlinear dynamic system by the linearized error equations when the dynamic system remains in B-i.e. each subsystem remains in its corresponding boundary layer B_i. A more convenient form of (20) is shown as follows:

$$[r_{ii}]\delta\ddot{q} + [s_{ii}]\delta\dot{q} + [t_{ii}]\delta q + [d_i(\dot{q}, q)] + [u_{i\varepsilon}^{\times}] = [0] \tag{21}$$

If there are missing terms such that the linearized error equation cannot be guaranteed in the form of(21) -with $[r_{ii}]$, $[s_{ii}]$, and $[t_{ii}]$ all in rank n—the control $u_{i\varepsilon}$; in (20) is used to supply the missing terms and to guarantee the form of (21). Lyapunov's direct method can then be applied to design the control law $u_{i\varepsilon}^{\times}$, which completes the tracking control and eliminates the control chattering. Even though the term $[d_i(\dot{q}, q)]$ in (21) is the coupling term between subsystems, (21) can still be treated as a linearized and decentralized system when $[d_i(\dot{q}, q)]$ is treated as a "general disturbance" to the system. Since the linearized system is also a decentralized system, the design of the ith control law $u_{i\varepsilon}^{\times}$ for the ith subsystem depends on that subsystem only. So, thc design of $u_{i\varepsilon}^{\times}$ can be treated separately for cach subsystem.

Since the design of the n control laws follows the same process, we need only show the design process once. Let us take the ith subsystem as an illustrative example. The ith linearized error equation that is used to design the ith control law $u_{i\varepsilon}^{\times}$ for the ith non-linear subsystem is shown as follows:

$$r_{ii}(q_d)\delta\ddot{q}_i + s_{ii}(\dot{q}_d, q_d)\delta\dot{q}_i + t_{ii}(\ddot{q}_d, \dot{q}_d, q_d)\delta q_i + d_i(\dot{q}, q) + u_{i\varepsilon}^{\times} = 0 \tag{22}$$

To represent the above equation in a state-space form, we define

$$x_1 = \delta q_i, \quad \dot{x}_1 = x_2$$

$$x_2 = \delta\dot{q}_i, \quad \dot{x}_2 = \delta\ddot{q}_i$$

$$X = \begin{bmatrix} x_1 \\ x_2 \end{bmatrix}$$

$$\dot{x}_2 = \frac{1}{r_{ii}(q_d)}[-s_{ii}(\dot{q}_d, q_d)x_2 - t_{ii}(\ddot{q}_d, \dot{q}_d, q_d)x_1 - d_i(\dot{q}, q) - u_{i\varepsilon}^{\times}]$$

The state-space form is represented as fellows:

$$\dot{X} = \begin{bmatrix} 0 & 1 \\ -\dfrac{t_{ii}(\ddot{q}_d, \dot{q}_d, q_d)}{r_{ii}(q_d)} & -\dfrac{s_{ii}(\dot{q}_d, q_d)}{r_{ii}(q_d)} \end{bmatrix} X + \begin{bmatrix} 0 & 0 \\ -\dfrac{d_i(\dot{q}, q)}{r_{ii}(q_d)} & -\dfrac{1}{r_{ii}(q_d)} \end{bmatrix} u_{i\varepsilon}^{\times}$$

Thc abbreviation of this state-space cquation is

$$\dot{X} = A_i X - B_i u_{i\varepsilon}^{\times} + D_i \tag{23}$$

The last term D_i is treated as the general disturbance to the subsystem. It includes part of system dynamics, modelling error, error correction term due to system approximation, interactions between subsystems, and disturbances to the subsystem. Since it is a feature of the robust control, we see that the general disturbance D_i does not affect the asymptotic stability of this error equation. So, the general disturbance term does nothing to the stability of (23), i.e. D_i has no effect on the performance of the system tracking.

4.6. Lyapunov's direct method

By Lyapunov's direct method, if a Lyapunov function of the ith subsystem V_i can be found such that $V_i > 0$ and $\dot{V}_i < 0$ then the ith subsystem approaches the origin asymptotically. Thus, if we define the Lyapunov function V_i of (23) as

$$V_i = X^T P X$$

and we find a symmetric positive definite matrix

$$P = \begin{bmatrix} a & b \\ b & c \end{bmatrix} > 0$$

with a, c > 0 and b 0, such that

$$\dot{V}_i = \dot{X}^T P X + X^T P \dot{X} < 0 \qquad (24)$$

then the ith subsystem approaches the origin asymptotically. In other words, we have

$$\delta\dot{q}_i \rightarrow 0, \quad \text{i.e. } \dot{q}_i \rightarrow \dot{q}_{id}$$
$$\delta q_i \rightarrow 0, \qquad q_i \rightarrow q_{id}$$

Thus, the tracking follows. This can always be achieved, since the control input $u_{i\varepsilon}^{\times}$ in (24) can always be chosen at will, such that the negative definite property of (24) is always guaranteed. We demonstrate this in the following.

The problem now reduces to the choice of the control law$u_{i\times}^{\times}$,such that the existence of $\dot{V}_i < 0$ is guaranteed. From (23) and (24),·we have

$$\begin{aligned} \dot{V}_i &= \dot{X}^T P X + X^T P \dot{X} \\ &= X^T(A_i^T P + P A_i)X - 2X^T P B_i u_{i\varepsilon}^{\times} + 2X^T P D_i \end{aligned}$$

If the control $u_{i\varepsilon}^{\times}$ is chosen as

$$u_{i\varepsilon}^{\times} > \frac{\text{sgn}\,(X^T P B_i)}{|2X^T P B_i| + \eta} |X^T(A_i^T P + P A_i)X + 2X^T P D_i|_{\max}$$

when $X^T PB > 0$ and

$$u_{i\varepsilon}^{\times} < \frac{\text{sgn}\,(X^T P B_i)}{|2X^T P B_i| + \eta} |X^T(A_i^T P + P A_i)X + 2X^T P D_i|_{\max}$$

when $X^TPB < 0$, where η is a small contant such that $|2X^TPB_i| + \eta \simeq |2X^TPB_i|$ and sgn (A) is the sign of A, then $\dot{V}_i < 0$ is guaranteed, and hence so is the tracking. The tracking of the whole system is completed by designing nu_{it}^{x}, to guarantee the tracking of the n subystems.

5. Conclusions

Several conclusions may be drawn regarding the robust tracking control scheme.

(i) A variable structure control scheme is used as a linearization and decentralization technique.

(ii) Since we can drive any non-linear system to an arbitrarily chosen linear system via the variable structure control scheme, it is obvious that such a scheme is a very powerful linearization technique. A linear time-varying error system is chosen to fit the robust tracking control scheme in this paper. It is determined via the perturbation method.

(iii) After linearization, the Lyapunov direct method can easily be applied to thesystem. It drives the linear time-varying error system to the origin, thus achieving tracking.

(iv) Since the control method switches from VSS to Lyapunov's direct method before the control chattering occurs, we can eliminate the control chattering caused by VSS.

(v) The design of robust control laws is based on two unequal criteria,

$$\frac{1}{2}\frac{dS^2}{dt} < 0 \quad \text{and} \quad \dot{V} < 0$$

These two criteria give great freedom in the design of robust control laws. The design principle is very simple and is always possible for putting the system under control.

(vi) More control effort may be needed by using this control scheme, although the simplicity of the design principle balances this out.

(vii) It has the capability of handling non-linear time-varying systems.

(viii) Two control flow charts are used to express the control scheme in a clearer way. One shows the control of a single system (Fig. 6) and the other shows the control of a system with interconnected subsystems (Fig. 7).

(ix) With respect to the comparison of the linearization technique developed in this paper and the familiar linearization technique, a boundary layer around the switching hyperplane $B = \{X: |S(X)| \leqslant \varepsilon\}$ is established to show the border where the switch of control and system takes place. We say that the non-linear system can be represented by a linearized system when the system enters the boundary layer B. The effectiveness of this statement is illustrated as follows.

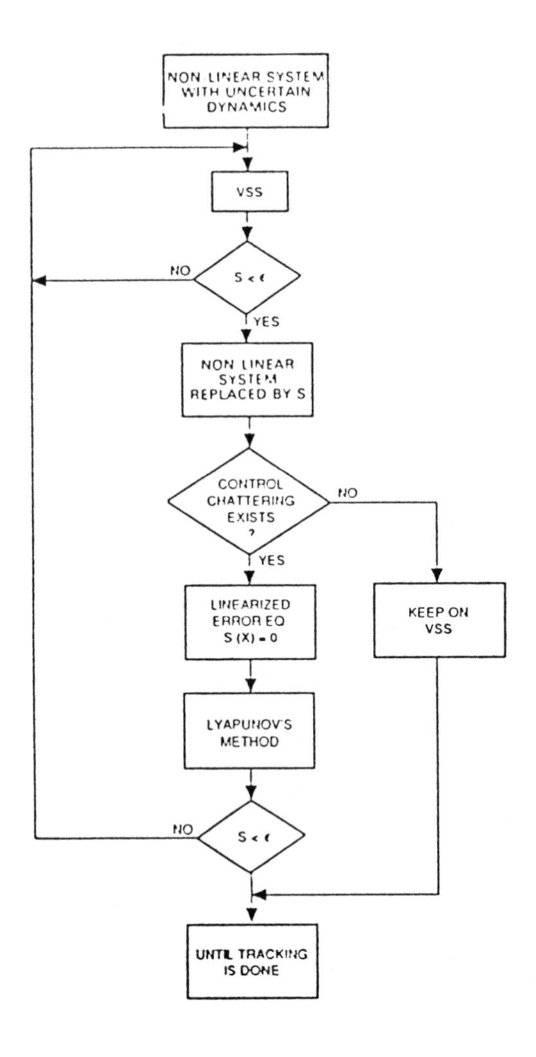

Figure 6. Control flow chart of single system.

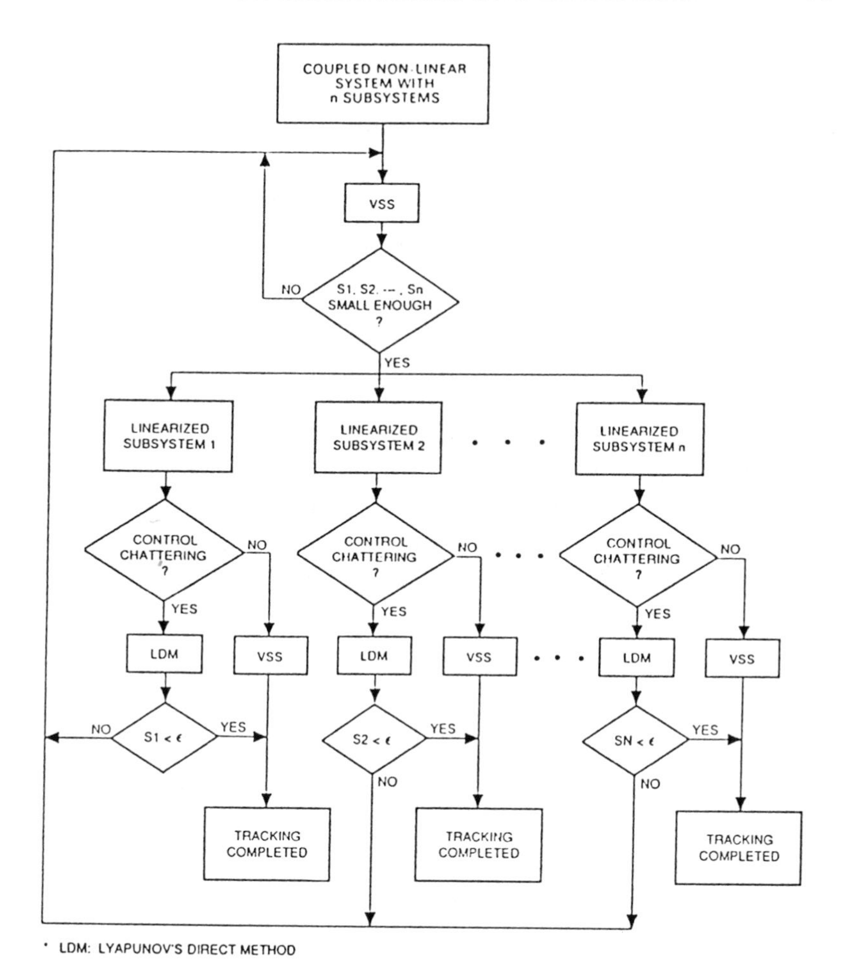

Figure 7. Control flow chart of system with N subsystems.

Since the switching hyperplane S(X) = 0 is also chosen as an error equation, the errors between the system and desired states should be small when the system enters the boundary layer B. Then, the non-linear system is replaced by a linear error equation and Lyapunov's direct method drives its error states to the origin. The error states are already small when Lyapunov's direct method is applied to the system - i.e. Lyapunov's direct method just brings those error states from the vicinity of the origin to the origin. Since it is a small motion, we can use a linearized system to replace the non-linear system, which is just the same as local linearization. The non-linear system can be viewed as a linear system if its motion is confined in a small local area. In this aper, the non-linear system is brought into the vicinity of the origin via VSS. Then system motion is confined within a small segment of the origin, and the non-linear system acts like a linear one in this area. Now we can see the similarity between this new linearization technique and local linearization i.e. they actually follow thc same idea.

Since we are dealing with a time-varying system, the origin is also time-varying - i.e. it is not fixed, and it shifts from time to time. However, the system is also corrected from time to time. The shift of origin should be small if the time interval between the two control actions is set to be small, and the system therefore stays in a small segment of origin, even when it is time-varying. The linearization is thus still effective.

Part II Application

1. Introduction

In general, physical systems are complex, highly coupled, non-linear, time-varying systems. It is usually hard to represent a physical system exactly by a mathematical model, due to the existence of uncertain dynamics, unmodelled dynamics, disturbances, and the complexity of the real system. A simple two-link manipulator (see Fig. 1) is used here as the system to be controlled. Using Lagrange's method, the system is described by two equation, (1) and (2). These are mathematical descriptions of link 1 and link 2, respectively. The dynamic equation of link 1 is

$$\begin{aligned} u_1 = {} & [(m_1 + m_2)d_1^2 + m_2 d_2^2 + 2m_2 d_2^2 + 2m_2 d_1 d_2 \cos \theta_2]\ddot{\theta}_1 \\ & + (m_2 d_2^2 + m_2 d_1 d_2 \cos \theta_2)\ddot{\theta}_2 - 2m_2 d_1 d_2 \sin (\theta_2)\, \dot{\theta}_1 \dot{\theta}_2 \\ & - m_2 d_1 d_2 \sin (\theta_2)\, \dot{\theta}_2^2 + (m_1 + m_2) g d_1 \sin \theta_1 \\ & + m_2 g d_2 \sin (\theta_1 + \theta_2) \end{aligned} \tag{1}$$

The dynamic equation of link 2 is

$$\begin{aligned} u_2 = {} & (m_2 d_2^2 + m_2 d_1 d_2 \cos \theta_2)\ddot{\theta}_1 + m_2 d_2^2 \ddot{\theta}_2 \\ & - 2m_2 d_1 d_2 \sin (\theta_2)\, \dot{\theta}_1 \dot{\theta}_2 - m_2 d_1 d_2 \sin (\theta_2)\, \dot{\theta}_1^2 \\ & + m_2 g d_2 \sin (\theta_1 + \theta_2) \end{aligned} \tag{2}$$

Accroding to (1) and (2), it is obvious that the dynamics of the two-link manipulator system vary with the variation of the load m_2 and the change of system configuration. This shows that uncertain dynamics do exists, even in this simple manipulator system. Since the friction at the joints and the time delays of the control inputs are not taken into account when we model the system, and the fact that there is no system in the real world with perfect rigidity, then unmodelled dynamics do exist. All of this shows us that uncertain and unmodelled dynamics are usually included in the model of the physical system. An error in the mathematical model of a real system always exists, even when we are trying to model a real system as precisely as possible.

For this paper's purposes, we assume that the system is perfecly modelled and insulated form its surroundings, such that no unmodelled dynamics and disturbances are taken into consideration. The unforecasted load m_2 and the system dynamics affected by the geometry of the manipulator are the only two kinds of system uncertainties considered here. Even though the problem has been simplified, the effectiveness of the robust tracking control scheme can still be tested.

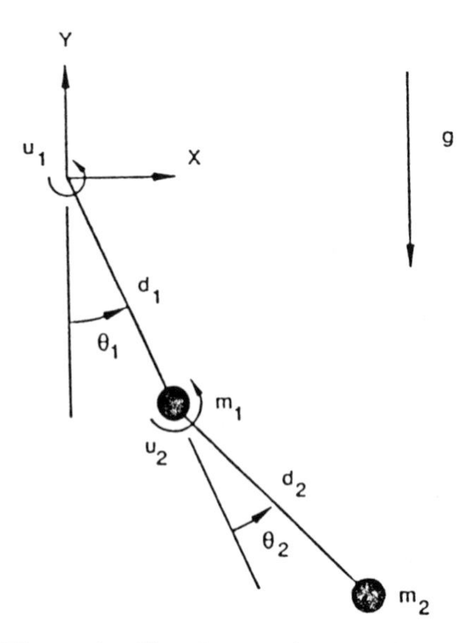

Figure 1 Simple two-link manipulator

2. Control scheme

The control laws designed the following process.

(a) Deriving variational equations

Obtain the variational equation for each link of the non-linear two-link manipulator system by the perturmation method. If we represent (1) and (2) generally by $H_1(\ddot{\theta}, \dot{\theta}, \theta, u_1) = 0$ and $H_2(\ddot{\theta}, \dot{\theta}, \theta, u_2) = 0$, with $(\ddot{\theta}_d, \dot{\theta}_d, \theta_d, u_d)$ as the desired path we are going to track, then two variational equations of the non-linear two-link manipulator system may be derived as

$$\left(\frac{\partial H_1}{\partial \ddot{\theta}_1}\right)_d \delta\ddot{\theta}_1 + \left(\frac{\partial H_1}{\partial \dot{\theta}_1}\right)_d \delta\dot{\theta}_1 + \left(\frac{\partial H_1}{\partial \theta_1}\right)_d \delta\theta_1 + \left(\frac{\partial H_1}{\partial u_1}\right)_d \delta u_1 = 0$$

and

$$\left(\frac{\partial H_2}{\partial \ddot{\theta}_2}\right)_d \delta\ddot{\theta}_2 + \left(\frac{\partial H_2}{\partial \dot{\theta}_2}\right)_d \delta\dot{\theta}_2 + \left(\frac{\partial H_2}{\partial \theta_2}\right)_d \delta\theta_2 + \left(\frac{\partial H_2}{\partial u_2}\right)_d \delta u_2 = 0$$

where

$$\delta\ddot{\theta}_i = \ddot{\theta}_i - \ddot{\theta}_{id}, \quad \delta\dot{\theta}_i = \dot{\theta}_i - \dot{\theta}_{id}$$

$$\delta\theta_i = \theta_i - \theta_{id}, \quad \delta u_i = u_i - u_{id}, \quad i = 1, 2$$

(b) Choosing switching hyperplanes

According to the two derived variational equations, two switching hyperlanes $S_1(\dot{\theta}, \theta) = 0$ and $S_2(\dot{\theta}, \theta) = 0$ are chosen such that two linearizaed error equation can be obtained from them. These two linearized error equations are chosen as the modified version of the variational equations of the system. They are used to represent the non-linear system approximately when the two links stay repectively at the vicinity of the two switching hyperplanes.

(c) Variable structure control

Driving the first link, $H_1(\ddot{\theta}, \dot{\theta}, \theta, u_1) = 0$ approaches the first switching hyperplane $S_1(\dot{\theta}, \theta) = 0$ by choosing a control law u_i such that it guarantees the existence of the sliding condition

$$\frac{1}{2}\frac{dS_1^2}{dt} < 0$$

The design of u_i is based on the method of variable structure systems (VSS).The control law u_2 is designed in a similar way, such that the second link is driven towards the second switching hyperplane. Control chattering around the switching hyperplane is the main drawback of VSS.

(d) System linearization and decentralization

The first link can then be represented approximately by the first switching hyperplane, wheneever the first link is kept in the vicinity of the first switching hyperplane. It also induces the first linearized error equation to represent the first link approximately, such that Lyapunov's direct method can easily be applied. The second linearized error equation is derived in a similar way, to represent the second link approximately. Since the two linearized error equations are chosen independently of each other, decentralization is also achieved when linearization is achieved.

(e) Lyapunov's direct method

According to Laypunov's direct method, two linearized error equations (two modified variational equations) that are derived from the switching hyperplanes are driven asymptotically to the origin by a suitable choice of the control laws $u_{i\varepsilon}, i = 1, 2$. It therefore follows the etracking control of the non-linear system, since the error states (the difference between the system states and the desired states) are all zero at origin.

(f) Completion of tracking

The tracking of the two-link manipulator system is achieved if each kink achieves its own tracking.

(g) Smoothing control chattering

The variable structure control method results in control chattering at the switching hyperplane. The control chattering at the first switching hyperplane is eliminated by switching the control law of the first link from the one derived by the method of VSS to the one derived by Lyapunov's direct method, whenever the first link trajectory enters a thin boundary layer of the first switching hyperplane $S_1(\dot{\theta}, \theta) = 0$. The thin boundary layer is defined as $B_1 = \{(\dot{\theta}, \theta) : |S_1(\dot{\theta}, \theta)| \leqslant \varepsilon\}$, where ε is a small positive constant. The control chattering at the second switching hyperplane can also be eliminated in similar way. the thin boundary layer of the second switching hyperplane is defined as $B_2 = \{(\dot{\theta}, \theta) : |S_2(\dot{\theta}, \theta)| \leqslant \varepsilon\}$.

(h) Robustness

The robustness of the tracting control is assured by choosing controls u_i and $u_{i\varepsilon}$ such that the sliding condition

$$\frac{1}{2}\frac{dS_i^2}{dt} < 0$$

exists and the time rate of Lyapunov's function is negative definite ($\dot{V}_i < 0$) ,i = 1,2. Since all the system uncertainties are on the left-hand side of these two criteria, then, if we know the range of variation of the system uncertainties, the controls can be chosen to guarantee the negative definite property of these two criteria and to overcome the effect of system uncertainties on the system.Thus, the effect of uncertainties on the system stability is eliminated. Robustness of the compensated system is achieved by applying this kind of control technique to it.

3. Robust tracking control laws

3.1 Deriving variational equations

The basic idea of this tracking control algorithm is to drive a linearized error equation (modified variational equation) of a non-linear system to the origin using Lyapunov's direct method. In doing this, the non-linear system should be first driven to the linearized error equation by the theory of VSS. Since the linearized error equation is obtained from the switching hyperplane, the selection of the switching hyperplane plays a key role in this control algorithm.

The switching hyperplane is chosen according to (15) of Wang and Leondes (1989). This choice is based on two important facts:

(a) the derivation of the linearized error equation from the switching hyperplane ; and
(b) the realization of the control law.

The process to determine the switching hyperplane depends mainly upon the trial-and-error method. Based on the derived variational equation, we set up two end equations - the initial switching hyperplane and the initial linearized error equation - and match these two together by continous modification, until the linearized error equation can be obtained from the switching hyperplane and the control laws, which are designed according to the switching hyperplane and linearized error equation, can be realized in the real world.

The variational equations of link 1 and link 2 (Fig. 1) are derived via the perturbation method. For link 1,it is

$$\begin{aligned}[(m_1+m_2)d_1^2+m_2d_2^2+2m_2d_1d_2\cos\theta_{2d}]\delta\ddot{\theta}_1-2m_2d_1d_2\sin(\theta_{2d})\,\dot{\theta}_{2d}\delta\dot{\theta}_1\\ +[(m_1+m_2)gd_1\cos(\theta_{1d})+m_2gd_2\cos(\theta_{1d}+\theta_{2d})]\delta\theta_1=\delta u_1\end{aligned} \tag{3}$$

For link 2, it is

$$\begin{aligned}m_2d_2^2\delta\ddot{\theta}_2-2m_2d_1d_2\sin(\theta_{2d})\,\dot{\theta}_{1d}\delta\dot{\theta}_2\\ -m_2d_1d_2\sin(\theta_{2d})\,\ddot{\theta}_{1d}\delta\theta_2-2m_2d_1d_2\dot{\theta}_{1d}\dot{\theta}_{2d}\cos(\theta_{2d})\,\delta\theta_2\\ -m_2d_1d_2\dot{\theta}_{1d}^2\cos(\theta_{2d})\,\delta\theta_2+m_2gd_2\cos(\theta_{1d}+\theta_{2d})\,\delta\theta_2=\delta u_2\end{aligned} \tag{4}$$

3.2 Choosing switching hyperplanes

Based on the principle described by (15) of Wang and Leondes (1989) and the two variational equations (3) and (4), the switching hyperplanes of links 1 and 2 are chosen accordingly. For link 1,it is chosen as

$$[(m_1 + m_2)d_1^2 + m_2 d_2^2 + 2m_2 d_1 d_2 \cos \theta_2]\delta\dot{\theta}_1$$
$$+ [m_2 d_2^2 + m_2 d_1 d_2 \cos \theta_2]\delta\dot{\theta}_2$$
$$+ [(m_1 + m_{2m})d_1^2 + m_{2m} d_2^2 + 2m_{2m} d_1 d_2 \cos \theta_{2d}]\delta\dot{\theta}_1$$
$$+ 2m_{2m} d_1 d_2 \sin (\theta_{2d})\, \dot{\theta}_{2d} \delta\theta_1 = S_1(\Theta) = 0$$

Owing to the consideration of realization of the control law, the switching hyperplane $S_1(\Theta) = 0$ is modified as follows:

$$[(m_1 + m_{2m})d_1^2 + m_{2m} d_2^2 + 2m_{2m} d_1 d_2 \cos \theta_{2d}]\delta\dot{\theta}_1$$
$$+ [(m_1 + m_2)d_1^2 + m_2 d_2^2 + 2m_2 d_1 d_2 \cos \theta_2]\delta\dot{\theta}_1 + W_{f1}\delta\theta_1$$
$$+ [m_2 d_2^2 + m_2 d_1 d_2 \cos \theta_2]\delta\dot{\theta}_2 = S_1(\Theta) = 0 \tag{5}$$

where W_{f1}, a weighting factor determined in the simulation process, is a positive constant.

The switching hyperplane of the second link is chosen in a similar way. It is shown as follow:

$$m_2 d_2^2 \delta\dot{\theta}_2 + (m_2 d_2^2 + m_2 d_1 d_2 \cos \theta_2)\delta\dot{\theta}_1$$
$$+ m_{2m} d_2^2 \delta\dot{\theta}_2 + 2m_{2m} d_1 d_2 \sin (\theta_{2d})\dot{\theta}_{1d}\delta\theta_2$$
$$+ W_{f2}\delta\theta_2 = S_2(\Theta) = 0 \tag{6}$$

where W_{f1}, another weighting factor determined in the simulation process, is also a positive constant. The two weighting factors, W_{f1} and W_{f2}, are chosen such that values of error terms - e.g. $\delta\theta_1, \delta\theta_2, \ldots$ - are all at the same level when the control switch takes place. In doing this, thee controls do not jump sharply from one level to another at the moment of control switching. The controls are also more realizable if W_{f1} and W_{f2} are suitably chosen.

3.3 Variable structure control

Based on the theory of VSS, the derivation of the control law u_i, which assures the attractiveness of the switching hyperplane $S_i(\Theta) = 0$, is the main focus of this section.

3.3.1 *The first link.* Now consider the derivation of the control u_i for the first link, which attracts the non-linear first link subsystem to the selected switching hyperplane $S_1(\Theta) = 0$. The dynamic system of link 1 is related to the switching hyperplane $S_1(\Theta) = 0$ — see (5) — by introducting the dynamic equation (1) into $\dot{S}_1(\Theta) = 0$. We then have

$$\begin{aligned}
&[(m_1 + m_{2m})d_1^2 + m_{2m}d_2^2 + 2m_{2m}d_1 d_2 \cos \theta_{2d}]\delta\ddot{\theta}_1 \\
&\quad - 2m_{2m}d_1 d_2 \sin(\theta_{2d})\, \dot{\theta}_{2d}\delta\dot{\theta}_1 \\
&\quad - [(m_1 + m_2)d_1^2 + m_2 d_2^2 + 2m_2 d_1 d_2 \cos \theta_2]\ddot{\theta}_{1d} \\
&\quad + 2m_2 d_1 d_2 \sin(\theta_2)\, \dot{\theta}_{1d}\dot{\theta}_2 + W_{f1}\delta\dot{\theta}_1 \\
&\quad -(m_2 d_2^2 + m_2 d_1 d_2 \cos \theta_2)\ddot{\theta}_{2d} \\
&\quad + m_2 d_1 d_2 \sin(\theta_2)\, \dot{\theta}_{2d}\dot{\theta}_2 \\
&\quad + u_1 - (m_1 + m_2)gd_1 \sin \theta_1 - m_2 g d_2 \sin(\theta_1 + \theta_2) = \dot{S}_1(\Theta) = 0
\end{aligned}$$

The non-linear first link subsystem is driven towards the switching hyperplane $S_1(\Theta) = 0$ by choosing the control u_i such that the sliding condition

$$\frac{1}{2}\frac{dS_1^2}{dt} < 0$$

is satisfied. This is achieved by choosing the law u_i as

$$\begin{aligned}
u_1 &= m_1 g d_1 \sin \theta_1 + 2m_{2m}d_1 d_2 \sin(\theta_{2d})\, \dot{\theta}_{2d}\delta\dot{\theta}_1 \\
&\quad - [(m_1 + m_{2m})d_1^2 + m_{2m}d_2^2 + 2m_{2m}d_1 d_2 \cos(\theta_{2d})]\delta\ddot{\theta}_1 \\
&\quad + m_1 d_1^2 \ddot{\theta}_{1d} - W_{f1}\delta\dot{\theta}_1 \\
&\quad - (\gamma_1 + 2m_{2m}d_1 d_2 |\dot{\theta}_2|\, |\dot{\theta}_{1d}|_{\max} + m_{2m}d_1 d_2 |\dot{\theta}_2|\, |\dot{\theta}_{2d}|_{\max}) \operatorname{sgn} S_1
\end{aligned} \tag{7}$$

with the positive constant γ_1 defined as

$$\begin{aligned}
\gamma_1 &> m_{2m}gd_1 + m_{2m}gd_2 + m_{2m}(d_1 + d_2)^2 |\ddot{\theta}_{1d}|_{\max} \\
&\quad + m_{2m}d_2(d_1 + d_2)|\ddot{\theta}_{2d}|_{\max}
\end{aligned} \tag{8}$$

Then, the sliding condition

$$\begin{aligned}
\frac{1}{2}\frac{dS_1^2}{dt} &= S_1 \dot{S}_1 \\
&= S_1\{-m_2 g d_1 \sin \theta_1 - m_2 g d_2 \sin(\theta_1 + \theta_2) \\
&\qquad - (m_2 d_1^2 + m_2 d_2^2 + 2m_2 d_1 d_2 \cos \theta_2)\ddot{\theta}_{1d} \\
&\qquad + 2m_2 d_1 d_2 \sin(\theta_2)\, \dot{\theta}_{1d}\dot{\theta}_2 - (m_2 d_2^2 + m_2 d_1 d_2 \cos \theta_2)\ddot{\theta}_{2d}
\end{aligned}$$

$$+ m_2 d_1 d_2 \sin(\theta_2)\, \dot\theta_{2d} \dot\theta_2 \}$$
$$- (\gamma_1 + 2m_{2m} d_1 d_2 |\dot\theta_2|\, |\dot\theta_{1d}|_{\max} + m_{2m} d_1 d_2 |\dot\theta_2|\, |\dot\theta_{2d}|_{\max}) |S_1| < 0$$

is thus guaranteed - as is the attractiveness of the switching hyperplane $S_1(\Theta) = 0$. Thus, the behaviour of the first link finally becomes similar to the behaviour of the switching hyperplane $S_1(\Theta) = 0$ when link 1 reaches $S_1(\Theta) = 0$.

3.3.2 *The sccond link* .Following the same argument, the control u_2 of the second link is chosen to satisfy the sliding condition

$$\frac{1}{2}\frac{dS_2^2}{dt} < 0$$

such that it attracts the non-lienar second link subsystem to the selected switching hyperplane $S_2(\Theta) = 0$—(6). This is achieved by choosing the control law u_2 as

$$u_2 = - m_{2m} d_2^2 \delta\ddot\theta_2 - 2m_{2m} d_1 d_2 \sin(\theta_{2d})\, \dot\theta_{1d} \delta\dot\theta_2 - W_{f2} \delta\dot\theta_2$$
$$- 2m_{2m} d_1 d_2 (\cos(\theta_{2d})\, \dot\theta_{1d} \dot\theta_{2d} + \sin(\theta_{2d})\, \ddot\theta_{1d}) \delta\theta_2$$
$$- (\gamma_2 + m_{2m} d_1 d_2 |\dot\theta_2|\, |\dot\theta_{1d}|_{\max}$$
$$+ m_{2m} d_1 d_2 |\dot\theta_1 (\dot\theta_1 + \dot\theta_2)|) \operatorname{sgn} S_2 \tag{9}$$

with the positive constant γ_2 defined as

$$\gamma_2 > m_{2m} d_2^2 |\ddot\theta_{2d}|_{\max} + m_{2m} d_2 (d_1 + d_2) |\ddot\theta_{1d}|_{\max} + m_{2m} g d_2 \tag{10}$$

Then, the sliding condition

$$\frac{1}{2}\frac{dS_2^2}{dt} = S_2 \dot S_2$$
$$= S_2 \{ - m_2 d_2^2 \ddot\theta_{2d} - (m_2 d_2^2 + m_2 d_1 d_2 \cos\theta_2) \ddot\theta_{1d} - m_2 g d_2 \sin(\theta_1 + \theta_2)$$
$$+ m_2 d_1 d_2 \sin(\theta_2)\, \dot\theta_{1d} \dot\theta_2 + m_2 d_1 d_2 \sin(\theta_2)\, \dot\theta_1 (\dot\theta_1 + \dot\theta_2) \}$$
$$- (\gamma_2 + m_{2m} d_1 d_2 |\dot\theta_2|\, |\dot\theta_{1d}|_{\max} + m_{2m} d_1 d_2 |\dot\theta_1 (\dot\theta_1 + \dot\theta_2)|) |S_2| < 0$$

is this guaranteed - as is the attractiveness of the switchung hyperplane $S_2(\Theta) = 0$ The behaviour of the second link finally becomes similar to the behaviour of the switching hyperplane $S_2(\Theta) = 0$ when link 2 reaches $S_2(\Theta) = 0$.

3.3.3 *Summary*. Thus, the non-linear first link subsystem is attracted towards the switching hyperplane $S_1(\Theta) = 0$ by applying the control u_1 (7) to the first link, with γ_1 defined as in (8). The second link subsystem is attracted towards the switching hyperplane $S_2(\Theta) = 0$ by applying the control law u_2 (9) to the second link, with γ_2 defined as in (10).

The approaching speed of the non-linear system to $S(\Theta) = 0$ is determined by the choice of the constants γ_1 and γ_2. A larger γ_i gives a shorter reaching time. Therefore, there exists a compromise between the reaching time and the available controls, and γ_1 and are chosen such that the two links approach repectively to the two switching hyperplanes at the same speed. Thus, the control switch between VSS and Lyapunov's direct mehod acts at the same time for both subsystems.

3.4 Elimination of control chattering

The main drawback of variable structure control algorithm is the existence of control chattering around the switching hyperplane $S(\Theta) = 0$. Since the control chattering is not desired, and is realizable in the real world, it should be, and can be, eliminated by switching the control method from VSS to Lyapunov's direct method when the system is in the vicinity of the switching hyperplane.

Two boundary layers, $B_1 = \{\Theta : |S_1(\Theta)| < \varepsilon\}$ centred at the switching hyperplane $S_1(\Theta) = 0$ and $B_2 = \{\Theta : |S_2(\Theta)| \leqslant \varepsilon\}$ centred at the switching hyperplane $S_2(\Theta) = 0$, are established, to serve as the borders of the control switch,such that the control chattering can be eliminated. The control u_1 (7) is applied to link 1 to keep the attractiveness of the switching hyperplane $S_1(\Theta) = 0$ whenever $|S_1(\Theta)| > \varepsilon$. The control u_2 (9) is applied to link 2 to keep the attractiveness of the switching hyperplane $S_2(\Theta) = 0$ whenever $|S_2(\Theta)| > \varepsilon$. control u_1 then switches to $u_{1\varepsilon}$ when the system trajectory of link 1 enters the boundary layer B_1. Likewise, the control u_2 switches to $u_{2\varepsilon}$ when the system trajectory of link 2 enters the boundary layer B_2. The design of the control laws $u_{1\varepsilon}$ and $u_{2\varepsilon}$ lies in the fact that

(a) the linearized error equations that are used to represent the dynamic system approximately can be obtained in the form of (22) of Wang and Leondes (1989), and

(b) the linearized error equations can be driven asymptotically to the origin using Lyapunov's direct method.

Owing to the switch of control laws, we eliminate the control chattering that are caused by applying the control law u_1 and u_2 to the system only.

In other wards, the robust tracking control is achieved by two kinds of controls, u_i and $u_{i\varepsilon}$, $i = 1, 2$. The controls u_i retain the attractiveness of the boundary layer B_i, and finally cause the non-linear system to be described approximately by the switching hyperplanes. The controls $u_{i\varepsilon}$ are used to guarantee the form of the linearized error equations and to force them to the origin whenever the system stays inside the boundary layer. The tracking is thus complete, and the control chattering eliminated. The control laws $u_{i\varepsilon}$ are determined in § 3.6.

3.5 System linearization and decentralization

The non-linear two-link manipulator system can be represented approximately by two linearized error equations, when two links are trapped respectively in two thin boundary layers B_1 and B_2. Thus, the non-linear system can be represented approximately by a linearized system when the system is in the boundary layer. This greatly eases the difficulties in the design of a control law for a non-linear system, since the design of control law is now based on an approximate linearized system.

3.5.1 *Linearized error equation of the first link.* We begin with the equation of the switching hyperplane $S_1(\Theta) = 0$ -(5) - to obtain the linearized error equation of link 1. This linearized error equation is the replacement of the non-linear first link subsystem when link 1 is inside the boundary layer B_1. Form (1) and $\dot{S}_1(\Theta) = 0$, we relate the first link with $S_1(\Theta) = 0$ by

$$\begin{aligned}
&[(m_1 + m_{2m})d_1^2 + m_{2m}d_2^2 + 2m_{2m}d_1 d_2 \cos \theta_{2d}]\delta\ddot{\theta}_1 \\
&- 2m_{2m}d_1 d_2 \sin(\theta_{2d})\, \dot{\theta}_{2d}\delta\dot{\theta}_1 + W_{f1}\,\delta\dot{\theta}_1 \\
&- [(m_1 + m_2)d_1^2 + m_2 d_2^2 + 2m_2 d_1 d_2 \cos \theta_2]\ddot{\theta}_{1d} \\
&- (m_2 d_2^2 + m_2 d_1 d_2 \cos \theta_2)\ddot{\theta}_{2d} \\
&+ m_2 d_1 d_2 \sin(\theta_2)\, \dot{\theta}_2(2\dot{\theta}_{1d} + \dot{\theta}_{2d}) \\
&+ u_{1\varepsilon} - (m_1 + m_2)g d_1 \sin(\theta_1) - m_2 g d_2 \sin(\theta_1 + \theta_2) = \dot{S}_1(\Theta) = 0
\end{aligned}$$

The control law $u_{1\varepsilon}$ is chosen to form the linearized error equation in the same form as (22) of Wang and Leondes (1989), and to guarantee that this non-linear subsystem tracks the desired path. Since we do not have a $\delta\theta_1$ term in the above equation, it should be supplied by the control $u_{1\varepsilon}$ to avoid degeneracy of the system matrix when the linearized error equation is represented in the state-space form. Thus, the linearized error equation is formed by choosing $u_{1\varepsilon}$ as

$$u_{1\varepsilon} = [(m_1 + m_{2m})gd_1 + m_{2m}gd_2]\delta\theta_1 - u_{1\varepsilon}^{\times} \tag{11}$$

Then, we have the first linearized error equation as

$$\begin{aligned}
&[(m_1 + m_{2m})d_1^2 + m_{2m}d_2^2 + 2m_{2m}d_1 d_2 \cos \theta_{2d}]\delta\ddot{\theta}_1 \\
&\quad - 2m_{2m}d_1 d_2 \sin (\theta_{2d})\, \dot{\theta}_{2d}\delta\dot{\theta}_1 + W_{f1}\delta\dot{\theta}_1 \\
&\quad - (m_1 + m_2)gd_1 \sin (\theta_1) - m_2 gd_2 \sin (\theta_1 + \theta_2) \\
&\quad - [(m_1 + m_2)d_1^2 + m_2 d_2^2 + 2m_2 d_1 d_2 \cos \theta_2]\ddot{\theta}_{1d} \\
&\quad - (m_2 d_2^2 + m_2 d_1 d_2 \cos \theta_2)\ddot{\theta}_{2d} \\
&\quad + m_2 d_1 d_2 \sin (\theta_2)\, \dot{\theta}_2(2\dot{\theta}_{1d} + \dot{\theta}_{2d}) \\
&\quad + [(m_1 + m_{2m})gd_1 + m_{2m}gd_2]\delta\theta_1 + O_1(\varepsilon) - u_{1\varepsilon}^{\times} = 0
\end{aligned} \tag{12}$$

The correction term $O_1(\varepsilon)$ is added to correct the error due to the approximation of the non-linear subsystem by the switching hyperplane $S_1(\Theta) = 0$.

3.5.2. *Linearized error equation of the second link.* The linearized error equation of the second link is determined in the same way as link 1. It is represented as

$$\begin{aligned}
&m_{2m}d_2^2\delta\ddot{\theta}_2 + 2m_{2m}d_1 d_2 \sin (\theta_{2d})\, \dot{\theta}_{1d}\delta\dot{\theta}_2 + W_{f2}\delta\dot{\theta}_2 \\
&\quad + 2m_{2m}d_1 d_2(\cos (\theta_{2d})\, \dot{\theta}_{1d}\dot{\theta}_{2d} + \sin (\theta_{2d})\, \ddot{\theta}_{1d})\delta\theta_2 \\
&\quad - m_2 d_2^2\ddot{\theta}_{2d} - (m_2 d_2^2 + m_2 d_1 d_2 \cos \theta_2)\ddot{\theta}_{1d} \\
&\quad + m_2 d_1 d_2 \sin (\theta_2)\, \dot{\theta}_2\dot{\theta}_{1d} \\
&\quad + m_2 d_1 d_2 \sin (\theta_2)\, \dot{\theta}_1(\dot{\theta}_1 + \dot{\theta}_2) \\
&\quad - m_2 gd_2 \sin (\theta_1 + \theta_2) + u_{2\varepsilon} = \dot{S}_2(\Theta) = 0
\end{aligned} \tag{13}$$

There is no missing term in this case. The control law $u_{2\varepsilon}$ is the control that applies to the second link when it is in B_2.

3.6. Lyapunov's direct method

The system motion consider in this section is the motion inside the boundary layer only. Equation (12) and (13) are used to represent approximately the non-linear system. Lyapunov's direct method is the control scheme used for design of our control laws.

3.6.1. *The first link.* Owing to the format of Lyapunov's direct method, we need to represent (12) in the state-space form. In order to do this, we set

$$J_1 = (m_1 + m_{2m})d_1^2 + m_{2m}d_2^2 + 2m_{2m}d_1 d_2 \cos \theta_{2d}$$

$$x_1 = \delta\theta_1, \quad \dot{x}_1 = x_2$$

$$x_2 = \delta\dot{\theta}_1, \quad \dot{x}_2 = \delta\ddot{\theta}_1$$

$$\dot{x}_2 = \frac{2m_2 d_1 d_2 \sin(\theta_{2d})\,\dot{\theta}_{2d} - W_{f1}}{J_1} x_2 - \frac{(m_1 + m_{2m})gd_1 + m_{2m}gd_2}{J_1} x_1 + \frac{u_{1\varepsilon}^{\times}}{J_1} - \frac{G_1 + O_1(\varepsilon)}{J_1}$$

The state-space form is then represented as follows:

$$\dot{X}_1 = \begin{bmatrix} 0 & 1 \\ -\dfrac{(m_1 - m_{2m})gd_1 + m_{2m}gd_2}{J_1} & \dfrac{2m_{2m}d_1 d_2 \sin(\theta_{2d})\,\dot{\theta}_{2d} - W_{f1}}{J_1} \end{bmatrix} X_1 + \begin{bmatrix} 0 \\ -\dfrac{G_1 + O_1(\varepsilon)}{J_1} \end{bmatrix} + \begin{bmatrix} 0 \\ \dfrac{1}{J_1} \end{bmatrix} u_{1\varepsilon}^{\times}$$

where

$$X_1 = \begin{bmatrix} x_1 \\ x_2 \end{bmatrix}$$

$$\begin{aligned} G_1 = &-(m_1 + m_2)gd_1 \sin(\theta_1) - m_2 g d_2 \sin(\theta_1 + \theta_2) \\ &- [(m_1 + m_2)d_1^2 + m_2 d_2^2 + 2m_2 d_1 d_2 \cos \theta_2]\ddot{\theta}_{1d} \\ &- (m_2 d_2^2 + m_2 d_1 d_2 \cos(\theta_2))\ddot{\theta}_{2d} \\ &+ m_2 d_1 d_2 \sin(\theta_2)\,\dot{\theta}_2(2\dot{\theta}_{1d} + \dot{\theta}_{2d}) \end{aligned}$$

The abbreviation of the state-space equation is set as

$$\dot{X}_1 = A_1 X_1 + B_1 u_{1\varepsilon}^{\times} + D_1 \tag{14}$$

The last term, D_1, can be treated as a general disturbance to the system. Owing to the properties of the robust control, we see that the general disturbancce D_1 does not affect the asymptotic stability of the system. Thus, the general disturbance does nothing to the stability of the system.

By Lyapunov's direct method, if a Lyapunov function V_1 can be found, such that $V_1 > 0$ and $\dot{V}_1 < 0$, then the system approaches the origin asymptiotically. Thus, if the Lyapunov function V_1 is defined as

$$V_1 = X_1^{\mathrm{T}} P X_1$$

and we can find a symmetric positive definite matrix

$$P = \begin{bmatrix} a & b \\ b & c \end{bmatrix} > 0$$

with

$$a, c > 0, \quad b \geqslant 0$$

such that

$$\dot{V}_1 = \dot{X}_1^{\mathrm{T}} P X_1 + X_1^{\mathrm{T}} P \dot{X}_1 < 0 \tag{15}$$

then the system approaches the origin asymptotically. In other words, we have

$$\delta\dot{\theta}_1 \to 0, \quad \text{i.e. } \dot{\theta}_1 \to \dot{\theta}_{1d}$$

$$\delta\theta_1 \to 0, \qquad \theta_1 \to \theta_{1d}$$

Thus, the tracking follows.

Now the problem boils down to the choice of the control law $u_{1\varepsilon}^{\times}$, such that the existence of $\dot{V}_1 < 0$ is guaranteed. From (14) and (15), we have

$$\dot{V}_1 = X_1^{\mathrm{T}}(A_1^{\mathrm{T}} P + P A_1) X_1 + u_{1\varepsilon}^{\times} B_1^{\mathrm{T}} P X_1 + X_1^{\mathrm{T}} P B_1 u_{1\varepsilon}^{\times} + D_1^{\mathrm{T}} P X_1 + X_1^{\mathrm{T}} P D_1$$

$$= -\frac{2}{J_1} b[(m_1 + m_{2m}) g d_1 + m_{2m} g d_2] x_1^2$$

$$+ \frac{2}{J_1}\{J_1 a - c[(m_1 + m_{2m}) g d_1 + m_{2m} g d_2]$$

$$
\begin{aligned}
&+ 2bm_{2m}d_1 d_2 \sin(\theta_{2d})\,\dot{\theta}_{2d} - 2bW_{f1}\}x_1 x_2 \\
&+ \frac{2}{J_1}[J_1 b + 2cm_{2m}d_1 d_2 \sin(\theta_{2d})\,\dot{\theta}_{2d} - 2cW_{f1}]x_2^2 \\
&+ \frac{2}{J_1}(bx_1 + cx_2) \\
&\times \{(m_1 + m_2)gd_1 \sin\theta_1 + m_2 g d_2 \sin(\theta_1 + \theta_2) \\
&\quad + [(m_1 + m_2)d_1^2 + m_2 d_2^2 + 2m_2 d_1 d_2 \cos\theta_2]\ddot{\theta}_{1d} \\
&\quad + (m_2 d_2^2 + m_2 d_1 d_2 \cos\theta_2)\ddot{\theta}_{2d} \\
&\quad - m_2 d_1 d_2 \sin(\theta_2)\dot{\theta}_2(2\dot{\theta}_{1d} + \dot{\theta}_{2d}) + O_1(\varepsilon)\} \\
&+ \frac{2}{J_1}(bx_1 + cx_2)u_{1\varepsilon}^{\times}
\end{aligned}
\tag{16}
$$

If the control law $u_{1\varepsilon}^{\times}$ is chosen as

$$
\begin{aligned}
u_{1\varepsilon}^{\times} = &- K_1 \frac{x_2^2 + |x_1 x_2|}{|bx_1 + cx_2| + \eta}\,\mathrm{sgn}\,(bx_1 + cx_2) - m_1 g d_1 \sin\theta_1 - m_1 d_1^2 \ddot{\theta}_{1d} \\
&- \{m_{2m} g d_1 |\sin\theta_1| + m_{2m} g d_2 |\sin(\theta_1 + \theta_2)| \\
&\quad + [m_{2m}(d_1^2 + d_2^2) + 2m_{2m} d_1 d_2 |\cos\theta_2|]|\ddot{\theta}_{1d}| \\
&\quad + (m_{2m} d_2^2 + m_{2m} d_1 d_2 |\cos\theta_2|)|\ddot{\theta}_{2d}| + |O_1(\varepsilon)|_{\max} \\
&\quad + m_{2m} d_1 d_2 |\sin\theta_2|\,|\dot{\theta}_2|\,|2\dot{\theta}_{1d} + \dot{\theta}_{2d}|\} \frac{bx_1 + cx_2}{|bx_1 + cx_2| + \eta}
\end{aligned}
$$

with the positive constant K_1 chosen as

$$
\begin{aligned}
K_1 = \max\{&J_{1m} b + 2cm_{2m} d_1 d_2 |\dot{\theta}_{2d}|_{\max} + 2cW_{f1}; \\
&J_{1m} a + c[(m_1 + m_{2m})gd_1 + m_{2m} g d_2] \\
&+ 2bm_{2m} d_1 d_2 |\dot{\theta}_{2d}|_{\max} + 2bW_{f1}\}
\end{aligned}
$$

where η is a very small constant

$$
J_{1m} = (m_1 + m_{2m})d_1^2 + m_{2m} d_2^2 + 2m_{2m} d_1 d_2
$$

and

m_{2m} = maximum load of the manipulator + mass of link 2

Then $\dot{V} < 0$ is guaranteed, as is the tracking.

3.6.2 *The second link.* The control law $u_{2\varepsilon}$, which is chosen to ensure the tracking of link 2, is determined using the same process as that for link 1.We start with the second linearized error equation (13). In order to represent (13) in the state-space form, we set

$$J_2 = m_{2m} d_2^2$$

$$x_3 = \delta\theta_2, \quad \dot{x}_3 = x_4$$

$$x_4 = \delta\dot{\theta}_2, \quad \dot{x}_4 = \delta\ddot{\theta}_2$$

$$\begin{aligned}
\dot{x}_4 = &-\frac{2}{J_2} m_{2m} d_1 d_2 \sin(\theta_{2d})\, \dot{\theta}_{1d} x_4 - W_{f2} x_4 \\
&-\frac{2}{J_2} m_{2m} d_1 d_2 [\cos(\theta_{2d})\dot{\theta}_{1d}\dot{\theta}_{2d} + \sin(\theta_{2d})\ddot{\theta}_{1d}] x_3 \\
&-\frac{u_{2\varepsilon}}{J_2} + \frac{1}{J_2}[(m_2 d_2^2 + m_2 d_1 d_2 \cos\theta_2)\ddot{\theta}_{1d} \\
&\qquad - m_2 d_1 d_2 \sin(\theta_2)\, \dot{\theta}_2 \dot{\theta}_{1d} + m_2 d_2^2 \ddot{\theta}_{2d} \\
&\qquad - m_2 d_1 d_2 \sin(\theta_2) \dot{\theta}_1 (\dot{\theta}_1 + \dot{\theta}_2) \\
&\qquad + m_2 g d_2 \sin(\theta_1 + \theta_2) + O_2(\varepsilon)]
\end{aligned}$$

The state-space form is then represented as follows:

$$\dot{X}_2 = \begin{bmatrix} 0 & 1 \\ -\dfrac{2m_{2m} d_1 d_2 [\cos(\theta_{2d})\dot{\theta}_{1d}\dot{\theta}_{2d} + \sin(\theta_{2d})\,\ddot{\theta}_{1d}]}{J_2} & -\dfrac{2m_{2m} d_1 d_2 \sin(\theta_{2d})\,\dot{\theta}_{1d}}{J_2} - W_{f2} \end{bmatrix} X_2 + \begin{bmatrix} 0 \\ \dfrac{G_2 + Q_2(\varepsilon)}{J_2} \end{bmatrix} + \begin{bmatrix} 0 \\ -\dfrac{1}{J_2} \end{bmatrix} u_{2\varepsilon} \tag{17}$$

where

$$X_2 = \begin{bmatrix} x_3 \\ x_4 \end{bmatrix}$$

$$G_2 = (m_2 d_2^2 + m_2 d_1 d_2 \cos \theta_2)\ddot{\theta}_{1d} - m_2 d_1 d_2 \sin(\theta_2)\, \dot{\theta}_2 \dot{\theta}_{1d}$$
$$+ m_2 g d_2 \sin(\theta_1 + \theta_2) + m_2 d_2^2 \ddot{\theta}_{2d} - m_2 d_1 d_2 \sin(\theta_2)\, \dot{\theta}_1 (\dot{\theta}_1 + \dot{\theta}_2)$$

The abbreviation of (17) is set as

$$\dot{X}_2 = A_2 X_2 + B_2 u_{2\varepsilon} + D_2 \tag{18}$$

The last term, D_2, is treated as a general disturbance to the system. Owing to the robust control scheme, the general disturbance does not affect the asymptotic stability of the system. Thus, D_2 does nothing to the stability of the system.

By applying Lyapunov's direct method, the Lyapuniv function V_2 is defined as

$$V_2 = X_2^T P X_2$$

Again, if we can find a symmetric positive definite matrix

$$P = \begin{bmatrix} a & b \\ b & c \end{bmatrix} > 0$$

with

$$a, c > 0, \quad b \geqslant 0$$

such that

$$\dot{V}_2 = \dot{X}_2^T P X_2 + X_2^T P \dot{X}_2 < 0 \tag{19}$$

then the system approaches the origin asymptotically. In other words, we have

$$\delta\dot{\theta}_2 \to 0, \quad \text{i.e. } \dot{\theta}_2 \to \dot{\theta}_{2d}$$
$$\delta\theta_2 \to 0, \qquad \theta_2 \to \theta_{2d}$$

Thus, the tracking of link 2 follows.

Now, the problem reduces to the choice of the control law $u_{2\varepsilon}$, such that the existence of $\dot{V}_2 < 0$ is guaranteed. Form (18) and (19), we have

$$\dot{V}_2 = X_2^T(A_2^T P + P A_2) X_2 + u_{2\varepsilon} B_2^T P X_2 + X_2^T P B_2 u_{2\varepsilon} + D_2^T P X_2 + X_2^T P D_2$$

$$
\begin{aligned}
&= -4b\frac{d_1}{d_2}[\cos(\theta_{2d})\,\dot\theta_{1d}\dot\theta_{2d} + \sin(\theta_{2d})\,\ddot\theta_{1d}]x_3^2 \\
&\quad + \Big\{2a - 4b\frac{d_1}{d_2}\sin(\theta_{2d})\,\dot\theta_{1d} - 2bW_{f2} \\
&\qquad - 4c\frac{d_1}{d_2}[\cos(\theta_{2d})\,\dot\theta_{1d}\dot\theta_{2d} + \sin(\theta_{2d})\,\ddot\theta_{1d}]\Big\}x_3x_4 \\
&\quad + \left[2b - 4c\frac{d_1}{d_2}\sin(\theta_{2d})\,\dot\theta_{1d} - 2cW_{f2}\right]x_4^2 \\
&\quad + \frac{2}{J_2}[(m_2d_2^2 + m_2d_1d_2\cos\theta_2)\,\ddot\theta_{1d} - m_2d_1d_2\sin(\theta_2)\,\dot\theta_2\dot\theta_{1d} \\
&\qquad + m_2gd_2\sin(\theta_1+\theta_2) + m_2d_2^2\ddot\theta_{2d} \\
&\qquad - m_2d_1d_2\sin(\theta_2)\,\dot\theta_1(\dot\theta_1+\dot\theta_2) + O_2(\varepsilon)](bx_3+cx_4) \\
&\quad - \frac{2}{J_2}(bx_3+cx_4)u_{2\varepsilon}
\end{aligned} \tag{20}
$$

If the control law $u_{2\varepsilon}$ is chosen as

$$
\begin{aligned}
u_{2\varepsilon} &= J_2K_2\frac{x_3^2 + W_ex_4^2 + |x_3x_4|}{|bx_3+cx_4|+\eta}\operatorname{sgn}(bx_3+cx_4) \\
&\quad + [(m_{2m}d_2^2 + m_{2m}d_1d_2|\cos\theta_2|)|\ddot\theta_{1d}| + m_{2m}d_1d_2|\sin(\theta_2)|\,|\dot\theta_2|\,|\dot\theta_{1d}| \\
&\qquad + m_{2m}gd_2|\sin(\theta_1+\theta_2)| + m_{2m}d_2^2|\ddot\theta_{2d}| \\
&\qquad + m_{2m}d_1d_2|\dot\theta_1(\dot\theta_1+\dot\theta_2)| + |O_2(\varepsilon)|_{\max}] \times \frac{bx_3+cx_4}{|bx_3+cx_4|+\eta}
\end{aligned}
$$

with the positive constant K_2 chosen as

$$
\begin{aligned}
K_2 = \max\Big\{&2b\frac{d_1}{d_2}(|\dot\theta_{1d}|_{\max}|\dot\theta_{2d}|_{\max} + |\ddot\theta_{1d}|_{\max});\ a + 2b\frac{d_1}{d_2}|\dot\theta_{1d}|_{\max} \\
&+ 2c\frac{d_1}{d_2}(|\dot\theta_{1d}|_{\max}|\dot\theta_{2d}|_{\max} + |\ddot\theta_{1d}|_{\max}) + bW_{f2};\ b + 2c\frac{d_1}{d_2}|\dot\theta_{1d}|_{\max}\Big\}
\end{aligned}
$$

where η is a very small constant, $J_2 = m_{2m}d_2^2$, W_e and W_{f2} are weighting factors, and

m_{2m} = maximum load of the manipulator + mass of link 2

then $\dot V_2 < 0$ is guaranteed, as is the tracking.

3.7. Conclusion

Two kinds of control u_i and $u_{i\varepsilon}$, i = 1, 2, are used for the purposes of the robust tracking control. For the first link, they are

$$\begin{aligned} u_1 = {} & m_1 g d_1 \sin\theta_1 + 2m_{2m} d_1 d_2 \sin(\theta_{2d})\,\dot\theta_{2d}\delta\dot\theta_1 - W_{f1}\delta\dot\theta_1 \\ & - [(m_1 + m_{2m})d_1^2 + m_{2m}d_2^2 + 2m_{2m}d_1 d_2 \cos(\theta_{2d})]\delta\ddot\theta_1 \\ & + m_1 d_1^2 \ddot\theta_{1d} - (\gamma_1 + 2m_{2m}d_1 d_2 |\dot\theta_2|\,|\dot\theta_{1d}|_{\max} \\ & \qquad + m_{2m} d_1 d_2 |\dot\theta_2|\,|\dot\theta_{2d}|_{\max}) \operatorname{sgn} S_1 \end{aligned} \tag{21}$$

and

$$\begin{aligned} u_{1\varepsilon} = {} & [(m_1 + m_{2m})gd_1 + m_{2m} g d_2]\delta\theta_1 \\ & + K_1 \frac{\delta\dot\theta_1^2 + |\delta\theta_1\,\delta\dot\theta_1|}{|b\delta\theta_1 + c\delta\dot\theta_1| + \eta} \operatorname{sgn}(b\delta\theta_1 + c\delta\dot\theta_1) \\ & + m_1 g d_1 \sin\theta_1 + m_1 d_1^2 \ddot\theta_{1d} \\ & + \{m_{2m} g d_1 |\sin\theta_1| + m_{2m} g d_2 |\sin(\theta_1 + \theta_2)| \\ & \quad + [m_{2m}(d_1^2 + d_2^2) + 2m_{2m} d_1 d_2 |\cos\theta_2|]|\ddot\theta_{1d}| \\ & \quad + (m_{2m} d_2^2 + m_{2m} d_1 d_2 |\cos\theta_2|)|\ddot\theta_{2d}| + |O_1(\varepsilon)|_{\max} \\ & \quad + m_{2m} d_1 d_2 |\sin\theta_2|\,|\dot\theta_2|\,|2\dot\theta_{1d} + \dot\theta_{2d}|\} \times \frac{b\delta\theta_1 + c\delta\dot\theta_1}{|b\delta\theta_1 + c\delta\dot\theta_1| + \eta} \end{aligned} \tag{22}$$

where

$$\begin{aligned} \gamma_1 > {} & m_{2m} g d_1 + m_{2m} g d_2 + m_{2m}(d_1 + d_2)^2 |\ddot\theta_{1d}|_{\max} \\ & + m_{2m} d_2 (d_1 + d_2) |\ddot\theta_{2d}|_{\max} \end{aligned}$$

where W_{f1} is a weighting factor. η is a very small constant such that

$$|b\delta\theta_1 + c\delta\dot\theta_1| \simeq |b\delta\theta_1 + c\delta\dot\theta_1| + \eta$$

$$\begin{aligned} K_1 = \max\{ & J_{1m} b + 2c m_{2m} d_1 d_2 |\dot\theta_{2d}|_{\max} + 2cW_{f1}; \\ & J_{1m} a + c[(m_1 + m_{2m})gd_1 + m_{2m} g d_2] \\ & + 2b m_{2m} d_1 d_2 |\dot\theta_{2d}|_{\max} + 2bW_{f1}\} \end{aligned}$$

$$J_{1m} = (m_1 + m_{2m})d_1^2 + m_{2m} d_2^2 + 2m_{2m} d_1 d_2$$

m_{2m} = maximum load of the manipulator + mass of link 2

and sgn A denotes the sign of A.

For the second link, they are

$$\begin{aligned} u_2 = & -m_{2m} d_2^2 \delta\ddot{\theta}_2 - 2m_{2m} d_1 d_2 \sin(\theta_{2d})\, \dot{\theta}_{1d} \delta\dot{\theta}_2 - W_{f2} \delta\dot{\theta}_2 \\ & - 2m_{2m} d_1 d_2 (\cos(\theta_{2d})\, \dot{\theta}_{1d} \dot{\theta}_{2d} + \sin(\theta_{2d}) \ddot{\theta}_{1d}) \delta\theta_2 \\ & - (\gamma_2 + m_{2m} d_1 d_2 |\dot{\theta}_2|\, |\dot{\theta}_{1d}|_{\max} + m_{2m} d_1 d_2 |\dot{\theta}_1 (\dot{\theta}_1 + \dot{\theta}_2)|) \operatorname{sgn} S_2 \end{aligned} \tag{23}$$

and

$$\begin{aligned} u_{2\varepsilon} = & J_2 K_2 \frac{\delta\theta_2^2 + W_e \delta\dot{\theta}_2^2 + |\delta\theta_2 \delta\dot{\theta}_2|}{|b\delta\theta_2 + c\delta\dot{\theta}_2| + \eta} \operatorname{sgn}(b\delta\theta_2 + c\delta\dot{\theta}_2) \\ & + [(m_{2m} d_2^2 + m_{2m} d_1 d_2 |\cos \theta_2|) |\ddot{\theta}_{1d}| \\ & \quad + m_{2m} d_1 d_2 |\sin(\theta_2)|\, |\dot{\theta}_2|\, |\dot{\theta}_{1d}| \\ & \quad + m_{2m} g d_2 |\sin(\theta_1 + \theta_2)| + m_{2m} d_2^2 |\ddot{\theta}_{2d}| \\ & \quad + m_{2m} d_1 d_2 |\dot{\theta}_1 (\dot{\theta}_1 + \dot{\theta}_2)| + |O_2(\varepsilon)|_{\max}] \times \frac{b\delta\theta_2 + c\delta\dot{\theta}_2}{|b\delta\theta_2 + c\delta\dot{\theta}_2| + \eta} \end{aligned} \tag{24}$$

where

$$\begin{aligned} K_2 = \max \Big\{ & 2b \frac{d_1}{d_2} (|\dot{\theta}_{1d}|_{\max} |\dot{\theta}_{2d}|_{\max} + |\ddot{\theta}_{1d}|_{\max});\ a + 2b \frac{d_1}{d_2} |\dot{\theta}_{1d}|_{\max} \\ & + 2c \frac{d_1}{d_2} (|\dot{\theta}_{1d}|_{\max} |\dot{\theta}_{2d}|_{\max} + |\ddot{\theta}_{1d}|_{\max}) + bW_{f2};\ b + 2c \frac{d_1}{d_2} |\dot{\theta}_{1d}|_{\max} \Big\} \end{aligned}$$

$$J_2 = m_{2m} d_2^2$$

$$\gamma_2 > m_{2m} d_2^2 |\ddot{\theta}_{2d}|_{\max} + m_{2m} d_2 (d_1 + d_2) |\ddot{\theta}_{1d}|_{\max} + m_{2m} g d_2$$

and W_{f2} and W_2 are weighting factors.

The control u_i guaranteed the attractiveness of the boundary layer $B_i = \{\Theta : |S_i(\Theta)| \leqslant \varepsilon\}$. The control $u_{i\varepsilon}$ guarantees the asymptotic stability of the ith linearized error equation, and eliminates the control chattering around $S_i(\Theta) = 0$. The boundary of $B_i = \{\Theta : |S_i(\Theta)| \leqslant \varepsilon\}$ the control switch.

The control u_i is applied to the ith non-linear subsystem whenever the non-linear subsystem is out of the boundary layer B_i. This guarantees that the compensated subsystem finally enters the boundayr layer B_i, and is represented approximately by the switching hyperplane S_i. By a suitable choice of the switching hyperplane, the compensated non-linear subsystem can be represented approximately by an expected linearized error equation.

The control $u_{i\varepsilon}$ is then applied to the ith linearized error equation,which is the replacement of the non-linear ith subsystem whenever the subsystem is in the boundary layer B_i. The ith linearized error equation is then driven to the origin by the control $u_{i\varepsilon}$.Thus, we complete the tracking control of the ith subsystem. The tracking of the two-link manipulator system succeeds when each link completes its own tracking, repectively. Since the system we are dealing with is a time-varying system, we have a time-vayring origin. However, the proposed robust tracking control scheme still works successfully.

A more precise approximatation of the compensated non-linear system to the switching hyperplane is obtained by choosing a smaller boundary layer thickness ε. The control chattering occurs when ε is zero, or ε is too small to prevent the system crossing the boundary of B_i and entering the other side of the switching hyperplane. The latter case occurs when a limited control only is available - this is a general case in the real word. Thus, a compromise between the precision of the approximation and the control chattering exists in the design of ε when the control u is fixed.

4. Computer simulation

The simulation results of the robust tracking control of a simple -two link manipulator system are shown in this section. The computer used is an IBM 3090 mainframe.

4.1 Desired path

The desired paths, $(\theta_{id}, \dot{\theta}_{id}, \ddot{\theta}_{id})$, for $i = 1, 2,$ which are chosen to be tracked separately by links 1 and 2 are defined as follows. For link 1, the desired path is chosen as

$$\begin{aligned} \theta_{1d} &= 2{\cdot}4(1 - \exp(-t/n)) \\ \dot{\theta}_{1d} &= \frac{1}{n} 2{\cdot}4 \exp(-t/n) \\ \ddot{\theta}_{1d} &= -\frac{1}{n^2} 2{\cdot}4 \exp(-t/n) \end{aligned} \tag{25}$$

The desired path for link 2 is chosen as

$$\begin{aligned} \theta_{2d} &= 1{\cdot}2(1 - \exp(-t/n)) \\ \dot{\theta}_{2d} &= \frac{1}{n} 1{\cdot}2 \exp(-t/n) \end{aligned}$$

$$\ddot{\theta}_{2d} = -\frac{1}{n^2} 1{\cdot}2 \exp(-t/n) \tag{26}$$

Two kinds of system motions are defined. For n = 1, the system tracks a fast-moving path, and is called " a system with fast motion ". For n = 2, the system tracks a slow moving path, and is called " a system with slow motion".

4.2 System parameters

The system parameters used for the simulation are chosen as follows:

$$m_1 = 1{\cdot}5 \text{ kg}, \quad m_2 = 1{\cdot}0 \sim 1{\cdot}5 \text{ kg}$$

$$d_1 = 0{\cdot}3 \text{ m}, \quad d_2 = 0{\cdot}3 \text{ m}$$

$$W_{f1} = 90, \quad W_{f2} = 30$$

$$\gamma_1 = m_{2m} g d_1 + m_{2m} g d_2 + m_{2m}(d_1 + d_2)^2 |\ddot{\theta}_{1d}|_{\max} + m_{2m}(d_1 + d_2)|\ddot{\theta}_{2d}|_{\max} + 23$$

$$\gamma_2 = m_{2m} g d_2 + m_{2m} d_2^2 |\ddot{\theta}_{2d}|_{\max} + m_{2m} d_2 (d_1 + d_2)|\ddot{\theta}_{1d}|_{\max} + 1$$

$$\eta = 0{\cdot}01, \quad W_e = 15{\cdot}0$$

$$a = 6000{\cdot}0, \quad b = 0{\cdot}0, \quad c = 100{\cdot}0$$

$$|O_1(\varepsilon)|_{\max} = 0{\cdot}4, \quad |O_2(\varepsilon)|_{\max} = 0{\cdot}4$$

4.3 Dynamic system and controls

The dynamic equations used for simulation are described by (1) and (2). The controls u_1 and u_2 are designed for links 1 and 2 according to the VSS method, and are shown as (21) and (23) when the system is outside the boundary layer B.u_1 and u_2 are replaced by $u_{1\varepsilon}$ and $u_{2\varepsilon}$, defined by (22) and (24), when the system is inside the boundary layer B.The thinkness ε of the boundary layer B is chosen to be 0.3. The controls u_1 and $u_{1\varepsilon}$ are applied to the first link, so that link 1 tracks the desired path (25). The second link is driven to track (26) by applying u_2 and $u_{2\varepsilon}$ on it.

All the measurements needed to form the controls are generated by finding the solution of a set of first-order ordinary differential equation of the form $\dot{X} = F(X)$, which is the state-space representation of the two-link manipulator system. An IMSL subroutine named DVERK is called to solve the differential equations.

4.4. Cases studied in simulation

Four cases are studied in the simulation, to verify the property of robustness of the control algorithm. Two sets of desired paths are set to tracked by the manipulator. One set, (25) and (26) with n = 1, is call the desired path with fast motion. The second one, (25) and (26) with n = 2, is called the desired path with slow motion. The varying payload, with knowledge of its variation range, is used to represent the uncertainty of the system. The full load and empty load of the manipulator are represented by setting two extreme values of m_2 as 1.5 kg and 1.0 kg. The combination of these two sets of desired paths and two extreme payloads forms the four different cases that are studied in the simulation:

(a) Case 1: system with full load (m_2 = 1.5 kg) tracks the desired path with fast motion;
(b) Case 2: system with no load (m_2 = 1.0 kg) tracks the desired path with fast motion;
(c) Case 3: system with full load tracks the desired path with slow motion;
(d) Case 4: system with no load tracks the desired path with slow motion.

4.5. Simulation results and conclusions

The simulation results are presented in six kinds graph. They are

(i) values of S_1 and S_2 with respect to time,
(ii) the position tracking with respect to time,
(iii) the velocity tracking with respect to time,
(iv) the position tracking error with respect to time,
(v) the velocity tracking error with respect to time, and
(vi) the controls with respect to time.

The last three types of graph of Cases 1 and 4, shown by Fig.2, are used to shown the feature of robustness of the compensated system.

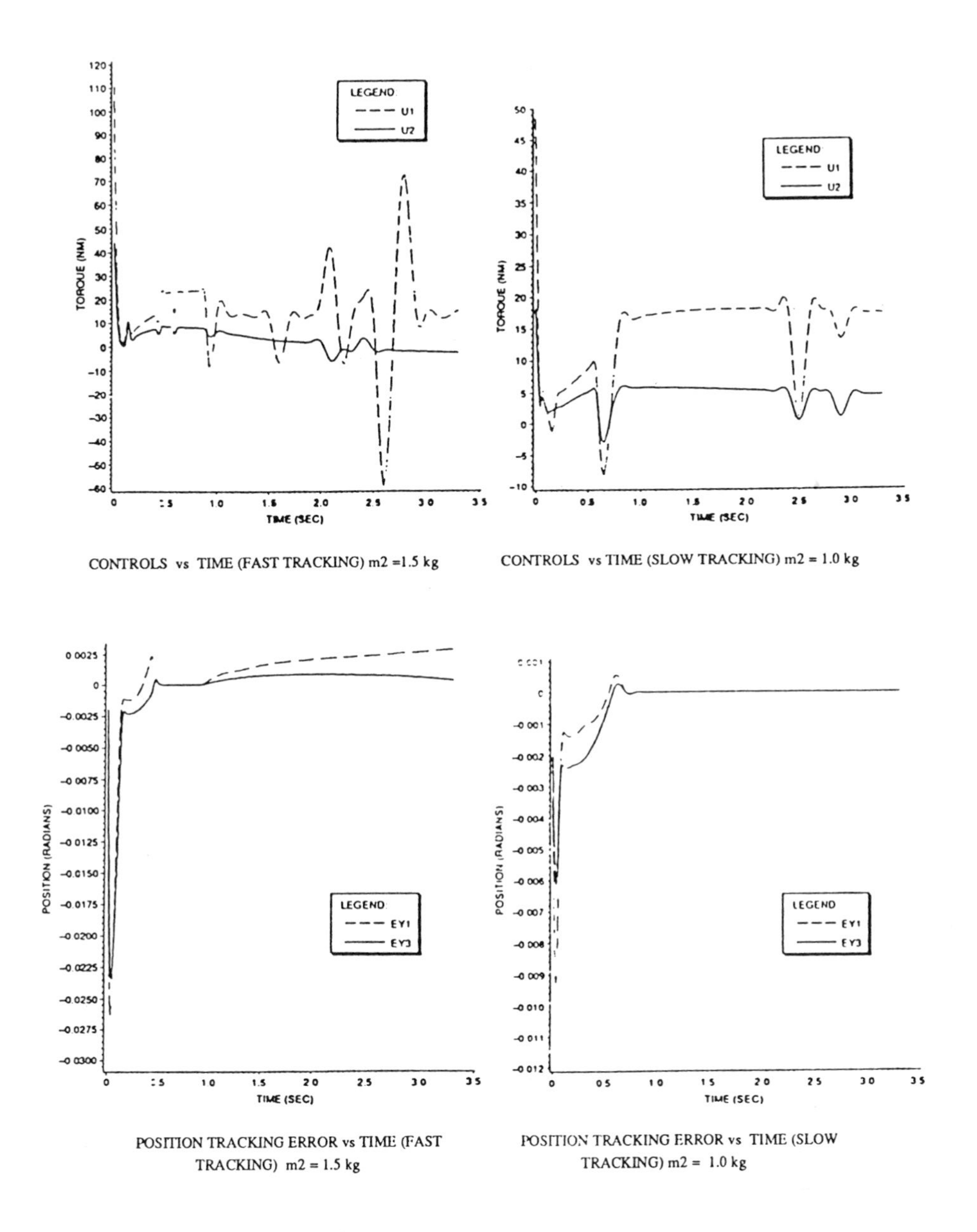

Figure 2. Robustness of compensated system

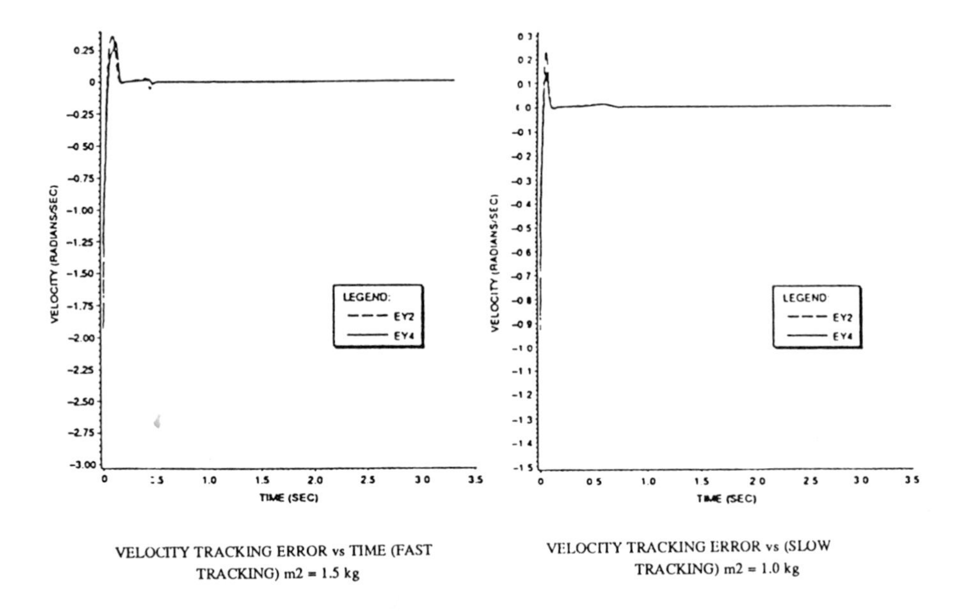

VELOCITY TRACKING ERROR vs TIME (FAST TRACKING) m2 = 1.5 kg

VELOCITY TRACKING ERROR vs (SLOW TRACKING) m2 = 1.0 kg

Figure 2 (continued)

Several conclusions are drawn from the simulation results, and these match with the engineering common sense. More control effort needs to be used for a system with a heavier load, and more control effect is also needed for a system that tracks a fast moving model than one that tracks a slow moving model.By applying our robust tracking control algorithm to the system, there is no singificant degradation on the system performance, even if we do not have a priori knowledge of the system payload and configuration. Thus, the robustness of the compensated system is guaranteed by applying this tracking control algorithm to it. The control chattering (see Fig. 3) is also eliminated by this tracking control algorithm.

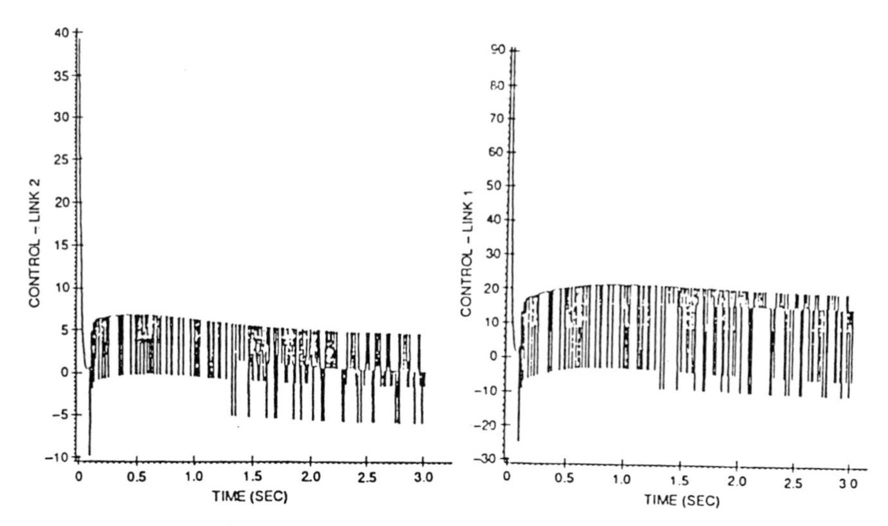

Figure 3 Control chattering caused by VSS.

References

AMBROSION, G., CELENTANO, G., and GAROFALO, F., 1985, I.E.E.E. Trans. Autom. Control, 30, 275.

BALESTRINO, A., DEMARIA, G., and ZINOBER, A. S., 1984, Automatica, 20, 559.

BARMISH, B. R., CORLESS, M., and LEITMANN, G., 1983 b, SIAM J. Control Optim., 21, 246.

BARMISH, B. R., PETERSEN, R., and FEUER, A., 1983a, Automatica, 19, 523

BARTOLINI, G., and ZOLEZZI, T., 1985, I.E.E.E. Trans. Autom. Control, 30, 681.

CORLESS, M. J., and LEITMANN, G., 1981, I.E.E.E. Trans. Autom. Control, 26, 1139.

DRAZENOVIC, B., 1969, Automatica, 5, 287.

GUO, T. H., and KOIVO, A. J., 1984, J. Robot. Systems, 1, 141.

GUTMAN, S., 1979, I.E.E.E. Trans. Autom. Control, 24, 437.

GUTMAN, S., and PALMOR, Z., 1982, SIAM J. Control Optim., 20, 850.

ITKIS, U., 1976, Control Systems of Variable Structure (New York: Wiley).

MORGAN, R. G., and OZGUNER, U., 1985, I.E.E.E. J. Robot. Automn, 1, 57.

NICOSIA, S., and TOMEI, P., 1984, Automatica, 20, 635.

RICHTER, S., LEFEBURE, S., and DECARLO, R., 1982, I.E.E.E. Trans. Autom. Control, 27, 492.

SINGH, S. N., 1985, I.E.E.E. Trans. Autom. Control, 30, 1099.

SLOTINE, J. J., 1984, Int. J. Control, 40, 421; 1985, Int. J. Robot. Res., 4, 49.

SLOTINE, J. J., and SASTRY, S. S., 1983, Int. J. Control, 38, 465.

UTKIN, V. I., 1971, Avtomat. Telemekh., 42; 1972, Ibid., 51; 1977, I.E.E.E. Trans. Autom. Control, 22, 212.

YOUNG, K. K. D. 1978, I.E.E.E. Trans. Autom. Control, 23, 1079.

ADAPTIVE ROBUST CONTROL OF UNCERTAIN SYSTEMS

Y.H. Chen* **J.S. Chen**

The George W. Woodruff School of Mechanical Engineering
Georgia Institute of Technology
Atlanta, Georgia 30332-0405

I. Introduction

The control design approach adopted here falls into the *deterministic* framework (Leitmann 1983). The features of this framework are best described in terms of its description on uncertainty and its performance requirement.

First, although the uncertainty may be *nonlinear* (in terms of its influence on the system) and (possibly fast) *time-varying*, its statistical information is never assumed. In turn, all is required is the knowledge of the set which the uncertainty value may lie within. This bounding set is characterized via certain deterministic properties (such as compactness).

Second, the system performance is described in a deterministic manner. For example, one of the many properties that we will impose on the system performance is called uniform ultimate boundedness (P3 of Definition 1). Loosely speaking, the system trajectory will eventually enter a ball around the nominal equilibrium position, whose size can be calculated by the designer, and stay there thereafter. This is guaranteed

* Author to whom the correspondence should be addressed.

CONTROL AND DYNAMIC SYSTEMS, VOL. 50

regardless of the actual behavior of the uncertainty.

The major development of control system design by using this deterministic approach can be divided into three stages.

Representing work in the first stage should at least include Gutman (1979) and Leitmann (1979). Under certain structural conditions, a class of feedback control scheme was proposed. The controls render the uncertain systems asymptotically stable. However, the control function is discontinuous and its practical implementation, which is always under the limitation of actuators, may result in an undesirable "chattering".

Facing this difficulty, the second stage of development focuses on *continuous-type* saturation control (Leitmann 1981, Corless and Leitmann 1981) and non-saturation control (Barmish *et al.* 1983). The latter may be reduced to a linear controller if the uncertain system is linear. Although this continuous design removes chattering concern, it still requires the knowledge of the bound of uncertainty. It is sometimes possible that with an insufficient physical understanding of the uncertainty, the bound may not be known. A similar difficulty is that if the uncertainty is poorly studied in advance, the control engineer is likely to choose a conservative estimation of the bound for safety reason. It is possible that the control magnitude may become excessively large and actuator saturation arises.

The third stage work intends to relax the *a priori* requirement of the bound of uncertainty. Adaptive robust control, which uses adaptive scheme to track the bound on-line, was proposed. Corless and Leitmann (1983, 1984) first constructed a class of (continuous) saturation-type adaptive robust controls which are based on full state information and assume no measurement noise. The bound of uncertainty is considered to be an unknown parameter (vector). An adaptive scheme is constructed to update the information of this bound. The robust control then uses this updated information to generate control signal. The uncertain systems under considerations are nonlinear with matched uncertainty. Chen (1989, 1990) proposes a new design which allows all admissible (non-adaptive) robust controls, as constructed in Chen (1988), to be converted to their adaptive versions. Moreover, the adaptive scheme in Corless and Leitmann (1983, 1984) is modified to tolerate mismatched uncertainty and measurement noise. The tradeoff is that the state $x(t)$ does not converge to zero (while it does so in Corless and Leitmann 1983, 1984). Furthermore, the measurement noise considered in Chen (1989, 1990) should be "smooth".

In this work, we will first show the basic framework under which non-adaptive robust control is designed. We then extend the design to adaptive robust controls under various settings, which include mismatched uncertainty and output matrix uncertainty.

The work reported here serves two purposes. First, it intends to provide a complete (though not comprehensive) description of the theoretical basis under which some major adaptive robust control designs are derived. Theoreticians are able to design other alternatives for their specific usages. Second, it intends to demonstrate its applications through two examples, one simple and one complex. Practitioners are able to follow the procedure to solve their own problems.

The organization of this work is outlined as follows. After a thorough description of the system, uncertainty, and performance (Section II), non-adaptive robust control based on full state information is presented (Section III). The cone-boundedness feature of the uncertainty bound and the presence of measurement noise are the main theme in this section. As the output rather than the full state is available, the non-adaptive robust control is redesigned (Section IV). If the bound of uncertainty is unknown, adaptive version of the robust output feedback control is proposed (Section V). So far all designs (both non-adaptive and adaptive) are based on the matching condition. Mismatched uncertainty case is explored in Section VI. Variation of output matrix is studied in Section VII. Section VIII provides a simple illustrative example to demonstrate the use of the previous controllers. Section IX removes the cone-boundedness requirement on the uncertainty bound, which all previous designs must meet. Section X provides a more complex example to illustrate the use in Section IX. Section XI summarizes the main ideas and provides some further elaborations. This section is especially useful for practitioners. Section XII concludes the work.

II. Uncertain dynamical systems and problem statement

Consider the following uncertain system

$$\dot{x}(t) = Ax(t) + Bu(t) + e(x(t), \sigma(t), u(t), t), \tag{1}$$

$$x(t_0) = x_0,$$

where $t \in \mathbf{R}$ is the independent variable, $x(t) \in \mathbf{R}^n$ is the state, $u(t) \in \mathbf{R}^m$ is the control, and $\sigma(t) \in \mathbf{R}^s$ is the uncertain parameter. The "nominal" (i.e., the portion without uncertainty) system matrix A and the "nominal" input matrix B are of appropriate dimensions and are both *known*. The *unknown* function $\sigma(\cdot) : \mathbf{R} \to \Sigma \subset \mathbf{R}^s$ is Lebesgue measurable (see the Appendix for a definition; readers not familiar with this may also assume it continuous). with Σ compact (that is, closed and bounded). The bounding set Σ may be *known* or *unknown*. The sections to follow will clearly indicate which case they refer to. The function $e(\cdot) : \mathbf{R}^n \times \Sigma \times \mathbf{R}^m \times \mathbf{R} \to \mathbf{R}^n$ is Caratheodory (see the Appendix for a definition; readers not familiar with this may also assume it continuous). For simplicity, we shall assume that the nominal system matrix A is Hurwitz (i.e., all eigenvalues of A have negative real parts). This may be replaced by a stabilizability condition of (A, B) (i.e., there exists a gain matrix K such that $A - BK$ is Hurwitz).

Assumption 1: *There exists a function* $\hat{e}(\cdot) : \mathbf{R}^n \times \Sigma \times \mathbf{R}^m \times \mathbf{R} \to \mathbf{R}^m$ *such that for all* $x \in \mathbf{R}^n$, $\sigma \in \Sigma$, $u \in \mathbf{R}^m$, *and* $t \in \mathbf{R}$,

$$e(x(t), \sigma(t), u(t), t) = B\ \hat{e}(x(t), \sigma(t), u(t), t). \tag{2}$$

◇◇

This condition, which implies that e should lie within the range space of B, is generally recognized as the *matching condition*.

For a dynamic plant, the expression (1) is not unique. In practice, given a system, say $\dot{x} = f(x, \sigma, u, t)$, one may first *choose* the nominal portions A and B. The other part is then lumped into $e := f - Ax - Bu$. Hence that A and B are both *known* is by no means a restriction in the present setup. However, the matching condition (2) imposes a condition on how the uncertainty $e(x, \sigma, u, t)$ enters the system. Some discussions of the matching condition are in order.

The purpose of matching condition is to serve as an indication of the uncertainty *direction* under which the uncertainty enters the system. Mismatched case arises as the range space of B is too "narrow" and needs to be "expanded". Consider, for example, that

$$e = \begin{bmatrix} 0 & \sigma_2 \\ \sigma_1 & \sigma_3 \end{bmatrix}, \qquad B = \begin{bmatrix} 1 \\ 0 \end{bmatrix}, \tag{3}$$

where $\sigma = [\sigma_1 \sigma_2 \sigma_3]^T$. This is a single input case. Matching condition is not met. If, however, one applies an extra control and hence

$$B = \begin{bmatrix} 1 & 0 \\ 0 & 1 \end{bmatrix}, \tag{4}$$

then the matching condition is met. Hence the matching condition may be regarded as a criteria which can be managed in the initial design stage. Mismatched case simply means a "cripple" design. The matching condition in fact comes with no surprise since otherwise one is not able to use the control to target to the uncertainty from the "opposite" (which is, of course, the most effective) direction. We also mention that this pre-condition on the uncertain system is in general weaker than some of the well accepted conditions in the society such as the conicity assumption for small gain method or vanishing assumption of u in e (Safonov 1980).

We will relax this matching condition later in Section VI.

Definition 1: *Consider the dynamic system*

$$\dot{z}(t) = g(z(t), t), \tag{5}$$

$$z(t_0) = z_0,$$

where $z \in \mathbf{R}^q$ and $g(\cdot) : \mathbf{R}^q \times \mathbf{R} \to \mathbf{R}^q$. The system is practically stable if and only if there exists a constant $\underline{d} > 0$ such that the following properties hold:

(P1) Existence of Solutions. Given $(z_0, t_0) \in \mathbf{R}^q \times \mathbf{R}$, there exists a solution $z(\cdot) : [t_0, \infty) \to \mathbf{R}^q$ for $z(t_0) = z_0$.

(P2) Uniform Boundedness. Given any $\bar{r} > 0$ and $\|z_0\| \leq \bar{r}$, there exist a constant $d(\bar{r}) > 0$ such that $\|z(t)\| \leq d(\bar{r})$ for all $t \in [t_0, \infty)$.

(P3) Uniform Ultimate Boundedness. For each $\bar{d} > \underline{d}$ and $\bar{r} > 0$, given any $z(\cdot) : [t_0, \infty) \to \mathbf{R}^q$, $z(t_0) = z_0$, with $\|z_0\| \leq \bar{r}$, there is a finite time $T(\bar{d}, \bar{r}) \geq 0$ such that $\|z(t)\| \leq \bar{d}$ for all $t \geq t_0 + T(\bar{d}, \bar{r})$.

(P4) Generalized Uniform Stability. For each $\bar{d} > \underline{d}$, there is a constant $r(\bar{d}) > 0$ such that $\|z_0\| \leq r(\bar{d})$ implies that $\|z(t)\| \leq \bar{d}$ for all $t \geq t_0$. ◇◇

Vector norms are taken to be euclidean and matrix norms are the corresponding induced ones (see, e.g., Desoer and Vidyasagar 1975); and thus $\|\Xi\|^2 = \lambda_M(\Xi^T \Xi)$ where Ξ is a real matrix, $\lambda_M(\cdot)$ ($\lambda_m(\cdot)$) denotes the maximum (minimum) eigenvalue of the designated matrix (if all its eigenvalues are real).

Some elaborations of P1-P4 are in order. Existence of solutions (P1) is imposed to assure the other three properties P2-P3, which all refer to solution properties, are meaningful.

Uniform boundedness property (P2) states that the solution will remain within a ball, whose size may be dependent on the initial ball (the ball in which the initial state lies) but not on the initial time t_0. In terms of the system performance, uniform boundedness assures no large transient overshoot.

Uniform ultimate boundedness property (P3) states that the state will eventually enter a target ball around the origin after a finite time and remains there thereafter. The required time for entering the ball are dependent on the size of the initial ball and the target ball but not on the initial time t_0. In terms of the system performance, this provides a description on the steady state performance.

The combination of P2 and P3 then gives a complete description on the system performance, both initially and afterward.

Generalized uniform stability (P4) states that a prescribed performance, in terms of the size of the ball where the state lies within for *all* time, can always be achieved provided the initial ball is properly chosen.

Note that P1-P4 are not to be confused with the *practical stabilizability* in Barmish *et al.* (1983). The constant $\underline{d} > 0$ in there is determined by the designer. The $\underline{d} > 0$ here is determined by the system and the applied control.

Note also that P2 is different from *uniform stability* in the sense of Lyapunov (Hahn 1967) which provides the choice for the initial ball *after* (not *before*) the ball for $z(t)$ is chosen. In other word, the P2 scenario does not claim the following: For any $\bar{d} > 0$, there exist a constant $r(\bar{d}) > 0$ such that $\|z(t)\| \leq \bar{d}$ for all $\|z_0\| \leq r(\bar{d})$ and $t \in [t_0, \infty)$.

Note that $\underline{d}$ determines the lower limit of the performance dictated by P3 and P4.

III. Non-adaptive state feedback robust control

We first consider the situation that measurement noise arises:

$$y_1(t) = x(t) + \omega_1(t), \tag{6}$$

where $y_1(t) \in \mathbf{R}^n$ and $\omega_1(\cdot) : \mathbf{R} \to \Omega_1 \subset \mathbf{R}^n$ is Lebesgue measurable with Ω_1 compact and *known*.

Assumption 2: *The bounding set Σ is known. Furthermore, there are known constants $k_1 \in [0, \infty)$, $k_2 \in [0, \infty)$, and $k_3 \in [0, 1)$ such that for all $x \in \mathbf{R}^n$, $\sigma \in \Sigma$, $u \in \mathbf{R}^m$, and $t \in \mathbf{R}$,*

$$\|\hat{e}(x, \sigma, u, t)\| \leq k_1 + k_2\|x\| + k_3\|u\|. \tag{7}$$

◇◇

The is in fact the *cone-boundedness* condition for the uncertainty bound (Barmish and Leitmann 1982). The admissible plants that satisfy this condition may be nonlinear in general. However, its intrinsic feature may be viewed more clearly by referring to the following linear uncertain system (Chen 1986, Chen and Leitmann 1987):

$$\dot{x}(t) = [A + \Delta A(\sigma(t))]x(t) + [B + \Delta B(\sigma(t))]u(t) + Cv(t), \quad (8)$$

where $\sigma(t) \in \Sigma$, $v(t) \in V$ (Σ and V are compact bounding sets) and $\Delta A(\sigma)$, $\Delta B(\sigma)$, and C satisfy the following matching condition:

$$\Delta A(\sigma) = BD(\sigma), \qquad \Delta B(\sigma) = BE(\sigma), \qquad C = BF. \quad (9)$$

Equation (7) is met if we choose k_1, k_2, and k_3 such that

$$k_1 \geq \max_{v \in V} \|Fv\|, \qquad k_2 \geq \max_{\sigma \in \Sigma} \|D(\sigma)\|, \qquad k_3 \geq \max_{\sigma \in \Sigma} \|E(\sigma)\|. \quad (10)$$

This clearly shows the following: The plant (1) can be nonlinear but its uncertainty bound should "act linearly". The restriction of $k_3 \in [0, 1)$ is interpreted as follows: The uncertainty in the input channel cannot be too severe to reverse the direction of control action; otherwise one is not able to tell if the control is in the desired direction.

Theorem 1: *Under Assumptions 1 and 2, there exists a linear robust control*

$$u(t) = -\gamma B^T P y_1(t) \quad (11)$$

such that the controlled system of (1) is practically stable. Here the constant gain γ *is chosen such that*

$$\gamma \geq \frac{k_2^2}{2(1-k_3)C_1\lambda_m(Q)}, \quad (12)$$

where the constant $C_1 \in (0, 1)$ *is a design parameter and the matrix* $Q > 0$.

Proof: Throughout the proofs, arguments are sometimes omitted for simplicity. Using elementary results from continuous functions and measurable functions, one can show that the right-hand side of the controlled system of (1) is Caratheodory. By Coddington and Levinson (1955) and Hale (1980), the solution exists almost everywhere for a finite time interval.

Let the Lyapunov function candidate be

$$v(x) = x^T P x, \tag{13}$$

where matrix $P > 0$ is the solution of the Lyapunov equation

$$A^T P + PA + Q = 0. \tag{14}$$

For a given $\sigma(\cdot)$ and $\omega_1(\cdot)$, the time derivative of v along the controlled system trajectory is given by

$$\begin{aligned}
\dot{v} &= 2x^T P[Ax - B\gamma B^T P y_1 + e] \\
&= 2x^T PAx - 2\gamma x^T PBB^T Px - 2\gamma x^T PBB^T P\omega_1 + 2x^T PB\hat{e} \\
&\le -x^T Qx - 2\gamma\|B^T Px\|^2 + 2\gamma\|B^T Px\|\|B^T P\omega_1\| \\
&\quad + 2\|B^T Px\|\left[k_1 + k_2\|x\| + k_3\|\gamma B^T P(x+\omega_1)\|\right] \\
&\le -\lambda_m(Q)\|x\|^2 - 2\gamma(1-k_3)\|B^T Px\|^2 \\
&\quad + \left[2\gamma(1+k_3)\|B^T P\omega_1\| + 2k_1 + 2k_2\|x\|\right]\|B^T Px\|.
\end{aligned} \tag{15}$$

The second and the third terms in the last inequality form a quadratic expression for $\|B^T Px\|$. A maximization analysis of this quadratic expression yields

$$\begin{aligned}
\dot{v} &\le -\lambda_m(Q)\|x\|^2 + \frac{\left[\gamma(1+k_3)\|B^T P\omega_1\| + k_1 + k_2\|x\|\right]^2}{2\gamma(1-k_3)} \\
&= -\alpha_1\|x\|^2 + \alpha_{12}\|x\| + \alpha_{13},
\end{aligned} \tag{16}$$

where

$$\alpha_1 := \lambda_m(Q) - \frac{k_2^2}{2\gamma(1-k_3)}, \tag{17}$$

$$\alpha_{12} := \frac{\left[\gamma(1+k_3)\bar{\omega}_1 + k_1\right]k_2}{\gamma(1-k_3)}, \tag{18}$$

$$\alpha_{13} := \frac{\left[\gamma(1+k_3)\bar{\omega}_1 + k_1\right]^2}{2\gamma(1-k_3)}, \tag{19}$$

$$\bar{\omega}_1 := \max_{\omega_1 \in \Omega_1} \|B^T P\omega_1\|. \tag{20}$$

Note that $k_3 \in [0,1)$ which implies $1 - k_3 > 0$. α_1 can always be made positive if γ is sufficient large (such as (12)). Choosing the constant gain γ according to (12), we have

$$\dot{v} \le -(1 - C_1)\lambda_m(Q)\|x\|^2 + \alpha_{12}\|x\| + \alpha_{13}. \tag{21}$$

As the design parameter $C_1 \in (0,1)$, the first term on the right is always negative. Therefore, $\dot{v}$ is negative definite for a sufficiently large $\|x\|$. Properties P1-P2 then follows upon using standard arguments (Corless and Leitmann 1981) and letting

$$d(\bar{r}) = \left[\frac{\lambda_M(P)}{\lambda_m(P)}\right]^{1/2} R \qquad \text{if } \bar{r} \le R, \tag{22}$$

$$d(\bar{r}) = \left[\frac{\lambda_M(P)}{\lambda_m(P)}\right]^{1/2} \bar{r} \qquad \text{if } \bar{r} > R, \tag{23}$$

where

$$R := \frac{\alpha_{12} + \sqrt{\alpha_{12}^2 + 4(1 - C_1)\lambda_m(Q)\alpha_{13}}}{2(1 - C_1)\lambda_m(Q)}. \tag{24}$$

P3 and P4 also follow by taking

$$\underline{d} = \left[\frac{\lambda_M(P)}{\lambda_m(P)}\right]^{1/2} R, \tag{25}$$

$$T(\bar{d},\bar{r}) = \begin{cases} 0 \quad \text{if } \bar{r} \le (\lambda_m(P)/\lambda_M(P))^{1/2}\bar{d}, \\ [\lambda_M(P)\bar{r}^2 - \lambda_m(P)\bar{d}^2]/[(1 - C_1)\lambda_m(Q)\lambda_m(P)\lambda_M^{-1}(P)\bar{d}^2 \\ \quad - \alpha_{12}\lambda_m^{1/2}(P)\lambda_M^{-1/2}(P)\bar{d} - \alpha_{13}] \qquad \text{otherwise,} \end{cases} \tag{26}$$

$$r(\bar{d}) = \left[\frac{\lambda_M(P)}{\lambda_m(P)}\right]^{1/2} \bar{d}. \tag{27}$$

◇◇

Note that the lower bound for γ does not depend on k_1 because the effect of k_1 is equivalent to an external disturbance which does not affect practical stability (recall that it only requires the existence of a constant $\underline{d}$). Nevertheless, k_1 affects the magnitude of $\underline{d}$ for ultimate boundedness.

From the designer's point of view, however, one usually determines a desirable performance and then translates it in terms of $\underline{d}$. This then dictates the value for R which needs to be achieved by the robust control. The designer's responsibility then becomes to choose an appropriate gain γ (which is in term related to its design parameter) for a given R.

If there is no measurement noise (hence $\omega_1 = 0$, $\bar{\omega}_1 = 0$) and $k_2 \neq 0$, the relation between R and the design parameter C_1 can be calculated from (24):

$$C_1 \leq \left[1 + \frac{k_1}{k_2 R}\right]^{-2}. \tag{28}$$

We see that a smaller R demands a smaller C_1 which renders high control gain (refer to (12)). The above equation can be utilized to estimate the required control gain for a given ultimate boundedness region. If measurement noise arises, the relationship between C_1 and R can be also configured, which leads to the required gain for a specific R.

As $\omega_1 = 0$ and $k_2 = 0$, from (21) and the expressions for α_{12} and α_{22}, the γ value required for a specific R is:

$$\gamma \geq \frac{k_1^2}{2(1 - k_3)\lambda_m(Q) R^2}.$$

From the expressions for R, α_{12}, and α_{13}, we see that if there are no measurement noise, R approaches zero when γ approach infinity. This means that the ball of ultimate boundedness can be made arbitrary small. On the other hand, R is lower bounded by a finite constant if measurement noise arises. This brings no surprise since if one can not even tell where the state is due to noise corruption, there is certainly no way to adjust the control value to drive the state in a very precisely way.

When there are no measurement error and "equivalent external input" ($\omega_1 = 0$, $k_1 = 0$), both α_{12} and α_{12} are equal to zero. Therefore, $R = 0$ ($\underline{d} = 0$) and the origin is asymptotically stable.

The choice of the design parameter C_1 is dictated by the tradeoff (determined by control engineers) between the control magnitude and the performance requirement. A smaller C_1 results in a smaller $\underline{d}$ (hence better performance) but a larger γ (hence more control effort).

IV. Non-adaptive output feedback robust control

We next consider the case as full state information is not available. Suppose that the measured output $y_2(t) \in \mathbf{R}^p$ is given by:

$$y_2(t) = Hx(t) + \omega_2(t), \tag{29}$$

where $H \in \mathbf{R}^{p\times n}$ is the output matrix and $\omega_2(\cdot) : \mathbf{R} \to \Omega_2 \in \mathbf{R}^p$ is the measurement noise. The bounding set Ω_2 is compact and known. The following additional assumption is needed.

Assumption 3: *There exists a matrix $\Psi \in \mathbf{R}^{m\times p}$ such that the transfer function matrix*

$$T(s) = \Psi H(sI - A)^{-1}B \tag{30}$$

is strictly positive real (SPR) (Narendra and Taylor 1973). Thus there exists a matrix pair $(P, Q) > 0$ for the Lyapunov equation $A^T P + PA + Q = 0$ such that

$$\Psi H = B^T P. \tag{31}$$

$\diamond\diamond$

This assumption will be relaxed later in Section VII.

Theorem 2. *Under Assumptions 1-3, there exists a linear robust control*

$$u(t) = -\gamma\Psi y_2(t) \tag{32}$$

such that the controlled system of (1) is practically stable. Here the constant gain γ is chosen such that

$$\gamma \geq \frac{k_2^2}{2(1-k_3)C_1\lambda_m(Q)}. \tag{33}$$

where the constant $C_1 \in (0, 1)$ is a design parameter and the matrix $Q > 0$.

Proof: Consider the Lyapunov function candidate

$$v(x) = x^T Px, \tag{34}$$

where matrix P is the solution of the Lyapunov equation (14). For a given $\sigma(\cdot)$ and $\omega_2(\cdot)$, the time derivative of v along the controlled system trajectory is given by

$$\begin{aligned}
\dot{v} &= 2x^T P[Ax - B\gamma\Psi y_2 + e] \\
&= 2x^T PAx - 2\gamma x^T PB\Psi(Hx + \omega_2) + 2x^T PB\hat{e} \\
&\leq -x^T Qx - 2\gamma\|B^T Px\|^2 + 2\gamma\|B^T Px\|\|\Psi\omega_2\| \\
&\quad + 2\|B^T Px\|\left[k_1 + k_2\|x\| + k_3\|\gamma\Psi(Hx + \omega_2)\|\right] \\
&\leq -\lambda_m(Q)\|x\|^2 - 2\gamma(1-k_3)\|B^T Px\|^2 \\
&\quad + \left[2(1+k_3)\gamma\|\Psi\omega_2\| + 2k_1 + 2k_2\|x\|\right]\|B^T Px\| \\
&\leq -\lambda_m(Q)\|x\|^2 + \frac{\left[\gamma(1+k_3)\|\Psi\omega_2\| + k_1 + k_2\|x\|\right]^2}{2\gamma(1-k_3)} \\
&= -\alpha_1\|x\|^2 + \alpha_{22}\|x\| + \alpha_{23},
\end{aligned} \tag{35}$$

where α_1 is given as (17) and

$$\alpha_{22} := \frac{[\gamma(1+k_3)\bar{\omega}_2 + k_1]\, k_2}{\gamma(1-k_3)}, \tag{36}$$

$$\alpha_{23} := \frac{[\gamma(1+k_3)\bar{\omega}_2 + k_1]^2}{2\gamma(1-k_3)}, \tag{37}$$

$$\bar{\omega}_2 := \max_{\omega_2 \in \Omega_2} \|\Psi\omega_2\|. \tag{38}$$

Choosing γ according to (0), we have

$$\dot{v} \leq -(1-C_1)\lambda_m(Q)\|x\|^2 + \alpha_{22}\|x\| + \alpha_{23} \tag{39}$$

Applying the similar augments with the proof of Theorem 1, it can be shown that $d(\bar{r})$, $\underline{d}$, $T(\bar{d}, \bar{r})$, $r(\bar{d})$ are the same as (22)-(23), and (25)-(27). The only difference is in R:

$$R := \frac{\alpha_{22} + \sqrt{\alpha_{22}^2 + 4(1-C_1)\lambda_m(Q)\alpha_{23}}}{2(1-C_1)\lambda_m(Q)}. \tag{40}$$

$\diamond\diamond$

In the special case that full state is available, we can take H as $n \times n$ identity matrix. For every Lyapunov equation solution pair $(P, Q) > 0$, there is a corresponding Ψ such that Assumption 3 is satisfied (simply let $\Psi = B^T P$). Therefore, Theorem 1 can be regarded as a special case of Theorem 2.

If Assumption 3 is not met, we can rewrite (29) as

$$y_2(t) = (\bar{H} + \Delta H)x(t) + \omega_2(t), \tag{41}$$

where $\bar{H}$ is chosen such that the following transfer function is SPR:

$$\bar{T}(s) = \Psi\bar{H}(sI - A)^{-1}B.$$

Here the output matrix "deterioration" ΔH can be treated as an "uncertainty" in output matrix. (There is certainly a slight abuse of terminology here for ΔH is actually known. However, it is to be *treated* as a part of uncertainty.) The practical stability of the controlled system for this case will be discussed in latter section.

V. Adaptive output feedback robust control

To implement the control (32) in the absence of the knowledge of bounds of uncertainty (i.e., k_1, k_2, and k_3), it is considered necessary that an adaptive scheme for γ be designed. In order to set up the frame, Assumption 2 needs to be modified as follows.

Assumption 4: *The bounding set Σ is unknown. However, there are (unknown) constants k_1 and $k_2 \in [0, \infty)$ such that for all $x \in \mathbf{R}^n$, $\sigma \in \Sigma$, $u \in \mathbf{R}^m$, and $t \in \mathbf{R}$,*

$$\|\hat{e}(x, \sigma, u, t)\| \leq k_1 + k_2\|x\|. \tag{42}$$

$\diamond\diamond$

Theorem 3: *Suppose that the uncertain system (1) with output equation (29) satisfies Assumptions 1, 3, and 4. Consider the control*

$$u(t) = -\hat{\gamma}(t)\Psi y_2(t), \tag{43}$$

$$\dot{\hat{\gamma}}(t) = l_1\|\Psi y_2(t)\|^2 - l_2\hat{\gamma}(t) - l_3\|\Psi y_2(t)\|\hat{\gamma}(t). \tag{44}$$

System (1) under this control is practically stable.

Proof: Let us consider the following Lyapunov function candidate

$$v(x, \hat{\gamma}) = \xi_1 v_1(x) + \xi_2 v_2(\hat{\gamma}), \tag{45}$$

$$v_1(x) = x^T P x, \qquad v_2(\hat{\gamma}) = \frac{1}{2}(\hat{\gamma} - \gamma)(\hat{\gamma} - \gamma), \tag{46}$$

where matrix P satisfies Lyapunov equation (14) and ξ_1, ξ_2 are positive constants to be decided later. For a given $\sigma(\cdot)$ and $\omega_2(\cdot)$, the time derivative of v_1 along the controlled system trajectory is given by

$$\begin{aligned}\dot{v}_1(x) =& 2x^T PAx - 2\hat{\gamma}x^T PB\Psi y_2 + 2x^T PB\hat{e} \\ =& 2x^T PAx - 2\gamma x^T PB\Psi y_2 + 2x^T PB\hat{e} - 2(\hat{\gamma} - \gamma)x^T PB\Psi y_2.\end{aligned} \tag{47}$$

From Assumption 3, we have

$$\begin{aligned}x^T PB =& x^T H^T \Psi^T \\ =& (y_2 - w_2)^T \Psi^T.\end{aligned} \tag{48}$$

Choose γ as in the last section. By the result from the previous section and the above equation, (47) can be written as

$$\begin{aligned}\dot{v}_1(x) \leq &-(1-C_1)\lambda_m(Q)\|x\|^2 + \alpha_{32}\|x\| + \alpha_{33}\\ &-2(\hat{\gamma}-\gamma)y_2^T\Psi^T\Psi y_2 + 2(\hat{\gamma}-\gamma)\omega_2^T\Psi^T\Psi y_2\\ \leq &-(1-C_1)\lambda_m(Q)\|x\|^2 + \alpha_{32}\|x\| + \alpha_{33}\\ &-2(\hat{\gamma}-\gamma)\|\Psi y_2\|^2 + 2\|\hat{\gamma}-\gamma\|\|\Psi\omega_2\|\|\Psi y_2\|\\ \leq &-(1-C_1)\lambda_m(Q)\|x\|^2 + \alpha_{32}\|x\| + \alpha_{33}\\ &-2(\hat{\gamma}-\gamma)\|\Psi y_2\|^2 + 2\bar{\omega}_2\|\hat{\gamma}-\gamma\|\|\Psi y_2\|,\end{aligned} \tag{49}$$

where $\bar{\omega}_2$ is given in (38) and

$$\alpha_{32} := \frac{[\gamma\bar{\omega}_2 + k_1]\,k_2}{\gamma}, \tag{50}$$

$$\alpha_{33} := \frac{[\gamma\bar{\omega}_2 + k_1]^2}{2\gamma}. \tag{51}$$

For a given $\sigma(\cdot)$ and $\omega_2(\cdot)$, the time derivative of v_2 along any solution of the controlled system is given by

$$\begin{aligned}\dot{v}_2(\hat{\gamma}) =&(\hat{\gamma}-\gamma)\dot{\hat{\gamma}}\\ =&(\hat{\gamma}-\gamma)\left[l_1\|\Psi y_2\|^2 - l_2\hat{\gamma} - l_3\hat{\gamma}\|\Psi y_2\|\right]\\ =&(\hat{\gamma}-\gamma)l_1\|\Psi y_2\|^2 - (\hat{\gamma}-\gamma)l_2(\hat{\gamma}-\gamma) - (\hat{\gamma}-\gamma)l_2\gamma\\ &-(\hat{\gamma}-\gamma)l_3(\hat{\gamma}-\gamma)\|\Psi y_2\| - (\hat{\gamma}-\gamma)l_3\gamma\|\Psi y_2\|\\ \leq& l_1(\hat{\gamma}-\gamma)\|\Psi y_2\|^2 - l_2\|\hat{\gamma}-\gamma\|^2 + l_2\gamma\|\hat{\gamma}-\gamma\|\\ &-l_3\|\hat{\gamma}-\gamma\|^2\|\Psi y_2\| + l_3\gamma\|\hat{\gamma}-\gamma\|\|\Psi y_2\|.\end{aligned} \tag{52}$$

Combining last two relationships, the total time derivative of v is then

$$\begin{aligned}\dot{v}(x,\hat{\gamma}) \leq &-\xi_1(1-C_1)\lambda_m(Q)\|x\|^2 + \xi_1\alpha_{32}\|x\| + \xi_1\alpha_{33}\\ &-2\xi_1(\hat{\gamma}-\gamma)\|\Psi y_2\|^2 + 2\bar{\omega}_2\xi_1\|\hat{\gamma}-\gamma\|\|\Psi y_2\|\\ &+\xi_2 l_1(\hat{\gamma}-\gamma)\|\Psi y_2\|^2 - \xi_2 l_2\|\hat{\gamma}-\gamma\|^2 + \xi_2 l_2\gamma\|\hat{\gamma}-\gamma\|\\ &-\xi_2 l_3\|\hat{\gamma}-\gamma\|^2\|\Psi y_2\| + \xi_2 l_3\gamma\|\hat{\gamma}-\gamma\|\|\Psi y_2\|.\end{aligned} \tag{53}$$

For any given l_1, we now choose ξ_1 and ξ_2 such that

$$2\xi_1 = \xi_2 l_1. \tag{54}$$

Moreover, we observe that

$$\begin{aligned} &\left[(2\xi_1\bar{\omega}_2 + \xi_2 l_3\gamma)\|\hat{\gamma} - \gamma\| - \xi_2 l_3\|\hat{\gamma} - \gamma\|^2\right]\|\Psi y_2\| \\ &\quad \leq \left[\frac{(2\xi_1\bar{\omega}_2 + \xi_2 l_3\gamma)^2}{4\xi_2 l_3}\right]\|\Psi y_2\| \\ &\quad := \alpha_3\|\Psi y_2\| \\ &\quad \leq \alpha_3\|\Psi H\|\|x\| + \alpha_3\|\Psi\omega_2\| \\ &\quad \leq \alpha_3\|\Psi H\|\|x\| + \alpha_3\bar{\omega}_2. \end{aligned} \tag{55}$$

The total time derivative $\dot{v}$ can be reduced to

$$\begin{aligned} \dot{v}(x,\hat{\gamma}) \leq &- \xi_1(1 - C_1)\lambda_m(Q)\|x\|^2 + (\xi_1\alpha_{32} + \alpha_3\|\Psi H\|)\|x\| + \xi_1\alpha_{33} \\ &+ \alpha_3\bar{\omega}_2 - \xi_2 l_2\|\hat{\gamma} - \gamma\|^2 + \xi_2 l_2\gamma\|\hat{\gamma} - \gamma\|. \end{aligned} \tag{56}$$

Let $\eta = [\, x^T \quad \hat{\gamma} - \gamma\,]^T$ and hence $\|\eta\|^2 = \|x\|^2 + \|\hat{\gamma} - \gamma\|^2$. Observing the following

$$\begin{aligned} ab + cd &= [\, a \quad c\,][\, b \quad d\,]^T \\ &\leq (a^2 + c^2)^{1/2}(b^2 + d^2)^{1/2}, \end{aligned} \tag{57}$$

where a, b, c, $d \in \mathbf{R}$, we then have

$$\dot{v}(x,\hat{\gamma}) \leq -\alpha_4\|\eta\|^2 + \alpha_5\|\eta\| + \alpha_6, \tag{58}$$

where

$$\alpha_4 = \min\{\xi_1(1 - C_1)\lambda_m(Q),\ \xi_2 l_2\}, \tag{59}$$

$$\alpha_5 = \left[(\xi_1\alpha_{32} + \alpha_3\|\Psi H\|)^2 + (\xi_2 l_2\gamma)^2\right]^{1/2}, \tag{60}$$

$$\alpha_6 = \xi_1\alpha_{33} + \alpha_3\bar{\omega}_2. \tag{61}$$

Since ξ_1 and ξ_2 are related according to (54), we can in turn write

$$\alpha_4 = \xi_2\ \min\left\{\frac{l_1}{2}(1 - C_1)\lambda_m(Q),\ l_2\right\}, \tag{62}$$

$$\alpha_5 = \left[(\xi_2 l_1\alpha_{32}/2 + \alpha_3\|\Psi H\|)^2 + (\xi_2 l_2\gamma)^2\right]^{1/2}. \tag{63}$$

The Lyapunov derivative $\dot{v}(x,\hat{\gamma})$ is negative definite for a sufficiently large $\|\eta\|$ since $C_1 \in (0,1)$. P1-P4 then follow upon using standard arguments and letting

$$d(\bar{r}) = \left[\frac{\gamma_2}{\gamma_1}\right]^{1/2} R \quad \text{if } \bar{r} \leq R, \tag{64}$$

$$d(\bar{r}) = \left[\frac{\gamma_2}{\gamma_1}\right]^{1/2} \bar{r} \quad \text{if } \bar{r} > R, \tag{65}$$

where

$$\gamma_1 = \min \left\{ \xi_1 \lambda_m(P), \frac{\xi_2}{2} \right\} = \frac{\xi_2}{2} \min \left\{ l_1 \lambda_m(P), 1 \right\}, \tag{66}$$

$$\gamma_2 = \max \left\{ \xi_1 \lambda_M(P), \frac{\xi_2}{2} \right\} = \frac{\xi_2}{2} \max \left\{ l_1 \lambda_M(P), 1 \right\}, \tag{67}$$

$$R = \frac{1}{2\alpha_4} \left[\alpha_5 + (\alpha_5^2 + 4\alpha_4\alpha_6)^{1/2} \right], \tag{68}$$

$$\underline{d} = \left[\frac{\gamma_2}{\gamma_1}\right]^{1/2} R, \tag{69}$$

$$T(\bar{d}, \bar{r}) = \begin{cases} 0 \quad \text{if } \bar{r} \leq (\gamma_1/\gamma_2)^{1/2} \bar{d}, \\ \dfrac{\gamma_2 \bar{r}^2 - \gamma_1 \bar{d}^2}{\alpha_4 \gamma_1 \gamma_2^{-1} \bar{d}^2 - \alpha_5 \gamma_1^{1/2} \gamma_2^{-1/2} \bar{d} - \alpha_6} \quad \text{otherwise,} \end{cases} \tag{70}$$

$$r(\bar{d}) = \left[\frac{\gamma_2}{\gamma_1}\right]^{1/2} \bar{d}. \tag{71}$$

◇◇

The following is the motivation of the adaptive scheme (44). First, a mechanism should be provided such that the scheme can "act" more strongly as the system performance is considered "poor". Second, the scheme should be able to adjust its strategy as the system performance becomes "satisfactory". In terms of the above adaptive scheme, the first term grows rapidly as $\|\Psi y_2\|$ increases (which reflects the "poor" system performance). This will result in a high increase rate of $\hat{\gamma}$ and therefore a high strength of the feedback. The second and third terms determine the decreasing rate. They will become significant as $\|\Psi y_2\|$ is small (i.e., the system performance is "satisfactory") and $\hat{\gamma}$ is large (i.e., the control action is excessive).

Following the term used in Ioannou and Kokotovic (1983), the above adaptive scheme belongs to the so-called "leakage type" (the second and third terms denote the leakage) with output dependent leakage.

We notice the difference between the *adaptive robust control* in the present work and the *robust adaptive control* (see, e.g., Kreisselmeier and Anderson 1986, Narendra and Annaswamy 1986, and Peterson and Narendra 1982). The former is more concerned with compensating time-varying uncertainty and nonlinearity by using information related to the bound. The latter is more concerned with improving the robustness of certain adaptive controls against certain extra uncertainty (such as unmodelled dynamics, time-varying input disturbance, etc.).

As the remark after Theorem 2, the control in Theorem 3 can also be used for full state feedback control. We simply set $H = I^{n \times n}$ and $\Psi = B^T P$.

VI. Robustness against mismatched uncertainty

Equation (2) of Assumption 1 restricts the uncertainty to lie within the range space of B. Many physical designs may result in the settings where Assumption 1 is violated. Thus it is of practical importance that the proposed control is robust against the mismatched part of uncertainty, i.e., the portion of the uncertainty that lies outside the range space of B.

Consider the system (1). Now choose the mappings $\hat{e}(\cdot) : \mathbf{R}^n \times \Sigma \times \mathbf{R}^m \times \mathbf{R} \to \mathbf{R}^m$ and $\tilde{e}(\cdot) : \mathbf{R}^n \times \Sigma \times \mathbf{R}^m \times \mathbf{R} \to \mathbf{R}^n$ such that for all $x \in \mathbf{R}^n$, $\sigma \in \Sigma$, $u \in \mathbf{R}^m$, and $t \in \mathbf{R}$

$$e(x, \sigma, u, t) = B\hat{e}(x, \sigma, u, t) + \tilde{e}(x, \sigma, u, t). \tag{72}$$

It can be seen easily that (2) is now a special case of last equation as $\tilde{e}(\cdot) = 0$. Moreover, the decomposition of e into $\hat{e}$ and $\tilde{e}$ is non-unique.

Assumption 5: *The bounding set Σ is unknown. However, there are (unknown) constants δ_1, $\delta_2 \geq 0$ such that for all $(x, \sigma, u, t) \in \mathbf{R}^n \times \Sigma \times \mathbf{R}^m \times \mathbf{R}$*

$$\|\tilde{e}(x, \sigma, u, t)\| \leq \delta_1 + \delta_2 \|x\|. \tag{73}$$

$\diamond\diamond$

Theorem 4: *Consider the uncertain system (1) with output (29) is subject to the control (43) with update law of $\hat{\gamma}$ given by (44). Let the function $e(\cdot)$ be decomposed into $\hat{e}(\cdot)$ and $\tilde{e}(\cdot)$ as shown in (72) with $\hat{e}(\cdot)$ and $\tilde{e}(\cdot)$ satisfying (42) and (73), respectively. Then there exists a constant $\underline{d} > 0$ such that the controlled system possesses properties P1 - P4 provided*

$$\delta_2 < \frac{\lambda_m(Q)}{2\lambda_M(P)}. \tag{74}$$

Proof: By substituting (43) and (72) into (1), we have

$$\dot{x} = Ax - B\hat{\gamma}\Psi y_2 + B\hat{e} + \tilde{e}. \tag{75}$$

Let us adopt the same Lyapunov function candidate v in (45). Upon performing similar analysis as (47)-(56) and incorporating (73), the time derivative of v along any solution of the last equation is given by

$$\begin{aligned}\dot{v}(x,\hat{\gamma}) \leq & -\xi_1(1-C_1)\lambda_m(Q)\|x\|^2 + (\xi_1\alpha_{32} + \alpha_3\|\Psi H\|)\|x\| + \xi_1\alpha_{33} \\ & + 2\xi_1\lambda_M(P)\delta_2\|x\|^2 + 2\xi_1\lambda_M(P)\delta_1\|x\| \\ & + \alpha_3\bar{\omega}_2 - \xi_2 l_2\|\hat{\gamma}-\gamma\|^2 + \xi_2 l_2\gamma\|\hat{\gamma}-\gamma\|.\end{aligned} \tag{76}$$

After some algebraic manipulation, it turns out that

$$\dot{v} \leq -\tilde{\alpha}_4\|\eta\|^2 + \tilde{\alpha}_5\|\eta\| + \tilde{\alpha}_6, \tag{77}$$

where

$$\tilde{\alpha}_4 = \xi_2 \min\left\{\frac{l_1}{2}\left[(1-C_1)\lambda_m(Q) - 2\lambda_M(P)\delta_2\right],\quad l_2\right\}, \tag{78}$$

$$\tilde{\alpha}_5 = \left[(\xi_2 l_1\alpha_{32}/2 + \alpha_3\|\Psi H\| + \xi_2 l_1\lambda_M(P)\delta_1)^2 + (\xi_2 l_2\gamma)^2\right]^{1/2}, \tag{79}$$

$$\tilde{\alpha}_6 = \alpha_6. \tag{80}$$

Notice that $\tilde{\alpha}_4$ is positive if (74) holds and C_1 approaches zero. P1 - P4 are concluded by taking $d(\bar{r})$, $\underline{d}$, $T(\bar{d},\bar{r})$, and $r(\bar{d})$ as shown in (64)-(71) but replacing α_4 and α_5 by $\tilde{\alpha}_4$ and $\tilde{\alpha}_5$. ◇◇

With minor modifications, Theorem 4 can also be applied to the (non-adaptive) robust control law mentioned in Theorems 1 and 2. In other words, a similar robustness argument against mismatched uncertainty holds as the non-adaptive robust (either full state feedback or output feedback) control is applied.

Note that (74) gives the upper bound of δ_2 which the controlled system can tolerate. This bound depends on the ratio $\lambda_m(Q)/\lambda_M(P)$. This ratio is only dependent on the nominal portion of the uncertain system and can in fact be maximized for a given A by taking $Q = I$ (Patel and Toda 1980). On the other hand, the closed-loop system (75) is able to withstand *any* δ_1 at the expense of increasing the size $\bar{d}$ of the ultimate boundedness ball.

VII. Uncertainty in the output matrix

The previous robust controllers, both non-adaptive and adaptive, are designed based on the assumption that the output matrix H is constant and known. A more general case should consider the uncertainty of the output matrix. The output matrix uncertainty can come from two possible sources. One is intrinsic to the system. The other is artificially created by the designer such that the nominal system satisfies the SPR requirement (recall Assumption 3). For simplicity, we shall only consider the system (8) with the matching condition (9). Similar work can be generalize to system (1) with the matching condition (2). Consider the output equation as

$$y_2(t) = [H + \Delta H(\beta(t))]x(t) + \omega_2(t), \tag{81}$$

where $\beta(t) \in \mathbf{R}^t$ is the uncertain parameter. The function $\beta(\cdot) : \mathbf{R} \to \Lambda \subset \mathbf{R}^t$ is Lebesgue measurable with the set Λ compact . Moreover, the function $\Delta H(\cdot)$ is continuous.

Theorem 5: *(Chen 1987) Suppose that the uncertain system (8) with output (81) satisfies the matching condition (9) and the SPR assumption (30). Consider the non-adaptive robust output feedback control*

$$u(t) = -\gamma \Psi y_2(t), \tag{82}$$

$$\gamma = \frac{\max_{\sigma\in\Sigma} \|D(\sigma)\|^2}{2(1 - \max_{\sigma\in\Sigma} \|E(\sigma)\|)C_1\lambda_m(Q)}. \tag{83}$$

where the constant $C_1 \in (0,1)$ and $Q > 0$. Then the system under this control is practical stable if either (i) or (ii) is met: (i)

$$\min_{\sigma\in\Sigma,\beta\in\Lambda} \lambda_m(\Theta(\sigma,\beta)) \geq 0, \tag{84}$$

(ii)

$$\min_{\sigma\in\Sigma,\beta\in\Lambda} \lambda_m(\Theta(\sigma,\beta)) < 0, \tag{85}$$

but with

$$\max_{\sigma\in\Sigma} \|D(\sigma)\| < \lambda_m(Q) \left[\frac{2C_1(1-C_1)(1-\max_{\sigma\in\Sigma}\|E(\sigma)\|)}{-\min_{\sigma\in\Sigma,\beta\in\Lambda} \lambda_m(\Theta(\sigma,\beta))}\right]^{1/2}. \tag{86}$$

Here the test matrix $\Theta(\sigma,\beta)$ is defined as

$$\Theta(\sigma,\beta) := PB[I+E(\sigma)]\Psi\Delta H(\beta) + \Delta H^T(\beta)\Psi^T[I+E(\sigma)]^T B^T P \tag{87}$$

Proof: Take the Lyapunov function candidate as

$$v(x) = x^T P x, \tag{88}$$

where $P > 0$ satisfies the Lyapunov equation. Choose k_1, k_2, and k_3 as follows:

$$k_1 = \max_{v \in V} \|Fv\|, \qquad k_2 = \max_{\sigma \in \Sigma} \|D(\sigma)\|, \qquad k_3 = \max_{\sigma \in \Sigma} \|E(\sigma)\|. \tag{89}$$

For a given uncertainty realization, the time derivative of v along the controlled system trajectory is given by (recall (35))

$$\begin{aligned} \dot{v} \leq & -\lambda_m(Q)\|x\|^2 + \frac{\left[\gamma(1+k_3)\|\Psi\omega_2\| + k_1 + k_2\|x\|\right]^2}{2\gamma(1-k_3)} \\ & - 2\gamma x^T PB(I+E)\Psi\Delta Hx. \end{aligned} \tag{90}$$

Note that

$$\begin{aligned} & -2\gamma x^T PB(I+E)\Psi\Delta Hx \\ & = -\gamma x^T \left[PB(I+E)\Psi\Delta H + \Delta H^T \Psi^T (I+E)^T B^T P\right] x \\ & \leq -\gamma\lambda_m(\Theta)\|x\|^2. \end{aligned} \tag{91}$$

Therefore, it can be shown that

$$\begin{aligned} \dot{v} & \leq -\lambda_m(Q)\|x\|^2 + \frac{\left[\gamma(1+k_3)\|\Psi\omega_2\| + k_1 + k_2\|x\|\right]^2}{2\gamma(1-k_3)} - \gamma\lambda_m(\Theta)\|x\|^2 \\ & \leq -\bar{\alpha}_1\|x\|^2 + \alpha_{22}\|x\| + \alpha_{23}, \end{aligned} \tag{92}$$

where α_{22} and α_{23} are defined in (36,37) and

$$\bar{\alpha}_1 := \lambda_m(Q) - \frac{k_2^2}{2\gamma(1-k_3)} + \gamma\lambda_m(\Theta). \tag{93}$$

Substituting in the value for γ, we have

$$\bar{\alpha}_1 = (1 - C_1)\lambda_m(Q) + \frac{k_2^2 \lambda_m(\Theta)}{2(1-k_3)C_1\lambda_m(Q)}. \tag{94}$$

As $\lambda_m(\Theta) \geq 0$, $\bar{\alpha}_1$ is always positive. The system is practical stable. As $\lambda_m(\Theta) < 0$, $\bar{\alpha}_1$ is still positive as long as (86) is satisfied. ◇◇

Roughly speaking, the test matrix $\Theta(\sigma,\beta)$ is to check if B^TP and $\Psi\Delta H(\beta)$ are "co-direction". If they are (hence (84)), the direction of the control is not distorted by $\Psi\Delta H(\beta)$ and the practical stability is preserved. On the other hand, if the direction of the control is distorted by $\Delta H(\beta)$, it is necessary that (86) is met. In the special case that $\Delta H(\beta)\equiv 0$ for all $\beta\in\Lambda$, $\lambda_m(\Theta(\sigma,\beta))\equiv 0$ for all σ and β. Therefore, it is a special case of (84). The conditions (84) and (85) are meaningful only when Σ and Λ are both known.

Theorem 6: *(Chen 1991) Suppose the uncertain system (8) with output (81) satisfies the matching condition (9) and the SPR assumption (30). Consider the adaptive robust control*

$$u(t) = -\hat{\gamma}\Psi y_2(t), \tag{95}$$

where

$$\dot{\hat{\gamma}}(t) = l_1\|\Psi y_2(t)\|^2 - l_2\hat{\gamma}(t) - l_3\|\Psi y_2(t)\|^2\hat{\gamma}(t), \tag{96}$$

and l_1, l_2, and l_3 are positive constants. Then the system under this control is practically stable provided

(i) The ratio l_3/l_1 is sufficiently small; and

(ii) $\max_{\beta\in\Lambda}\|\Psi\Delta H(\beta)\|$ is also sufficiently small. ◇◇

Note that the update law in (96) is different from the update law in (44) and there is no restrictions on the choice of l_2.

VIII. Application to a simple pendulum

Let us consider a simple pendulum system shown in Figure 1. (Corless and Leitmann 1981 and Barmish *et al.* 1983). The length of the pendulum is l. It is subject to a control moment $u(\cdot)$ (per unit moment of inertia) and an uncertain acceleration $q(\cdot)$, with $|q(t)|\le \rho l$ for all $t\in\mathbf{R}$. Let x_1 and x_2 denote the angle and the angular velocity, respectively. The equation of motion is given by

$$\dot{x}_1(t) = x_2(t), \tag{97}$$

$$\dot{x}_2(t) = -\frac{g\,\sin x_1(t)}{l} + u(t) - \frac{q(t)\,\cos x_1(t)}{l}, \tag{98}$$

where g is the gravity constant. For simplicity, set $l = g$. System (97,98) is nonlinear but can be put into the form of (1) by posing the following linear nominally system:

$$\dot{x}(t) = Ax(t) + Bu(t), \tag{99}$$

where

$$A = \begin{bmatrix} 0 & 1 \\ -1 & -1 \end{bmatrix}, \quad B = \begin{bmatrix} 0 \\ 1 \end{bmatrix}. \tag{100}$$

The uncertain system (97,98) can be rewritten as

$$\dot{x}(t) = Ax(t) + Bu(t) + [\ f(x(t)) - Ax(t)\ ], \tag{101}$$

where

$$f(x) = \begin{bmatrix} x_2 \\ -\sin x_1 - q\ \cos x_1/l \end{bmatrix}. \tag{102}$$

In other words, we have $e(x) = f(x) - Ax$. Assumptions 1 and 2 are met by taking

$$\hat{e} = [-\sin x_1 - q\ \cos x_1/l + x_1 + x_2], \tag{103}$$

$$k_1 = 1 + \rho, \qquad k_2 = (1^2 + 1^2)^{1/2} = \sqrt{2}, \qquad k_3 = 0 \tag{104}$$

Note that $|-\sin x_1| \le 1$, $|q\ \cos x_1/l| \le \rho l/l = \rho$.

If full state is available, robust state feedback control can be applied. The Lyapunov matrix P is obtained by solving the Lyapunov equation (notice that A is Hurwitz). This yields (by choosing $Q = 2I$)

$$P = \begin{bmatrix} 3 & 1 \\ 1 & 2 \end{bmatrix}. \tag{105}$$

From (11),

$$u = -\gamma [1 \quad 2] \begin{bmatrix} y_{11} \\ y_{12} \end{bmatrix}, \tag{106}$$

where y_{11} and y_{12} are measured states. Taking C_1 in (12) as 0.85, we need

$$\gamma \ge \frac{\sqrt{2}^2}{2 \cdot (1 - 0) \cdot 0.85 \cdot 2} = 0.59. \tag{107}$$

Next consider output feedback case where:

$$y_2(t) = [1 \quad 2]\, x(t) + w_2(t). \tag{108}$$

Choosing $\Psi = 1$, Assumption 4 is satisfied by taking the same P as (105). The non-adaptive robust output feedback control is given by, recalling (32),

$$u(t) = -\gamma\, y_2(t), \tag{109}$$

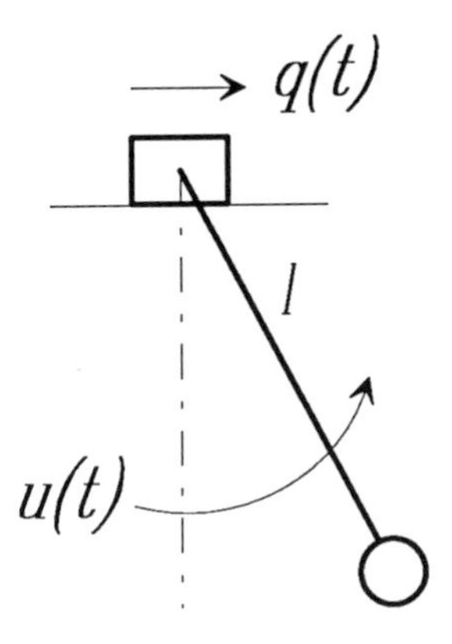

Figure 1. Example: A simple pendulum

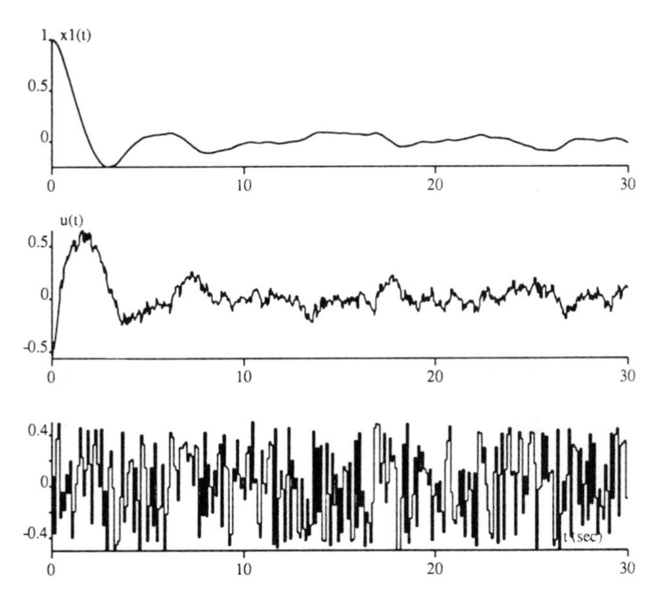

Figure 2. Robust output feedback control

a. Angular displacement (x_1) history

b. Control (u) history

c. Disturbance (q/l) history

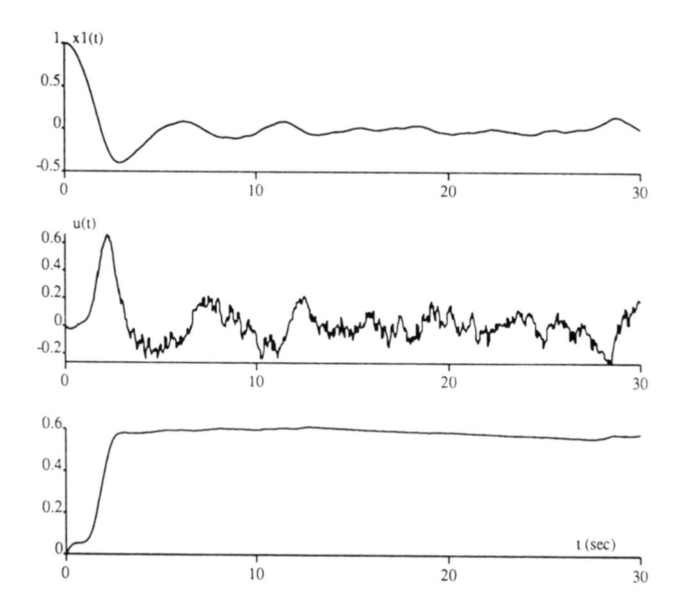

Figure 3. Adaptive robust output feedback control

a. Angular displacement (x_1) history

b. Control (u) history

c. Adaptive parameter ($\hat{\gamma}$) history

where γ needs to satisfy (33).

For the same output equation (108), the adaptive robust output feedback control is given by, recalling (43,44),

$$u(t) = -\hat{\gamma}(t)\, y_2(t), \tag{110}$$

$$\dot{\hat{\gamma}}(t) = l_1|y_2(t)|^2 - l_2\hat{\gamma}(t) - l_3\hat{\gamma}(t)|y_2(t)|. \tag{111}$$

Figure 2 shows the simulation results for non-adaptive robust output feedback control (109) where γ is chosen as 0.6. In this and the following simulations, initial condition is $x(0) = [1 \quad 0]^T$. Both $q(t)$ and $w_2(t)$ were chosen to be rectangularly distributed random numbers in the interval of ± 0.1 and ± 0.5 (therefore $\rho = 0.5$). Figure 3 shows the response of the adaptive robust output feedback control (110,111). The system response is comparable to the use of control (109). The adaptive parameter $\hat{\gamma}$ increases mostly in the beginning and then maintains steady. This means that the adaptive scheme (111) halts its action after the time when the performance is considered satisfactory.

IX. Adaptive robust control with nonlinear bound

The class of uncertainty considered so far can be characterized as the *cone-boundedness* class (recall (7)). The advantage for the control design is its linear structure with respect to the state (or output). However, it is certainty possible that a more general uncertainty may emerge in a broader application domain. In this section, we shall consider such case. The uncertainty bound only needs to be *concave* (to be clarified in Assumption 7). A much general uncertainty bound can be included in this class. This work is due to Corless and Leitmann (1983, 1984).

Consider the following class of uncertain systems:

$$\dot{x}(t) = Ax(t) + \Delta f(x(t), \sigma(t), t) + [B(x(t), t) + \Delta B(x(t), \sigma(t), t)]u(t), \tag{112}$$

where $t \in \mathbf{R}$, $x(t) \in \mathbf{R}^n$ is the state, $\sigma(t) \in \Sigma \subset \mathbf{R}^s$ is the uncertain parameter, and $u(t) \in \mathbf{R}^m$ is the control. The matrices A, $\Delta f(x, \sigma, t)$, $B(x, t)$, and $\Delta B(x, \sigma, t)$ are of appropriate dimensions. The matrix A and the function $B(\cdot)$ are both *known*. The functions $\Delta f(\cdot)$ and $\Delta B(\cdot)$ are *uncertain*. That is, they are not assumed known but are assumed to satisfy certain conditions (to be stated in Assumption 7). We now state the following assumptions.

Assumption 6: *The matrix A is Hurwitz.*

Assumption 7: *(1) The functions $\Delta f(\cdot)$, $B(\cdot)$, and $\Delta B(\cdot)$ are Caratheodory.*

(2) There exist uncertain functions $h(\cdot)$ and $E(\cdot)$ such that for all $(x, \sigma, t) \in \mathbf{R}^n \times \Sigma \times \mathbf{R}$

$$\Delta f(x(t), \sigma(t), t) = B(x(t), t) h(x, \sigma(t), t), \tag{113}$$

$$\Delta B(x(t), \sigma(t), t) = B(x(t), t) E(x, \sigma(t), t). \tag{114}$$

(3) There exists a (unknown) constant λ such that for all $(x, \sigma, t) \in \mathbf{R}^n \times \Sigma \times \mathbf{R}$,

$$\frac{1}{2} \min_{\sigma \in \Sigma} \lambda_m \left[E^T(x, \sigma, t) + E(x, \sigma, t) \right] \geq \lambda > -1. \tag{115}$$

(4) There exist an unknown constant vector $\beta \in (0, \infty)^q$ and a continuous function (known or unknown) $\rho(\cdot) : \mathbf{R}^n \times \mathbf{R} \times (0, \infty)^q$ such that for all $(x, \sigma, t) \in \mathbf{R}^n \times \Sigma \times \mathbf{R}$,

$$\max_{\sigma \in \Sigma} \| h(x, \sigma, t) \| \leq \rho(x, t, \beta). \tag{116}$$

The function $\rho(\cdot)$ is strongly Caratheodory (see Appendix for a definition).

(5) For each $(x, t) \in \mathbf{R}^n \times \mathbf{R}$, the function $\rho(x, t, \cdot) : (0, \infty)^q \to \mathbf{R}_+$ is C^1, concave, and non-decreasing with respect to each coordinate of its argument, β. Here the concave property means that for any β^1, $\beta^2 \in (0, \infty)^q$,

$$\rho(x, t, \beta^1) - \rho(x, t, \beta^2) \leq \frac{\partial \rho}{\partial \beta}(x, t, \beta^2)(\beta^1 - \beta^2). \tag{117}$$

The function $\partial \rho / \partial \beta$ is strongly Caratheodory. ◇◇

Assumption 6 imposes a condition on the *nominal* portion of the system (112). Notice that for a given system, the way of choosing A and $\Delta f(x, \sigma, t)$, etc. is not unique. This thus suggests the way of decomposing the uncertain system (112) (that is, one first "chooses" a matrix A and then lumps the rest of the system portion in $\Delta f(x, \sigma, t)$).

The condition on the constant λ assures that the adaptive robust control proposed later for the system (112) can "act" in the desired direction. That is, the control direction is not deteriorated by the uncertain portion $E(x, t)$. However, it is possible that the control magnitude is affected by the uncertain portion $E(x, t)$. The positive constant vector β can be interpreted as is related to the bound of $\Delta f(x,\sigma,t)$ and $\Delta B(x,\sigma,t)$. However, the relationship may not be direct. Notice that the dimension of β (i.e., q) can be different from that of $\Delta f(x,\sigma,t)$, etc. That β is unknown reflects the bound is unknown. However, certain properties (shown in Assumption 7(5)) on how the system *depends on* β are known.

We now propose the following class of adaptive robust controls:

$$u(t) = p(x(t), t, \hat{\beta}(t), \epsilon(t)), \tag{118}$$

$$p(x,t,\hat{\beta},\epsilon) = -\frac{\mu(x,t,\hat{\beta})}{\|\mu(x,t,\hat{\beta})\|}\rho(x,t,\hat{\beta}) \qquad \text{if } \|\mu(x,t,\hat{\beta})\| > \epsilon, \tag{119}$$

$$p(x,t,\hat{\beta},\epsilon) = -\frac{\mu(x,t,\hat{\beta})}{\epsilon}\rho(x,t,\hat{\beta}) \qquad \text{if } \|\mu(x,t,\hat{\beta})\| \leq \epsilon, \tag{120}$$

$$\dot{\hat{\beta}}(t) = L\frac{\partial \rho^T}{\partial \beta}(x(t), t, \hat{\beta}(t))\|\nu(x(t), t)\|, \tag{121}$$

$$\dot{\epsilon}(t) = -l\epsilon(t), \tag{122}$$

$$\hat{\beta}(t_0) \in (0,\infty)^q, \qquad \epsilon(t_0) \in (0,\infty), \tag{123}$$

where $\nu(x,t) := B(x,t)^T Px$, $P > 0$ is the solution of the Lyapunov equation $A^T P + PA + Q = 0$, $Q > 0$, $\mu(x,t,\hat{\beta}) := \nu(x,t)\rho(x,t,\hat{\beta})$, $L \in \mathbf{R}^{q\times q}$ is diagonal with positive elements, and $l > 0$.

The controlled system and the adaptive scheme can then be expressed as follows:

$$\begin{aligned} \dot{x}(t) =& Ax(t) + \Delta f(x(t), \sigma(t), t) \\ &+ [B(x(t), t) + \Delta B(x(t), \sigma(t), t)]p(x(t), t, \hat{\beta}(t), \epsilon(t)), \end{aligned} \tag{124}$$

$$\dot{\hat{\beta}}(t) = L\frac{\partial \rho^T}{\partial \beta}(x(t), t, \hat{\beta}(t))\|\nu(x(t), t)\|, \tag{125}$$

$$\dot{\epsilon}(t) = -l\epsilon(t). \tag{126}$$

The resulting controlled system performance is described as follows. We first define the parameter "estimate"† vector

$$\hat{\xi}(t) = (\hat{\beta}(t)^T \ \epsilon(t))^T, \tag{127}$$

---

† This is not to be interpreted as $\hat{\xi}(t) \to \xi$. All this means is that $\hat{\beta}$ is used in the control scheme (118) in place of β.

$$\hat{\xi}(t_0) = (\hat{\beta}(t_0)^T \ \epsilon(t_0))^T \in (0, \infty)^{q+1}, \tag{128}$$

and the parameter vector

$$\xi := (\beta^T \ 0)^T. \tag{129}$$

Theorem 7: *Consider the dynamic system described by (112). Suppose that Assumptions 6 and 7 are met. As the adaptive robust control (118-123) is applied to (112), the resulting controlled system and the adaptive scheme can be described by (124-126) and has the following properties.*

(i) Existence of solutions. For each $(x_0, t_0, \hat{\xi}(t_0)) \in \mathbf{R}^n \times \mathbf{R} \times (0, \infty)^{q+1}$ *there exists a solution* $(x(\cdot), \hat{\xi}(\cdot)) : [t_0, t_1) \to \mathbf{R}^n \times (0, \infty)^{q+1}$ *of (124-126) with* $(x(t_0), \hat{\xi}(t_0)) = (x_0, \hat{\xi}_0)$.

(ii) Uniform stability of $(0, \xi)$. *For each* $\eta > 0$ *there exists* $\varsigma > 0$ *such that if* $(x(\cdot), \hat{\xi}(\cdot))$ *is any solution of (124-126) with* $\|x(t_0)\|$, $\|\hat{\xi}(t_0) - \xi\| < \varsigma$ *then* $\|x(t)\|$, $\|\hat{\xi}(t) - \xi\| < \eta$ *for all* $t \in [t_0, t_1)$.

(iii) Uniform boundedness of solutions. For each r_1, $r_2 > 0$ *there exist* $d_1(r_1, r_2)$, $d_2(r_1, r_2) \geq 0$ *such that if* $(x(\cdot), \hat{\xi}(\cdot))$ *is any solution of (124-126) with* $\|x(t_0)\| \leq r_1$ *and* $\|\hat{\xi}(t_0) - \xi\| \leq r_2$ *then* $\|x(t)\| \leq d_1(r_1, r_2)$ *and* $\|\hat{\xi}(t) - \xi\| \leq d_2(r_1, r_2)$ *for all* $t \in [t_0, t_1)$.

(iv) Extension of solutions. Every solution of (124-126) can be extended into a solution defined on $[t_0, \infty)$.

(v) Convergence of $x(\cdot)$ *to zero. If* $(x(\cdot), \hat{\xi}(\cdot)) : [t_0, \infty) \to \mathbf{R}^n \times (0, \infty)^{q+1}$ *is a solution of (124-126) then*

$$\lim_{t \to \infty} x(t) = 0. \tag{130}$$

Proof: See Corless and Leitmann (1984). ◇◇

If one considers the cone-boundedness property (7) as a *linear* structure of the bound, the current configuration of the bound, which is depicted by β in $\rho(x, t, \beta)$, as a *nonlinear* structure.

The adaptive robust control is of saturation type. The direction of the control is pre-specified to be $-\mu$. The magnitude of the control is however determined by whether the state variable x is outside of the saturation region $\|\mu\| = \epsilon$ (hence (119)) or inside (hence (120)). The control is designed without knowing the uncertainty. It is *guaranteed* that the state $x(t)$ converges to zero.

There are certain design parameters involved. The value of L determines the rate of adaptation. A larger value of L implies a faster learning rate. It can be shown that if the initial condition $\hat{\xi}(t_0)$ is chosen positive (which is required in this adaptive scheme) then the parameter "estimate" vector $\hat{\xi}(t)$ remains positive for all $t \in [t_0, \infty)$ (Corless and Leitmann 1984). This fits the physical interpretation of $\hat{\beta}$ (which is a part of $\hat{\xi}$) that it is related to the bound of the uncertainty. The parameter $\epsilon(t)$ is governed by (122) which is decoupled with the state variable $x(t)$ and the adaptive parameter $\hat{\beta}(t)$. The value of ϵ determines the size of the saturation region $\|\mu\| = \epsilon$ for the adaptive robust control (118). In fact (122) shows that $\epsilon(t)$ converges to zero asymptotically. The choice of the initial condition $\epsilon(t_0)$ and the constant l is arbitrary (as long as they are positive). Hence one can manipulate the value of ϵ (at least initially) and hence the size of the saturation region in practical application.

X. Application to a single zone HVAC system

Heating, ventilation, and air-conditioning (HVAC) systems are comprised of a large number of subsystems, each of which may exhibit time-varying and/or nonlinear characteristics. For example, a detailed description of the dynamics of a typical five-zone commercial HVAC system requires on the order of 1,000 differential and algebraic equations (Kelly *et al.* 1984). Furthermore, the parameters of this dynamical description generally vary with load, weather, and building occupancy. These complexities suggest that the use of some simple control schemes (such as the on-off control which many HVAC systems are still using) may not be appropriate for some of the new load-management technologies and systems. A brief overview on the HVAC control development is in order. A system model includes a complete set of HVAC components. Stoecker (1976) modeled an HVAC system with polynomial expressions whose coefficients were determined through experimental or on-site performance data. This formulation is quite valuable for estimating the steady-state operation of an HVAC system. A dynamical model was developed by Thompson and Chen (1979) which included transfer function expressions for various HVAC components. These components were strung together to model an HVAC system. Thompson (1981) later modified the thermostat module. Although these researchers developed a digital simulation scheme to identify energy sensitive parameters, they never studied the effects of system dynamics. Mehta (1984) described the concept of a rational model, which includes the dynamic interaction between the HVAC system and the heating/cooling loads. This approach had been suggested by an earlier successful experimental validation (Mehta and Woods 1980) of HVAC models, obtained by linking proper modular blocks. Kaya (1976, 1979, 1981) and Kaya *et al.* (1982) tackled the problem of the optimal control formulation. Temperature, humidity, and air velocity were considered as three major comfort variables, and the comfort condition was treated as a region. A two-step optimization procedure (static and dynamic) was discussed. Sud (1984) discussed a three-step optimization procedure which included operational modes and control hierarchies. Schumann (1980) presented a simple air-conditioning system using a parameter-adaptive deadbeat controller and a parameter-adaptive optimal state controller. There were also works devoted to the parameter estimation issue which was considered as a major step toward the use of adaptive control. Diderrich and Kelly (1984) described the use of Kalman filtering methods for the failure detection of HVAC sensors. Forrester and Wepfer (1984) and Li and Wepfer (1985) applied off-line least square estimation schemes to data taken from a large commercial office building and developed load prediction algorithm. Later Li and Wepfer (1987) also developed an on-line recursive estimation methods for a multi-input multi-output HVAC system.

This section considers the control issue for an HVAC system which possesses modelling uncertainty and nonlinearity. The uncertainty may be due to thermal storage effect, heat and moisture generation (by the people), and outside temperature and humidity. These may be unpredictable for a generic room. It is difficult in practice to collect statistical information of the uncertainty. This is since buildings may have different situations (for loading, materials, etc.) as well as different managerial policies. Statistical information, if it is to be reliable, must be gathered for each special application and very likely needs to be updated for each period. The cost involved in this collection procedure and the resulting controller adjustment may become a concern for builders. An alternative view is to consider the uncertainty to be *bounded* but the bound is *unknown*. This falls directly into the framework currently addressed. The adaptive robust control introduced in the last section is capable of compensating the uncertainty. In order to simplify the formulation and hence to emphasize the feature of the control algorithm, only a single-zone HVAC system is considered in this section. However, the extension to multi-zone HVAC system is straightforward. This part of work is also reported in Chen and Lee (1991).

A. Modelling

A single zone HVAC system in a generic room is considered. The system is constructed by a direct application of the energy conservation principle. Assumptions adopted here for the modelling include ideal gas behavior, perfect mixing, negligible radiative heat transfer, and constant pressure.

Conservation of energy leads to (Li and Wepfer 1987):

$$\begin{pmatrix}\text{energy}\\ \text{stored}\\ \text{in room}\end{pmatrix} = \begin{pmatrix}\text{energy in}\\ \text{via air}\\ \text{supply}\end{pmatrix} + \begin{pmatrix}\text{heat con-}\\ \text{duction}\\ \text{via walls}\end{pmatrix} + \begin{pmatrix}\text{heat in}\\ \text{due to}\\ \text{occupants}\end{pmatrix} - \begin{pmatrix}\text{energy}\\ \text{loss via}\\ \text{return air}\end{pmatrix} \tag{131}$$

Converting the above relationships to symbols (with some terms combined) yields:

$$\frac{dT}{dt} = -\frac{h_{fg} + c_{pw}T}{(V\rho)_{rm}c_p}\left[(F\rho)_e(W_e - W) + \dot{m}_O\right] + \frac{(F\rho)_e(h_e - h)}{(V\rho)_{rm}c_p} + \frac{UA(T_O - T)}{(V\rho)_{rm}c_p} + \frac{\dot{Q}_O}{(V\rho)_{rm}c_p} \tag{132}$$

where all mass-specific quantities are given per kilogram of dry air. Conservation of mass of moisture in the air leads to:

$$\begin{pmatrix}\text{moisture}\\ \text{increase}\\ \text{in air}\end{pmatrix} = \begin{pmatrix}\text{moisture}\\ \text{in via}\\ \text{supply air}\end{pmatrix} + \begin{pmatrix}\text{moisture}\\ \text{in via}\\ \text{occupants}\end{pmatrix} - \begin{pmatrix}\text{moisture}\\ \text{out via}\\ \text{return air}\end{pmatrix} \tag{133}$$

In terms of symbols this becomes:

$$\frac{dW}{dt} = \frac{(F\rho)_e}{(V\rho)_{rm}}(W_e - W) + \frac{\dot{m}_O}{(V\rho)_{rm}} \tag{134}$$

The following two relationships hold:

$$c_p = \sum_i c_{pi} \tag{135}$$

$$\rho = \frac{P}{R(T+273)}\frac{0.622}{(0.622+W)} \sim \text{constant (humid air density)} \quad (136)$$

Here in (135) the heat capacity is the summation of that of air, wall, furniture, equipment, etc.

According to the ASHRAE standards (ASHRAE 1981), the comfort region (in terms of temperature and humidity) may be approximated as shown in Figure 4. Based on this, we now define the following state variables:

$$x_1 = T^*, \qquad T^* = \frac{T - T_{med}}{T_{max} - T_{min}} \quad (137)$$

$$x_2 = W^*, \qquad W^* = \frac{W - W_{med}}{W_{max} - W_{min}} \quad (138)$$

$$u_1 = T_e^*, \qquad T_e^* = \frac{T_e - T_{med}}{T_{max} - T_{min}} \quad (139)$$

$$u_2 = W_e^*, \qquad W_e^* = \frac{W_e - W_{med}}{W_{max} - W_{min}} \quad (140)$$

where "max" and "min" refer to the upper and lower boundaries of the comfort region as shown in Figure 4, $T_{med} = \frac{1}{2}(T_{max} + T_{min})$, and $W_{med} = \frac{1}{2}(W_{max} + W_{min})$. Then (132) and (134) can be converted to the following state space form:

$$\begin{bmatrix} \dot{x}_1 \\ \dot{x}_2 \end{bmatrix} = \begin{bmatrix} -(\alpha_1 + 1) & \gamma\Delta W \frac{T_{med}}{\Delta T} + \alpha_2 \\ 0 & -1 \end{bmatrix} \begin{bmatrix} x_1 \\ x_2 \end{bmatrix}$$
$$+ \begin{bmatrix} 1 & -\alpha_2 - \gamma\Delta W(x_1 + \frac{T_{med}}{\Delta T}) \\ 0 & 1 \end{bmatrix} \begin{bmatrix} u_1 \\ u_2 \end{bmatrix} \quad (141)$$
$$+ \begin{bmatrix} \alpha_1 & 1 & -\alpha_2 - \gamma\Delta W(x_1 + \frac{T_{med}}{\Delta T}) \\ 0 & 0 & 1 \end{bmatrix} \begin{bmatrix} u_3 \\ u_4 \\ u_5 \end{bmatrix} + \begin{bmatrix} \gamma\Delta W x_1 x_2 \\ 0 \end{bmatrix}$$

where $\dot{x}_1 = dx_1/dt^*$, $\dot{x}_2 = dx_2/dt^*$, $t^* = tF_e/V_{rm}$, $\Delta T = T_{max} - T_{min}$, $\Delta W = W_{max} - W_{min}$,

$$\alpha_1 = \frac{UA}{\rho c_p F_e}, \qquad \alpha_2 = \frac{h_{fg}\Delta W}{c_p \Delta T}, \qquad \gamma = \frac{c_{pw}}{c_p}, \quad (142)$$

$$u_3 = \frac{T_0 - T_{med}}{\Delta T}, \qquad u_4 = \frac{\dot{Q}_0}{c_p F_e \Delta T}, \qquad u_5 = \frac{\dot{m}_0}{F_e \rho \Delta W}. \quad (143)$$

This is a nonlinear system. Moreover, the control u_2 is coupled with the state x_1. In practice for a generic room the thermal storage effects (which may be due to wall, equipment, and furniture, etc.) should not be ignored. However, these are difficult to model in a very precise way. Hence it is realistic to face the fact that the model (141) possesses certain degree of uncertainty. We shall treat the heat capacity c_p as an uncertain parameter. To be more specific, let

$$c_p = c_p^{nominal} + \Delta c_p(t) \tag{144}$$

where the nominal value of c_p is a known constant. However, the uncertain portion $\Delta c_p(t)$ is time-varying and unpredictable. It is reasonable to assume that $\Delta c_p(t)$ is *bounded.* The bound is however *unknown.* In addition to c_p, there are other uncertainties in the model. These include u_3, u_4, and u_5. Here u_3 is determined by the outside temperature T_0 which may be varying due to weather change. The other two inputs u_4 and u_5 are related to the heat and moisture generation (through $\dot{Q}_0$ and $\dot{m}_0$). We shall also treat u_3, u_4, and u_5 as uncertainties. These are time-varying and their changes are unpredictable. However, it is again realistic to assume that the variations (around certain nominal values) are *bound.* The bound is however *unknown.*

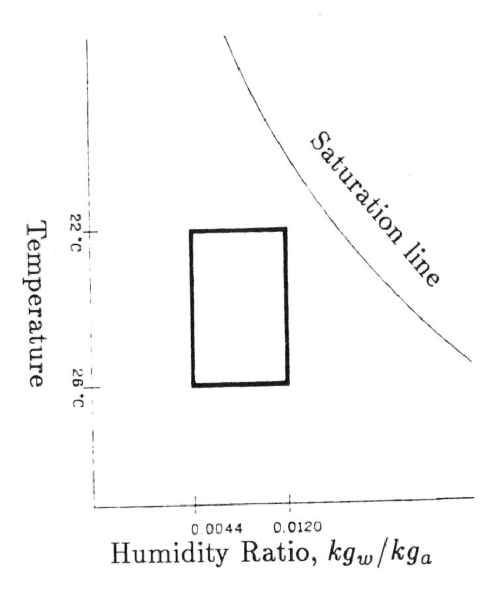

Figure 4: The comfort zone

B. Control

We now apply the adaptive robust control to the single zone HVAC system. Assume both room temperature and humidity measurement are available. We first decompose the system (141) into the nominal and uncertain portions. Based on the decomposition of c_p performed in (144), we are able to decompose α_1, α_2, and γ such that

$$\alpha_1 = \alpha_1^{nominal} + \Delta\alpha_1(t), \tag{145}$$

$$\alpha_1^{nominal} = \frac{UA}{\rho c_p^{nominal} F_e}, \tag{146}$$

$$\Delta\alpha_1(t) = \frac{UA}{\rho F_e} \frac{-\Delta c_p(t)}{(c_p^{nominal} + \Delta c_p(t))c_p^{nominal}}, \tag{147}$$

$$\alpha_2 = \alpha_2^{nominal} + \Delta\alpha_2(t), \tag{148}$$

$$\alpha_2^{nominal} = \frac{h_{fg}\Delta W}{\Delta T c_p^{nominal}}, \tag{149}$$

$$\Delta\alpha_2(t) = \frac{h_{fg}\Delta W}{\Delta T} \frac{-\Delta c_p(t)}{(c_p^{nominal} + \Delta c_p(t))c_p^{nominal}}, \tag{150}$$

$$\gamma = \gamma^{nominal} + \Delta\gamma(t), \tag{151}$$

$$\gamma^{nominal} = \frac{c_{pw}\Delta W}{c_p^{nominal}}, \qquad \Delta\gamma(t) = \frac{c_{pw}\Delta W(-\Delta c_p(t))}{(c_p^{nominal} + \Delta c_p(t))c_p^{nominal}}. \tag{152}$$

Notice again that $\Delta c_p(t)$ is *unknown*. No deterministic or statistical information is assumed. Comparing (141) with (112), we choose

$$A = \begin{bmatrix} -(\alpha_1^{nominal} + 1) & \gamma^{nominal}\Delta W \frac{T_{med}}{\Delta T} + \alpha_2 \\ 0 & -1 \end{bmatrix}, \tag{153}$$

$$\Delta f(x,t) = \begin{bmatrix} -\Delta\alpha_1(t) & \Delta\gamma(t)\Delta W \frac{T_{med}}{\Delta T} + \Delta\alpha_2(t) \\ 0 & 0 \end{bmatrix} \begin{bmatrix} x_1 \\ x_2 \end{bmatrix} \tag{154}$$

$$+ \begin{bmatrix} \alpha_1 & 1 & -\alpha_2 - \gamma\Delta W(x_1 + \frac{T_{med}}{\Delta T}) \\ 0 & 0 & 1 \end{bmatrix} \begin{bmatrix} u_3 \\ u_4 \\ u_5 \end{bmatrix} + \begin{bmatrix} \gamma\Delta W x_1 x_2 \\ 0 \end{bmatrix},$$

$$B(x,t) = \begin{bmatrix} 1 & -\alpha_2^{nominal} - \gamma^{nominal}\Delta W(x_1 + \frac{T_{med}}{\Delta T}) \\ 0 & 1 \end{bmatrix}, \tag{155}$$

$$\Delta B(x,t) = \begin{bmatrix} 0 & -\Delta\alpha_2(t) - \Delta\gamma(t)\Delta W(x_1 + \frac{T_{med}}{\Delta T}) \\ 0 & 0 \end{bmatrix}. \tag{156}$$

In order to implement the control, we need to show the satisfactions of Assumptions 6 and 7. The matrix A is Hurwitz. The matching condition (113,114) is met since the square matrix B in (155) is non-singular (hence one may choose $h = B^{-1}\Delta f$ and $E = B^{-1}\Delta B$). Assumption 7(3) is verified by restricting the operating condition. This is since E depends on x and in practice one only has to satisfy (115) for all system solution $x(\cdot)$. Next we need to perform the bounding analysis on $h(x,t)$. It is sufficient to show that $\Delta f(x,t)$ satisfies Assumption 7(4,5) since $\|h(x,t)\| \leq \|B^{-1}\|\|\Delta f(x,t)\|$ and $\|B^{-1}\| = 1$ (although B depends on x_1). We first show that the first two terms on the right-hand side of (154) are cone-bounded. That is, there exist (unknown) constants β_2 and β_3 such that

$$\|\Psi_1 x + \Psi_2(x)\bar{u}\| \leq \beta_2\|x\| + \beta_3, \tag{157}$$

where

$$\Psi_1 x = \begin{bmatrix} -\Delta\alpha_1(t) & \Delta\gamma(t)\Delta W(\frac{T_{med}}{\Delta T} + \Delta\alpha_2(t)) \\ 0 & 0 \end{bmatrix} \begin{bmatrix} x_1 \\ x_2 \end{bmatrix}, \tag{158}$$

$$\Psi_2(x) = \begin{bmatrix} \alpha_1 & 1 & -\alpha_2 - \gamma\Delta W(x_1 + \frac{T_{med}}{\Delta T}) \\ 0 & 0 & 1 \end{bmatrix}, \tag{159}$$

$$\bar{u} = \begin{bmatrix} u_3 \\ u_4 \\ u_5 \end{bmatrix}. \tag{160}$$

It is clear that

$$\|\Psi_1 x\| \leq \|\Psi_1\|\|x\|. \tag{161}$$

We separate the matrix $\Psi_2(x)$ into two parts:

$$\begin{aligned} \Psi_2(x) &= \begin{bmatrix} \alpha_1 & 1 & -\alpha_2 - \gamma\Delta W\frac{T_{med}}{\Delta T} \\ 0 & 0 & 1 \end{bmatrix} \\ &\quad + \begin{bmatrix} 0 & 0 & -\gamma\Delta W \\ 0 & 0 & 0 \end{bmatrix} \begin{bmatrix} 0 & 0 & 0 \\ 0 & 0 & x_2 \\ 0 & 0 & x_1 \end{bmatrix} \\ &= \Psi_{21} + \Psi_{22} \begin{bmatrix} 0 & 0 & 0 \\ 0 & 0 & x_2 \\ 0 & 0 & x_1 \end{bmatrix} \end{aligned} \tag{162}$$

Then a simple bounding analysis yields

$$\|\Psi_2(x)\bar{u}\| \leq \max \|\bar{u}\|(\|\Psi_{21}\| + \|\Psi_{22}\|\|x\|) \tag{163}$$

The cone-boundedness property (157) is proven. In fact, we can choose the constants β_2 and β_3 to be

$$\beta_2 = \|\Psi_1\| + \max \|\bar{u}\|\|\Psi_{22}\| \tag{164}$$

$$\beta_3 = \max \|\bar{u}\|\|\Psi_{21}\| \tag{165}$$

The third term on the right-hand side of (154) is bounded by $\|x\|^2$ since $2|x_1 x_2| \leq x_1^2 + x_2^2 = \|x\|^2$ and hence

$$\|\Psi_3(x)\| \leq \max|\gamma \Delta W|\|x\|^2 := \beta_1 \|x\|^2, \tag{166}$$

where

$$\Psi_3(x) = \begin{bmatrix} \gamma \Delta W x_1 x_2 \\ 0 \end{bmatrix}. \tag{167}$$

Combining (157) and (166) we conclude that

$$\|h(x,t)\| \leq \beta_1 \|x\|^2 + \beta_2 \|x\| + \beta_3 := \rho(x,t,\beta) \tag{168}$$

It is then easy to check that Assumption 7(4,5) is met. The adaptive scheme is now constructed as follows:

$$\dot{\hat{\beta}}_1(t) = L_1 \|\nu(x)\|\|x\|^2 \tag{169}$$

$$\dot{\hat{\beta}}_2(t) = L_2 \|\nu(x)\|\|x\| \tag{170}$$

$$\dot{\hat{\beta}}_3(t) = L_3 \|\nu(x)\| \tag{171}$$

where L_1, L_2, and L_3 are the diagonal elements of the matrix L. The adaptive robust control for the single zone HVAC system is given by (118) with

$$\rho(x,t,\beta) = \beta_1 \|x\|^2 + \beta_2 \|x\| + \beta_3 \tag{172}$$

where $\beta = [\beta_1 \ \beta_2 \ \beta_3]^T$.

Computer simulations are performed for system analysis. The parameters for the single zone prototype are summarized in Tables 1 and 2. The uncertain parameters (i.e., $\dot{Q}_0(t)$, $\dot{m}_0(t)$, $T_0(t)$, and $c_p(t)$) are decomposed with their nominal values given in Table 1. Their uncertain portions are given in the following form for simulation purposes:

$$\Delta \dot{Q}_0(t) = a_1 + b_1 \; sin \left[\frac{2\pi t}{c_1}\right] + d_1 \; norm(t) \tag{173}$$

$$\Delta \dot{m}_0(t) = a_2 + b_2 \; sin \left[\frac{2\pi t}{c_2}\right] + d_2 \; norm(t) \tag{174}$$

$$\Delta T_0(t) = a_3 + b_3 \; sin \left[\frac{2\pi t}{c_3}\right] + d_3 \; norm(t) \tag{175}$$

$$\Delta c_p(t) = a_4 + b_4 \; sin \left[\frac{2\pi t}{c_4}\right] + d_4 \left[rect(t) - 0.5\right] \tag{176}$$

where $norm(t)$ is a random number with mean equal to zero and standard deviation equal to 1 and $rect(t)$ is a random number with rectangular distribution in the interval $[0, \; 1]$. The purpose of using these forms of functions for the uncertainty in simulations is to consider the combinations of various practical situations, including constant uncertainty, high frequency periodic uncertainty, and random uncertainty. The numerical values of the parameters a_1, etc. which are adopted for simulations are summarized in Table 2. Moreover, we take the outside temperature $T_0 = 35^oC$, the heat gain $\dot{Q}_0 = 75 \; W$ hourly, and the moisture load $\dot{m}_0 = 3 \times 10^{-5} \; kg/s$ hourly.

For comparison purpose, an on-off control (which many current HVAC systems are still using) is also implemented for the single zone system under the same uncertainty. The control is given in the following form:

$$u_1(t) = \left\{ \begin{array}{l} -u_{1max} \text{ if } x_1 > \delta_1 \\ 0 \text{ if } -\delta_1 \leq x_1 \leq \delta_1 \\ u_{1max} \text{ if } x_1 < -\delta_1 \end{array} \right\}, \tag{177}$$

$$u_2(t) = \left\{ \begin{array}{l} -u_{2max} \text{ if } x_2 > \delta_2 \\ 0 \text{ if } -\delta_2 \leq x_2 \leq \delta_2 \\ u_{2max} \text{ if } x_2 < -\delta_2 \end{array} \right\}, \tag{178}$$

where u_{imax}, $i = 1, 2$, is the maximum control magnitude, $[-\delta_i, \; \delta_i]$ is the dead-zone (in terms of the control action). For simulation purpose, we take $u_{1max} = 500$, $u_{2max} = 300$, $\delta_1 = 0.5$, and $\delta_2 = 0.5$. The other parameters chosen for the adaptive robust control (118) are: $L_1 = L_2 = L_3 = 1$, $l = 0.01$, $\epsilon(0) = 10$, $\hat{\beta}_1(0) = \hat{\beta}_2(0) = \hat{\beta}_3(0) = 10$.

A	cross-sectional area (m^2)
e	subscript, supply air
F	volumetric flow rate (m^3/s)
h_{fg}	latent heat of water (J/kg)
h	enthalpy $(J/kg - dry\ air)$
$\dot{m}_0$	internal moisture load (kg/s)
P	pressure (Pa)
$\dot{Q}_0$	internal heat load (W)
R	specific gas constant (J/kg^oK)
rm	subscript, room
t	time (s)
T	temperature (^oC)
U	heat transfer coeff. $(W/m^{2o}K)$
V	volume (m^3)
W	moisture/dry air ratio by mass
ρ	air density

Type	Room
U value, wall $(W/m^2\ K)$	1.42
U value, window $(W/m^2\ K)$	6.42
UA value (W/K)	22.78
h_{fg} (J/kg)	2.501E+06
P (MPa)	0.101
ΔT (K)	4
ΔW (kg_w/kg_a)	0.0076
Work Intensity	Light
V_{rm} (m^3)	27
F_e (m^3/s)	9.44E-03
c_{pa} $(J/kg\ K)$	1005
c_{pw} $(J/kg\ K)$	1820
R_{air} $(J/kg\ K)$	287
ρ_{air} (kg/m^3)	1.2
$100t^*$ Time scale (s)	28.60

Table 1: Descriptive Parameters for the Single-zone Prototype

		Index Number			
		1	2	3	4
Parameter Name	a	3	0.00001	2	300
	b	10	0.00001	8	300
	c	0.01	1	1	1
	d	1	0.00001	1	30

Table 2: Parameters Used for Uncertainty

The initial conditions of the system are chosen to be $x_1(0) = 6$ and $x_2(0) = 4$. This significant deviation from the comfort region is intended to test the recovery capability of the control system in a severe situation. Figures 5-10 depict the system and control performance. The time axis is scaled such that it is equivalent to $100tF_e/V_{rm}$. In other words, a unit time in the figure is equivalent to 28.60 seconds based on the prototype parameters in Table 1. Figure 5 is the state performance under no control. Figure 6 is due to the use of the on-off control. Figures 7-10 are due the use of the adaptive robust control. It is interesting to note that the system performance due to the control (177,178) has a very high temperature overshoot (shown as curve 1 in Figure 6). This is mainly due to the coupling of the control u_2 with x_1 (as shown in (141)). However, the steady state performance is rather satisfactory. On the other hand, the system performance due to the adaptive robust control shows a much less overshoot (Figure 9). The steady state performance has certain oscillations. Due to the practical need for room comfort, it is usually more important to be able to maintain small overshoot than to have a slight improvement in the steady state performance. A human body can not always tell the difference of 1°C (which is about the difference between the steady state values of Figures 6 and 7). However, the overshoot in Figure 6 certainly indicates a significant discomfort. This comparison in fact also suggests the practical need for a realistic room temperature-humidity control system. It is more important that the control system is *robust* against the uncertainty (in the sense of maintaining small overshoot) than showing a slight improvement in the steady state performance.

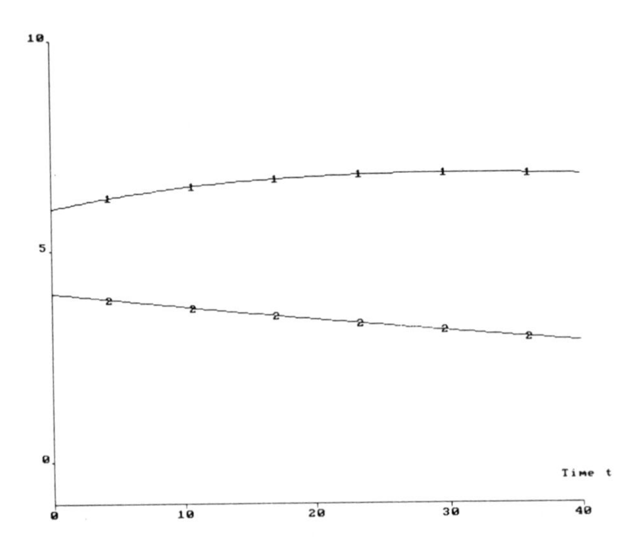

Figure 5: System performance (x_1 and x_2), without control

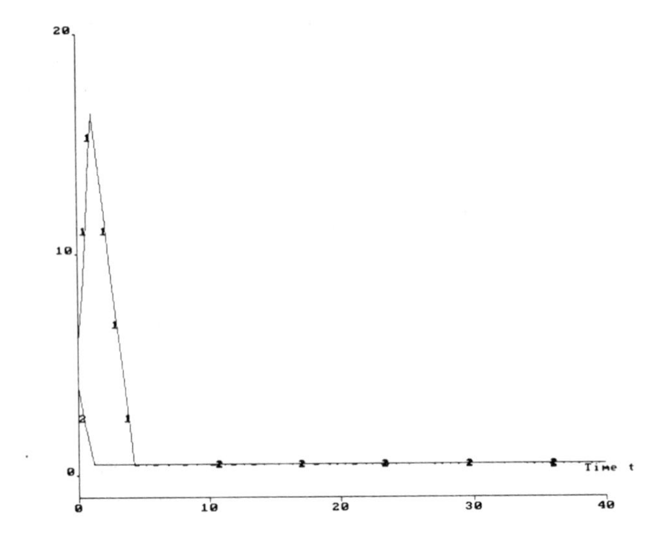

Figure 6: System performance (x_1 and x_2), under on-off control

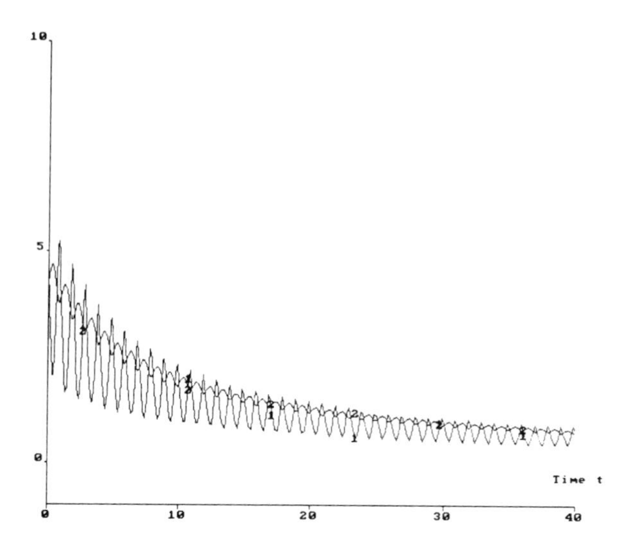

Figure 7: System performance (x_1 and x_2), under adaptive robust control

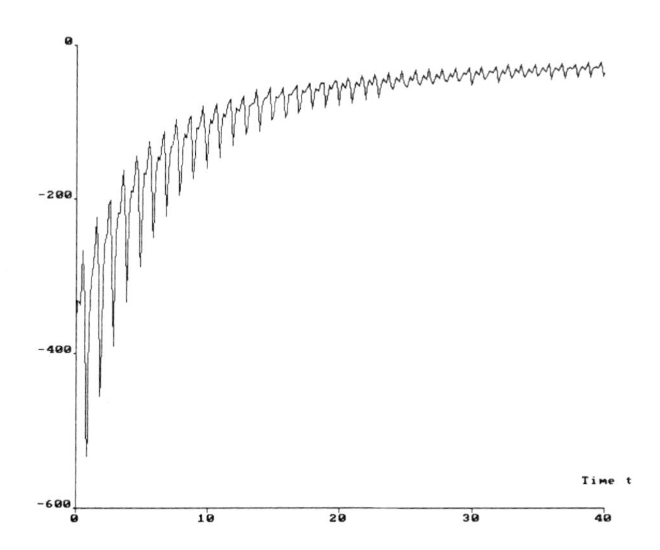

Figure 8: Control history (u_1), adaptive robust control

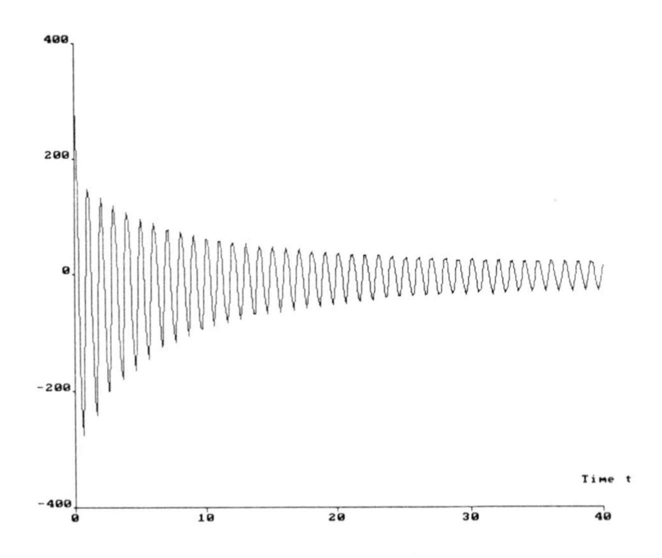

Figure 9: Control history (u_2), adaptive robust control

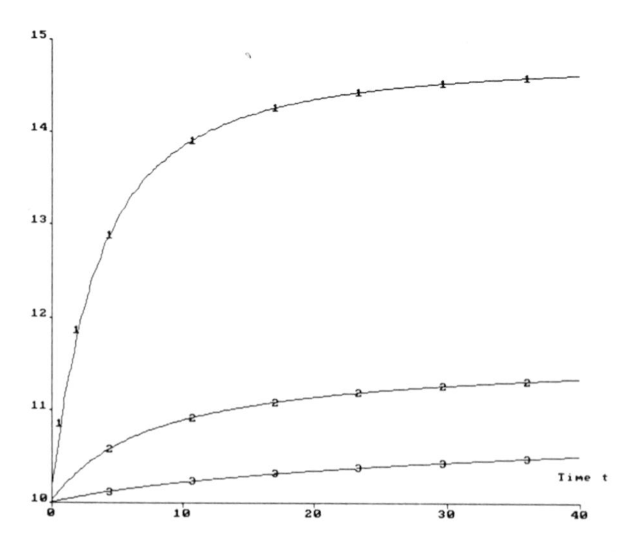

Figure 10: Adaptive parameters history ($\hat{\beta}_1$, $\hat{\beta}_2$, $\hat{\beta}_3$)

XI. User's summary

We adopt the so-called deterministic approach for control design as well as uncertainty treatment. This approach treats the uncertainty, which may vary with time, by only using its possible bound. So far, the approach is one of the very few alternatives in the control society that are able to handle *nonlinear* systems with *(possibly fast) time varying* uncertainty. We emphasize that the procedure we used to tackle the control problem *starts with* a given uncertainty bound. In other words, the designer is first given the bound. Then an appropriate control is posed for it. This differs from some other commonly seen procedure where one studies how much uncertainty that a given controlled system can tolerate. That is, one starts with a control, not the uncertainty.

The control design involves two layers of concepts. First, one makes sure if a (non-adaptive) robust control exists as the uncertainty bound is known. This control may not be feasible if the bound is not available. However, this *existence inquiry* assures that a follow up effort, namely, the quest for an adaptive robust control, is not in vain. Second, if the (non-adaptive) robust control exists (that is, if one knows how to control the uncertain system once the bound is given), then a corresponding adaptive robust control can be applied. The first layer effort is well justified for it simply says the adaptive robust control certainly does not make something out of nothing.

The design steps for an adaptive robust output feedback controller are summarized as follows:

(i) Write down the mathematical model of the physical system. Identify the uncertainty and estimate its bound.

(ii) Separate the system equations into nominal part and uncertain part. Try every effort to satisfy the matching condition (2) and the SPR condition (30).

(iii) If the SPR condition is met and the output matrix does not have "uncertainty" (be *real* or *artificial*), as addressed in (41) and the beginning of Section VII, check whether the cone-boundedness condition (42) is satisfied. If yes, design the controller according to Theorem 3 ((43),(44)) based on the *matched part* of the uncertainty. If the cone-boundedness condition is not satisfied, we need full state feedback. Check whether Assumption 7 ((113)-(117)) is satisfied. If the answer is yes, use the controller described in (118-123).

If the uncertainty has mismatched part, the controlled system based on (43,44) is practically stable if (74) is satisfied.

(iv) If SPR condition is met but the output matrix has "uncertainty", design the controller according to Theorem 6 ((95),(96)). Remember to check the two conditions in Theorem 6.

Note that full state feedback can always be used to satisfy the SPR condition.

XII. Conclusions

Adaptive robust control has its origin from (non-adaptive) robust control where uncertainty bound is the primary design base. An adaptive scheme is then incorporated into this (non-adaptive) robust control. Therefore the control is first considered robust and then made adaptive. This is not to be confused with *robust adaptive control* where an adaptive control is made robust (Narendra and Annaswamy 1989).

In the adaptive robust control, the adaptive scheme generates the signal which is equivalently treated as the uncertainty bound. This *equivalent principle* is implicitly applied to all designs. However, caution must be made that there is neither proof nor intention to claim that the signal value is *equal* to the bound (stated at least in the limiting sense). In fact, chasing the bound value is not of too much significance in the current setting. The uncertainty we consider here is unknown and unpredictable in its nature. Therefore its maximum value always has the potential to become even larger as time proceeds. In other words, it is even difficult for one to tell what the bound is by keeping track of the past history of the uncertainty value. Despite this, the current adaptive robust control is able to generate an appropriate signal which in turn results in satisfactory performance. This is made possible by using another information of the system, namely, the functional dependence property of the bound (such as cone-boundedness or convexity).

Acknowledgement

This research was supported in part by the National Science Foundation under Grant MSS-9014714.

Appendix

A function $a(\cdot) : \mathbf{R} \to \mathbf{R}^s$ is Lebesgue measurable if it is continuous everywhere *except* on a set of measure zero, i.e., except at an atmost countable number of points.

Let D be a subset of $\mathbf{R}^q$. A function $f(\cdot) : D \times \mathbf{R} \to \mathbf{R}^s$ is Caratheodory if and only if for each $t \in \mathbf{R}$, $f(\cdot, t)$ is continuous; for each $z \in D$, $f(z, \cdot)$ is Lebesgue measurable; and for each compact subset $\mathcal{C}$ of $D \times \mathbf{R}$, there exists a Lebesgue integrable function $M_{\mathcal{C}}(\cdot)$ such that for all $(z,t) \in \mathcal{C}$, $\|f(z,t)\| \leq M_{\mathcal{C}}(t)$.

A function $f(\cdot) : D \times \mathbf{R} \to \mathbf{R}^s$ is strongly Caratheodory if and only if it is Caratheodory with $M_{\mathcal{C}}(\cdot)$ replaced by a constant $M_{\mathcal{C}}$.

References

ASHRAE, (1981). *ASHRAE Handbook - 1981 Fundamentals*, American Society of Heating, Refrigerating and Air-Conditioning Engineers, Inc., Atlanta, GA, pp. 8.27.

Barmish, B.R., Corless, M., and Leitmann, G. (1983). "A new class of stabilizing controllers for uncertain dynamical systems", *SIAM J. Control Optimization*, **21**, pp. 246-255.

Barmish, B.R., and Leitmann, G. (1982). "On ultimate boundedness control of uncertain systems in the absence of matching conditions", *IEEE Trans. Automatic Control*, **AC-27**, pp. 153-158.

Chen, Y.H. (1986). "On the deterministic performance of uncertain dynamical systems", *International J. Control*, **43**, pp. 1557-1579.

Chen, Y.H. (1987). "Robust output feedback controller: direct design", *International J. Control*, **46**, pp. 1083-1091.

Chen, Y.H. (1988). "Design of robust controllers for uncertain dynamical systems", *IEEE Trans. Automatic Control*, **AC-33**, pp. 487-491.

Chen, Y.H. (1989). "Modified adaptive robust control system design", *International J. Control*, **49**, pp. 1869-1882.

Chen, Y.H. (1990). "Robust control system design: non-adaptive versus adaptive", *International J. Control*, **51**, pp. 1457-1477.

Chen, Y.H. (1991). "Adaptive robust control system design: Using information related to the bound of uncertainty", *Control-Theory and Advanced Technology*, **7**, pp. 31-53.

Chen, Y.H., Lee, K.M. and Wepfer, W.J. (1990). "Adaptive robust control scheme applied to a single-zone HVAC system", *America Control Conference*, San Diego, CA, pp. 1076-1081.

Chen, Y.H., and Leitmann, G. (1987). "Robustness of uncertain systems in the absence of matching assumptions", *International J. Control*, **45**, pp. 1527-1542.

Coddington, E.A., and Levinson, N. (1955). *Theory of Ordinary Differential Equations*, McGraw-Hill, New York, NY.

Corless, M.J., and Leitmann, G. (1981). "Continuous state feedback guaranteeing uniform ultimate boundedness for uncertain dynamic systems", *IEEE Trans. Automatic Control*, **AC-26**, pp. 1139-1143.

Corless, M., and Leitmann, G. (1983). "Adaptive control of systems containing uncertain functions and unknown functions with uncertain bounds", *J. Optimization Theory Applications*, **41**, pp. 155-168.

Corless, M., and Leitmann, G. (1984). "Adaptive control for uncertain dynamical systems", In A. Blaquiere and G. Leitmann (Eds.), *Dynamical Systems and Microphysics: Control Theory and Mechanics*, Academic Press, New York, NY.

Desoer, C.A., and Vidyasagar, M. (1975). *Feedback Systems: Input-Output Properties*, Academic Press, New York, NY.

Diderrich, G.T., and Kelly, R.M. (1984). "Estimating and correcting sensor data in a chiller system: An application of Kalman filter theory", *ASHRAE Trans.*, **90**, Part 2B, pp. 511-522.

Forrester, J.R., and Wepfer, W.J. (1984). "Formulation of a load prediction algorithm for a large commercial building", *ASHRAE Trans.*, **90**, Part 2B, pp. 536-551.

Gutman, S. (1979). "Uncertain dynamical systems - a Lyapunov min-max approach", *IEEE Trans. Automatic Control*, **AC-24**, pp. 437-443.

Hahn, W. (1967). *Stability of Motion*, Spring-Verlag, New York, NY.

Hale, J.K. (1980). *Ordinary Differential Equations*, Krieger, Huntington.

Ioannou, P.A., and Kokotovic, P.V. (1983). *Adaptive Systems with Reduced Models*, Springer-Verlag, New York, NY.

Kaya, A. (1976). "Analytical techniques for controller design", *ASHRAE Journal*, **18**, pp. 35-39.

Kaya, A. (1979). "Modelling of an environmental space for optimum control of energy use", *Seventh IFAC World Congress*, Helsinki, Finland, pp. 327-334.

Kaya, A. (1981). "Optimum control of HVAC system to save energy", *Eighth IFAC World Congress*, Kyoto, Japan, pp. 3231-3240.

Kaya, A., Chen, C.S., and Raina, S. (1982). "Optimum control policies to minimize energy use in HVAC systems", *ASHRAE Trans.*, **88**, Part 2, pp. 235-248.

Kelly, G., Park, C., Clark, D.R., and May, W.B. (1984). "HVACSIM+, A dynamic building-HVAC-control systems simulation program", *Workshop on HVAC Controls, Modeling, and Simulation*, Atlanta, GA.

Kreisselmeier, G., and Anderson, B.D.O. (1986). "Robust model reference adaptive control", *IEEE Trans. Automatic Control*, **AC-31**, pp. 127-133.

Leitmann, G. (1979). "Guaranteed asymptotic stability for some linear systems with bounded uncertainties", *J. Dynamic Systems, Measurement, and Control*, **101**, pp. 212-216.

Leitmann, G. (1981). "On the efficacy of nonlinear control in uncertain systems", *J. Dynamic Systems, Measurement, and Control*, **103**, pp. 95-102.

Leitmann, G. (1983). "Deterministic control of uncertain systems", *R & D Newsletter on Robotics and Artificial Intelligence*, **1**, pp. 1-12.

Li, X.M., and Wepfer, W.J. (1985). "Implementation of adaptive control to building HVAC systems", *Eighth World Energy Engineering Congress*, pp. 143-152.

Li, X.M., and Wepfer, W.J. (1987). "Recursive estimation methods applied to a single-zone HVAC system", *ASHRAE Trans.*, **93**, Part 1, pp. 1814-1829.

Mehta, D.P., (1984). "Modeling of environmental control components", *Workshop on HVAC Controls, Modeling, and Simulation*, Atlanta, GA.

Mehta, D.P., and Woods, J.E. (1980). "An Experimental Validation of a rational Model for Dynamic response of Buildings", *ASHRAE Trans.*, **86**, Part 2, pp. 497-520.

Narendra, K.S., and Annaswamy, A.M. (1987). "A new adaptive law for robust adaptive control without persistent excitation", *IEEE Trans. Automatic Control*, **AC-32**, pp. 134-145.

Narendra, K.S., and Annaswamy, A.M. (1989). *Stable Adaptive Systems*, Prentice-Hall, New York, NY.

Narendra, K.S., and Taylor, J.H. (1973). *Frequency Domain Criteria for Absolute Stability*, Academy Press, New York, NY.

Patel, R.V., and Toda, M. (1980). "Quantitative measures of robustness for multivariable systems", *Joint Automatic Control Conference*, San Francisco, CA, TP8-A.

Peterson, B.B., and Narendra, K.S. (1982). "Bounded error adaptive control", *IEEE Trans. Automatic Control*, **AC-27**, pp. 1161-1168.

Safonov, M.G. (1980). *Stability and robustness of multivariable feedback systems*, MIT Press, Cambridge, MA.

Schumann, R. (1980). "Digital parameter adaptive control of an air conditioning plant", *Fifth IFAC/IFIP Conference on Digital Computer Applications*, Dusseldorf, FRG.

Stoecker, W.F. (1976). *Procedures for Simulating the Performance of Components and Systems for Energy Calculations*, 3rd ed., ASHRAE, Atlanta, GA.

Sud, I. (1984). "Development of a simulation technique for evaluating control strategies for minimum energy usage", *Workshop on HVAC Controls, Modeling, and Simulations*, Atlanta, GA.

Thompson, J.G., and Chen, P.N.T. (1979). "Digital simulation of the effect of room and control system dynamics on energy consumption", *ASHRAE Trans.*, **85**, Part 2, pp. 222-237.

Thompson, J.G., (1981). "The effect of room and control systems dynamics on energy consumption", *ASHRAE Trans.*, **87**, Part 2, pp. 883-896.

Robustness Techniques in Nonlinear Systems with Applications to Manipulator Adaptive Control *

Nader Sadegh, Assistant Professor

The George W. Woodruff School of Mechanical Engineering
The Georgia Institute of Technology
Atlanta, Georgia, 30332-0405

I Introduction

The main advantage of using adaptive control for robotic applications is the ability of the controller to tune itself automatically to configuration changes and non-linearities associated with the manipulator dynamics. One of the drawbacks of adaptive control, however, is the possible loss of robustness to unmodelled dynamics and bounded disturbances [1,2].

In this chapter, we study the robustness of a direct adaptive control law for robot manipulators introduced by Sadegh and Horowitz [1]. This control law, called the Desired Compensation Adaptive Law (DCAL), utilizes the desired, instead of the actual, trajectory information for the computations of the feedforward control input, and the parameter adaptation algorithm. The DCAL has several advantages over the adaptive controllers that use the actual trajectory information [3,4]: 1) it is more efficient computationally, 2) under

*This work was supported by the National Science Foundation under grant MSS-8910427.

some mild conditions, it is exponentially convergent, and 3) it is more robust to input and measurement disturbances [1]. In fact, if the desired trajectories are known in advance, most of the DCAL's computations can be performed off-line and the results stored in memory. Consequently, the on-line computational complexity of the algorithm is reduced to that of a conventional linear controller. The DCAL has been implemented on several robotic systems, including the Berkeley SCARA robot [5] and an IBM 7545 industrial robot [6].

The main contribution of the present work is twofold. First, It is shown that the DCAL is exponentially stable provided a persistent excitation condition on the desired trajectory is satisfied; and, second, under a less restrictive semi–persistent excitation condition, the robustness of the DCAL to a class of unmodelled dynamics and bounded input disturbances is fully analyzed. To evaluate the robust performance of the DCAL, and to illustrate the design procedure, the implementation details of the DCAL on an industrial robot (IBM 7545) are also provided.

This chapter is organized as follows: Section II gives some preliminary results on the exponential stability of a class of nonlinear systems and its relation to robust stability. In Section III the dynamic equation of the manipulator and its class of unmodelled dynamics is introduced. Section IV reviews the details of the DCAL algorithm, and section V presents the pertaining exponential and robust stability results. The implementation details of the DCAL are included in section VI. Finally, conclusions are drawn in section VII.

The following notations are freely used in the paper: $x := y$ means x is defined to be y; $\forall$ means "for all"; $\exists$ means "there exist(s)". $\mathbb{R}^+$ = set of nonnegative real numbers; $\mathbb{R}^n$ = n dimensional Euclidean space; $\mathbb{R}^{m \times n}$ = space of $m \times n$ matrices of real numbers. For $\mathbf{x} \in \mathbb{R}^n$, x_i denotes the i-th component of $\mathbf{x}$; $|\mathbf{x}| := \sqrt{x_1^2 + \cdots + x_n^2}$; For $\mathbf{A} \in \mathbb{R}^{m \times n}$, $\mathbf{A}^T$ means transpose of $\mathbf{A}$; $\|\mathbf{A}\|$ is the matrix norm induced by $|.|$, i.e., $\|\mathbf{A}\| := \sup_{\mathbf{x} \in \mathbb{R}^n} |\mathbf{A}\mathbf{x}|/|x|$.

II Unmodelled Dynamics and Robust Stability

In this section we introduce the class of unmodelled dynamics considered in this chapter followed by a robust stability result. This result is subsequently used to analyze the robustness of the manipulator adaptive controller to be introduced. The class of unmodelled dynamics considered here consists of $\mathcal{L}_p$ input–output stable operators. The basic definitions and notations for $\mathcal{L}_p$ stability used here are, in most parts, adopted from Desoer and Vidyasagar [7].

Let $\mathcal{L}$ denote the set of Lebesgue measurable functions $f : [0, \infty) \rightarrow \mathbb{R}^n$, and define the function space $\mathcal{L}_{p_T}$ to be the set of all $f \in \mathcal{L}$ such that $||f||_{p_T} < \infty$ with the pseudo–norm $||f||_{p_T}$ given by

$$||f||_{p_T} := \begin{cases} \left[\int_0^T |f(t)|^p \, dt\right]^{1/p} & \text{if } p \in [1, \infty) \\ \text{ess sup}_{t\in[0,T]} \; |f(t)| & \text{if } p = \infty \end{cases} \tag{1}$$

where $|.|$ is the usual Euclidean norm, and "ess sup" denotes the essential supremum [7]. Also, let

$$\mathcal{L}_p := \{f \in \mathcal{L} : ||f||_p < \infty\}, \quad \text{with} \quad ||f||_p := \lim_{T\rightarrow\infty} ||f||_{p_T}$$

Definition 1 *An operator $G : \mathcal{L}_{p_T} \rightarrow \mathcal{L}$ is said to be $\mathcal{L}_p$* stable *$(\in \mathcal{L}_p^s)$ if there exist $K_p^G > 0$ and $d_{p_T}^G > 0$ such that*

$$||G(f)||_{p_T} \leq K_p^G \; ||f||_{p_T} + d_{p_T}^G, \quad \forall \, T > 0 \tag{2}$$

The scalar K_p^G is called the $\mathcal{L}_p$ gain of G.
G is said to be asymptotically stable *$(\in \mathcal{L}_a^s)$ if $G \in \mathcal{L}_\infty^s$, and there exist K_a^G and $d_a^G > 0$ such that*

$$\overline{\lim_{t\rightarrow\infty}} \; |G(f)(t)| \leq K_a^G \; \overline{\lim_{t\rightarrow\infty}} |f(t)| + d_a^G, \quad \forall f \in \mathcal{L}_\infty \tag{3}$$

where $\overline{\lim}$ denotes the upper limit. The scalar K_a^G is called the $\mathcal{L}_a$ gain of G.

Remarks

i) If G is an asymptotically stable linear operator, then it is $\mathcal{L}_p$ stable for all $p \in [1, \infty]$.

ii) The positive scalars $d_{p_T}^G$ and d_a^G can be regarded as the $\mathcal{L}_{p_T}$ upper bound and upper limit, respectively, of the input disturbances associated with G.

For the remainder of this section, we study the robustness properties of dynamical systems of the form

$$\begin{aligned} \dot{\mathbf{x}}(t) &= F(\mathbf{x}(t), \mathbf{u}, t) \\ \mathbf{u} &= G(\mathbf{x}) \end{aligned} \tag{4}$$

where $\mathbf{x} \in R^n$ is the state variable, t is the "time" variable, $F(\mathbf{x}, \mathbf{u}, t)$ is piecewise continuous in t and continuous in $\mathbf{x}$ and $\mathbf{u}$, $F(0, 0, t) = 0$, $\forall t \geq 0$, $G \in \mathcal{L}_p^s \cap \mathcal{L}_a^s$. The system $\dot{\mathbf{x}} = f(\mathbf{x}, t) := F(\mathbf{x}, 0, t)$ is referred to as the *unperturbed* system. The input $\mathbf{u} = G(\mathbf{x})$ includes the unmodelled dynamics and input disturbances that affect the unperturbed system $\dot{\mathbf{x}} = f(\mathbf{x}, t)$, respectively. We assume that the system in Eq. (4) satisfies the following:

Assumption 2

i) Dynamical system in Eq. (4) always possesses a unique solution over $[0, \infty)$.

ii) $f(\mathbf{x}, t)$ is continuously differentiable in $\mathbf{x}$. Moreover, letting $\mathbf{A}(t) := \frac{\partial f}{\partial \mathbf{x}}|_{\mathbf{x}=0}$, then

$$\lim_{|\mathbf{x}| \to 0} \frac{|f(\mathbf{x}, t) - \mathbf{A}(t)\mathbf{x}|}{|\mathbf{x}|} = 0 \quad \text{uniformly in } t$$

iii) $F(\mathbf{x}, \mathbf{u}, t)$ is Lipschitz in $\mathbf{u}$: there exist h and $l > 0$, such that

$$|F(\mathbf{x}, \mathbf{u}, t) - f(\mathbf{x}, t)| \leq l|\mathbf{u}|, \qquad \forall\, \mathbf{x} \in B_h$$

where B_h denotes the open ball of radius h centered at zero.

We now define the (local) exponential stability ([2], p. 24) of the unperturbed system $\dot{\mathbf{x}} = f(\mathbf{x}, u)$, which, as we shall see, is closely related to the robust stability of the perturbed system in Eq. (4):

Definition 3 *The system $\dot{\mathbf{x}} = f(\mathbf{x}, t)$ is said to be (locally) exponentially stable at $\mathbf{x} = 0$ if there exist m, α, and $h > 0$ such that $\mathbf{x}(t)$ satisfies*

$$|\mathbf{x}(t)| \leq m\, e^{-\alpha(t-t_0)}\, |\mathbf{x}(t_0)|$$

for all $\mathbf{x} \in B_h$, $t \geq t_0 \geq 0$.

The following theorem, adopted from Sastry and Bodson [2], relates exponential stability to the existence of a special Lyapunov function.

Theorem 4 *Consider the system $\dot{\mathbf{x}} = f(\mathbf{x}, t)$ satisfying assumption 2. The following statements are equivalent:*

a) *$\dot{\mathbf{x}} = f(\mathbf{x}, t)$ is exponentially stable at $\mathbf{x} = 0$.*
b) *There exists a function $v(\mathbf{x}, t)$, and some positive constants h', α_1–α_4, such that $\forall\ \mathbf{x} \in B_{h'}$,*

$$\alpha_1^2 |\mathbf{x}|^2 \leq v(\mathbf{x}, t) \leq \alpha_2^2 |\mathbf{x}|^2 \tag{5}$$

$$\frac{\partial v(\mathbf{x}, t)}{\partial t} + \frac{\partial v(t, \mathbf{x})}{\partial \mathbf{x}} f(\mathbf{x}, t) \leq -\alpha_3 |\mathbf{x}|^2 \tag{6}$$

$$\left| \frac{\partial v(\mathbf{x}, t)}{\partial \mathbf{x}} \right| \leq \alpha_4 |\mathbf{x}| \tag{7}$$

The following lemma shows that the exponential stability of a nonlinear system may be deduced from the uniform asymptotic stability of its linearized model, and vice versa.

Lemma 5 *The system $\dot{\mathbf{x}} = f(\mathbf{x}, u)$ satisfying assumption 2–ii is exponentially stable at $\mathbf{x} = 0$ if and only if $\dot{\mathbf{x}} = \mathbf{A}(t)\mathbf{x}$ is uniformly asymptotically stable, where $\mathbf{A}(t) := (\partial f / \partial \mathbf{x})|_{\mathbf{x}=0}$.*

Proof See appendix A.

The preceding lemma shows that the exponential stability of a nonlinear system is directly related to its linearized model. The next series of results are useful in establishing the exponential stability of a linear time varying system. First, we have the definition of uniform complete observability [2]:

Definition 6 *The linear time varying system* $\dot{\mathbf{x}}(t) = \mathbf{A}(t)\mathbf{x}(t)$, $\mathbf{y}(t) = \mathbf{C}(t)\mathbf{x}(t)$, *or simply the pair* $(\mathbf{A}, \mathbf{C})$, *is said to be uniformly completely observable (UCO) if there exists* s, β_1, $\beta_2 > 0$ *such that observability Grammian*

$$N(t_0, t_1) := \int_{t_0}^{t_1} \Phi^T(\tau, t_0)\mathbf{C}^T(\tau)\mathbf{C}(\tau)\Phi(\tau, t_0)\ d\tau$$

satisfies

$$\beta_1 \mathbf{I} \geq N(t_0, t_0 + s) \geq \beta_2 \mathbf{I}$$

$\forall t_0 \geq 0$, *where* $\phi(t, \tau)$ *is the state transition matrix corresponding to* $\mathbf{A}(t)$.

Remark If the system $\dot{x} = \mathbf{A}\mathbf{x}$, $\mathbf{y} = \mathbf{C}\mathbf{x}$ is UCO, then it follows that there exist s, β_1, and $\beta_2 > 0$ such that

$$\beta_1 |\mathbf{x}(t_0)|^2 \geq \int_{t_0}^{t_0+s} |\mathbf{y}(t)|^2\ dt \geq \beta_2 |\mathbf{x}(t_0)|^2$$

for all $t_0 \geq 0$. In particular, $\int_{t_0}^{t_0+s} |\mathbf{y}(t)|^2 dt$ approaches zero if and only if the initial state $\mathbf{x}(t_0)$ approaches zero.

The following theorem is extremely useful in investigating the exponential stability of adaptive systems:

Theorem 7 *Consider the linear time varying system* $\dot{x}(t) = \mathbf{A}(t)\mathbf{x}(t)$ *partitioned as*

$$\frac{d}{dt}\begin{bmatrix} \mathbf{x}_1(t) \\ \mathbf{x}_2(t) \end{bmatrix} = \begin{bmatrix} \mathbf{A}_{11}(t) & \mathbf{A}_{12}(t) \\ \mathbf{A}_{21}(t) & \mathbf{A}_{22}(t) \end{bmatrix} \begin{bmatrix} \mathbf{x}_1(t) \\ \mathbf{x}_2(t) \end{bmatrix}$$

where the elements of $\mathbf{A}(t)$ *are bounded and uniformly continuous functions of* t. *Suppose that there exist a bounded strictly positive definite matrix* $\mathbf{P}(t)$ *and scalar* $\gamma > 0$ *such that*

$$\frac{d}{dt}\left[\mathbf{x}^T(t)\mathbf{P}(t)\mathbf{x}(t)\right]\big|_{\dot{\mathbf{x}}=\mathbf{A}\mathbf{x}} \leq -\gamma|\mathbf{x}_1(t)|^2 \tag{8}$$

Then, $\dot{\mathbf{x}} = \mathbf{A}(t)\mathbf{x}$ *is exponentially stable if the pair* $(\mathbf{A}_{22}, \mathbf{A}_{12})$ *is UCO.*

Proof We first claim that if $(\mathbf{A}_{22}, \mathbf{A}_{12})$ is UCO for some $s > 0$, as in definition 6, then the system $\dot{\mathbf{x}} = \mathbf{A}\mathbf{x}$, $\mathbf{y} = \mathbf{x}_1$ is also UCO for the same s. The proof of this claim proceeds by contradiction. If the system $\dot{\mathbf{x}} = \mathbf{A}\mathbf{x}$, $\mathbf{y} = \mathbf{x}_1$ is not UCO, then by definition, $\lambda_{\min}[N(t_0, t_0 + s)] \to 0$ as $t_0 \to \infty$. This implies that for every $t_0 \geq 0$, there exists a $\mathbf{z}(t_0) \in \mathbb{R}^n$ bounded away from zero such that

$$||\mathbf{x}_1||_{t_0,t_0+s} := \int_{t_0}^{t_0+s} |\mathbf{x}_1(t)|^2 \to 0$$

as $t_0 \to \infty$, where $\mathbf{x}(t) = [\mathbf{x}_1(t)\ \mathbf{x}_2(t)]^T = \Phi(t, t_0)\mathbf{z}(t_0)$. To complete the proof, we need the following technical lemma.

Lemma 8 *Let* $f : \mathbb{R}^+ \to \mathbb{R}^n$ *be uniformly continuous. Define*

$$g_\delta(t) := \int_t^{t+\delta} f(\tau)d\tau$$

Then

$$\lim_{t\to\infty} ||g_\delta||_{t,t+s} \to 0 \quad \forall\ 0 < \delta \leq s \iff \lim_{t\to\infty} ||f||_{t,t+s} \to 0$$

Proof See appendix A.

From the first state equation, $\forall\ t_0 \geq 0$, and $0 < \delta \leq s$, we have

$$\mathbf{x}_1(t_0 + \delta) - \mathbf{x}_1(t_0) = \int_{t_0}^{t_0+\delta} \mathbf{A}_{11}(t)\mathbf{x}_1(t)\ dt + g_\delta(t_0)$$

where

$$g_\delta(t_0) := \int_{t_0}^{t_0+\delta} \mathbf{y}_2(t)\, dt, \qquad \mathbf{y}_2(t) := \mathbf{A}_{12}(t)\mathbf{x}_2(t)$$

Since $\|\mathbf{x}_1\|_{t_0,t_0+s} \to 0$ as $t \to \infty$, by the "only if" part of the lemma we have that

$$\lim_{t_0\to\infty} \|g_\delta\|_{t_0,t_0+s} = 0$$

As $A_{12}(t)$ is bounded and uniformly continuous, and $\mathbf{x}(t)$ and $\dot{\mathbf{x}}(t)$ are bounded, it follows that $\mathbf{y}_2(t)$ is uniformly continuous. Thus by the "if" part of the lemma, $\|\mathbf{y}_2\|_{t_0,t_0+s}$ goes to zero as $t_0 \to \infty$. From the second state equation, we have

$$\mathbf{x}_2(t) = \Phi_{22}(t,t_0)\mathbf{x}_2(t_0) + \int_{t_0}^{t} \Phi_{22}(\tau,t_0)\mathbf{A}_{21}(\tau)\mathbf{x}_1(\tau)\, d\tau$$

so that

$$\lim_{t_0\to\infty} |\mathbf{x}_2(t) - \Phi_{22}(t,t_0)\mathbf{x}_2(t_0)| = 0$$

$\forall t \in [t_0, t_0+s]$, where $\Phi_{22}(t,\tau)$ is the state transition of $\mathbf{A}_{22}(t)$. It now follows that

$$\lim_{t_0\to\infty} \int_{t_0}^{t_0+s} |\mathbf{A}_{12}(t)\Phi_{22}(t,t_0)\mathbf{x}_2(t_0)|^2\, dt = 0$$

for some $\mathbf{x}_2(t_0)$ that is bounded away from zero. This clearly contradicts the UCO assumption made about the pair (A_{22}, A_{12}). Thus the system $\dot{\mathbf{x}} = \mathbf{A}\mathbf{x}$, $\mathbf{y} = \mathbf{x}_1$ is UCO, and $\exists\ \beta_1$, and $\beta_2 > 0$ such that

$$\beta_1|\mathbf{x}(t_0)|^2 \geq \|\mathbf{x}_1\|_{t_0,t_0+s}\, dt \geq \beta_2|\mathbf{x}(t_0)|^2$$

for all $t_0 \geq 0$. To complete the proof, let $v(\mathbf{x},t) := \mathbf{x}^T\mathbf{P}(t)\mathbf{x}$. Integrating Eq. (8) from $t_k := sk$ to t_{k+1} along the trajectories of $\dot{\mathbf{x}} = \mathbf{A}(t)\mathbf{x}$ gives

$$v_{k+1} - v_k \leq -\gamma\|\mathbf{x}_1\|^2_{t_k,t_{k+1}} \leq -\gamma\beta_2|\mathbf{x}(t_k)| \leq -\gamma\beta_2\underline{\lambda}_{\mathrm{P}} v_k$$

where $v_k = v(\mathbf{x}(t_k), t_k)$, and $\underline{\lambda}_{\mathrm{P}}$ is the infimum of the smallest eigenvalue of $\mathbf{P}(t)$ over all $t \geq 0$. Thus $v_{k+1} \leq \sigma v_k$ for some $0 < \sigma < 1$, and the system is exponentially stable. $\square$.

We begin the robustness analysis with a basic definition:

Definition 9 *The system given in Eq. (4) is said to be* $\mathcal{L}_p$ robust–stable *if there exist positive scalars* $\overline{K}$, $\overline{K}_p$, $\overline{K}_a$, $\overline{d}$, *and* r_0, *such that for all* $G \in \mathcal{L}_p^s \cap \mathcal{L}_a^s$ *satisfying Eqs. (2) and (3) with* $K_\infty^G \leq \overline{K}$, $K_p^G \leq \overline{K}_p$, $d_\infty^G \leq \overline{d}$, $K_a^G \leq \overline{K}_a$, *and for all (initial conditions)* $\mathbf{x}(0)$ *satisfying* $|\mathbf{x}(0)| < r_0$, *we have* $\|\mathbf{x}\|_\infty < \infty$,

$$\|x\|_{p_T} \leq a_1 d_{p_T}^G + a_2, \quad \forall T \in [0, \infty) \tag{9}$$

for some positive scalars a_1, $a_2 > 0$, *and*

$$\overline{\lim_{t \to \infty}} \, |\mathbf{x}(t)| \leq a_1' \; d_a^G \tag{10}$$

for some $a_1' > 0$.

Remark The first requirement of the $\mathcal{L}_p$ robust–stability states that the state $\mathbf{x}(t)$ is a bounded function of time provided $|\mathbf{x}(0)|$ is chosen sufficiently small. The second requirement, which is not usually included in the existing robustness analyses, is very important from a practical point of view. It requires that the asymptotic value of $|\mathbf{x}(t)|$ be bounded by that of the input disturbance, d_a^G. In particular, if $d_a^G = 0$, then $\mathbf{x}(t)$ must also converge to zero asymptotically.

The following theorem presents the main result of this section.

Theorem 10 *Consider the dynamical system given by Eq. (4) satisfying assumption 2. If the unperturbed system* $\dot{\mathbf{x}} = f(\mathbf{x}, t)$ *is exponentially stable at* $\mathbf{x} = 0$, *then the perturbed system in Eq. (4) is* $\mathcal{L}_p$ *robust–stable for* $p \in [2, \infty]$.

Proof By theorem (4) there exists a function $v(\mathbf{x}, t)$ that satisfies Eqs. (5), (6), and (7). By the Lipschitz assumption on $F(\mathbf{x}, \mathbf{u}, t)$, there exist h' and $l > 0$ such that we can write $F(\mathbf{x}, \mathbf{u}, t) = f(\mathbf{x}, t) + F'(\mathbf{x}, \mathbf{u}, t)$ with $|F'(\mathbf{x}, \mathbf{u}, t)| \leq l|\mathbf{u}|$, $\forall \; \mathbf{x} \in B_{h'}$. Evaluating dv/dt along the trajectories of Eq. (4), and using the properties of v, we have

$$\frac{dv(\mathbf{x}, t)}{dt} = \frac{\partial v(\mathbf{x}, t)}{\partial t} + \frac{\partial v(\mathbf{x}, t)}{\partial \mathbf{x}} [f(\mathbf{x}, t) + F'(\mathbf{x}, \mathbf{u}, t)]$$

$$\begin{aligned} &\leq -\alpha_3|\mathbf{x}|^2 + l\alpha_4|\mathbf{x}||\mathbf{u}| \\ &\leq -\frac{\alpha_3}{2}\left[|\mathbf{x}|^2 - \left(\frac{l\alpha_4}{\alpha_3}\right)^2|\mathbf{u}|^2\right], \quad \forall \mathbf{x} \in B_h \cap B_{h'} \end{aligned} \tag{11}$$

The last inequality follows from the identity $-2|\mathbf{x}|^2 + 2b|\mathbf{x}||\mathbf{u}| \leq -|\mathbf{x}|^2 + b^2|\mathbf{u}|^2$. Defining $w(t) := v(\mathbf{x}(t), t)$, from the property of $v(\mathbf{x}, t)$ given by Eq. (5), we obtain

$$\begin{aligned} \frac{d}{dt}w^2(t) &\leq -\gamma\left[w^2(t) - \alpha_1^2\beta^2|\mathbf{u}(t)|^2\right] \\ \mathbf{u} &= G(\mathbf{x}) \end{aligned} \tag{12}$$

for all $w(t) < r := \min\{h, h'\}\alpha_1$, where

$$\gamma = \frac{\alpha_3}{2\alpha_2^2}, \quad \text{and} \quad \beta = \frac{\alpha_2\alpha_4 l}{\alpha_1\alpha_3}$$

Multiplying both sides of Eq. (12) by $e^{\gamma t}$, we have

$$\frac{d}{dt}\left[e^{\gamma t}w^2(t)\right] \leq \gamma\alpha_1^2\beta^2 e^{\gamma t}|\mathbf{u}(t)|^2$$

Integrating this equation from t_0 to t, we obtain

$$w^2(t) \leq e^{-\gamma(t-t_0)}w^2(t_0) + \gamma\alpha_1^2\beta^2 \int_{t_0}^{t} e^{-\gamma(t-\tau)}\,|\mathbf{u}(\tau)|^2 d\tau \tag{13}$$

We now show that by choosing $K_\infty^G \leq \overline{K} < \sqrt{2}/2\beta$ and $d_{\infty_T}^G \leq \overline{d} < r/\rho$ with

$$\rho := \frac{\alpha_1\beta\sqrt{2}}{\sqrt{1 - 2\beta^2\overline{K}^2}}$$

then

$$w(0) < r \implies w(t) < r, \quad \forall t \geq 0$$

which guarantees that

$$|\mathbf{x}(0)| < r_0 := \min\{h, h'\}\frac{\alpha_1}{\alpha_2} \implies \|\mathbf{x}\|_\infty < \min\{h, h'\}$$

Toward this end, we consider the following cases:

Case i: $w(0) \leq \rho\bar{d} < r$

We claim that $w(t) \leq \rho\bar{d}$ for all $t \geq 0$. If this is not the case, then by the continuity of $w(t)$, it follows that there exist $0 \leq t_0 < T$ such that

$$w(0) \leq w(t_0) = \rho\bar{d} < w(T) < r$$

and $w(T) = \|w\|_{\infty T}$. Setting $t = T$ in Eq. (13), and using that

$$\left[\sup_{t\in[0,T]} |G(\mathbf{x})(t)|\right]^2 \leq 2\left[\overline{K}^2|\mathbf{x}(T)|^2 + \bar{d}^2\right]$$

we obtain

$$\begin{aligned} w^2(T) &\leq \sigma w^2(t_0) + 2\beta^2(1-\sigma)\left[\overline{K}w^2(T) + \alpha_1^2\bar{d}^2\right] \\ &\leq 2\sigma\left[\frac{w^2(t_0)}{2} - \alpha_1^2\beta^2\bar{d}^2 - \beta^2\overline{K}^2w^2(T)\right] \\ &\quad +2\beta^2\overline{K}^2w^2(T) + 2\alpha_1^2\beta^2\bar{d}^2 \end{aligned} \tag{14}$$

where $\sigma = e^{-\gamma(T-t_0)} < 1$. Since $w(t_0) = \rho\bar{d}$, and $w(T) > \rho\bar{d}$, Eq. (15) implies that

$$\begin{aligned} (1 - 2\beta^2\overline{K}^2)w(T) &\leq 2\sigma\left[\rho^2\beta^2\overline{K}^2\bar{d}^2 - \beta^2\overline{K}^2w^2(T)\right] + 2\alpha_1^2\beta^2\bar{d}^2 \\ &\leq 2\alpha_1^2\beta^2\bar{d}^2 \end{aligned}$$

which is a contradiction since $w(T) > \rho\bar{d}$.

Case ii: $\rho\bar{d} < w(0) < r$

In this case we claim that $w(t) \leq w(0)$ for all $t \geq 0$. If not, as in the preceding case, there exist $0 \leq t_0 < T$ such that $w(t_0) = w(0)$, $w(T) > w(0)$, and $w(T) = \|w\|_{\infty T}$. Setting $t = T$ in Eq. (13), similarly to the preceding case, we obtain

$$\begin{aligned} w^2(T) &\leq \sigma\left[(1 - 2\beta^2\overline{K}^2)w^2(t_0) - 2\alpha_1^2\beta^2\bar{d}^2\right] \\ &\quad +2\beta^2\overline{K}^2w^2(T) + 2\alpha_1^2\beta^2\bar{d}^2 \\ &\leq (1 - 2\beta^2\overline{K}^2)w^2(t_0) + 2\beta^2\overline{K}^2w^2(T) \end{aligned}$$

The last inequality implies that $w(T) \leq w(0)$, which is a contradiction. Thus in any case $w(t) < r$ for all $t \geq 0$. This proves that under the above conditions, $||\mathbf{x}||_\infty < \infty$.

We now prove that if, in addition to the above conditions for G, and $\mathbf{x}(0)$, $K_p^G \leq \overline{K}_p < 1/\beta$, then Eq. (9) of definition 9 is satisfied. Taking the $L_{p/2}$ norm of both sides of Eq. (13) with $t_0 = 0$, $p \in [2, \infty]$, and simplifying the convolution norm using the well known identity $||w * u||_{p_T} \leq ||w||_1 ||u||_{p_T}$ [7], we obtain

$$\begin{aligned} ||w||_{p_T} &\leq w(0)\sqrt[p]{\frac{2}{\gamma p}} + \alpha_1\beta\left[||G(\mathbf{x})||_{p_T}\right] \\ &\leq w(0)\sqrt[p]{\frac{2}{\gamma p}} + \alpha_1\beta\left[\overline{K}_p||\mathbf{x}||_{p_T} + d^G_{p_T}\right] \end{aligned}$$

Since $\beta\overline{K}_p < 1$ and $||w||_{p_T} \geq \alpha_1||\mathbf{x}||_{p_T}$, we conclude that

$$||\mathbf{x}||_{p_T} \leq a_1 d^G_{p_T} + a_2$$

where

$$a_1 = \frac{\beta}{1 - \beta\overline{K}_p}, \quad \text{and} \quad a_2 = \frac{w(0)}{(1 - \beta\overline{K}_p)\alpha_1}\sqrt[p]{\frac{2}{\gamma p}}$$

Finally, we claim that Eq. (10) of definition 9 holds with $K_a^G \leq \overline{K}_a < 1/\beta$. Taking the upper limit of Eq. (13) yields

$$\overline{\lim_{t\to\infty}}\, w(t) \leq \beta\overline{K}_a \overline{\lim_{t\to\infty}}\, w(t) + \alpha_1\beta d_a^G$$

which proves Eq. (10) with $a_1' = \beta/(1 - \beta\overline{K}_a) > 0$. □

III Dynamics of a Robot Manipulator

A manipulator is defined to be a kinematic chain of rigid links connected through cylindrical or spherical joints. Furthermore, it is assumed that each degree of freedom of the manipulator is powered by an independent torque source. Using the Lagrangian formulation, the equations of motion of an n-th degree of freedom manipulator may be expressed by

$$\mathbf{M}(\mathbf{q})\ddot{\mathbf{q}} + [\mathbf{C}(\mathbf{q},\dot{\mathbf{q}}) + \mathbf{V}]\,\dot{\mathbf{q}} + \mathbf{g}(\mathbf{q}) = \mathbf{T}_{\text{act}} \tag{15}$$

where $\mathbf{q} \in \mathbb{R}^n$ is the vector of generalized (joint) positions, $\dot{\mathbf{q}} \in \mathbb{R}^n$ is the $n \times 1$ vector of generalized (joint) velocities, $\mathbf{M}(\mathbf{q}) \in \mathbb{R}^{n\times n}$ is an $n \times n$ symmetric, bounded, positive definite and C^∞ matrix function, also called the generalized inertia matrix, $\mathbf{V}$ is the diagonal matrix of *viscous* joint friction coefficients, $\mathbf{g}(\mathbf{q})$ is a smooth vector field due to gravity and other conservative (i.e., position dependent) forces, and $\mathbf{T}_{\text{act}} \in R^n$ represents the vector of net actuator torques applied to the manipulator.

The ij–th element of $\mathbf{C}$, c_{ij}, also known as the Christoffel symbol, is given by

$$c_{ij} = \frac{1}{2}\sum_k \dot{q}_k \left[\frac{\partial m_{ij}}{\partial q_k} + \frac{\partial m_{ik}}{\partial q_j} - \frac{\partial m_{jk}}{\partial q_i}\right] \tag{16}$$

where m_{ij} is the ij–th element of $\mathbf{M}$, q_j is the j–th element of $\mathbf{q}$, and $\dot{q}_k$ is the k–th element of $\dot{\mathbf{q}}$. See [1,9] for details.

The term $\mathbf{C}(\mathbf{q},\dot{\mathbf{q}})\dot{\mathbf{q}}$ is also known in the robotics literature as the Coriolis and centripetal acceleration vector. The following proposition reveals an important relationship between the inertia matrix and the Coriolis vector.

Proposition 11 *Given any C^1 vector function of time* $\mathbf{v}(.) : \mathbb{R}^+ \to \mathbb{R}^n$

$$\frac{d}{dt}\mathbf{v}^T\mathbf{M}(\mathbf{q})\mathbf{v} = 2\mathbf{v}^T\left[\mathbf{M}(\mathbf{q})\frac{d}{dt}\mathbf{v} + \mathbf{C}(\mathbf{q},\dot{\mathbf{q}})\mathbf{v}\right] \tag{17}$$

Proof The proposition follows using the well known fact that the matrix $\left[\frac{d}{dt}\mathbf{M}(\mathbf{q}) - 2\mathbf{C}(\mathbf{q},\dot{\mathbf{q}})\right]$ is *skew–symmetric* (see [1] for a proof). □

We assume the following model for the actuator torque vector $\mathbf{T}_{\text{act}}$:

$$\mathbf{T}_{\text{act}} = (1 + \tilde{G})\mathbf{u} \tag{18}$$

where $\mathbf{u} \in \mathbb{R}^n$ is the vector of control inputs to the n actuators, and $\tilde{G} \in \mathcal{L}_p^s \cap \mathcal{L}_a^s$, $p \in [2, \infty]$, represents the actuator unmodelled dynamics and disturbances. More specifically, $\exists\ K_p^{\tilde{G}}$, $K_a^{\tilde{G}}$, $d_{pT}^{\tilde{G}}$, and $d_a^{\tilde{G}} > 0$ such that

$$\|\tilde{G}(\mathbf{u})\|_{pT} \leq K_p^{\tilde{G}} \|\mathbf{u}\|_{pT} + d_{pT}^{\tilde{G}} \tag{19}$$

$$\overline{\lim_{t\to\infty}}\ |\tilde{G}(\mathbf{u})| \leq K_a^{\tilde{G}}\ \overline{\lim_{t\to\infty}}\ |\mathbf{u}| + d_a^{\tilde{G}} \tag{20}$$

Through the remainder of the chapter we *assume* that Eq. (15) has a unique solution given any continuous control law $\mathbf{u} = g(\mathbf{q}, t)$.

IV The Desired Compensation Adaptive Law (DCAL)

In this section we will discuss the motion control of robotic manipulators using an adaptive computed torque controller referred to as the Desired Compensation Adaptive Law (DCAL) [1]. This controller, utilizes the desired trajectory information, instead of the actual ones, for feedforward compensation and adaptation. The objective of the DCAL is to force both the position and velocity of the manipulator, $\mathbf{q}(t)$ and $\dot{\mathbf{q}}(t)$, to asymptotically converge to a prescribed C^2 (i.e., twice continuously differentiable) function of time $\mathbf{q}_d(t)$ and its derivative $\dot{\mathbf{q}}_d(t)$, respectively. Throughout the chapter, $\mathbf{q}_d(t)$, $\dot{\mathbf{q}}_d(t)$ and $\ddot{\mathbf{q}}_d(t)$, called the *desired* position, velocity and acceleration, respectively, are assumed to be a bounded function of t.

In order to estimate only constant parameters in the adaptation law, we use the reparametrization technique [3,1] by assuming that the left side of Eq. 15 can be expressed as a product of a *known* matrix function and an *unknown* parameter vector. More precisely,

$$\mathbf{M}(\mathbf{q}) + [\mathbf{C}(\mathbf{q}, \dot{\mathbf{q}}) + \mathbf{V}]\, \dot{\mathbf{q}} + \mathbf{g}(\dot{\mathbf{q}}) = \mathbf{W}(\mathbf{q}, \dot{\mathbf{q}}, \ddot{\mathbf{q}})\mathbf{\Theta} \tag{21}$$

where the matrix $\mathbf{W}(.,.,.)$ is a known function of the joint position, velocities and accelerations and $\boldsymbol{\Theta}$ is a constant vector of the manipulator inertial parameters. For future references, we define $\mathbf{W}_d(t) := \mathbf{W}(\mathbf{q}_d, \dot{\mathbf{q}}_d, \ddot{\mathbf{q}}_d)$ so that

$$\mathbf{W}_d(t)\ \boldsymbol{\Theta} = \mathbf{M}(\mathbf{q}_d)\frac{d}{dt}\dot{\mathbf{q}}_d + [\mathbf{C}(\mathbf{q}_d, \dot{\mathbf{q}}_d) + \mathbf{V}]\,\dot{\mathbf{q}}_d + \mathbf{g}(\mathbf{q}_d) \tag{22}$$

The control law consists of three main parts: 1) a proportional derivative (PD) feedback controller, 2) a non–linear feedback compensation, both of which have fixed or predetermined gains, and 3) an adaptive feedforward compensator. The PD control action, which utilizes an inner velocity and an outer position loop, together with the non–linear controller is used to guarantee the exponential stability of the overall system. The adaptive feedforward compensator provides the required feedforward torque input for trajectory following purposes by estimating the unknown parameter $\boldsymbol{\Theta}$ on–line. To formulate the control law, we first need to define the trajectory following position and velocity errors. The tracking position error, $\mathbf{e}(t)$, is defined by

$$\mathbf{e}(t) := \mathbf{q}(t) - \mathbf{q}_d(t) \tag{23}$$

For stability purposes, the reference velocity input to the inner velocity loop is set to

$$\dot{\mathbf{q}}_r(t) := \dot{\mathbf{q}}_d(t) - \lambda_P \mathbf{e}, \qquad \lambda > 0 \tag{24}$$

Defining the error between the reference and actual velocity to be

$$\mathbf{e}_r(t) := \dot{\mathbf{q}}(t) - \dot{\mathbf{q}}_r(t) = \dot{\mathbf{e}}(t) + \lambda_P \mathbf{e}(t),$$

the control law, which determines the actuator input, is given by the following set of equations:

$$\mathbf{u} = -\mathbf{K}_P\mathbf{e} - \mathbf{K}_D\mathbf{e}_r - \mathbf{u}_n(\mathbf{e}_r, \mathbf{e}) + \mathbf{W}_d(t)\hat{\boldsymbol{\Theta}} \tag{25}$$

$$\frac{d}{dt}\hat{\boldsymbol{\Theta}} = -\mathbf{K}_I\mathbf{W}_d(t)\ \mathbf{e}_r(t), \qquad \mathbf{K}_I > 0 \tag{26}$$

where $\mathbf{K}_P$ and $\mathbf{K}_D > 0$ are the PD gain matrices, $\mathbf{K}_I > 0$ is the adaptation gain, and $\mathbf{u}_n$ is the nonlinear feedback term given by

$$\mathbf{u}_n(\mathbf{e}_r, \mathbf{e}) = \sigma_n \left|\mathbf{e}\right|^2 \mathbf{e}_r, \qquad \sigma_n > 0 \tag{27}$$

Remark The control law in Eqs. (25) and (26) contains the desired trajectory quantities in both the feedforward compensation term and the parameter adaptation law. Therefore, the signal $\mathbf{W}_d(t)$ may be calculated and stored off-line, resulting in a significant reduction in the number of on-line computations, when compared with the conventional computed torque method. In addition, since the matrix $\mathbf{W}_d(t)$ is not contaminated with noise, the error signal $\mathbf{e}_r$ is uncorrelated with the adaptation signal, hence, avoiding the infinite gain phenomenon mentioned in Rohrs. [10].

The following theorem [1] shows that, in the absence of unmodelled effects, the DCAL forces the tracking error to zero asymptotically.

Theorem 12 *Applying the adaptive control law given by Eqs. (25) and (26) to the unperturbed (i.e., $\tilde{G}(\mathbf{u}) = 0$) manipulator system given by Eq. (15) forces both $\mathbf{e}$ and $\dot{\mathbf{e}}$ to converge to zero asymptotically starting from any initial conditions provided the control gains $\mathbf{K}_P$, $\mathbf{K}_D$ and σ_n are chosen sufficiently large.*

V Exponential Stability of the DCAL

In this section we will show that under a persistent excitation (PE) assumption on the desired trajectory of the manipulator, the closed loop system resulting form the DCAL is (locally) exponentially stable. The exponential stability, as was shown previously, leads to robust–stability in the presence of unmodelled dynamics and input disturbances. Furthermore, we will show that the robust–stability of the DCAL is preserved under a less restrictive assumption, called the semi–persistent excitation (SPE) [11] condition. The PE condition relevant to the DCAL is defined below:

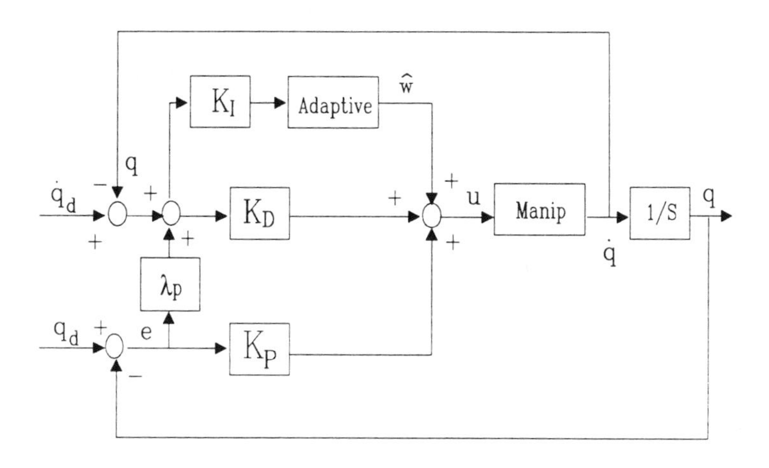

Figure 1: Overall DCAL Block Diagram.

Definition 13 *A matrix function* $\mathbf{W} : \mathbb{R}^{+} \rightarrow \mathbb{R}^{n \times m}$ $(n \leq m)$ *is said to be persistently exciting* (PE) *if the pair* $(0, \mathbf{W})$ *is UCO, that is, there exist positive scalars* β_1, β_2, *and s such that*

$$\beta_1 \mathbf{I} \geq \int_t^{t+s} \mathbf{W}_d^T(\tau)\mathbf{W}_d(\tau)\, d\tau \ \geq \beta_2 \mathbf{I} \tag{28}$$

for all $t > 0$.

For simplicity in the present analysis, we omit the nonlinear control term $\mathbf{u}_n$ from the DCAL (i.e., set $\sigma_n = 0$ in Eq. 27) , which was used to guarantee global stability. See Fig. (1) for an overall block diagram of the DCAL. The exponential stability results that we obtain here, with or without $\mathbf{u}_n$, are local in the sense of definition (3). Applying the control law Eq. (25) to the manipulator system whose dynamics is governed by Eq. (15), we obtain the following error equation:

$$\begin{aligned} \mathbf{M}(\mathbf{q})\frac{d}{dt}\mathbf{e}_r &= -\mathbf{C}(\mathbf{q},\dot{\mathbf{q}})\mathbf{e}_r - \mathbf{K}_D\mathbf{e}_r - \mathbf{K}_P\mathbf{e} \\ &\quad -\eta(\mathbf{e},\mathbf{e}_r,t) + \tilde{\mathbf{u}} + \mathbf{W}_d(t)\tilde{\boldsymbol{\Theta}} \\ \frac{d}{dt}\mathbf{e} &= \mathbf{e}_r - \lambda_P\mathbf{e} \\ \frac{d}{dt}\tilde{\boldsymbol{\Theta}} &= -\mathbf{K}_I\mathbf{W}_d^T\,\mathbf{e}_r \end{aligned} \tag{29}$$

where $\tilde{\boldsymbol{\Theta}} = \hat{\boldsymbol{\Theta}} - \boldsymbol{\Theta}$,

$$\eta(\mathbf{e},\mathbf{e}_r,t) = \mathbf{M}(\mathbf{q})\ddot{\mathbf{q}}_r + \mathbf{C}(\mathbf{q},\dot{\mathbf{q}})\dot{\mathbf{q}}_r + \mathbf{g}(\mathbf{q}) - \mathbf{W}_d(t)\boldsymbol{\Theta} \tag{30}$$

and $\tilde{\mathbf{u}} = \tilde{G}(\mathbf{u})$ represents the unmodelled term. See [1] for a more detailed derivation. Defining the error state vector as

$$\mathbf{x} := \begin{bmatrix} {\mathbf{e}_r}^T & \mathbf{e}^T & \tilde{\boldsymbol{\Theta}}^T \end{bmatrix}^T \tag{31}$$

Eq. (29) can be expressed in the standard state form

$$\dot{\mathbf{x}} = F(\mathbf{x},\tilde{\mathbf{u}},t) \tag{32}$$

where $F(\mathbf{x},\tilde{\mathbf{u}},t)$ is given by

$$\begin{bmatrix} \mathbf{M}^{-1}(\mathbf{q}_d + \mathbf{e})\left[-\mathbf{K}'_D(\mathbf{x},t)\mathbf{e}_r - \mathbf{K}_P\mathbf{e} - \eta(\mathbf{e}_r,\mathbf{e},t) + \tilde{\mathbf{u}} + \mathbf{W}_d(t)\tilde{\boldsymbol{\Theta}}\right] \\ \mathbf{e}_r - \lambda_P\mathbf{e} \\ -\mathbf{K}_I\mathbf{W}_d^T(t)\mathbf{e}_r \end{bmatrix}$$

where $\mathbf{K}'_D(\mathbf{x},t) = \mathbf{K}_D + \mathbf{C}(\mathbf{q}_d + \mathbf{e}, \dot{\mathbf{q}}_d + \mathbf{e}_r + \lambda_P\mathbf{e})$.

To study the stability properties of the error system, we linearize Eq. (32) about the desired trajectory. The following lemma will be used to linearize the term $\eta(\mathbf{e}_r,\mathbf{e},t)$ in the above equation.

Lemma 14 *The term $\eta(\mathbf{e}_r,\mathbf{e},t)$ given by Eq. (30) is continuously differentiable with respect to $\mathbf{e}_r$ and $\mathbf{e}$, and $\eta(0,0,t) = 0$ for all $t \geq 0$. Moreover, the partial derivatives*

$$\Delta_1(t) := \frac{\partial\eta}{\partial\mathbf{e}}(0,0,t), \quad \text{and} \quad \Delta_2(t) := \frac{\partial\eta}{\partial\mathbf{e}_r}(0,0,t)$$

are uniformly bounded, i.e., $\exists\ \overline{\eta}_1,\ \overline{\eta}_2 > 0$ such that $\|\Delta_1(t)\| < \overline{\Delta}_1$ and $\|\Delta_2(t)\| < \overline{\Delta}_2$ for all $t \geq 0$.

Proof See appendix A.

The following theorem shows that the DCAL closed system is exponentially stable when the regression matrix $\mathbf{W}_d(t)$ is PE:

Theorem 15 *Consider the manipulator dynamics given by Eq. (15) with $\tilde{G}(\mathbf{u}) = 0$ subject to the adaptive control law given by Eqs. (25) and (26). If $\mathbf{W}_d(t)$ is uniformly continuous and PE, then there exists a choice of control gains $\mathbf{K}_P$ and $\mathbf{K}_D$ for which the resulting closed loop system given by Eq. (32) is exponentially stable.*

Proof Denote the unperturbed closed loop system (i.e., $\tilde{\mathbf{u}} = 0$), Eq. (32), by $\dot{\mathbf{x}} = \mathbf{f}(\mathbf{x}, t) := F(\mathbf{x}, 0, t)$. Linearizing this system about the equilibrium state $\mathbf{x} = 0$, we obtain the linear time varying differential equation $\dot{\mathbf{x}} = \mathbf{A}(t)\mathbf{x}$ with $\mathbf{A}(t)$ given by

$$\begin{bmatrix} -\mathbf{M}^{-1}(\mathbf{q}_d)\left[\mathbf{K}_D + \mathbf{C}(\mathbf{q}_d, \dot{\mathbf{q}}_d) + \Delta_1\right] & -\mathbf{M}^{-1}(\mathbf{q}_d)\left[\mathbf{K}_P + \Delta_2\right] & -\mathbf{M}^{-1}(\mathbf{q}_d)\mathbf{W}_d \\ \mathbf{I} & -\lambda_P \mathbf{I} & 0 \\ -\mathbf{K}_I \mathbf{W}_d^T & 0 & 0 \end{bmatrix}$$

As the desired position, velocity and acceleration are assumed to be uniformly bounded, it can be easily verified that

$$\lim_{|\mathbf{x}| \to 0} \frac{|f(\mathbf{x}, t) - \mathbf{A}(t)\mathbf{x}|}{|\mathbf{x}|} = 0 \quad \text{uniformly in } t$$

Thus assumption 2–*ii* is satisfied. We now prove that $\dot{\mathbf{x}} = \mathbf{A}(t)\mathbf{x}$ is exponentially stable. Define the Lyapunov function candidate $v(\mathbf{x}, t) = \frac{1}{2}\mathbf{x}^T\mathbf{P}(t)\mathbf{x}$, where $\mathbf{P}(t) = \mathrm{diag}(\mathbf{M}(\mathbf{q}_d(t)), \mathbf{K}_P, \mathbf{K}_I^{-1})$. By proposition 11, we have that

$$\frac{d}{dt}v(t, x) = \mathbf{e}_r{}^T\left[\mathbf{M}(\mathbf{q}_d)\frac{d}{dt}\mathbf{e}_r + \mathbf{C}(\mathbf{q}_d, \dot{\mathbf{q}}_d)\mathbf{e}_r\right] + \mathbf{e}^T\mathbf{K}_P\frac{d}{dt}\mathbf{e} + \tilde{\mathbf{\Theta}}^T\mathbf{K}_I^{-1}\frac{d}{dt}\tilde{\mathbf{\Theta}}$$

Evaluating this derivative along the trajectories of $\dot{\mathbf{x}} = \mathbf{A}(t)\mathbf{x}$, we obtain

$$\frac{d}{dt}v(t, x) = -\mathbf{e}_r{}^T\left(\mathbf{K}_D + \Delta_1\right)\mathbf{e}_r - \lambda_P\mathbf{e}^T\mathbf{K}_P\mathbf{e} - \mathbf{e}_r{}^T\Delta_2\mathbf{e}$$

Since Δ_1 and Δ_2 are uniformly bounded, by choosing $\mathbf{K}_D$ and $\mathbf{K}_P$ sufficiently large, there exists $\gamma > 0$ such that

$$\frac{d}{dt}v(t, x) \le -\gamma(|\mathbf{e}_r|^2 + |\mathbf{e}|^2)$$

Partitioning $\mathbf{x}$ as $[\mathbf{x}_1^T\ \mathbf{x}_2^T]^T$ with $\mathbf{x}_1^T := [\mathbf{e}_r{}^T\ \mathbf{e}^T]$ and $\mathbf{x}_2 := \tilde{\mathbf{\Theta}}$, then A is partitioned as $\begin{bmatrix} \mathbf{A}_{11} & \mathbf{A}_{12} \\ \mathbf{A}_{21} & \mathbf{A}_{22} \end{bmatrix}$ with $\mathbf{A}_{22} = 0$ and $\mathbf{A}_{12} = \begin{bmatrix} \mathbf{M}^{-1}(\mathbf{q}_d)\mathbf{W}_d \\ 0 \end{bmatrix}$. We now claim that $(\mathbf{A}_{22}, \mathbf{A}_{12})$ is UCO, that is, $\exists\ \beta_1', \beta_2'$ and $s > 0$ such that

$$\beta_1'\mathbf{I} \geq \int_{t_0}^{t_0+s} \mathbf{A}_{12}^T(t)\mathbf{A}_{12}(t)\ dt \geq \beta_2'\mathbf{I}, \quad \forall\ t_0 \geq 0$$

The inequality on the left hand side is automatically satisfied because of the boundedness of $\mathbf{W}_d(t)$. To prove the other inequality, by the hypothesis that $\mathbf{W}_d(t)$ is PE, we have

$$\int_{t_0}^{t_0+s} \mathbf{A}_{12}^T(t)\mathbf{A}_{12}(t)\ dt \geq \frac{1}{\overline{\lambda_M}^2} \int_{t_0}^{t_0+s} \mathbf{W}_d^T(t)\mathbf{W}_d(t) \geq \frac{\beta_1}{\overline{\lambda_M}^2}$$

where $\overline{\lambda_M}$ is the supremum of the largest eigenvalue of $\mathbf{M}(\mathbf{q}_d)$ over all $\mathbf{q}_d \in \mathbb{R}^n$. Therefore, the hypothesis of theorem 7 is satisfied and $\dot{\mathbf{x}} = \mathbf{A}(t)\mathbf{x}$ is exponentially stable. As $\dot{\mathbf{x}} = \mathbf{A}(t)\mathbf{x}$ is the linearized model of $\dot{\mathbf{x}} = \mathbf{f}(\mathbf{x}, t)$, the proof of the theorem is established by lemma 5. □

V-A Robustness Analysis

We now are in position to analyze the robustness of the DCAL presented above. We first show that the closed loop error dynamics of the DCAL has the same form as Eq. (4). Rewriting the control law in Eq. (25) as

$$\mathbf{u}(t) = \mathbf{K}(t)\mathbf{x}(t) + \mathbf{W}_d(t)\mathbf{\Theta}$$

with $\mathbf{K}(t) := \text{diag}(\mathbf{K}_P, \mathbf{K}_D, \mathbf{W}_d(t))$, and defining

$$G(\mathbf{x}) := \tilde{G}(\mathbf{Kx} + \mathbf{W}_d\mathbf{\Theta})$$

the unmodelled term in the error Eq. (29) can be expressed as $\tilde{\mathbf{u}} = G(\mathbf{x})$. Taking the $\mathcal{L}_p$ norm, and the upper limit of $G(\mathbf{x})$, it is easily seen that $G \in \mathcal{L}_p^a \cap \mathcal{L}_a^s$, and

$$\begin{aligned} \|G(\mathbf{x})\|_{pT} &\leq K_p^G \|\mathbf{x}\|_{pT} + d_{pT}^G \\ \overline{\lim_{t\to\infty}} |G(\mathbf{x})| &\leq K_a^G \overline{\lim_{t\to\infty}} |\mathbf{x}| + d_a^G \end{aligned}$$

where

$$\begin{aligned} K_p^G &= K_p^{\tilde{G}} \sup_{t\geq 0} \|\mathbf{K}(t)\| \\ K_a^G &= K_a^{\tilde{G}} \overline{\lim_{t\to\infty}} \|\mathbf{K}(t)\| \\ d_{pT}^G &= K_p^{\tilde{G}} \|\mathbf{W}_d \boldsymbol{\Theta}\|_{pT} + d_{pT}^{\tilde{G}} \\ d_a^G &= K_a^{\tilde{G}} \overline{\lim_{t\to\infty}} |\mathbf{W}_d(t)\boldsymbol{\Theta}| + d_a^{\tilde{G}} \end{aligned}$$

Based on this reformulation, the perturbed error Eq. (32) becomes

$$\begin{aligned} \frac{d}{dt}\mathbf{x} &= F(\mathbf{x}, \tilde{\mathbf{u}}, t) = f(\mathbf{x}, t) + \mathbf{M}^{-1}(\mathbf{q}_d + \mathbf{e})\tilde{\mathbf{u}} \\ \tilde{\mathbf{u}}(t) &= G(\mathbf{x}) \end{aligned} \tag{33}$$

where $F(\mathbf{x}, \tilde{\mathbf{u}}, t)$ is as in Eq. (32), and $f(\mathbf{x}, t) = F(\mathbf{x}, 0, t)$. As was shown in the proof of theorem 15, $f(\mathbf{x}, t)$ satisfies assumption 2–*ii*. Moreover, as the inertia matrix $M(\mathbf{x})$ is bounded for all $\mathbf{x} \in \mathbb{R}^n$, the Lipschitz condition in assumption 2–*iii* holds globally with $l = 1/\overline{\lambda}_M$, where $\overline{\lambda}_M$ is as defined in the proof of theorem 15. Thus the error Eq. (33) is of the same form as Eq. (4), and satisfies assumption 2. As the unperturbed system $\dot{\mathbf{x}} = \mathbf{f}(\mathbf{x}, t)$ is exponentially stable under the hypothesis of theorem 15, theorem 10 leads to the following result:

Theorem 16 *Consider the manipulator dynamics given by Eq. (15) with $\tilde{G}_u \in \mathcal{L}_p^s \cap \mathcal{L}_a^s$, for some $p \in [2, \infty]$, subject to the adaptive control law given by Eqs. (25) and (26). If $W_d(t)$ is uniformly continuous and PE, then there exists a choice of control gains $\mathbf{K}_P$ and $\mathbf{K}_D$ for which the resulting closed loop system given by Eq. (33) is $\mathcal{L}_p$ robust–stable.*

V-B Relaxation of Persistent Excitation Condition

In this subsection, we show that even if the $\mathbf{W}_d(t)$ matrix is not PE, it is still possible to preserve the robust stability of the DCAL. The less restrictive condition required for robust stability is defined below:

Definition 17 *A matrix function* $\mathbf{W}_d : \mathbb{R}^+ \rightarrow \mathbb{R}^{n\times m}$ $(n \geq m)$ *is said to be* Semi-Persistently Exciting (SPE) *if there exist a symmetric positive semi-definite matrix* $\mathbf{S}$ *and constants* $s > 0$ *and* $\beta_1 \geq \beta_2 > 0$ *such* $\forall\ t \geq 0$,

$$\beta_1 \mathbf{S} \geq \int_t^{t+s} \mathbf{W}_d^T(\tau)\mathbf{W}_d(\tau)\, d\tau \ \geq \beta_2 \mathbf{S} \tag{34}$$

Remark The SPE condition holds when $\mathbf{W}_d(t)$ is periodic, or even constant. For example, if $\mathbf{W}_d$ is constant, then $\mathbf{S}$ can simply taken to be $\mathbf{W}_d^T\mathbf{W}_d$ with $\beta_2 = 1$, and $s = 1$.

For the remainder of this section, we assume the following:

Assumption 18 *The* $\mathbf{W}_d(t)$ *matrix function in the DCAL control law, Eqs. (25) and (26),*

i) is bounded and uniformly continuous as a function of t*;*
ii) is SPE, and satisfies Eq. (34) for a symmetric positive semi-definite matrix $\mathbf{S} \in \mathbb{R}^{m\times m}$ *of rank* $r \leq m$.

It is well known that $\mathbf{S}$ as prescribed in the above assumption has the following Singular Value Decomposition (SVD) [12]:

$$\mathbf{S} = \begin{bmatrix} \mathbf{U}_{\mathcal{R}} & \mathbf{U}_{\mathcal{N}} \end{bmatrix} \begin{bmatrix} \mathbf{\Lambda}_r & 0 \\ 0 & 0 \end{bmatrix} \begin{bmatrix} \mathbf{U}_{\mathcal{R}}^T \\ \mathbf{U}_{\mathcal{N}}^T \end{bmatrix} = \mathbf{U}_{\mathcal{R}}\mathbf{\Lambda}_r\mathbf{U}_{\mathcal{R}}^T \tag{35}$$

where $[\mathbf{U}_{\mathcal{R}}\ \mathbf{U}_{\mathcal{N}}] \in \mathbb{R}^{m\times m}$ is an orthogonal matrix with $\mathbf{U}_{\mathcal{R}} \in \mathbb{R}^{m\times r}$, $\mathbf{U}_{\mathcal{N}} \in \mathbb{R}^{m\times(m-r)}$, and $\mathbf{\Lambda}_r = \mathrm{diag}(\lambda_1, \ldots, \lambda_r)$ with $\lambda_1 \geq \ldots \lambda_r > 0$. Moreover, $\mathbf{U}_{\mathcal{R}}\mathbf{U}_{\mathcal{R}}^T$ is the orthogonal projection matrix onto the range space of $\mathbf{S}$, denoted by $\mathcal{R}(\mathbf{S})$, and $\mathbf{U}_{\mathcal{N}}\mathbf{U}_{\mathcal{N}}^T$ is the orthogonal projection matrix onto the null space of $\mathbf{S}$, denoted by $\mathcal{N}(\mathbf{S})$.

Lemma 19 *Let $\mathbf{W}_d(t)$ satisfy assumption 18. Then*

$$\mathbf{W}_d(t)\mathbf{U}_{\mathcal{N}}\mathbf{U}_{\mathcal{N}}^T = 0, \quad \text{and} \quad \mathbf{W}_d(t)\mathbf{U}_{\mathcal{R}}\mathbf{U}_{\mathcal{R}}^T = \mathbf{W}_d(t)$$

for all $t \geq 0$.

Proof See appendix A.

Let $\tilde{\mathbf{\Theta}}_{\mathcal{R}}$ and $\tilde{\mathbf{\Theta}}_{\mathcal{N}}$ denote the orthogonal components of $\tilde{\mathbf{\Theta}}$ along $\mathcal{R}(\mathbf{S})$ and $\mathcal{N}(\mathbf{S})$, respectively. Since $\mathbf{U}_{\mathcal{R}}\mathbf{U}_{\mathcal{R}}^T$ and $\mathbf{U}_{\mathcal{N}}\mathbf{U}_{\mathcal{N}}^T$ are orthogonal projection matrices, we have

$$\tilde{\mathbf{\Theta}}_{\mathcal{R}} = \mathbf{U}_{\mathcal{R}}^T\tilde{\mathbf{\Theta}}, \quad \text{and} \quad \tilde{\mathbf{\Theta}}_{\mathcal{N}} = \mathbf{U}_{\mathcal{N}}^T\tilde{\mathbf{\Theta}} \tag{36}$$

As we shall see, $\tilde{\mathbf{\Theta}}_{\mathcal{R}}$ is the only component of the parameter error vector that affects the response of the control system.

Using the fact that

$$\mathbf{W}_d(t) = \mathbf{W}_d(t)\mathbf{U}_{\mathcal{R}}\mathbf{U}_{\mathcal{R}}^T, \quad \text{and} \quad \mathbf{U}_{\mathcal{N}}^T\mathbf{W}_d^T(t) = 0,$$

from lemma 19, the error dynamics of the DCAL in Eq. (29) can be rewritten as

$$\begin{aligned} \mathbf{M}(\mathbf{q})\frac{d}{dt}\mathbf{e}_r &= -\mathbf{C}(\mathbf{q},\dot{\mathbf{q}})\mathbf{e}_r - \mathbf{K}_D\mathbf{e}_r - \mathbf{K}_P\mathbf{e} \\ &\quad -\eta(\mathbf{e},\mathbf{e}_r,t) + \tilde{\mathbf{u}} + \overline{\mathbf{W}}_d\,\tilde{\mathbf{\Theta}}_{\mathcal{R}} \\ \frac{d}{dt}\mathbf{e} &= \mathbf{e}_r - \lambda_P\,\mathbf{e} \\ \frac{d}{dt}\tilde{\mathbf{\Theta}}_{\mathcal{R}} &= -\overline{\mathbf{K}}_I\,\overline{\mathbf{W}}_d^T\,\mathbf{e}_r \\ \frac{d}{dt}\tilde{\mathbf{\Theta}}_{\mathcal{N}} &= 0 \end{aligned} \tag{37}$$

where $\overline{\mathbf{W}}_d(t) = \mathbf{W}_d(t)\mathbf{U}_{\mathcal{R}}$ and $\overline{\mathbf{K}}_I = \mathbf{U}_{\mathcal{R}}^T\mathbf{K}_I\mathbf{U}_{\mathcal{R}}$. Since $\mathbf{S} = \mathbf{U}_{\mathcal{R}}\mathbf{\Lambda}_r\mathbf{U}_{\mathcal{R}}^T$, multiplying Eq. (34) from the left and right side by $\mathbf{U}_{\mathcal{R}}^T$ and $\mathbf{U}_{\mathcal{R}}$, respectively, yields

$$\beta_1\lambda_1\mathbf{I} \geq \int_t^{t+s} \overline{\mathbf{W}}_d^T(\tau)\overline{\mathbf{W}}_d(\tau)\,d\tau \;\leq \beta_2\lambda_r\mathbf{I}$$

so that $\overline{\mathbf{W}}_d(t)$ is PE. From the above error equation, it is also seen that $\tilde{\mathbf{\Theta}}_{\mathcal{N}}$ plays no role in the closed lop system, as it remains constant for all time. Defining the state error as

$$\mathbf{x} := \left[\begin{array}{ccc} \mathbf{e}_r{}^T & \mathbf{e}^T & \tilde{\mathbf{\Theta}}_{\mathcal{R}}^T \end{array} \right]^T$$

it follows that

$$\dot{\mathbf{x}} = F(\mathbf{x}, \tilde{\mathbf{u}}, t) \tag{38}$$

where F is given by Eq. (33) with $\mathbf{W}_d(t)$ and $\mathbf{K}_I$ replaced by $\overline{\mathbf{W}}_d(t)$ and $\overline{\mathbf{K}}_I$, respectively. Thus both the exponential stability of the unperturbed system and the robust stability of the perturbed one remains intact. This result is summarized below:

Theorem 20 *Consider the manipulator dynamics given by Eq. (15) with $\tilde{G}_u \in \mathcal{L}_p \cap \mathcal{L}_\infty$ for some $p \in [2, \infty]$ subject to the adaptive control law given by Eqs. (25) and (26). If $\mathbf{W}_d(t)$ satisfies assumption 18, then there exists a choice of control gains $\mathbf{K}_P$ and $\mathbf{K}_D$ for which the resulting closed loop system given by Eq. (38) is $\mathcal{L}_p$ robust-stable.*

VI Implementation Details

The design procedure for the DCAL and its implementation on a two–link (SCARA) industrial robot (IBM 7545, Fig. 2) is now discussed. The specific form of the inertia matrix $\mathbf{M}(\mathbf{q})$, the Coriolis matrix $\mathbf{C}(\mathbf{q}, \dot{\mathbf{q}})$, the gravity vector $\mathbf{g}(\mathbf{q})$, and the viscous friction matrix $\mathbf{V}$, for the first two revolute links of the IBM 7545 manipulator in terms of the constant inertial parameters $\Theta_1 - \Theta_5$ are:

$$\mathbf{M}(\mathbf{q}) = \left[\begin{array}{cc} \Theta_1 + 2\Theta_3 \cos(q_2) & \Theta_2 + \Theta_3 \cos(q_2) \\ \Theta_2 + \Theta_3 \cos(q_2) & \Theta_2 \end{array} \right]$$

$$\mathbf{C}(\mathbf{q}, \dot{\mathbf{q}}) = \left[\begin{array}{cc} -\Theta_3 \dot{q}_2 \sin(q_2) & -\Theta_3(\dot{q}_1 + \dot{q}_2) \sin(q_2) \\ \Theta_3 \dot{q}_1 \sin(q_2) & 0 \end{array} \right]$$

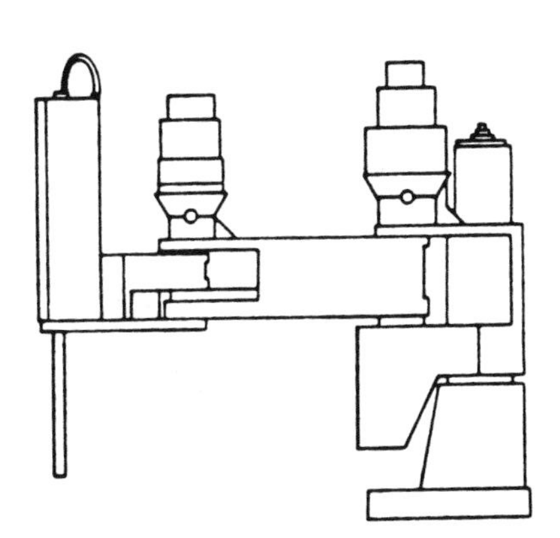

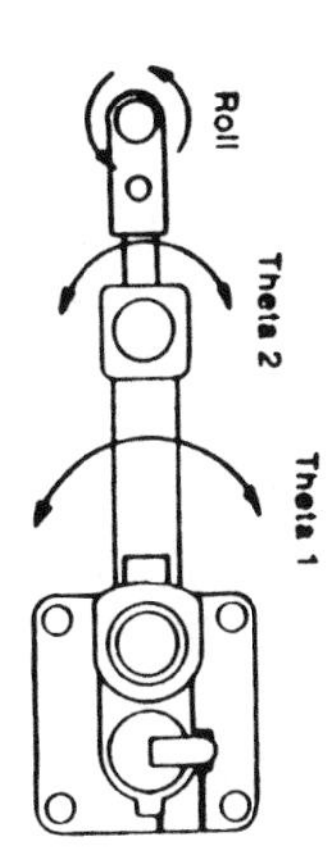

Figure 2: Schematic of the IBM 7545 SCARA Robot.

$\mathbf{g}(\mathbf{q}) = 0$, and $\mathbf{V} = \mathrm{diag}(\Theta_4, \Theta_5)$, where q_1 and q_2 are the position of the first and second joints, respectively. Based on these, the specific form of $\mathbf{W}_d$ using Eq. (22) is found to be

$$\begin{bmatrix} \ddot{q}_{d1} & \ddot{q}_{d2} & (2\ddot{q}_{d1} + \ddot{q}_{d2})\cos(q_{d2}) - (2\dot{q}_{d1} + \dot{q}_{d2})\dot{q}_{d2}\sin(q_{d2}) & \dot{q}_{d1} & 0 \\ 0 & \ddot{q}_{d1} + \ddot{q}_{d2} & \ddot{q}_{d1}\cos(q_{d2}) + \dot{q}_{d1}^2\sin(q_{d2}) & 0 & \ddot{q}_{d2} \end{bmatrix}$$

where q_{d1} and q_{d2} are the desired position of the first and second joints, respectively. The approximate value of the constant parameter vector, $\mathbf{\Theta}$, for the IBM 7545, which was determined by the DCAL adaptation algorithm, is

$$\mathbf{\Theta} = \begin{bmatrix} 0.45 & 0.22 & 0.08 & 1.80 & 1.40 \end{bmatrix}^T$$

The links of this robot are actuated by DC motors coupled to harmonic drives. The harmonic drives of the actuators introduce a

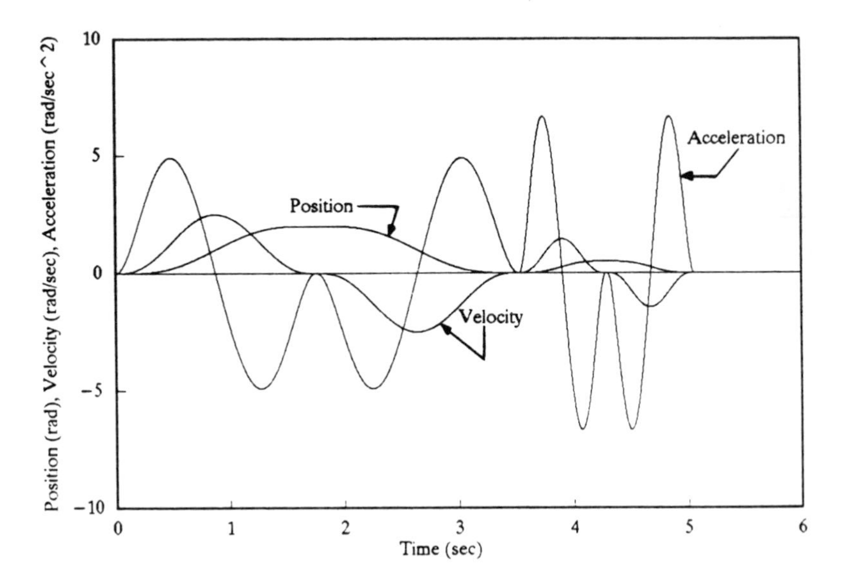

Figure 3: Desired Trajectory for each Revolute Joint.

significant degree of flexible unmodelled dynamics into the control system. In addition, a substantial amount of Coulomb friction is present at each joint. Both the flexibility and Coulomb friction are treated as unmodelled effects.

To make the DCAL digitally realizable, the following discrete time equivalent of the control law (Fig. 1) in Eqs. (25) and (26) was employed in the implementations:

$$\mathbf{u}(t_k) = -\mathbf{K}_P\mathbf{e}(t_k) - \mathbf{K}_D\mathbf{e}_r(t_k) + \mathbf{W}_d(t_k)\hat{\mathbf{\Theta}}(t_{\bar{k}}) \tag{39}$$

$$\hat{\mathbf{\Theta}}(t_{\bar{k}}) = \hat{\mathbf{\Theta}}(t_{\bar{k}-1}) - \mathbf{K}_I \sum_{i=k-3}^{k} \mathbf{W}_d^T(i)\, \mathbf{e}_r(i)\Delta t \tag{40}$$

where $k = 0, 1, 2, \ldots$, $t_k = k\Delta$, $\bar{k}$ is the integer part of $k/4$, and Δ is the sampling time, which was 2 milliseconds for the implementations. Note that the parameter vector $\hat{\mathbf{\Theta}}$ is updated every 8 milliseconds, which is 4 times slower than the updating rate for the feedback

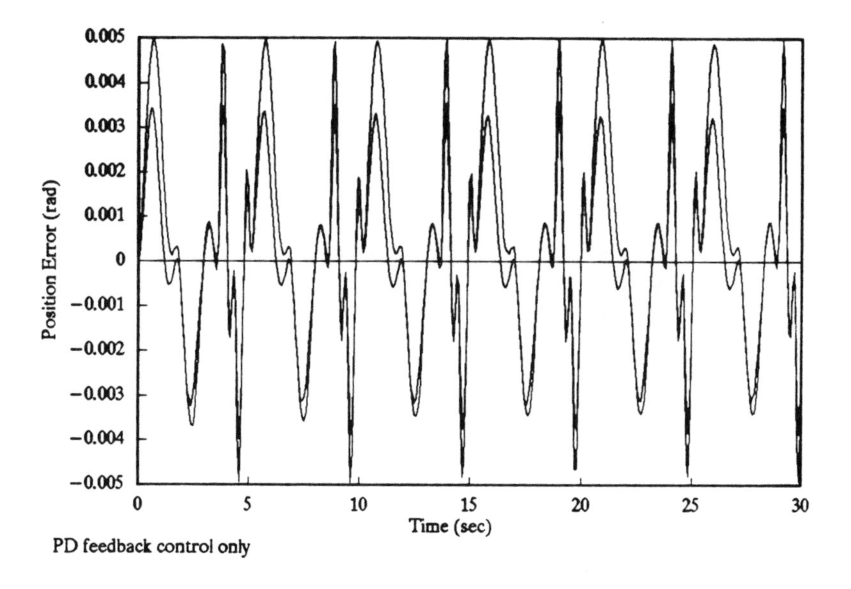

Figure 4: PD Feedback Position Error.

terms (i.e., $k = 4\bar{k}$). This is done to reduce the amount of on–line computations, and to average the error signal before using it in the adaptation algorithm. The following gains were used in the PD feedback portion of the control law above:

$$\begin{aligned} \mathbf{K}_P &= \text{diag}(800, 500) \\ \mathbf{K}_D &= \text{diag}(70, 30) \\ \lambda_P &= 10 \end{aligned}$$

These gains were determined using standard pole placement design based on the average value of the inertia matrix along the desired trajectory. The desired trajectory for each joint was prepared using a 7th order polynomial for the position variables (Fig. 3). The peak velocity in the desired trajectory corresponds to the maximum velocity specified by the manufacturer for the actuators. Therefore, this maximizes any non–linear dynamic effects and results in a very

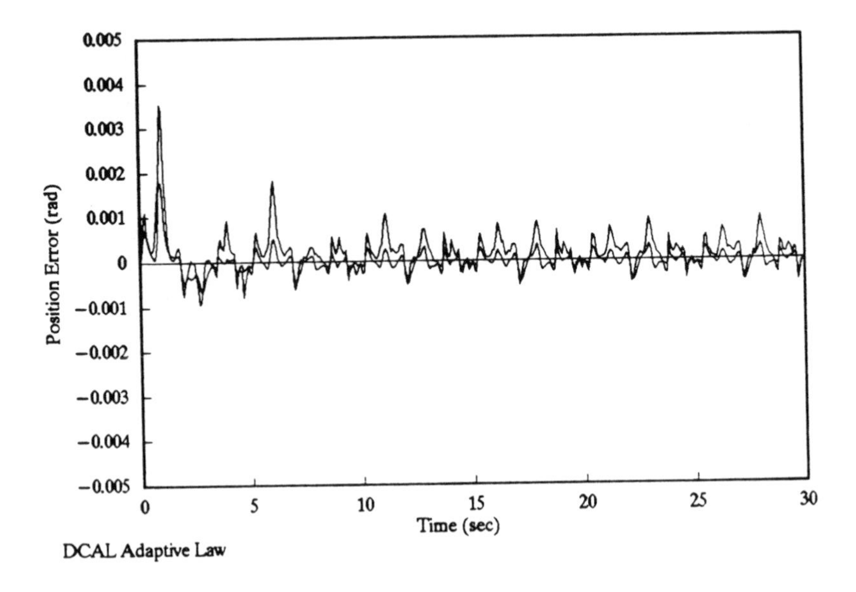

Figure 5: DCAL Position Error.

demanding desired task. The pure PD controller and the DCAL were both implemented. Figures (4–5) show position error over six cycles of the desired trajectory. The DCAL shows a fast convergence of less than 2 cycles and excellent steady state error as compared to the pure PD controller. The experimental results verifies the scheme is robust to the actuator unmodelled dynamics and input disturbances.

VII Conclusions

Some new results on the exponential and robust stability of a class of nonlinear systems was presented. It was shown that exponentially stability of the unperturbed system implies the robust stability of the perturbed one. These results were then applied to the robustness analysis of a direct adaptive controller for robot manipulators (DCAL). It was shown that if the regression matrix, $\mathbf{W}_d(t)$

is bounded, uniformly continuous, and PE, then the DCAL is exponentially and $\mathcal{L}_p$ robust stable. Moreover, using the preliminary results, it was shown that even if $\mathbf{W}_d(t)$ is semi–PE, the DCAL retains its robust stability. The implementation results of the DCAL demonstrated its robust performance in practical applications.

References

[1] Sadegh, N. and R. Horowitz, "Stability and Robustness Analysis of a Class of Adaptive Controllers for Robotic Manipulators," *International Journal of Robotics Research*, Vol. 9, No. 3, June 1990, pp. 74–92.

[2] Bodson, M. and S. Sastry, *Adaptive Control Stability, Convergence, and Robustness,* pp. 229–231, Prentice Hall Advanced Reference Series, 1989.

[3] Craig, J.J., P. Hsu, and S.S. Sastry, "Adaptive Control of Mechanical Manipulators," *The Int. Journal of Robotics Research,* vol. 6. no. 2, Aug. 1987, pp. 16–28.

[4] Slotine J-J. E. and W. Li, "Adaptive Manipulator Control: A Case Study," *IEEE trans. Automat. Contr.,* vol. 33, no. 11, Nov. 1988.

[5] Morando, A., R. Horowitz, and N. Sadegh, "Digital Implementation of Adaptive Control Algorithms for Robot Manipulators," *Proceedings of the 1989 Int. Conf. on Robotics and Automation,* vol. 3, pp. 1656–62, Scottsdate, Arizona, May 1989.

[6] Guglielmo, K. and N. Sadegh, "Experimental Evaluation of a New Robot Learning Controller," *Proceedings of the 1991 IEEE Int. Conf. on Robotics and Automation,* Sacramento, California, April 7–12, 1991.

[7] Desoer, C. A. and M. Vidyasagar, *Feedback Systems: Input–Output Properties,* Academic Press, 1975.

[8] Vidyasagar, M., *Nonlinear Systems Analysis,* Prentice–Hall, 1978.

[9] Sadegh, N., R. Horowitz, W.-W. Kao, and M. Tomizuka, "A Unified Approach to the Design of Adaptive and Repetitive Controllers for Robotic Manipulators," *Journal of Dynamic Systems, Measurement, and Control,* vol. 112, pp. 618–629, December 1990.

[10] Rohrs, C., "Adaptive Control in the Presence of Unmodelled Dynamics," *Lab. Inform. Decision Syst. LIDS–TH–1254, M.I.T.,* Cambridge, MA, 1982.

[11] Sadegh, N. and R. Horowitz, "An Exponentially Stable Adaptive Control Law for Robotic Manipulators," *IEEE Trans. on Robotics and Automation,* vol. 6, no. 4, pp. 491–496, August, 1990.

[12] Golub, G. H. and C. F. Vann Loan, *Matrix Computations,* Baltimore, MD: John Hopkins University Press, 1983.

A Appendix: Technical Proof of the Lemmas

Proof of lemma (5):
Here we only prove the "if" part. The "only if" proof may be found in Vidyasagar ([8], theorem 5.4.21, p. 188). If $\dot{\mathbf{x}} = f(\mathbf{x}, u)$ is exponentially stable, then by theorem (4) there exists a function $v(\mathbf{x}, t)$ that satisfies Eqs. (5), (6), and (7). Writing $f(\mathbf{x}, t) = \mathbf{A}(t)\mathbf{x} + f'(\mathbf{x}, t)$, by assumption (2), $\forall \varepsilon < 1$, there exists $\delta \leq h'$ such that if $|\mathbf{x}| < \delta$, then $|f'(\mathbf{x}, t)| < (\varepsilon\alpha_3/\alpha_4)|\mathbf{x}|$. Evaluating $v(\mathbf{x}, t)$ along the trajectories of $\dot{\mathbf{x}} = \mathbf{A}(t)\mathbf{x}$ gives

$$\begin{aligned}\frac{dv(\mathbf{x}(t), t)}{dt} &= \frac{\partial v(\mathbf{x}, t)}{\partial t} + \frac{\partial v(t, \mathbf{x})}{\partial \mathbf{x}}\mathbf{A}(t)\mathbf{x} \\ &= \frac{\partial v(\mathbf{x}, t)}{\partial t} + \frac{\partial v(t, \mathbf{x})}{\partial \mathbf{x}} f(\mathbf{x}, t) - \frac{\partial v(t, \mathbf{x})}{\partial \mathbf{x}} f'(\mathbf{x}, t)\end{aligned}$$

$$\leq \quad -\alpha_3|\mathbf{x}|^2 + \alpha_4|\mathbf{x}||f'(\mathbf{x},t)|, \quad \forall\, \mathbf{x} \in B_{h'} \tag{41}$$

where we used the properties of $v(\mathbf{x},t)$ to get Eq. (41). If $|\mathbf{x}| < \delta$, then dv/dt along the trajectories of $\dot{\mathbf{x}} = \mathbf{A}(t)\mathbf{x}$ satisfies

$$\frac{dv(\mathbf{x},t)}{dt} \leq -(1-\varepsilon)\alpha_3|\mathbf{x}|^2$$

Thus $\dot{\mathbf{x}} = \mathbf{A}(t)\mathbf{x}$ possesses a locally positive definite (l.p.d.f) decrescent Lyapunov function $v(\mathbf{x},t)$ such that $-dv(\mathbf{x},t)/dt$ is l.p.d.f. along its trajectories. By the Lyapunov theorem ([8], theorem 5.2.45, p. 153) the linearized system is uniformly asymptotically stable. □

Proof of lemma (8):
i) "$\Longleftarrow$" implication: we have

$$|g_\delta| \leq \int_t^{t+s} |f(\tau)|d\tau \leq \sqrt{s}||f||_{t,t+s}$$

so that $||g_\delta||_{t,t+s} \leq s||f||_{t,t+s}$. Thus $||g_\delta||_{t,t+s}$ goes to zero as $t \to \infty$.
"$\Longrightarrow$" implication: For a given $\varepsilon > 0$ choose $\delta > 0$ such that

$$|f(\tau) - f(t)| \leq \varepsilon, \quad \forall \tau \in [t, t+\delta], \quad \forall t \geq 0$$

Since $||g_\delta||_{t,t+s}$ goes to zero as t goes to infinity, $\exists\; T > 0$ such that for all $t \geq T$, $||g_\delta||_{t,t+s} \leq \delta\varepsilon$. Expressing g_δ as

$$g_\delta(t) = \delta f(t) + \int_t^{t+\delta} [f(\tau) - f(t)]\; d\tau$$

and taking the norm of both sides, we get

$$\delta||f||_{t,t+s} \leq ||g_\delta||_{t,t+s} + \delta s\varepsilon \leq \delta\varepsilon(1+s)$$

Thus $||f||_{t,t+s} \leq \varepsilon(1+s)$. Since ε is arbitrary, $||f||_{t,t+s}$ goes to zero as t goes to infinity. □

Proof of lemma (14):
First, from Eq. (30) it is easily seen that $\eta(0,0,t) = 0$, and that η

is continuously differentiable (in fact, C^{∞}) with respect to $\mathbf{e}$ and $\mathbf{e}_r$. Taking the partial derivative of η in Eq. (30) with respect to $\mathbf{e}$ and $\mathbf{e}_r$, respectively, we obtain

$$\begin{aligned}\Delta_1(t) &= \lambda_P^2 \mathbf{M}(\mathbf{q}_d) + \left[\frac{\partial \mathbf{M}(\mathbf{q})\ddot{\mathbf{q}}_d}{\partial \mathbf{q}} + \frac{\partial \mathbf{C}(\mathbf{q},\dot{\mathbf{q}}_d)\dot{\mathbf{q}}_d}{\partial \mathbf{q}} + \frac{\partial \mathbf{g}(\mathbf{q})}{\partial \mathbf{q}}\right]\Big|_{\mathbf{q}=\mathbf{q}_d} \\ \Delta_2(t) &= -\lambda_P \mathbf{M}(\mathbf{q}_d) + \frac{\partial \mathbf{C}(\mathbf{q}_d,\dot{\mathbf{q}})\dot{\mathbf{q}}_d}{\partial \dot{\mathbf{q}}}\Big|_{\dot{\mathbf{q}}=\dot{\mathbf{q}}_d}\end{aligned}$$

Since the $\mathbf{q}_d$, $\dot{\mathbf{q}}_d$, and $\ddot{\mathbf{q}}_d$ are uniformly bounded, the assertions of the lemma is proved. □

Proof of lemma (19):

Let $\mathbf{v}$ be an arbitrary vector belonging to $\mathcal{N}(\mathbf{S})$. Multiplying Eq. (34) from the left and right side by $\mathbf{v}^{\mathbf{T}}$ and $\mathbf{v}$, respectively, gives

$$\int_t^{t+s} |\mathbf{W}_d(\tau)\mathbf{v}|^2 \, \mathbf{d}\tau = \mathbf{0}, \qquad \forall\ \mathbf{t} \geq \mathbf{0}$$

Since $\mathbf{W}_d(t)$ is continuous, this equation implies that $\mathbf{W}_d(t)\mathbf{v} = \mathbf{0}$ for all $t \geq 0$. Thus $\mathbf{W}_d(t)\mathbf{U}_{\mathcal{N}}\mathbf{U}_{\mathcal{N}}^T = 0$ as columns of $\mathbf{U}_{\mathcal{N}}$ span $\mathcal{N}(\mathbf{S})$. Using that

$$\mathbf{U}_{\mathcal{R}}\mathbf{U}_{\mathcal{R}}^T + \mathbf{U}_{\mathcal{N}}\mathbf{U}_{\mathcal{N}}^T = \mathbf{I}$$

proves the second assertion. □

TECHNIQUES IN MODELING UNCERTAIN DYNAMICS FOR ROBUST CONTROL SYSTEM DESIGN

Altuğ İftar

Department of Electrical Engineering
University of Toronto
Toronto M5S 1A4, Canada

Ümit Özgüner

Department of Electrical Engineering
The Ohio State University
Columbus, Ohio 43210, U. S. A.

I. INTRODUCTION

One of the fundamental issues in feedback controller design is robustness. Since an exact model of a physical system would be very complicated if not impossible to obtain, the designer should base the controller design on a *nominal* model and should require the controller to perform satisfactorily under possible deviations from the nominal model. At this point, modeling of such deviations (*i.e.*, *uncertainties*) becomes a crucial issue. The uncertainties in a mathematical model of a physical system are basically due to two factors: unknown values of certain system parameters (*e.g.*, resistance of an electrical component) and totally unmodeled dynamics (*e.g.*, high frequency modes of a flexible structure). It has been generally accepted that state space models are more suitable for representing uncertainties due to parameter variations, while unmodeled dynamics can be represented easier in the frequency domain.

Various methods have been devised to test the stability of a system which is subject to parameter variations. A method based on vector Liapunov functions has been proposed by Šiljak [1]. Other Liapunov function based methods have been introduced by, among other researchers, Patel *et al.* [2], Yedavalli [3], Zhou and Khargonekar [4], and Keel *et al.* [5]. A frequency domain approach for systems represented in state space has been presented by Qiu and Davison [6]. Šiljak [7] has introduced a parameter space method based on Popov's stability criterion.

A number of approaches to determine whether a characteristic polynomial is Hurwitz invariant (*i.e.*, whether the roots are retained in the open left half complex plane) under parameter variations have also been studied. Barmish [8] has utilized the result of Kharitonov [9] to develop a test for perturbed systems. In his work [9], Kharitonov proved that it is necessary and sufficient to test only four polynomials to conclude Hurwitz invariance of a set of polynomials with independent coefficient variations. However, if the variations in coefficients of a polynomial are dependent (which is usually the case in control systems), Kharitonov's result becomes only sufficient and, hence, a degree of conservatism is involved. There has been some work to reduce such conservatism [10]. Alternative approaches have also been proposed. Biernachi *et al.* [11] have introduced a parameter space method to test Hurwitz invariance. Bartlett *et al.* [12] have shown that, if the family of all possible characteristic polynomials form a polytope in the coefficient space, then the exposed edges of the polytope determine whether the entire family is Hurwitz or not.

Special controller design methods for systems which are subject to parameter variations have also been proposed. Karmarkar and Šiljak [13] have developed a computer aided design approach to maximize the stability hypercube in parameter space. Ackermann [14] has introduced a pole placement approach. In [15], Bernstein and Hyland used the *Maximum Entropy/Optimal Projection* approach [16] to design a low–order robust controller for a system with uncertain parameters. This approach has been compared with some of the frequency domain approaches in [17]. Yedavalli [18] has proposed an optimal controller design approach with a cost function weighting measures of robustness and system performance. Keel *et al.* [5] have presented an approach which maximizes the radius of the stability hypersphere in parameter space. Bernstein and Haddad [19] have introduced a robust controller design methodology based on the Kharitonov's Theorem [9].

Much research has also been undertaken to establish frequency domain design strategies for uncertain systems (*e.g.*, [20]–[26]). However, most of these approaches are designed for systems with truely unstructured uncertainties and can not take advantage of all the available information. More specifically, these methods assume only a known upper bound on the magnitude of the possible perturbations represented in the frequency domain. However, in practice more information is generally available. For example, in the case of flexible structures with co–located actuation and sensing, it is known that the phase of the uncertain dynamics always lies between $0°$ and $180°$. Such additional information may relax the robust controller design problem considerably and would, in general, lead to a less conservative controller design.

Many physical systems possess both parameters with unknown exact values and uncertain dynamics. In [27], Wei and Yedavalli proposed a combined frequency domain and state space approach for such systems. However, such an approach would, in general, necessitate alternating design stages in state space and frequency domain. In [28], Boyd considered representing unstructured uncertainty in a structured form and, hence, combining both classes for a state space design. However, his approach is restricted to systems which can be transformed into the so–called standard form. Furthermore, only magnitude bounds on those blocks can be utilized in the design process. Some research has also been undertaken to represent structured uncertainty (which may be due to parameter variations) in the frequency domain. Horowitz [29] has proposed an approach based on constructing regions (called *templates*) on the complex plane to represent the plant variation for some frequency point. Once the templates are determined, stability–robustnees and acceptable performance bounds on the loop gain at the corresponding frequencies can be constructed; and (if possible) a controller can be designed to satisfy these bounds [29]. An alternative approach proposed by Sideris and Safonov uses a series of conformal mappings of the templates onto the unit disk to transform the structured uncertainty problem to one with unstructured uncertainty which can be solved using H^{∞}–theory or Nevanlinna–Pick interpolation techniques [30]. However, determination of the templates may be very tedious in many cases. Usually, a large number of frequency points must be considered for a useful representation of the plant parameter uncertainties. Furthermore, the approach proposed in [29] requires a number of trial and error designs and H^{∞} minimization suggested in [30] often results in a high dimensional controller.

In the present chapter, a unified approach in state space, which was first introduced in [31] and [32], is presented. This approach can be used for systems with both classes of uncertainties. Furthermore, the approach can utilize both phase and gain information about the uncertain dynamics. In Section II, it is demonstrated that, under mild conditions, it is possible to obtain a rational transfer function matrix (TFM), possibly parametrized by a finite dimensional vector, to represent uncertain dynamics and design a controller accordingly to satisfy stability and desired performance. The underlying idea here is to represent possibly very high dimensional uncertain dynamics in a relatively low order structured form. In Section III, the state space representation of such dynamics and the robust controller design problem is discussed. Parameter uncertainty can be combined with such a representation of uncertain dynamics at this stage. The presented approach is applied to a robust controller design problem for a flexible robot manipulator in Section IV. The system under consideration possesses both parameters with unknown exact values and unmodeled (high frequency) dynamics. Extension of the proposed approach to decentralized robust controller design is discussed in Section V. Section VI includes some concluding remarks.

In the sequel $\mathbf{C}$ denotes the space of complex numbers, $\mathbf{C}^+$ denotes the set of complex numbers with nonnegative real parts, $\mathbf{R}$ denotes the space of real numbers, $\mathbf{R}^k$ denotes the k dimensional real vector space, $\mathbf{R}^+$ denotes the set of nonnegative real numbers, I denotes the identity matrix of appropriate dimensions, $[a, b]$ denotes the closed interval of $\mathbf{R}$ from a to b, and $\overline{(\cdot)}$, $|\cdot|$, and $\angle(\cdot)$ denote, respectively, the complex conjugate, the magnitude, and the phase of $(\cdot)$.

II. REDUCED ORDER MODELS FOR UNCERTAIN DYNAMICS

The uncertain dynamics of a linear system, nominally modeled by $G(s)$, can be represented at the input by $\Delta_i(s)$, at the output by $\Delta_o(s)$, or additively by $\Delta_a(s)$ as shown in Figure 1. For brevity, here we consider only uncertainties represented at the output. Similar results can be proved for other types of representations along the same lines. Henceforth we drop the subscript "o" of $\Delta_o(s)$.

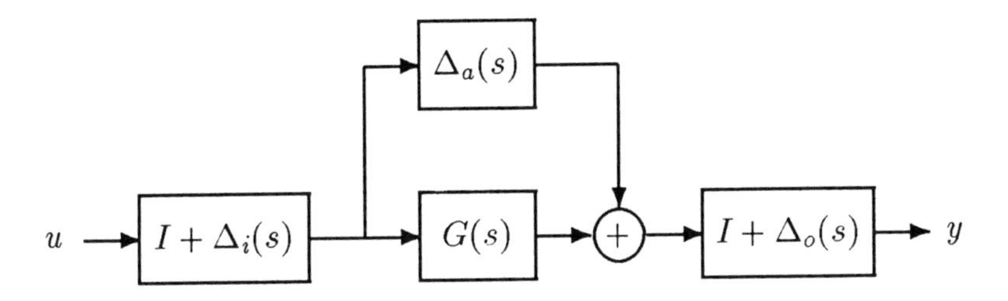

Figure 1: Representation of uncertain dynamics.

Although $\Delta(s)$ may not be known exactly, some information about it is generally available. Furthermore, $\Delta(s)$ may not be representable as a finite dimensional linear system with a rational TFM, or such a representation may require an unmanageably high dimensional model.

Our purpose is to determine a rational TFM $E(s;p)$ to replace $\Delta(s)$ in the model, such that a feedback controller $K(s)$ designed for such a model stabilizes the actual plant and achieves desired performance. Here $E(s;p)$ is possibly parametrized by a vector p which is taken from a parameter set Π. The following theorem demonstrates that once a suitable TFM and an associated parameter set are found to represent the uncertainties, such an approach would, in fact, guarantee the stability of the actual closed-loop system.

Theorem 1a (*Absolute Stability*): Suppose that $\Delta(s)$ is analytic in the closed right half complex plane $\mathbf{C}^+$, except possibly at some isolated singular points which constitute a total of m poles with due count of multiplicity.[1] Let $E(s;p)$ be a TFM which is rational in s and continuous in p and Π be a subset of a finite dimensional real vector space with the properties:

a) for all $\omega \in \mathbf{R}$ there exists a $p_\omega \in \Pi$ such that

$$E(j\omega; p_\omega) = \Delta(j\omega) \tag{1a}$$

if $\Delta(s)$ is analytic at $s = j\omega$ or

$$\lim_{r \to 0} E(j\omega + re^{j\theta}; p_\omega) = \lim_{r \to 0} \Delta(j\omega + re^{j\theta}) \tag{1b}$$

for all $\theta \in [-\frac{\pi}{2}, \frac{\pi}{2}]$, if $\Delta(s)$ is not analytic at $s = j\omega$,

b) there exists a $p_\infty \in \Pi$ such that

$$\lim_{r \to \infty} E(re^{j\theta}; p_\infty) = \lim_{r \to \infty} \Delta(re^{j\theta}) \tag{1c}$$

for all $\theta \in [-\frac{\pi}{2}, \frac{\pi}{2}]$, and

c) $E(s;p)$ has exactly m poles in $\mathbf{C}^+$ for all $p \in \Pi$.

For a given rational TFM $G(s)$, suppose that a rational TFM $K(s)$ is chosen such that the closed-loop TFM

$$T_E(s;p) \triangleq (I + E(s;p))G(s)K(s)[I + (I + E(s;p))G(s)K(s)]^{-1} \tag{2}$$

does not have any poles in $\mathbf{C}^+$ for all $p \in \Pi$. Also assume that $\det[I + (I + \Delta(s))G(s)K(s)] \neq 0$ on a dense subset of $\mathbf{C}$. Then, the true closed-loop TFM:

[1] Note that, one needs to know the total number of unstable poles of $\Delta(s)$. Their actual locations need not be known.

$$T_\Delta(s) \triangleq (I + \Delta(s))G(s)K(s)[I + (I + \Delta(s))G(s)K(s)]^{-1} \tag{3}$$

is analytic in $\mathbf{C}^+$.

Proof: Throughout in this proof we drop the arguments s and p for notational brevity. Let n denote the number of poles of $(I + \Delta)GK$ in $\mathbf{C}^+$. Then, since Δ and E have the same number of poles in $\mathbf{C}^+$, $(I + E)GK$ also has n poles in $\mathbf{C}^+$ for all $p \in \Pi$. Furthermore, since T_E does not have any poles in $\mathbf{C}^+$, by the multivariable Nyquist stability criterion [33], the map $\det[I + (I + E)GK]$, as s is varied on the standard Nyquist contour $\mathcal{D}$, encircles the origin $-n$ times for all $p \in \Pi$. By the hypothesis, Δ is analytic on $\mathbf{C}^+$, except possibly at isolated singular points. The same is also true for G and K, since they are rational TFMs. Therefore, $\det[I + (I + \Delta)GK]$ is analytic on $\mathbf{C}^+$, except possibly at isolated singular points. Furthermore, since $\det[I + (I + \Delta)GK] \neq 0$ on a dense subset of $\mathbf{C}$, T_Δ is well defined on such a set and we can apply the multivariable Nyquist stability criterion. Note that, the conditions a and b imply that the loci of $\det[I + (I + \Delta)GK]$, as s is varied on $\mathcal{D}$, is a subset of $\det[I + (I + E)GK]$ loci as s is varied on $\mathcal{D}$ and p is varied on Π (see Figure 2). Furthermore, since E is continuous in p, any member of the former set must encircle any point on the complex plane that is encircled by the latter set the same number of times. Thus, $det[I + (I + \Delta)GK]$ encircles the origin $-n$ times as s is varied on $\mathcal{D}$, which, by the multivariable Nyquist stability criterion, proves that T_Δ does not have any poles in $\mathbf{C}^+$. Hence, the result follows. $\square$

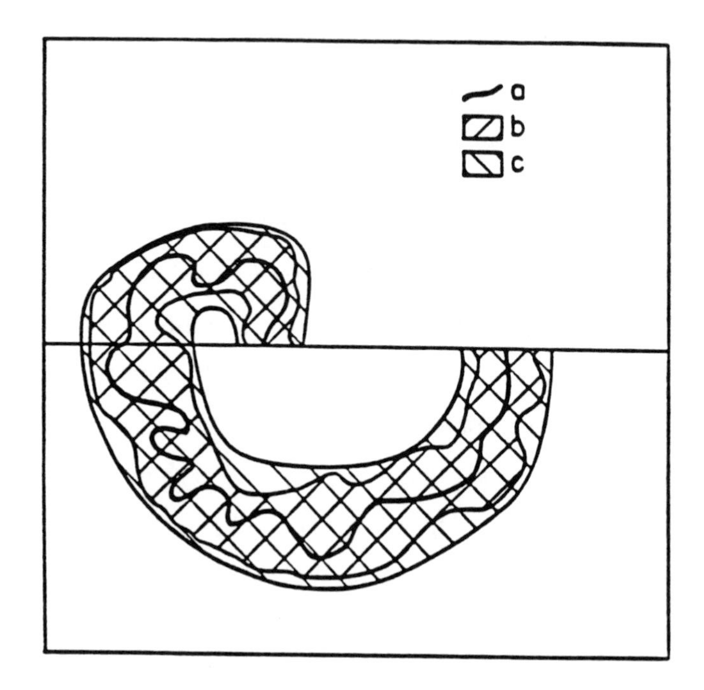

Figure 2: a) A representative $\det[I + (I + \Delta)GK]$ loci, b) region of all possible $\det[I + (I + \Delta)GK]$ loci, c) $\det[I + (I + E)GK]$ loci as s is varied on the half Nyquist contour and p is varied on Π.

Remark 1: Note that $T_\Delta(s)$ represents a physical system whenever $\Delta(s)$ does, since all other terms that appear in the right hand side of (3) are rational TFMs. Furthermore, $T_\Delta(s)$ represents the actual closed-loop system whenever $\Delta(s)$ is the TFM representing the actual unmodeled dynamics. Thus, whenever $T_\Delta(s)$ is analytic in $\mathbf{C}^+$, the actual closed-loop system is stable.

In many practical cases, absolute stability may not be sufficient.

Instead, for example, one may wish to confine all the closed-loop eigenvalues in a region $\mathcal{R}$ of the complex plane. Note that the imaginery axis and the right hand semi-circle of infinite radius together define the boundary of $\mathbf{C}^+$. Hence, if the conditions a and b of Theorem 1a are met, we say that $E(s,p)$ and $\Delta(s)$ are *matched* on the boundary of $\mathbf{C}^+$. Furthermore, we say that the two TFMs are matched on the boundary of a region $\mathcal{R}$, if the obvious generalizations of the conditions a and b of Theorem 1a hold. The following result can be proved as Theorem 1a, with an obvious modification of the Nyquist contour.

Theorem 1b (*Relative Stability*): Let $\mathcal{R}$ be an open subset of $\mathbf{C}$. Suppose that $\Delta(s)$ is analytic in $\mathcal{R}^c \triangleq \mathbf{C}\setminus\mathcal{R}$, except possibly at some isolated singular points which constitute a total of m poles with due count of multiplicity. Let $E(s;p)$ be a TFM which is rational in s and continuous in p and Π be a subset of a finite dimensional real vector space such that $E(s,p)$ and $\Delta(s)$ are matched on the boundary of $\mathcal{R}^c$ and $E(s,p)$ has exactly m poles in $\mathcal{R}^c$ for all $p \in \Pi$. For a given rational TFM $G(s)$, suppose that a rational TFM $K(s)$ is chosen such that the closed-loop TFM (2) does not have any poles in $\mathcal{R}^c$ for all $p \in \Pi$. Also assume that $\det[I + (I + \Delta(s))G(s)K(s)] \neq 0$ on a dense subset of $\mathbf{C}$. Then, the true closed-loop TFM (3) is analytic in $\mathcal{R}^c$. □

It is well known that most of the widely used performance measures (*e.g.*, disturbance rejection and steady state error) can be related to the return difference matrix [21]. For example, to achieve a certain degree of plant disturbance rejection at the output, one may require the return difference matrix to satisfy:

$$[I + (I + \Delta(j\omega))G(j\omega)K(j\omega)]^S \geq Q(\omega) \qquad \forall\omega \in \Omega\,, \tag{4}$$

provided that the left hand side is well defined.[2] Here $Q(\omega)$ is a positive definite matrix for all $\omega \in \Omega$, $\Omega \subset \mathbf{R}$ is the set of frequencies where disturbances are effective, for appropriately dimensioned matrices A and B, $A^S \triangleq A^H A$, A^H denotes the complex conjugate transpose of A, and $A \geq B$ means $A - B$ is positive semi–definite. The following theorem demonstrates that if a controller is designed to satisfy such a performance criterion for the *design model* (the model in which Δ is replaced by E), then the actual closed–loop system satisfies the same criterion.

Theorem 1c (*Good Performance*): Under the conditions of Theorem 1a, suppose that $K(s)$ is chosen such that

$$[I + (I + E(j\omega;p))G(j\omega)K(j\omega)]^S \geq Q(\omega) \tag{5}$$

for all $\omega \in \Omega$ and for all $p \in \Pi$. Then, (4) is satisfied, provided that the left hand side of (4) is well defined for all $\omega \in \Omega$.

Proof: If the left hand side of (4) is well defined, then $\Delta(s)$ is analytic at $s = j\omega$. By (1a), for such ω there exists a $p_\omega \in \Pi$ such that

[2]The more widely used criterion is defined as a lower bound on the singular values of the return difference matrix. However, note that (4) is a more general condition.

$$I + (I + E(j\omega; p_\omega))G(j\omega)K(j\omega) = I + (I + \Delta(j\omega))G(j\omega)K(j\omega).$$

Hence, the result follows. □

Under mild conditions on $\Delta(s)$, the existence of a TFM $E(s;p)$ and a parameter set Π, satisfying the conditions of the above theorems, can be ensured. To avoid notational complexity, we prove this only for the scalar case. The extension to the multivariable case is possible along similar lines. Furthermore, here we consider only absolute stability. A similar result can be proved for the relative stability case if the region $\mathcal{R}$ is symmetric about the real axis.

Theorem 2: Let

$$\Delta(s) = \frac{1}{d_u(s)}\Delta_s(s) \tag{6}$$

where $d_u(s)$ is an mth order monic polynomial with zeros in $\mathbf{C}^+$ and $\Delta_s(s)$ is analytic in $\mathbf{C}^+$ (*i.e.*, factor out the unstable poles of $\Delta(s)$ as $\frac{1}{d_u(s)}$). Furthermore, suppose that $\Delta(s)$ is such that $\Delta(j\omega) = \overline{\Delta(-j\omega)}$ for all $\omega \in \mathbf{R}$, and that $\lim_{r\to\infty} \Delta(re^{j\theta})$ is a finite real constant[3] for all $\theta \in [-\frac{\pi}{2}, \frac{\pi}{2}]$. Then, there exists a function $E(s;p)$, continuous in p and proper and rational in s, and a subset Π of a finite dimensional real vector space, such that conditions a, b, and c of Theorem 1a hold.

Proof: Let $p = (\alpha, \beta) = (\alpha_1, \ldots, \alpha_{11}, \beta_1, \ldots, \beta_m)$ and

$$E(s;p) = \frac{1}{e(s;\beta)}E_s(s;\alpha)$$

[3] The actual value of this constant need not be known.

where $e(s;\beta) = s^m + \beta_1 s^{m-1} + \cdots + \beta_{m-1}s + \beta_m$ and

$$E_s(s;\alpha) = \frac{\alpha_6 s^5 + \alpha_7 s^4 + \alpha_8 s^3 + \alpha_9 s^2 + \alpha_{10} s + \alpha_{11}}{s^5 + \alpha_1 s^4 + \alpha_2 s^3 + \alpha_3 s^2 + \alpha_4 s + \alpha_5} \, .$$

For any $\omega \in \mathbf{R}^+$, let $\mathcal{R}_u(\omega) \subset \mathbf{C}$ be the set of all possible values of $d_u(j\omega)$. Then, it is possible to find a subset $B(\omega)$ of $\mathbf{R}^m$, such that

(i) the loci of $e(j\omega;\beta)$ as β is varied over $B(\omega)$ contains $\mathcal{R}_u(\omega)$, and

(ii) $e(s,\beta)$ has m zeros in $\mathbf{C}^+$ for all $\beta \in B(\omega)$.

For any $\omega \in \mathbf{R}^+$, let $\mathcal{R}_s(\omega) \subset \mathbf{C}$ be the set of all possible values of $\Delta_s(j\omega)$.[4] Then, it is possible to find a subset $A(\omega)$ of $\mathbf{R}^{11}$, such that

(i) the loci of $E_s(j\omega;\alpha)$ as α is varied over $A(\omega)$ contains $\mathcal{R}_s(\omega)$, and

(ii) $E_s(s,\alpha)$ is analytic in $\mathbf{C}^+$ for all $\alpha \in A(\omega)$.

Let $\mathcal{R}_\infty \subset \mathbf{R}$ be the set of all possible values of $\lim_{r\to\infty} \Delta(re^{j\theta})$. Then, it is possible to find subsets A_∞ of $\mathbf{R}^{11}$ and B_∞ of $\mathbf{R}^m$, such that

(i) the loci of $\lim_{r\to\infty} E(re^{j\theta};\alpha,\beta)$ as α is varied over A_∞ and β is varied over B_∞ contains $\mathcal{R}_\infty$,

(ii) $E_s(s,\alpha)$ is analytic in $\mathbf{C}^+$ for all $\alpha \in A_\infty$, and

(iii) $e(s,\beta)$ has m zeros in $\mathbf{C}^+$ for all $\beta \in B_\infty$.

---

[4] $\mathcal{R}_s(\omega)$ is well defined since $\Delta_s(s)$ is analytic on the imaginary axis $j\mathbf{R} \subset \mathbf{C}^+$.

Let

$$\bar{A} \triangleq A_\infty \bigcup \left\{ \bigcup_{\omega \in \mathbf{R}+} A(\omega) \right\} ,$$

$$\bar{B} \triangleq B_\infty \bigcup \left\{ \bigcup_{\omega \in \mathbf{R}+} B(\omega) \right\} ,$$

and

$$\Pi \triangleq \bar{A} \times \bar{B} .$$

Then, $E(s;p)$ is continuous in p and proper and rational in s, Π is a subset of a finite dimensional real vector space ($\mathbf{R}^{m+11}$), and the conditions a, b, and c of Theorem 1a hold. □

Remark 2: In certain cases, depending on how much is known about $\Delta(s)$, the minimum required dimension of Π (*i.e.*, the number of parameters) can, in fact, be less than the number predicted in the above proof. On the other hand, in certain other cases, one may prefer to work with a higher order TFM and a higher dimensional parameter set to reduce the possible conservatism involved in representing $\Delta(s)$.

Remark 3: Note that $E(s;p)$ and $\Delta(s)$ may be matched by using a different $p \in \Pi$ at different points s on the boundary of $\mathbf{C}^+$ (or of $\mathcal{R}^c$ in the case of relative stability). However, this fact does not bring any restrictions in satisfying properties such as absolute stability, relative stability, or good performance of the actual system (see the proofs of Theorems 1a and 1c).

The following example illustrates the procedure of determining $E(s;p)$ and Π.

Example 1: Suppose, all that is known about $\Delta(s)$ is that it is analytic in $\mathbf{C}^+$, $\Delta(j\omega) = \overline{\Delta(-j\omega)}$, $\lim_{r\to\infty} \Delta(re^{j\theta}) = 0$ for all $\theta \in [-\frac{\pi}{2}, \frac{\pi}{2}]$,

$$\left|\frac{\delta_m \omega_1}{j\omega + \omega_1}\right| \leq |\Delta(j\omega)| \leq \left|\frac{\delta_M \omega_1}{j\omega + \omega_1}\right| \qquad \forall \omega \in \mathbf{R}^+ ,$$

and

$$\angle\left(\frac{1}{j\omega + \omega_1}\right) \leq \angle(\Delta(j\omega)) \leq \angle\left(\frac{1}{j\omega + \omega_2}\right) \qquad \forall \omega \in \mathbf{R}^+ ,$$

where $\delta_M > \delta_m > 0$ and $\omega_2 > \omega_1 > 0$ are known numbers. The gain and the phase of all possible $\Delta(s)$, together with a representative $\Delta(s)$ are depicted in Figure 3. Let us choose

$$E(s;p) = \frac{\epsilon}{s + \gamma} , \tag{7}$$

where $p = (\gamma, \epsilon)$. For each frequency $\omega \in \mathbf{R}^+$,

- by varying γ within the interval $[\omega_1, \omega_2]$ we can satisfy the phase condition $\angle(E(j\omega;p)) = \angle(\Delta(j\omega))$ and
- for a fixed $\gamma \in [\omega_1, \omega_2]$, we can meet the magnitude condition $|E(j\omega,p)| = |\Delta(j\omega)|$ by varying ϵ within the interval $[\delta_m \omega_1, \delta_M \gamma]$

for all possible $\Delta(s)$. Hence, we obtain

$$\Pi = \{(\gamma, \epsilon) \mid \omega_1 \leq \gamma \leq \omega_2 \ , \ \delta_m \omega_1 \leq \epsilon \leq \delta_M \gamma\} \tag{8}$$

which is a bounded subset of R^2 and is depicted in Figure 4. □

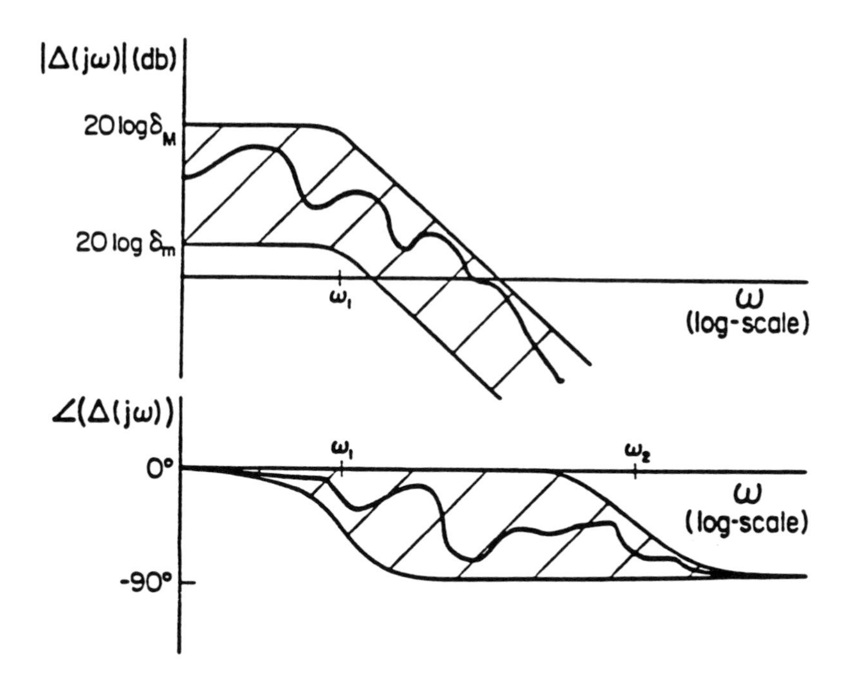

Figure 3: Regions of all possible magnitude and phase of $\Delta(j\omega)$ and a representative $\Delta(j\omega)$ for Example 1.

Remark 4: Note that, the tightest H^{∞} norm bound for $\Delta(s)$ described in the above example is δ_M. Suppose that a controller was to be designed, say to achieve robust stability, for a plant with uncertainty $\Delta(s)$ by using the H^{∞} approach [24]. Then, it would be necessary to design a controller to stabilize all the plants with uncertainty $E(s)$ satisfying

$$||E(s)||_{\infty} \leq \delta_M \ , \tag{9}$$

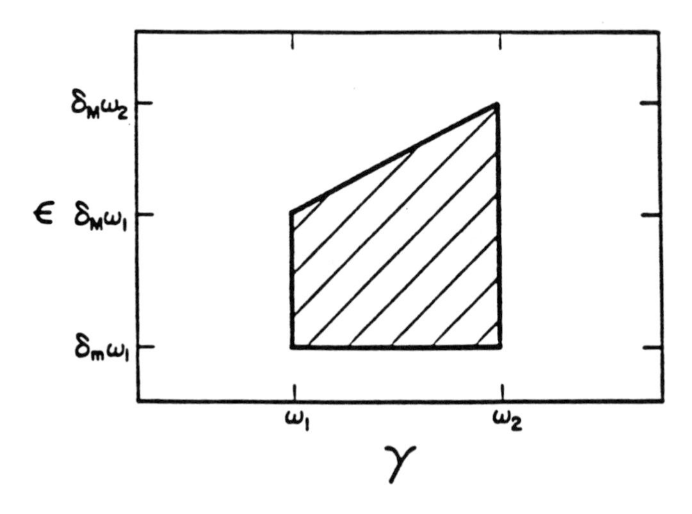

Figure 4: The parameter set Π for Example 1.

where $||\cdot||_\infty$ denotes the usual H^∞ norm. This, however, is a larger class than the one described by (7) and (8). Therefore, a controller designed by using the H^∞ approach would, in general, be more conservative than a controller designed using the present approach. It is, of course, possible to modify the nominal TFM and reduce the bound in (9); however, it is apparent that by using the norm bounds alone (as in the H^∞ approach) one can not get a tighter representation (of the actual uncertainty) than the representation obtained by the present approach.

III. STATE SPACE REPRESENTATION AND CONTROLLER DESIGN

Consider a system modeled by a nominal TFM $G(s;p_G)$, where $p_G \in \Pi_G$ is a vector denoting possible values of some system parameters. Suppose that the dynamics not modeled by G can be represented at the output by an uncertain TFM $\Delta(s)$ satisfying the conditions of Theorem 2. Then, one can obtain a rational TFM $E(s;p_E)$ and a corresponding parameter set Π_E, satisfying the conditions of Theorem 1a (or of Theorem 1b if relative, rather than absolute, stability is of concern), as discussed in the previous section.

Let $(A(p_G), B(p_G), C(p_G), D(p_G))$ be a realization for $G(s;p_G)$, *i.e.*, let

$$G(s;p_G) \triangleq C(p_G)(sI - A(p_G))^{-1}B(p_G) + D(p_G)\,.$$

Furthermore, let $(F(p_E), G(p_E), H(p_E), L(p_E))$ be a realization for $E(s;p_E)$. Then, we obtain the following state space model for the uncontrolled system

$$\begin{aligned} \frac{d}{dt}\begin{bmatrix} x \\ \xi \end{bmatrix} &= \begin{bmatrix} A & 0 \\ GC & F \end{bmatrix}\begin{bmatrix} x \\ \xi \end{bmatrix} + \begin{bmatrix} B \\ GD \end{bmatrix} u \\ y &= \begin{bmatrix} (I+L)C & H \end{bmatrix}\begin{bmatrix} x \\ \xi \end{bmatrix} + (I+L)Du\,. \end{aligned} \tag{10}$$

Suppose that a feedback controller $K(s)$, which has a realization

(A_K, B_K, C_K, D_K), is designed so that the closed–loop dynamics are described by

$$\frac{d}{dt}\begin{bmatrix} x \\ \xi \\ \eta \end{bmatrix} = A_{CL}\begin{bmatrix} x \\ \xi \\ \eta \end{bmatrix} \tag{11}$$

where

$$A_{CL} = \begin{bmatrix} A - BUD_K(I+L)C & -BUD_KH & BUC_K \\ GC - GDUD_K(I+L)C & F - GDUD_KH & GDUC_K \\ -B_K(I+L)C + B_K(I+L)DUD_K(I+L)C & -B_KH + B_K(I+L)DUD_KH & A_K - B_K(I+L)DUC_K \end{bmatrix}$$

where $U \triangleq (I + D_K(I+L)D)^{-1}$. The following result follows from Theorems 1a, 1b, and 1c.

Corollary 1: If $E(s;p_E)$ and Π_E are chosen to satisfy the conditions of Theorem 1a and the closed–loop system (11) is stable for all $p \triangleq (p_G, p_E) \in \Pi \triangleq \Pi_G \times \Pi_E$, then the actual closed–loop system, shown in Figure 5, is stable. Furthermore, if the closed–loop system (11) satisfies a performance criterion like (4) for all $p \in \Pi$, then the actual closed–loop system also satisfies the same criterion. If $E(s;p_E)$ and Π_E are chosen to satisfy the conditions of Theorem 1b and the closed–loop system (11) has eigenvalues only in the region $\mathcal{R}$ for all $p \in \Pi$, then the actual closed–loop system has the corresponding relative stability properties. □

Once a closed–loop system such as (11) is designed, one of the many existing state space methods, such as those mentioned in the

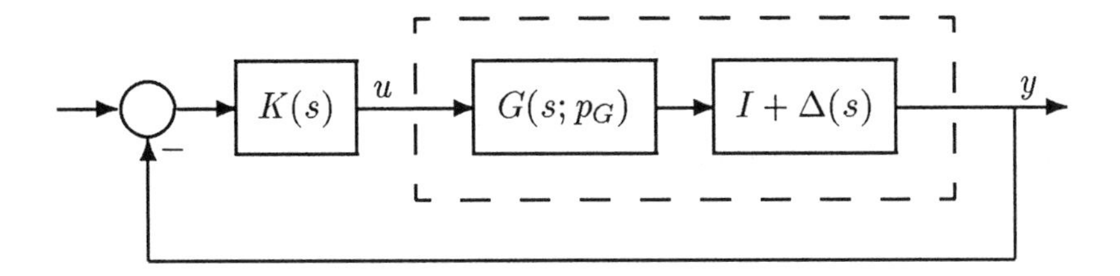

Figure 5: Closed–loop system.

introduction section of the present chapter, can be used to test the robust stability of this system. Alternatively, one of the robust controller design approaches, such as those mentioned in the same section, can be applied to the open–loop system (10) to obtain a robustly stable closed–loop system.

IV. APPLICATION TO A FLEXIBLE ROBOT MANIPULATOR

In this section, we consider the robust controller design problem for a one–link flexible robot manipulator on the horizontal plane [34]. The dynamics of the manipulator can be described by [35]:

$$\left[I_h + \int_0^L \rho x^2 dx \right] \ddot{\theta} = u - \int_0^L \rho x z_{tt} dx \tag{12a}$$

$$z_{tt} + \frac{EI_a}{\rho} z_{xxxx} = -x\ddot{\theta} \tag{12b}$$

where L, ρ, E, and I_a are, respectively, the length, linear mass density, Young's modulus, and area moment of inertia of the link; I_h is the inertia of the hub; θ is the angle between the "rigid–body" link and the reference frame; $z(x,t)$ is the perpendicular deviation of the link from the rigid–body shape at location x along the link (measured from the hub to the tip) and at time t; and u is the torque applied at the hub to control the motion of the link. Subscripts 'x' and 't' denote the partial derivatives with respect to x and the time variable t respectively. It is assumed that the link has uniform mass and cross–section. The axial displacement, torsion and shear effects are neglected. The parameters of the manipulator are:

$$E = 6.8944 \times 10^{10} \ \mathrm{N/m^2}\,, \qquad I_a = 7.7610 \times 10^{-9} \ \mathrm{m^4}\,,$$

$$\rho = 1.3750 \ \mathrm{kg/m}\,, \qquad L = 2.0000 \ \mathrm{m}\,, \qquad I_h = 0.9200 \ \mathrm{kg\ m^2}\,.$$

The boundary conditions are given by

$$z(0,t) = z_x(0,t) = z_{xx}(L,t) = z_{xxx}(L,t) = 0\,. \tag{13}$$

The manipulator is actuated by a constant field current DC motor which is described by:

$$R_a i + L_a \frac{di}{dt} + K_a \dot{\theta} = v \tag{14a}$$

$$u = K_a i \tag{14b}$$

where i is the armature current and v is the voltage applied to the armature circuit (the control input). The *nominal* values of the motor parameters are:

$$R_a^n = 1.05 \ \Omega\,, \qquad L_a^n = 0.0053 \ \mathrm{H}\,, \qquad K_a^n = 1.11 \ \mathrm{V\,s}\,.$$

It is assumed that the actual values of the armature resistance and the motor constant may be between R_a^m and R_a^M and K_a^m and K_a^M respectively, where

$$R_a^m = 1.00 \ \Omega\,, \qquad R_a^M = 1.10 \ \Omega\,,$$

$$K_a^m = 1.00 \ \mathrm{V\,s}\,, \qquad K_a^M = 1.22 \ \mathrm{V\,s}\,.$$

By substituting (12b) into (12a), we obtain a simplified description of the rigid–body dynamics:

$$I_h \ddot{\theta} = u + y \tag{15}$$

where

$$y = EI_a \int_0^L x z_{xxxx} dx = EI_a z_{xx}(0,t) \tag{16}$$

represents the effect of the flexible part on the rigid–body. By using (15), (12b) can be re–written as:

$$z_{tt} + \frac{EI_a}{\rho} z_{xxxx} = -\frac{x}{I_h}(u+y)\,. \tag{17}$$

The function $z(x,t)$ (which is assumed to be square integrable over $0 \leq x \leq L$) can be represented as:

$$z(x,t) = \sum_{i=1}^{\infty} \phi_i(x) \eta_i(t)\,, \tag{18}$$

where $\phi_i(x)$ $(i = 1, 2, ...)$ are the *mode shapes* [36] with corresponding *mode numbers* k_i $(i = 1, 2, ...)$, which are the solutions of the following frequency equation:

$$\cos(kL)\cosh(kL) = -1\,. \tag{19}$$

By substituting (18) into (17), multiplying both sides by $\phi_i(x)$, and integrating over $0 \leq x \leq L$, the following description of the "modal dynamics" is obtained:

$$\ddot{\eta}_i + \lambda_i \eta_i = b_i(u + y)\,, \qquad i = 1, 2, ... \tag{20}$$

where $\lambda_i = \frac{EI_a}{\rho} k_i^4$, and

$$b_i = \frac{1}{L}\left[-\frac{1}{I_h}\int_0^L x\phi_i(x)dx\right] = -\frac{2}{I_h L k_i^2}\,. \tag{21}$$

Moreover, by using (18), (16) can be re-written as:

$$y(t) = EI_a \sum_{i=1}^{\infty} \phi_i''(0)\eta_i(t) = \sum_{i=1}^{\infty} c_i \eta_i(t)\,, \tag{22}$$

where $c_i = 2EI_a k_i^2$.

Let

$$\eta \triangleq (\eta_1, \eta_2, ...)^T\,, \qquad b \triangleq (b_1, b_2, ...)^T\,,$$

$$c \triangleq (c_1, c_2, ...)\,, \qquad \Omega \triangleq \mathrm{diag}\left(\sqrt{\lambda_1}, \sqrt{\lambda_2}, ...\right)\,.$$

Then, the following description of the modal dynamics is obtained:

$$\ddot{\eta} = \left(-\Omega^2 + bc\right)\eta + bu \tag{23a}$$

$$y = c\eta\,. \tag{23b}$$

Although, the above derivation ignores any damping, all flexible structures have some inherent, typically small, damping. To account for this effect, we modify (23a)–(23b) as follows:

$$\ddot{\eta} = \left(-\Omega^2 + bc\right)\eta - 2Z\Omega\dot{\eta} + bu \tag{24a}$$

$$y = c\eta\,, \tag{24b}$$

where Z is the matrix of modal damping coefficients. Since (24a)–(24b) represents an infinite dimensional system, it may not be very useful for controller design. To overcome this difficulty, we undertake the approach presented in the earlier sections of the present chapter.

Typical Bode plots for (24a)–(24b) are shown in Figure 6. The phase plot indicates that a second order model is sufficient to represent these dynamics. We let:

$$E(s, p_E) = \frac{-K_p\omega_p^2}{s^2 + 2\zeta_p\omega_p s + \omega_p^2}\,, \qquad p_E = (\omega_p, \zeta_p, K_p) \in \Pi_E \tag{25}$$

be such a representation with:

$$\Pi_E = \{(\omega_p, \zeta_p, K_p) \mid \omega_m \leq \omega_p < \infty\ ,\ \zeta_m \leq \zeta_p \leq \zeta_M\ ,\ 0 \leq K_p \leq K_M\}\ , \tag{26}$$

where ω_m is a lower bound for the lowest frequency of the flexible structure and can be taken as $\sqrt{\lambda_1}$ (the interaction term bc shifts the

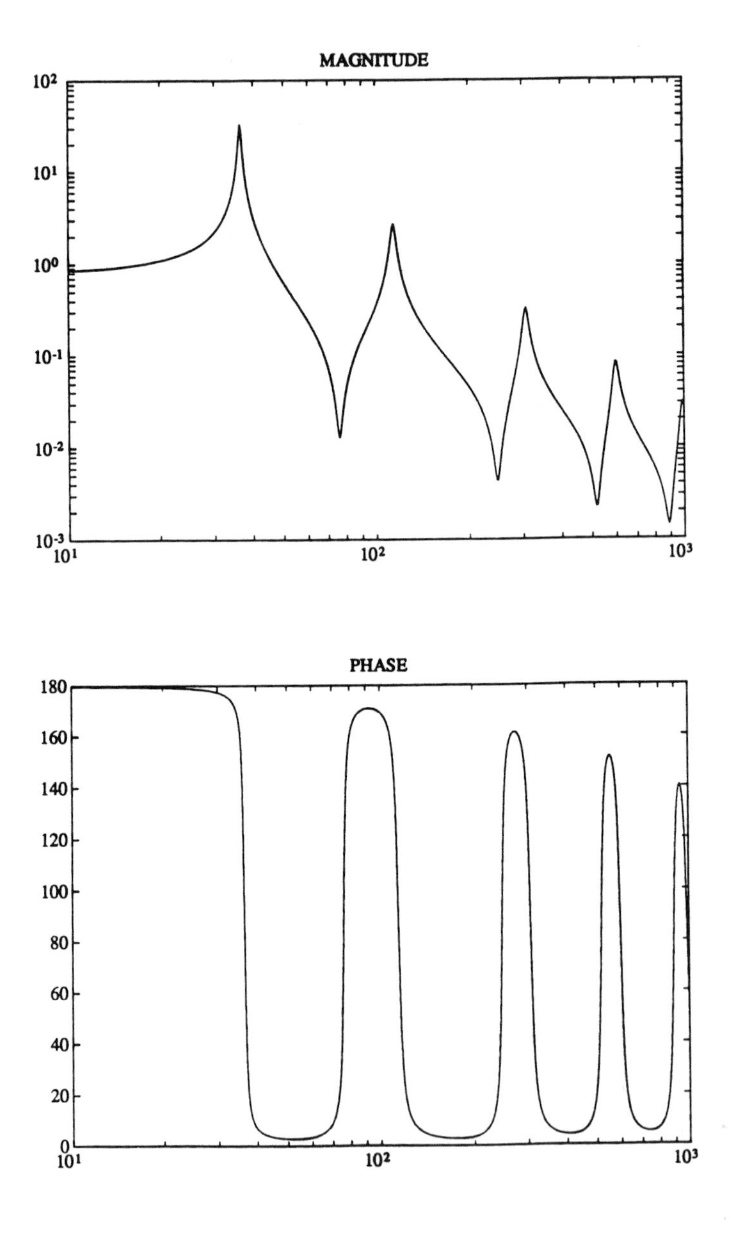

Figure 6: Typical Bode plots for the flexible dynamics.

frequencies up); ζ_m and ζ_M are, respectively, the lower and the upper bounds for the modal damping coefficients and, for a given physical structure, they can be determined experimentally; and K_M must be chosen such that

$$\max_{p\in\Pi} |E(j\omega, p)| \geq |\Delta(j\omega)|\,, \qquad \forall\omega \in \mathbf{R}\,, \tag{27}$$

where $\Delta(s)$ is the transfer function from u to y for the system (24a)–(24b). For the current example, the lower bound on ω_p is taken to be $\omega_m = 16$ rad/s, which is below the lowest structural frequency $\sqrt{\lambda_1} = 17.340$ rad/s; the minimum and the maximum damping ratios are assumed to be $\zeta_m = 0.01$ and $\zeta_M = 0.10$, respectively; and in order to satisfy (27), the upper bound on K_p is chosen as $K_M = 0.90265$ (see [34] for details).

A state space description of the design model can now be obtained as:

$$\dot{\xi} = A(p)\xi + Bv \tag{28}$$

where $\xi \triangleq \left(\theta, \dot{\theta}, \eta_E, \dot{\eta}_E, i\right)^T$, where $(\eta_E, \dot{\eta}_E)^T$ is the state of the pseudo system described by $E(s, p_E)$,

$$A(p) = \begin{bmatrix} 0 & 1 & 0 & 0 & 0 \\ 0 & 0 & \frac{\omega_p^2}{I_h} & 0 & \frac{K_a}{I_h} \\ 0 & 0 & 0 & 1 & 0 \\ 0 & 0 & -\omega_p^2 & -2\zeta_p\omega_p & -K_pK_a \\ 0 & -\frac{K_a}{L_a} & 0 & 0 & -\frac{R_a}{L_a} \end{bmatrix}, \qquad B = \begin{bmatrix} 0 \\ 0 \\ 0 \\ 0 \\ \frac{1}{L_a} \end{bmatrix},$$

and $p = (\omega_p, \zeta_p, K_p, R_a, K_a) \in \Pi$, where

$$\Pi = \{(\omega_p, \zeta_p, K_p, R_a, K_a) \mid \omega_m \le \omega_p < \infty \ , \ \zeta_m \le \zeta_p \le \zeta_M \ , \\ 0 \le K_p \le K_M \ , \ R_a^m \le R_a \le R_a^M \ , \ K_a^m \le K_a \le K_a^M \} \ . \quad (29)$$

In order to find a controller which will stabilize the overall system and which will achive satisfactory performance, let us consider the performance measure

$$J = \frac{1}{2} \int_{t_o}^{\infty} (\xi^T Q \xi + v^T v) dt \, , \quad (30)$$

for the design model (28), where in SI units $Q = \mathrm{diag}(100, 1, 1, 1, 1)$. For the purpose of optimization, the initial conditions of the system (28) are taken to be $\xi(t_0) = (\theta_0, 0, 0, 0, 0)^T$, where θ_0 is arbitrary. This choice corresponds to an initial condition of the physical manipulator being at rest without any deviations from the rigid body shape, and making an angle of θ_0 with the final desired position.

Our aim is to find a feedback gain matrix K^*, such that, when the feedback control law

$$v = K\xi \, , \quad (31)$$

with $K = K^*$, is applied to the system (28), the resulting closed–loop system is stable and, for some $p^* \in \Pi_*$,

$$J(K^*, p^*) = \min_{K \in \mathcal{K}} \max_{p \in \Pi_*} J(K, p) \, , \quad (32)$$

where Π_* is a specified bounded subset of Π,

$$\mathcal{K} \triangleq \{K \mid A(p) + BK \text{ is stable } \forall p \in \Pi\} \ , \quad (33)$$

and $J(K,p)$ is given by (30) with the control v as in (31). We take

$$\Pi_* = \{(\omega_p, \zeta_p, K_p, R_a, K_a) \mid \omega_m \leq \omega_p \leq \omega_M \ , \ \zeta_m \leq \zeta_p \leq \zeta_M \ , \\ 0 \leq K_p \leq K_M \ , \ R_a^m \leq R_a \leq R_a^M \ , \ K_a^m \leq K_a \leq K_a^M \} \ , \quad (34)$$

where $\omega_M = 90$ rad/s is between the first and the second modal frequencies of the flexible manipulator. Note that, although the optimization is carried out on the bounded set Π_*, the closed-loop system is required to be stable over the entire set Π.

The above optimization problem is solved by using the approach introduced in [37]. The optimal feedback gains in SI units are found to be:

$$K^* = \begin{bmatrix} -10.000 & -15.830 & -42.793 & -17.146 & -0.39276 \end{bmatrix} ;$$

the corresponding point in Π_* is $p^* = \begin{bmatrix} \omega_M & \zeta_m & K_M & R_a^M & K_a^m \end{bmatrix}$.

In order to apply the above optimal feedback to the actual system, note that the tip deviation $z_L \triangleq z(L,t)$ and its velocity $\dot{z}_L \triangleq z_t(L,t)$ can be approximately related to the state of the system (28) by

$$z_L \approx \frac{b_1 \phi_1(L)}{K_p} \eta_E \ , \qquad \dot{z}_L \approx \frac{b_1 \phi_1(L)}{K_p} \dot{\eta}_E \ .$$

Therefore, the feedback control law for the actual system is obtained as:

$$v = M^* q \ , \quad (35)$$

where $q = \left(\theta, \dot{\theta}, z_L, \dot{z}_L, i\right)^T$ and $M^* = K^* \cdot \text{diag}(1, 1, \alpha_1^*, \alpha_1^*, 1)$, where

$$\alpha_1^* = \frac{K_M}{b_1 \phi_1(L)} = -0.36498 \,.$$

For simulation purposes, the flexible part of the system is modelled by a 12th order modal model derived from (24a)–(24b), where the damping coefficient matrix is taken as $Z = 0.05I$. It is assumed that the objective is to slue the manipulator 30°. The resulting hub angle motion and the tip deviation motion, when the feedback control (35) is applied, are shown in Figure 7. It is observed that the hub angle converges to the reference smoothly and the tip deviation motion fades out very quickly.

To compare the proposed design, an alternative design approach, which does not take the flexible dynamics and the uncertainty of the motor parameters into account, is also considered. The design model in this case is

$$\frac{d}{dt}\begin{bmatrix} \theta \\ \dot{\theta} \\ i \end{bmatrix} = \begin{bmatrix} 0 & 1 & 0 \\ 0 & 0 & \frac{K_a^n}{I_h} \\ 0 & -\frac{K_a^n}{L_a^n} & -\frac{R_a^n}{L_a^n} \end{bmatrix} \begin{bmatrix} \theta \\ \dot{\theta} \\ i \end{bmatrix} + \begin{bmatrix} 0 \\ 0 \\ \frac{1}{L_a^n} \end{bmatrix} v \,, \tag{36}$$

and the feedback control law is chosen to minimize

$$J_s = \frac{1}{2}\int_{t_o}^{\infty} (100\,\theta^2 + \dot{\theta}^2 + i^2 + v^2) dt \,. \tag{37}$$

The resulting control law is

$$v = -10.000\,\theta - 4.0438\,\dot{\theta} - 0.41773\,i \,, \tag{38}$$

where all the measurements are taken in SI units. This control law is applied to the same simulation model considered above. The re-

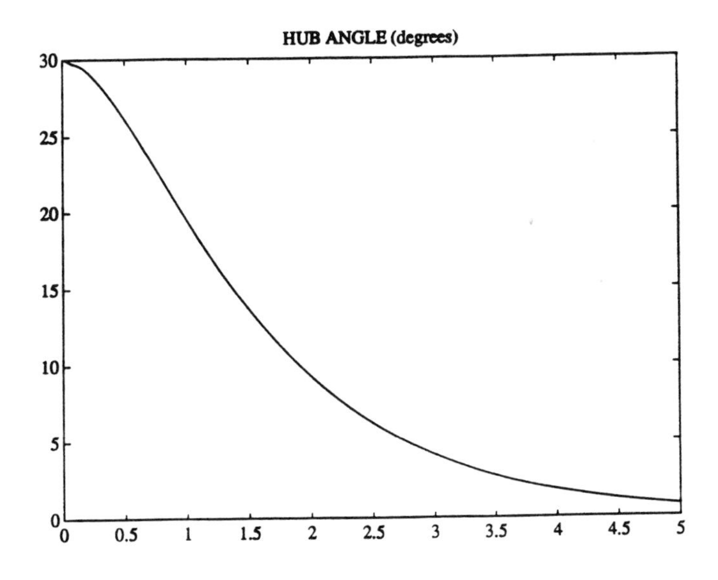

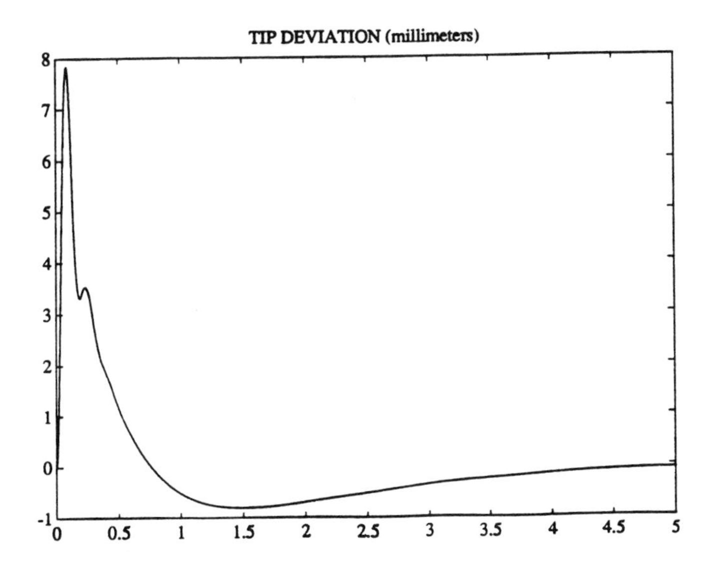

Figure 7: Simulation results for the proposed design.

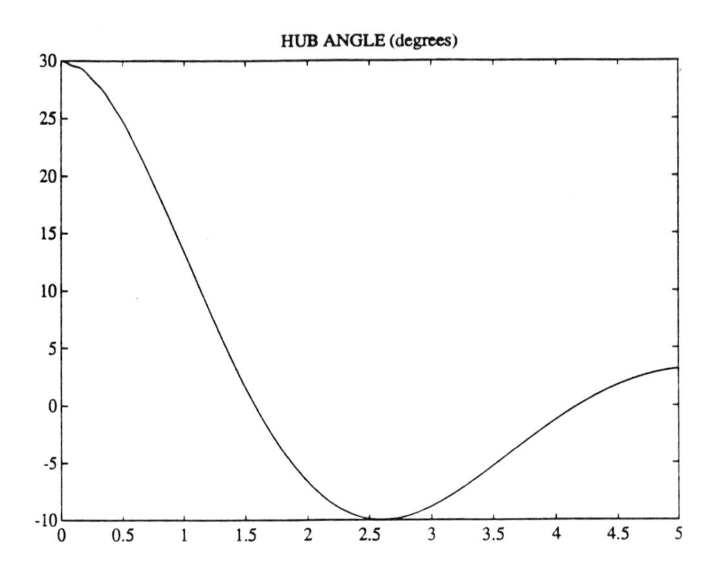

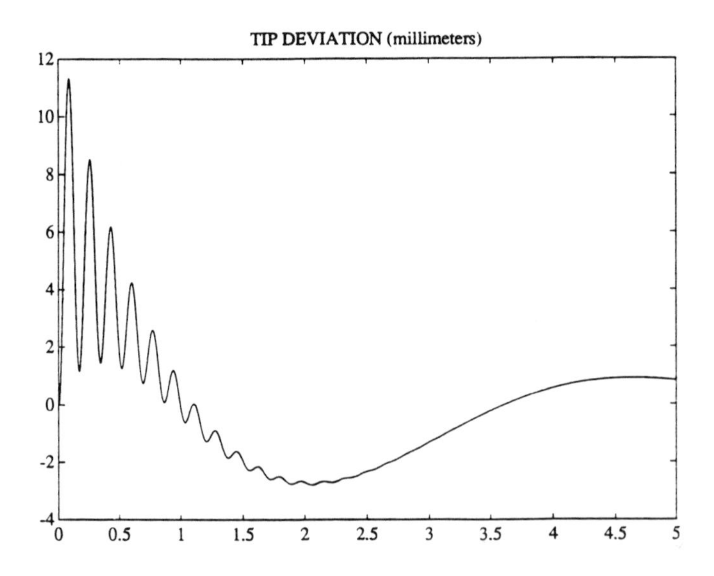

Figure 8: Simulation results for the alternative design.

sulting hub angle motion and the tip deviation motion are shown in Figure 8. By comparing these results with those depicted in Figure 7, the improvement obtained by using the proposed design is apparent.

V. DECENTRALIZED ROBUST CONTROLLER DESIGN

The approach presented in this chapter may also be useful in designing decentralized robust controllers for a large scale system. The approach is especially useful in a local controller design case where the control agents have limited knowledge of controller and subplant dynamics corresponding to other control agents [38]. To avoid notational complexity, here we consider a large scale system which is composed of only two subsystems and dynamic interconnections. After applying a certain *expansion* (see [39] for details), such a system can be described by:

$$\begin{array}{l} \dot{\hat{x}}_i = \hat{A}_i\hat{x}_i + \hat{A}_{ij}\hat{x}_j + \hat{B}_iu_i \\ y_i = \hat{C}_i\hat{x}_i \end{array} , \quad i,j = 1,2 \; , \quad j \neq i \; . \tag{39}$$

Suppose that a controller, described by:

$$\dot{z}_2 = F_2z_2 + G_2y_2 \tag{40a}$$

$$u_2 = H_2z_2 + K_2y_2 \; , \tag{40b}$$

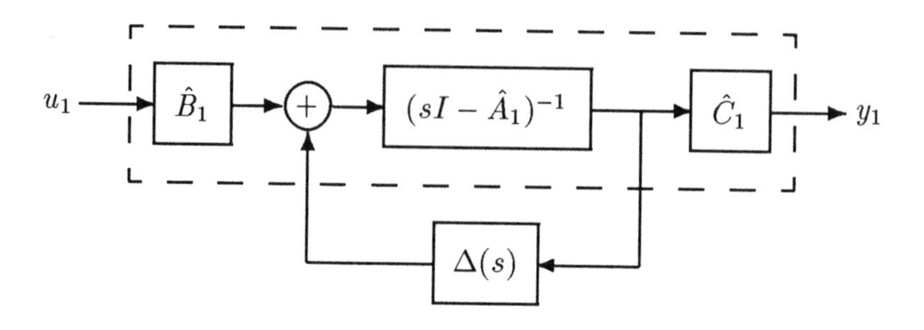

Figure 9: Design model for the first control agent.

is implemented by the second control agent. Then, the overall system dynamics can be described by:

$$\frac{d}{dt}\begin{bmatrix} \hat{x}_1 \\ \hat{x}_2 \\ z_2 \end{bmatrix} = \begin{bmatrix} \hat{A}_1 & \hat{A}_{12} & 0 \\ \hat{A}_{21} & \hat{A}_2 + \hat{B}_2 K_2 \hat{C}_2 & \hat{B}_2 H_2 \\ 0 & G_2 \hat{C}_2 & F_2 \end{bmatrix}\begin{bmatrix} \hat{x}_1 \\ \hat{x}_2 \\ z_2 \end{bmatrix} + \begin{bmatrix} \hat{B}_1 \\ 0 \\ 0 \end{bmatrix} u_1$$

$$y_1 = \hat{C}_1 \hat{x}_1 \, .$$

Note that the controller or even the subplant dynamics corresponding to the second control agent may not be known to the first agent. However, a certain amount of information, such as upper and lower bounds of the TFMs corresponding to these dynamics, may be available. By defining

$$\Delta(s) = \begin{bmatrix} \hat{A}_{12} & 0 \end{bmatrix} \left(sI - \begin{bmatrix} \hat{A}_2 + \hat{B}_2 K_2 \hat{C}_2 & \hat{B}_2 H_2 \\ G_2 \hat{C}_2 & F_2 \end{bmatrix} \right)^{-1} \begin{bmatrix} \hat{A}_{21} \\ 0 \end{bmatrix},$$

a design model, as depicted in Figure 9, is obtained for the first control agent.

By knowing certain bounds on $\Delta(s)$, a rational TFM $E(s;p)$ parametrized by a vector $p \in \Pi$ and a parameter set Π can be determined as explained in Section II. Let $(F(p), G(p), H(p), L(p))$ be a realization for $E(s;p)$. Then,

$$\frac{d}{dt}\begin{bmatrix} x \\ \xi \end{bmatrix} = \begin{bmatrix} \hat{A}_1 + L & H \\ G & F \end{bmatrix}\begin{bmatrix} x \\ \xi \end{bmatrix} + \begin{bmatrix} \hat{B}_1 \\ 0 \end{bmatrix} u_1 \tag{41a}$$

$$y_1 = \begin{bmatrix} \hat{C}_1 & 0 \end{bmatrix}\begin{bmatrix} x \\ \xi \end{bmatrix} \tag{41b}$$

is a design model for the first control agent. At this stage, any uncertainties in the first subsystem dynamics can also be incorporated in this model; and a robust controller for the first subsystem can be designed. A similar approach can be taken at the second station as well. By such an approach, overall stability and good performance of the resulting closed–loop system can be guaranteed.

VI. CONCLUSION

In this chapter, modeling of uncertain dynamics for robust controller design in state space has been discussed. It has been shown that, under mild conditions, it is possible to represent uncertain dynamics as a structured uncertainty with arbitrarily small conservatism. This representation is in the form of a rational TFM, possibly dependent on a parameter vector which varies over a subset of a finite dimensional real vector space. Furthermore, in many practical cases, uncertain dynamics can be represented by a relatively low order TFM, even if the actual dynamics are of very high order. The procedure of determining such a TFM has been discussed. It has been shown that a controller which stabilizes the nominal system, including such a representation of uncertain dynamics, also stabilizes the actual system. Furthermore, desired performance or relative stability can also be guaranteed.

Once a parameter dependent rational TFM is obtained to describe the uncertain dynamics, a parametrized state space design model for the overall system can be obtained. Any parameter uncertainties can also be incorporated in this model at this stage. Then, already existing methods can be used to design robust controllers. The presented approach gives a unified framework for the solution to the robust controller design problem for systems with both parameter uncertainties and uncertain dynamics. The approach has been successfully applied to a real life problem [40].

Decentralized robust controller design has also been discussed in this chapter. An approach to design local controllers when only limited information about rest of the closed–loop system is available to each control agent has been presented. If succesfully applied, the approach guarantees overall stability and desired performance for the closed–loop system. The presented approach may also be useful in many other diverse applications, such as model reduction.

REFERENCES

[1] D. D. Šiljak, *Large–Scale Dynamic Systems: Stability and Structure*. New York: North–Holland, 1978.

[2] R. V. Patel, M. Toda, and B. Sridhar, "Robustness of linear quadratic state feedback designs in the presence of system uncertainty," *IEEE Transactions on Automatic Control*, vol. AC–22, pp. 945–949, 1977.

[3] R. K. Yedavalli, "Perturbation bounds for robust stability in linear state space models," *International Journal of Control*, vol. 42, pp. 1507–1517, 1985.

[4] K. Zhou and P. P. Khargonekar, "Stability robustness bounds for linear state–space models with structured uncertainty," *IEEE Transactions on Automatic Control*, vol. AC–32, pp. 621–623, 1987.

[5] L. H. Keel, S. P. Bhattacharyya, and J. W. Howze, "Robust control with structured perturbations," *IEEE Transactions on Automatic Control*, vol. AC–33, pp. 68–78, 1988.

[6] L. Qiu and E. J. Davison, "New perturbation bounds for the robust stability of linear state space models," in *Proceedings of the IEEE Conference on Decision and Control*, pp. 751–755, Athens, Greece, December 1986.

[7] D. D. Šiljak, "Absolute stability and parameter sensitivity," *International Journal of Control*, vol. 8, pp. 279–283, 1968.

[8] B. R. Barmish, "Invariance of strict Hurwitz property for polynomials with perturbed coefficients," *IEEE Transactions on Au-*

tomatic Control, vol. AC–29, pp. 935–936, 1984.

[9] V. L. Kharitonov, "Asyptotic stability of an equilibrium position of a family of systems of linear differential equations," *Differentsialjnia Uravnenia*, vol. 14, pp. 2086–2088, 1978.

[10] B. R. Barmish, "A generalization of Kharitonov's four–polynomial concept for robust stability problems with linearly dependent coefficient perturbations," *IEEE Transactions on Automatic Control*, vol. AC–34, pp. 157–165, 1989.

[11] R. M. Biernacki, H. Hwang, and S. P. Bhattacharyya, "Robust stability with structured real parameter perturbations," *IEEE Transactions on Automatic Control*, vol. AC–32, pp. 495–506, 1987.

[12] A. C. Bartlett, C. V. Hollot, and H. Lin, "Root location of an entire polytope of polynomials: it suffices to check the edges," in *Proceedings of the American Control Conference*, pp. 1611–1616, Minneapolis, MN, June 1987.

[13] J. S. Karmarkar and D. D. Šiljak, "A computer aided design of robust regulators," in *Proceedings of the IFAC Workshop*, pp. 49–58, Denver, CO, June 1979.

[14] J. Ackermann, "Parameter space design of robust control systems," *IEEE Transactions on Automatic Control*, vol. AC–25, pp. 1058–1072, 1980.

[15] D. S. Bernstein and D. C. Hyland, "The Optimal Projection/Maximum Entropy approach to designing low–order, robust controllers for flexible structures," in *Proceedings of the IEEE Conference on Decision and Control*, pp. 745–752, Ft.

Lauderdale, FL, December 1985.

[16] D. C. Hyland and D. S. Bernstein, "The optimal projection equations for fixed-order dynamic compensation," *IEEE Transactions on Automatic Control*, vol. AC–29, pp. 1034–1037, 1984.

[17] M. Cheung and S. Yurkovich, "On the robustness of MEOP design versus asymptotic LQG synthesis," *IEEE Transactions on Automatic Control*, vol. AC–33, pp. 1061–1065, 1988.

[18] R. K. Yedavalli, "Dynamic compensator design for robust stability of linear uncertain systems," in *Proceedings of the IEEE Conference on Decision and Control*, pp. 34–36, Athens, Greece, December 1986.

[19] D. S. Bernstein and W. M. Haddad, "Robust controller synthesis using Kharitonov's theorem," in *Proceedings of the IEEE Conference on Decision and Control*, pp. 1222–1223, Honolulu, Hawaii, December 1990.

[20] J. C. Doyle and G. Stein, "Multivariable feedback design: Concepts for a classical/modern synthesis," *IEEE Transactions on Automatic Control*, vol. AC–26, pp. 4–16, 1981.

[21] M. G. Safonov, A. J. Laub, and G. L. Hartmann, "Feedback properties of multivariable systems: the role and use of return difference matrix," *IEEE Transactions on Automatic Control*, vol. AC–26, pp. 47–65, 1981.

[22] J. B. Cruz, J. S. Freudenberg, and D. P. Looze, "A relationship between sensitivity and stability of multivariable feedback systems," *IEEE Transactions on Automatic Control*, vol. AC–26, pp. 66–74, 1981.

[23] G. Zames and B. A. Francis, "Feedback, minimax sensitivity, and optimal robustness," *IEEE Transactions on Automatic Control*, vol. AC–28, pp. 585–600, 1983.

[24] B. A. Francis, J. W. Helton, and G. Zames, "$\mathcal{H}^{\infty}$–Optimal feedback controllers for linear multivariable systems," *IEEE Transactions on Automatic Control*, vol. AC–29, pp. 888–900, 1984.

[25] A. İftar and Ü. Özgüner, "Local LQG/LTR controller design for decentralized systems," *IEEE Transactions on Automatic Control*, vol. AC–32, pp. 926–930, 1987.

[26] H. Özbay and A. Tannenbaum, "On the structure of suboptimal $\mathcal{H}^{\infty}$ controllers in the sensitivity minimization problem for distributed stable plants," *Automatica*, vol. 27, pp. 293–305, 1991.

[27] K. H. Wei and R. K. Yedavalli, "Robust stabilizability for systems with both parameter variation and unstructured uncertainty," in *Proceedings of the IEEE Conference on Decision and Control*, pp. 2082–2087, Los Angeles, CA, December 1987.

[28] S. Boyd, "A note on parametric and nonparametric uncertainties in control systems," in *Proceedings of the American Control Conference*, pp. 1847–1849, Seattle, WA, June 1986.

[29] I. M. Horowitz, "Quantitative feedback theory," *Proceedings of the IEE, Part-D*, vol. 129, pp. 215–226, 1982.

[30] A. Sideris and M. G. Safonov, "A design algorithm for the robust synthesis of SISO feedback control systems using conformal maps and H_{∞}–theory," in *Proceedings of the American Control Conference*, pp. 1710–1715, Seattle, WA, June 1986.

[31] A. İftar and Ü. Özgüner, "Structured modeling of unstructured

uncertainties and robust controller design in state space," in *Proceedings of the ASME Winter Annual Meeting*, Paper No. 88-WA/DSC-12, Chicago, IL, November 1988.

[32] A. İftar and Ü. Özgüner, "Modeling of uncertain dynamics for robust controller design in state space," *Automatica*, vol. 27, pp. 141–146, 1991.

[33] A. G. J. MacFarlane, "Return–difference and return–ratio matrices and their use in analysis and design of multivariable feedback control systems," *Proceedings of the IEE*, vol. 117, pp. 2037–2049, 1970.

[34] A. İftar, "A robust controller design approach for flexible robot manipulators with actuator dynamics," in *Proceedings of the Bilkent International Conference on New Trends in Communication, Control, and Signal Processing*, pp. 932–938, Ankara, Turkey, July 1990.

[35] F. Khorrami, "Analysis of manipulators with flexible joints and links," in *Proceedings of the IEEE International Conference on Systems Engineering*, pp. 561–564, Dayton, OH, August 1989.

[36] S. Timoshenko, D. H. Young, and W. Weaver, *Vibration Problems in Enginnering*. New York: Wiley, 1974.

[37] A. İftar, "Robust optimal control for uncertain systems," in *Proceedings of the IEEE International Conference on Systems Engineering*, pp. 119–122, Dayton, OH, August 1989.

[38] A. İftar and Ü. Özgüner, "Representation of uncertain actuator dynamics and decentralized robust controller design," in *Proceedings of the Annual Allerton Conference on Communication,*

Control, and Computing, pp. 883–892, Monticello, IL, October 1988.

[39] A. İftar and Ü. Özgüner, "Decentralized LQG/LTR controller design for interconnected systems," in *Proceedings of the American Control Conference*, pp. 1682–1687, Minneapolis, MN, June 1987.

[40] A. İftar, F. Khorrami, and S. Jain, "A robust controller design approach and experiments on a sluing flexible structure," in *Proceedings of the IEEE International Conference on Systems, Man, and Cybernetics*, Charlottesville, VA, October 1991. (to appear).

Neoclassical Control Theory: A Functional Analysis Approach to Optimal Frequency Domain Controller Synthesis

A. M. Holohan

M. G. Safonov

Dept. of Electrical Engineering-Systems
University of Southern California
Los Angeles, CA 90089-2563

I. INTRODUCTION

In this chapter, the class of optimal k_m-synthesis (or μ-synthesis, robustness margin synthesis) problems involving rank one augmented plants is considered, where k_m is the Multivariable Stability Margin for systems with uncertain complex parameters. For this class, the optimal k_m-synthesis problem is shown to be convex in the Youla parameter. This class includes SISO control problems with robust stability and *robust* disturbance attenuation as the design specifications. Function space duality theory is used to establish an all-pass property. The optimal value of k_m is explicitly determined for stable and minimum phase plants.

It is fundamental that no real-life system can be modeled perfectly. Every physical system is uncertain to some degree, and frequently to a substantial degree. The effects of this uncertainty on the system's behavior can be dramatically reduced by the use of feedback. Indeed, the ability of feedback to counteract the effects of uncertainty is generally the primary motivation for its use. However, the application of excessive feedback usually causes a system to become unstable. This leaves the control engineer with the subtle problem of determining just how much feedback can safely be applied.

Early efforts to use feedback ran into exactly this difficulty, and it motivated much of the early development of control theory. The Maxwell-Routh approach grew out of the instability of Watt's governor for controlling the speed of steam engines. The Black-Bode-Nyquist approach was developed to deal with the instability of feedback amplifiers. These two theories led to the familiar root locus approach and lead-lag compensator approach respectively, and they are still the mainstay of much of industrial practice today. Both theories were developed to explain why feedback can lead to instability in practice, and how to avoid it.

Typically, control systems are designed and analyzed using a system model, arrived at by physical laws or perhaps from measured data. The successful controller design is expected to produce a stable closed-loop system, and to satisfy certain performance specifications. The presence of uncertainty impacts upon both of these requirements: The actual system may be unstable, even if the plant model predicts stability. Likewise, the model may erroneously predict that the performance specifications are met. Consequently, controller design procedures which rely heavily on the plant model and ignore its limitations are likely to produce controller designs which work well on paper but poorly if at all in practice. In other words, the very same uncertainty which motivates the application of feedback means that an analysis based on a single model of the system may very well fail to capture adequately an essential facet of the problem.

This suggests the desirability of a design methodology which treats the plant not as a *single* system, but rather as a *set* of systems, inside of which the actual system lies. In this perspective, the controller must stabilize not only the nominal model, but also a clearly defined set of systems, and likewise for the performance specifications. This approach to controller design can fairly be called *Robust Control Theory*. So a fundamental problem in system theory is to produce a controller design methodology which *ensures both stability and adequate performance in the presence of substantial uncertainty*. The *analysis problem* is then to determine if a given controller satisfies these requirements for a given set of plants. The *synthesis problem* is to find a controller which does so optimally.

In this paper, we propose to study a special case; namely, the optimal synthesis of SISO controllers using a formulation of the problem which is very much in the spirit of the classical frequency response methods, from the perspective of modern robust control theory. A satisfactory solution to the scalar case is a necessary precursor to a comprehensive and exact (non-conservative) theory capable of handling large complex problems and, in particular, multivariable problems.

As mentioned above, the most frequently used approaches are the root locus and the lead-lag compensator design methodologies. The centerpiece of both theories is an effective criterion for checking if a single linear system is stable. These are the familiar root location and Nyquist criteria respectively. They do not, however, provide practical tests for accurately checking the stability of families of systems (except in very restrictive cases). As a result, the practicing engineer must deal with the effects of uncertainty in a heuristic and approximate fashion, thus obliging him to err substantially on the side of safety. Although good working designs are routinely produced in this fashion, the result may be far from optimal, even in those cases where the uncertainty is easily characterized. Another important limitation of these methods is that they do not extend to multivariable problems. For heuristic reasoning to be reliable and not too far from optimum, the problem must be small and transparent. The essence of the difficulty seems to be the implicit treatment of the uncertainty in the plant model. Since the issue of plant uncertainty is central to both the objectives and difficulties of using feedback, it is reasonable to conclude that this uncertainty should be treated in an explicit way. With this in mind, the present study will consider the plant to be described as a family of systems, as follows. Each point of the plant's Nyquist diagram belongs to a disc with center $G_0(j\omega)$ and with radius $|V(j\omega)||G_0(j\omega)|$, so that $|V(j\omega)|$ represents the proportional uncertainty in the nominal model of the plant. With this setup, the plant may be thought of as being described by a "fuzzy" Nyquist curve. It is not difficult to see how an appropriate choice of $|V(j\omega)|$ can be arrived at from measured frequency response data in many typical engineering problems.

It has been argued above that the primary objective of using feedback is generally sensitivity reduction in one form or another, and a satisfactory controller must certainly yield a stable closed loop system. Thus motivated, we impose the specifications that the controller achieve (i) closed loop stability, and (ii) adequate disturbance attenuation (i.e. sensitivity reduction), *taking full account of plant uncertainty*, and propose to study the underlying *optimal synthesis* problem. See Section IV for the detailed problem description. This is an interesting problem in its own right, as it essentially corresponds to classical "Bode style" compensator design. It has a strong engineering motivation and is perhaps the simplest problem which captures the "$|S|$ versus $|T|$ tradeoff" [1].

The volume of literature on this problem is surprisingly modest. In [2], it is shown that robust disturbance attenuation is a much more stringent requirement than nominal sensitivity reduction in the multivariable case. In [3], the underlying robustness analysis problem for the scalar case is resolved, and qualitative results on the synthesis problem are presented. A closely related problem is that of optimally reducing the nominal sensitivity subject to the constraint that the system be robustly stable, and this problem has been studied in [4,5], both for the scalar case. In [4], bounds on the achievable performance are derived, while in [5] the problem is cast as the minimization of a non-convex function over the set of bounded real functions.

Several authors have considered the general robustness analysis problem for linear systems from a frequency domain perspective. This work led to the introduction of the Multivariable Stability Margin k_m [6] as a quantitative measure of the immunity of a multivariable control system to uncertainty. The problem of calculating this quantity has received considerable attention, e.g. [7-11]. While numerous procedures for calculating bounds on k_m have been suggested, there is still no fast reliable computational scheme for calculating k_m.

Only very special cases of the k_m-synthesis problem have been solved to date [12-17]. Indeed, the only k_m-synthesis problem involving complex uncertainties that has hitherto been solved is the case involving a single complex block. This is the now-classical one-block $\mathcal{H}^\infty$ problem [12]. In [18,19,20], a theory based on conformal mapping is described which is capable of handling certain k_m-synthesis problems, including, for example, gain margin maximization. This problem can be cast as an optimal k_m-synthesis problem with plant uncertainty described by a single real parameter. Synthesis procedures based on the inequality $\mu = k_m^{-1} \geq \overline{\sigma}(DMD^{-1})$ (see [8]) have been suggested [13-17]. However, the solutions produced by these methods may be local optima.

While the general synthesis problem has thus far defied solution, it is shown below that a certain class of such problems lead to an easy analysis problem and a *convex* synthesis problem. Note that the general problem (in both analysis and synthesis) is non-convex. This class includes the case where the plant is SISO, and the specifications are robust sensitivity reduction (i.e. robust disturbance attenuation) and robust stability, where the plant uncertainty at each frequency is described by a single complex unknown but bounded parameter. So there are realistic engineering problems which belong to this special case. Function space duality theory is used to establish an all-pass property, and to derive a pleasingly explicit expression for the optimal achievable value of k_m for the SISO stable minimum phase case.

II. BACKGROUND

This section describes three fundamental facts upon which our analysis rests, and defines the Multivariable Stability Margin and the Robustness Margin.

The Diagonal Perturbation Formulation:

Consider a finite-dimensional linear time-invariant system with n uncertain complex scalar parameters $\Delta_i(j\omega), \ \ i = 1, \ldots, n$ embedded in it. These parameters are uncertain but norm bounded. They are assumed to be stable and independent of each other. This will be a standing assumption throughout this paper. Safonov and Athans [21] note that these parameters can generally be isolated and placed in a diagonal matrix, using standard block diagram manipulations. That is, these parameters may be assumed to form a diagonal matrix $\Delta = \text{diag}\{\Delta_1, \ldots, \Delta_n\}$. The transfer function "seen" by Δ will be denoted by M. (The dependence on $s = j\omega$ is generally suppressed in our notation.) The resulting structure is illustrated in Figure 1.

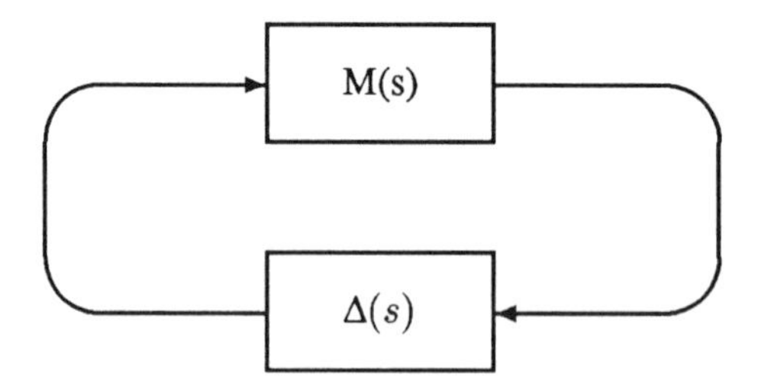

Figure 1

Any frequency dependence of the parameter bounds can be absorbed into the matrix M. So we may assume that the bounds are

$$|\Delta_i| \leq 1, \ \ i = 1, \ldots, n$$

The resulting set of Δ's will be denoted by $\mathcal{D}$, so that

$$\mathcal{D} := \{\Delta \in \mathcal{C}^{n \times n} | \overline{\sigma}(\Delta) \leq 1, \quad \Delta \text{ is diagonal}\}$$

Following Safonov [22,23], we say that the system is *nominally stable* if it is stable when $\Delta = 0$ and *robustly stable* (R.S.) if it is stable for all $\Delta \in \mathcal{D}$. These remarks also apply to uncertain complex blocks, where the Δ_i's may be matrix valued, and to uncertain real valued parameters. In this paper the uncertainties will always be complex blocks or complex scalars. See [10] for a recent treatment of the analysis problem in the presence of both real scalar and complex block uncertainties.

So most problems of this type can be rearranged into a standard format, a canonical form, as shown above. In this setup, a linear system with transfer function matrix M is perturbed multiplicatively by a diagonal uncertain matrix Δ.

The Robust Stability Theorem:

A basic result is the following [7,8].

Theorem 1 The system of Figure 1 is robustly stable if and only if it is nominally stable and

$$0 \neq \det(I + \Delta(j\omega)M(j\omega)) \;\; \forall \Delta \in \mathcal{D} \;\; \forall \omega \in \mathcal{R} \cup \{\infty\}$$

This result may be established straightforwardly from the multivariable Nyquist stability criterion and the continuity of the zeros of a polynomial to continuous changes in its coefficients.

The Robust Performance Theorem:

If a candidate design obeys a certain performance specification when $\Delta = 0$, it will be said to satisfy it *nominally*; if it does so for all $\Delta \in \mathcal{D}$, it will be said to satisfy it *robustly*. Doyle et al. [24] have shown how to embed a certain type of *robust* performance specification within the above framework by introducing an additional "fictitious" uncertainty Δ_{n+1}. Consider a performance specification that requires a transfer function from a system input u to a system output y to obey an infinity-norm upper bound. (The input and output involved may be multivariable.) Clearly, the problem can be set up as shown below.

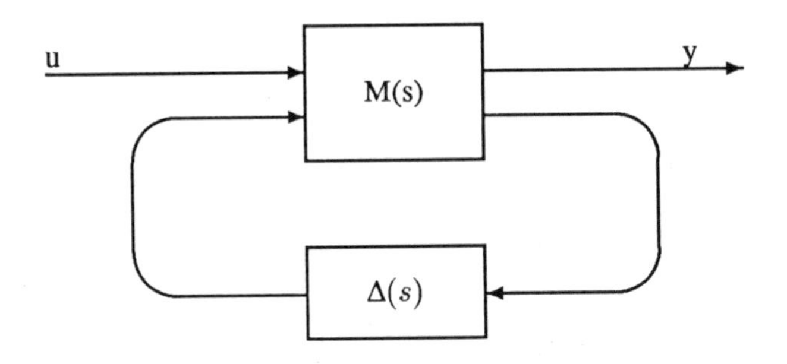

Figure 2

The matrix Δ contains the uncertainty terms on its diagonal, as described above. Introducing the new artificial uncertainty term Δ_{n+1} with its output connected to u and its input connected to y, it is easy to see that this extra uncertainty can then be placed on the diagonal of Δ as before. We say that the system model has been augmented with an additional "fictitious" uncertainty, and that Δ has been augmented by a performance block. The Robust Performance Theorem [24] then states that this modified system is robustly stable if and only if the original system is robustly stable and the closed loop transfer function from u to y has infinity norm less than one for all $\Delta \in \mathcal{D}$. So this more general problem can be reduced to a robust stability problem, to the original setup of Figure 1.

Note that the theorem does not apply to multiple specifications. That is, we may add only a single performance block as described above. However, the necessary extension to handle multiple specifications is straightforward.

The Robustness Margin and the Multivariable Stability Margin:

When assessing whether or not a candidate design is robustly stable, it is desirable to give the engineer more information than a mere yes/no answer. A quantitative measure of the degree of the robustness of a system is needed. This motivates the definition of the *Robustness Margin* K_m. (The notation r_{max} is often used.) It is a measure of the degree of robustness of a system, thereby indicating quantitatively the excess or shortfall in the level of robustness of a design. To define it, write the uncertainties as $K\Delta_i$, as shown in Figure 3, and continue to require $|\Delta_i| \leq 1$.

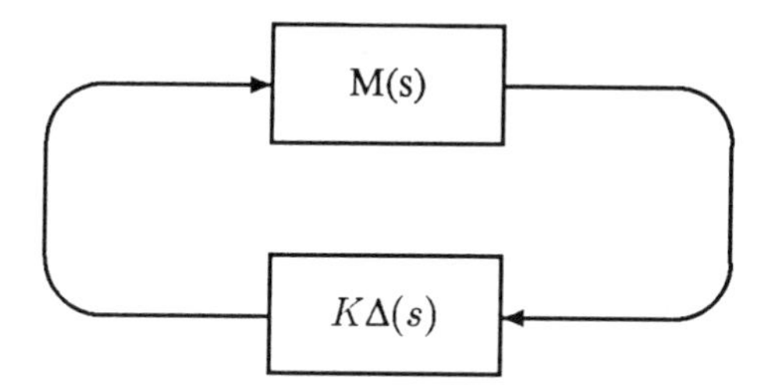

Figure 3

This approach is obviously equivalent to replacing the unity norm bound on the Δ_i's by the variable bound K. For the configuration of Figure 3, the *Robustness Margin* K_m is then defined as the minimum $K \geq 0$ for which the system fails to be robustly stable. Equivalently, it is the supremum of those values of $K \geq 0$ for which the system is robustly stable. It is zero if the system is not nominally stable (i.e. if $M(s)$ is not stable or if the system has unstable hidden modes), and is infinite if there is no Δ with the required diagonal structure which destabilizes the system, irrespective of its size (norm).

As noted above, when assessing whether or not a design is robustly stable, it is highly desirable to give the engineer as much information as possible. An algorithm which computes K_m only does not give the engineer insight on how to improve matters. An algorithm which produces, in addition, the (or a) smallest destabilizing Δ is clearly a good deal more useful. In a sense, it directs the engineer to the source of difficulty. Another natural approach is as follows. It is motivated by the desirability of giving the engineer robustness information on a frequency-by-frequency basis.

We introduced above the parameter K as a variable bound on the Δ_i's. That strategy amounts to floating the bounds on the uncertainties, *uniformly across* ω. An attractive idea is to float the bounds in a frequency dependent fashion. Replace each $\Delta_i(j\omega)$ by $\Delta_i(j\omega)k(\omega)$, and consider the question: How large can $k_m(\omega)$ be made at each frequency while maintaining robust stability? Thus motivated, define the *Multivariable Stability Margin* [6] to be

$$k_m(M) := \min_k \{k \geq 0 | \det(I + k\Delta M) = 0 \text{ for some } \Delta \in \mathcal{D}\}$$

(If no such k exists, take $k_m = +\infty$). So k_m is a mapping from the $n \times n$ complex matrices to the non-negative reals:

$$k_m : \mathcal{C}^{n \times n} \to \mathcal{R}^+$$

It is well defined for every $n \times n$ complex matrix. When applied to a transfer function matrix $M(j\omega)$, we obtain $k_m(M(j\omega))$, a function of ω, which will be abbreviated to $k_m(\omega)$, or just k_m. With this definition, the Robust Stability Theorem can be stated in the following equivalent way:

Theorem 1′ The system of Figure 1 is robustly stable if and only if it is nominally stable and

$$k_m(\omega) > 1 \;\; \forall \omega \in \mathcal{R} \cup \{\infty\}$$

Hence, for a stable $M(s)$,

$$K_m = \inf_{\omega} k_m(M(j\omega)) \quad \text{and} \quad R.S. \Leftrightarrow K_m > 1$$

(Our notation for K_m is motivated by the above expression.) Thus, one can think of robust stability as holding ($k_m > 1$), or failing to hold ($k_m \leq 1$) at a particular frequency. Strictly speaking, this is an abuse of terminology, but it is justified by the above theorem. For the case involving both complex blocks and real parameters, see [10].

Thus, $k_m(\omega)$ is a quantitative and exact measure of the immunity of a system, in both stability and performance, to (structured) uncertainty on a frequency-by-frequency basis. The Robust Stability Theorem establishes that this function is well defined. The reciprocal of k_m, $\mu = k_m^{-1}$, is called the *structured singular value* [8], and this notation is also widely used. The problem of the practical computation of k_m is an important unsolved problem, although many practical procedures for calculating bounds on k_m have been suggested, see for example [7,8,9,10,11].

In analyzing control systems, the frequency dependent function $k_m(\omega)$ contains more information than the scalar K_m, and generally provides greater insight.

III. THE RANK 1 CASE - ANALYSIS

In this section, we consider the special case where the matrix M has rank 1. As will become clear below, there are some interesting problems with a strong engineering motivation which belong to this special case.

Recall that all the Δ_i's are independent complex-valued scalars and are internally stable. Under these conditions, Fan and Tits [9] have shown that an explicit solution of the analysis problem is possible whenever the matrix M has rank 1.

Theorem 2 [9] If the matrix M has rank one, then

$$k_m^{-1} = \sum_{i=1}^{n} |m_{ii}| \qquad K_m^{-1} = \left\| \sum_{i=1}^{n} |m_{ii}| \right\|_{\infty}$$

where m_{ii} is the ith diagonal element of M.

Proof Expanding $\det(I + k\Delta M)$ leads to a function that is multilinear in the n variables Δ_i, $i = 1, \ldots, n$. Such a function has 2^n coefficients. In fact, the coefficients are all the diagonal minors of M. Now if M has rank 1, all minors are zero except the 1×1 minors. So all coefficients drop out except those involving a single Δ_i or none. That is, all the higher order terms drop out. Thus

$$\det(I + k\Delta M) = 1 + k\sum_{i=1}^{n} \Delta_i m_{ii}$$

Alternatively, note that if M has rank 1, then $M = uv^*$ for some column vectors u and v. Now

$$\det(I + k\Delta uv^*) = \det(1 + kv^*\Delta u) = 1 + k\sum_{i=1}^{n} \overline{v_i}\Delta_i u_i = 1 + k\sum_{i=1}^{n} \Delta_i m_{ii}$$

Clearly, this quantity is non-zero $\forall \Delta \in \mathcal{D}$ if and only if

$$1 > k\sum_{i=1}^{n} |m_{ii}|$$

and the theorem follows. □

This result amounts to a closed form expression for k_m in the rank 1 case. Of special interest is the quantity $K_m = \inf_\omega k_m$. For practical purposes, a plot of $k_m = \sum |m_{ii}|$ should suffice for this. In theory, k_m needs to be checked at an infinite number of ω's to obtain the exact value of K_m. A precise approach would be to bound the rate of change of $\sum |m_{ii}|$ with ω, which easily yields a rigorous approach. Alternatively, the problem is easily reduced to determining if a certain polynomial in ω has a real root (for each fixed K). Note that in the present case, k_m and K_m are clearly continuous in ω and in the elements of M, though this is not true in the general case [25].

Consider now the Youla Lemma [26], which parameterizes all stabilizing controllers in terms of a stable transfer function ($q(s)$ say) which will be referred to here as the Youla parameter. The matrix M will be affine in q. So each m_{ii} will be affine in the elements of q. Also, tha absolute value function $|.|$ obeys the triangle inequality. These observations establish the following:

Theorem 3 If the matrix M has rank 1, then k_m^{-1} and K_m^{-1} are convex in the elements of M and in the Youla parameter.

The immediate consequence of convexity, of course, is that it opens the door to reliable computational solutions when $\mathcal{H}^\infty$ is replaced by a finite dimensional "discretization". Thus, the method of [27] for example is applicable.

The requirement that $M(j\omega)$ have rank one for all $\omega \in \mathcal{R}$ is clearly a severe restriction. Nonetheless, it will be shown below that the robust sensitivity reduction problem for SISO plants has this structure.

IV. CLASSICAL COMPENSATOR DESIGN

In this section, we give the detailed problem description. We have chosen to cast the problem as an optimal multivariable stability margin synthesis problem to emphasize the mutual consistency and unification of the present problem with the k_m-formulation of modern robust control theory. The set up is shown in Figure 4 below.

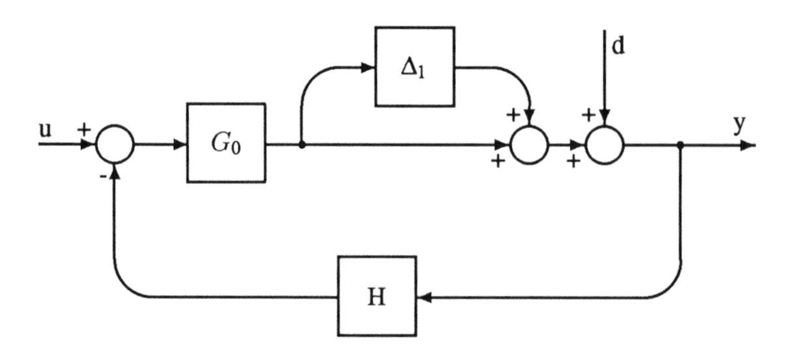

Figure 4

The nominal plant is G_0, the actual plant is $G = (1 + \Delta_1)G_0$, where Δ_1 represents the uncertainty in the plant model. The controller is H, d is a disturbance, y is the output and u is the command input. The nominal sensitivity and nominal complimentary sensitivity functions will, as usual, be denoted by S and T respectively, where

$$S = \frac{1}{1 + G_0 H}, \qquad T = \frac{G_0 H}{1 + G_0 H}$$

Our first specification is to ensure robust stability in the face of the plant uncertainty Δ_1 where $|\Delta_1| \leq K|V|$. Our second specification is a *robust* performance specification. To ensure adequate disturbance attenuation, we will require $|y/d| \leq |W^{-1}|/K$ for all $j\omega$ and *for all* Δ_1 obeying $|\Delta_1| \leq K|V|$.

The analysis problem is to calculate, for given H, the largest value of K for which the above conditions are satisfied. The synthesis problem is to determine an H which maximizes K subject to the above conditions. Note that increasing K from unity corresponds to tightening the specifications (i.e. demanding more), while decreasing K from unity corresponds to relaxing the specifications (i.e.

demanding less). It is fundamental that control specifications cannot be tightened arbitrarily. If K is made too large, there exists no controller that will meet the specifications. The above synthesis problem is to determine just how well, at best, one can do. That is, determine exactly how the fundamental limitations of feedback affect the achievable performance.

The Diagonal Perturbation Formulation:

The above problem is easily reformulated into the standard format discussed earlier. Firstly, a "fictitious" uncertain parameter Δ_2 is introduced to allow for the specification on $|y/d|$ as shown in Figure 5 below. The transfer function "seen" by Δ_2 is exactly $(1 + GH)^{-1}$ as expected. So requiring this transfer function to be small is the same as requiring robust stability for large Δ_2.

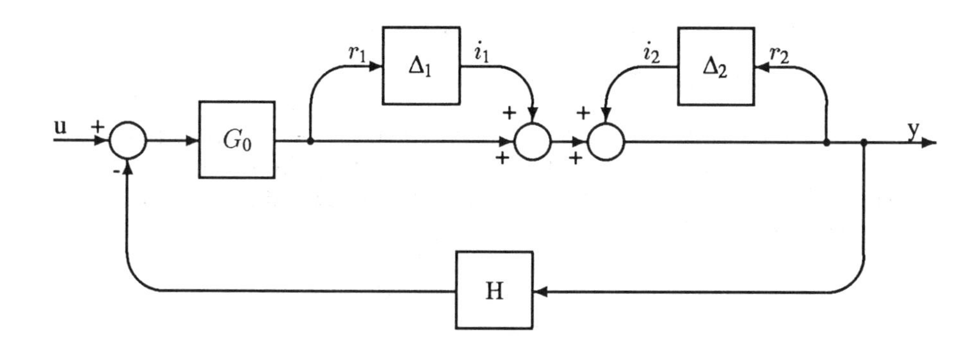

Figure 5

From Figure 5,

$$\begin{pmatrix} r_1 \\ r_2 \end{pmatrix} = \begin{pmatrix} -T & -T \\ S & S \end{pmatrix} \begin{pmatrix} i_1 \\ i_2 \end{pmatrix}$$

$$= M \begin{pmatrix} i_1 \\ i_2 \end{pmatrix} \text{ say.}$$

Secondly, weights are introduced to normalize the frequency dependence of the bounds on Δ_1 and Δ_2. Clearly,

$$|\Delta_1| \leq |V| \Leftrightarrow |\Delta_1 V^{-1}| = |\delta_1| \leq 1 \text{ where } \delta_1 = \Delta_1 V^{-1}, \text{ and}$$

$$|\Delta_2| \leq |W| \Leftrightarrow |\Delta_2 W^{-1}| = |\delta_2| \leq 1 \text{ where } \delta_2 = \Delta_2 W^{-1}.$$

So $\mathcal{D}$ and M become

$$\mathcal{D} = \{\delta \in C^{2\times 2} | \delta = \mathrm{diag}(\delta_1, \delta_2), |\delta_1|, |\delta_2| \leq 1\}$$

$$M = \begin{pmatrix} -VT & -VT \\ WS & WS \end{pmatrix}$$

Theorem 2 then yields

$$K_m^{-1} = \| \, |WS| + |VT| \, \|_\infty$$

which is the solution of the analysis problem.

In summary, our specifications are

- Robust Stability: $\forall \; |\Delta_1| \leq K|V|$
- Robust Performance: $|y/d| \leq K^{-1}|W^{-1}| \quad \forall \; j\omega \; \forall \; |\Delta_1| \leq K|V|$

and necessary and sufficient conditions for the above to hold are

- The nominal system is closed loop stable
- $K^{-1} > |VT| + |WS| \quad \forall \; \omega$

This result can easily be confirmed directly from the Nyquist criterion and simple geometry. It amounts to a closed form expression for K_m. Note that the matrix M above clearly has rank 1. The optimal achievable value of the robustness margin will be denoted by K_{mo}, so that

$$K_{mo} := \sup_H K_m(H)$$

where the supremum is taken over all stabilizing controllers.

The Standard $\mathcal{H}^\infty$ Approach:

The standard approach at present to tackling problems like the above is to minimize the infinity norm of the matrix M over all stabilizing controllers. This is a standard $\mathcal{H}^\infty$ problem. This approach is suboptimal. Specifically, the synthesis problem at hand is solved to within a factor of $\sqrt{2}$ by a mixed-sensitivity two-block infinity-norm problem:

Theorem 4 [3] The optimal robustness margin K_{mo} is bounded above and below by

$$\lambda \leq K_{mo}^{-1} \leq \sqrt{2} \; \lambda$$

$$\text{where} \quad \lambda := \inf_{q \in \mathcal{H}^\infty} \left\| \begin{matrix} WS \\ VT \end{matrix} \right\|_\infty \quad \text{and} \quad K_{mo}^{-1} := \inf_{q \in \mathcal{H}^\infty} K_m^{-1}$$

Proof For any two complex numbers a and b,

$$\overline{\sigma}^2 \begin{pmatrix} a \\ b \end{pmatrix} = |a|^2 + |b|^2 \leq (|a| + |b|)^2 \leq 2(|a|^2 + |b|^2) = 2\overline{\sigma}^2 \begin{pmatrix} a \\ b \end{pmatrix}$$

and the theorem follows immediately. □

This theorem, which appears in [3], is not true for the general rank 1 case, where weaker bounds apply. The above mixed-sensitivity two-block infinity-norm problem was originally posed and solved in [28,29].

V. A VECTOR SPACE FORMULATION FOR CLASSICAL COMPENSATOR DESIGN

In this and the following sections, we take some first steps towards an analytical solution to the above robust sensitivity reduction problem, a special case of the Multivariable Stability Margin synthesis problem. To this end, we first establish that the problem can be cast in the form: find a vector belonging to a subspace which minimizes the distance from this vector to another given vector. This allows us to apply function space duality theory, where there is a mature and powerful theory for problems with this structure. We follow Luenberger's treatment [30] closely. The methodology adopted here has been successfully applied to the scalar 1-block $\mathcal{H}^\infty$ problem [31,32], to optimal $\mathcal{L}^1$ problems [33,34], and to optimal $\mathcal{L}^2$ control problems [30]. It may be worthwhile remarking that functional analysis admits a unified treatment of these diverse control problems.

The Youla Lemma will be used to ensure closed loop stability and to express the problem in the desired form. It is assumed that the plant is rational and has neither poles nor zeros on the imaginary axis. Following [26], let

$$G_0 = \frac{n}{d} \qquad H = \frac{x}{y}$$

be stable coprime fractional representations of G_0 and H. The fact that n and d are coprime ensures the existence of a solution (u, v) to the Bezout identity

$$1 = nu + dv$$

The Youla Lemma [26] then states that the closed loop system is stable if and only if

$$x = -qd + u \qquad y = qn + v$$

for some stable proper $q(s)$. The transfer function $q(s)$, the Youla parameter, parameterizes the entire set of (closed loop) stabilizing controllers. Straightforward algebra then confirms that

$$S = d(qn + v) \qquad T = n(-qd + u)$$

Any stable and minimum phase factors in the plant can be cancelled by the controller, so we may assume that n and d are the Blaschke products corresponding to the plant's right half plane (RHP) zeros and poles respectively. The weights W and V are assumed to be rational, stable, minimum phase, and to have equal numerator and denominator degrees. Hence, the problem is to minimize

$$K_m^{-1} = \| \, |WS| + |VT| \, \|_\infty = \| \, |Wn^{-1}v + Wq| + |Vd^{-1}u - Vq| \, \|_\infty$$

over all stable proper $q(s)$.

Consider the vector space $\mathcal{L}^\infty \times \mathcal{L}^\infty$. ($\mathcal{L}^\infty$ is the vector space of the complex-valued essentially bounded measurable functions of $\omega \in \mathcal{R}$). The quantity

$$\| \, (a, b) \, \|_{\infty\times\infty} := \| \, |a| + |b| \, \|_\infty \quad \text{where } (a, b) \in \mathcal{L}^\infty \times \mathcal{L}^\infty$$

is a norm on this vector space. So the above problem is equivalent to minimizing

$$\| \, (Wq, -Vq) + (Wn^{-1}v, Vd^{-1}u) \, \|_{\infty\times\infty}$$

subject to q being proper and stable (i.e. $q \in \mathcal{H}^\infty$). Now the set of vectors in $\mathcal{L}^\infty \times \mathcal{L}^\infty$ which are of the form $(Wq, -Vq)$ for some $q \in \mathcal{H}^\infty$ constitute a vector space, N_0 say, which is a linear subspace of $\mathcal{L}^\infty \times \mathcal{L}^\infty$:

$$N_0 := \{(a, b) \in \mathcal{L}^\infty \times \mathcal{L}^\infty | a = Wq, b = -Vq, q \in \mathcal{H}^\infty\}$$

So the problem is now in the required form, viz. find a vector $(Wq, -Vq)$ in the subspace N_0 that best approximates a given vector $(Wn^{-1}v, Vd^{-1}u)$. We have established the following:

Theorem 5 If the matrix M has rank 1, then K_m^{-1} is a vector space norm, and the optimal K_m-synthesis problem is that of determining the minimum distance from a point to a subspace. For instance, the synthesis problem of Section IV is

$$\begin{aligned} K_{mo}^{-1} &= \inf_{q\in\mathcal{H}^\infty} \| \, |WS| + |VT| \, \|_\infty \\ &= \inf_{(\alpha,\beta)\in N_0} \| \, (\alpha, \beta) + (Wn^{-1}v, Vd^{-1}u) \, \|_{\infty\times\infty} \end{aligned} \tag{1}$$

There is a well developed theory for handling such problems. To this we now turn.

VI. ABSTRACT DUALITY THEORY

In this section, we give a brief overview of the relevant duality theory, and state the results that will be used.

Consider a normed vector space X with coefficients in the field of complex numbers $\mathcal{C}$. This space will be referred to as the primal space and its norm denoted by $\| \, . \, \|$. A *linear functional* f is a linear map from X to $\mathcal{C}$

$$f : X \to \mathcal{C}$$

It is clear that these can be multiplied by scalars and added together. Define its norm to be

$$\| f \| := \sup_{x \in X, x \neq 0} \frac{|f(x)|}{\| x \|} \tag{2}$$

Although the same notation will be used for the norm of vectors in X and for the norm of linear functionals over X, this will not cause ambiguity since the argument of the norm function identifies which norm is intended. The functional f will be said to be *bounded* if the above supremum exists (i.e. if it's $< \infty$). It is not difficult to see that the set of all bounded linear functionals over the primal space X forms a normed vector space [30]. This space of bounded linear functionals will be called the *dual space* of X, and will be denoted by X^d. The value of $y \in X^d$ at $x \in X$ will subsequently be denoted by $< x, y >$. A dual space is always a Banach space. The closed unit ball in a normed vector space X will be denoted by BX, i.e.

$$BX := \{x \in X | \; \| x \| \leq 1\}$$

Note that the definition of the dual norm implies that for any $x \in X$ and any $y \in X^d$

$$| < x, y > | \quad \leq \quad \| x \| \quad \| y \| \tag{3}$$

The vectors $x \in X$ and $y \in X^d$ will be said to be *aligned* if

$$< x, y > \quad = \quad \| x \| \quad \| y \|$$

Note that if equality holds in eq. (3) for x and y, then y can be multipied by a scalar α of magnitude one so that x and αy are aligned. Given a subspace M of X, we define the set $M^\perp$, the *orthogonal complement* of M, as follows:

$$M^\perp := \{y \in X^d | < x, y > = 0 \quad \forall x \in M\}$$

This set is a closed subspace of X^d.

Via the Hahn-Banach Theorem, the following two theorems can be established (see, for example, [30]).

Theorem 6 Suppose that X is a normed vector space with subspace M, and that $x \in X$. Then

$$\inf_{m \in M} \| x - m \| = \max_{y \in BM^\perp} | < x, y > |$$

and the maximum on the right is attained for some $y_o \in M^\perp$. If the infimum is attained, say by m_o, then there is a maximizing y_o which is aligned with $x - m_o$.

Theorem 7 Suppose that X is a normed vector space with subspace N, that its dual space is X^d, and that $y \in X^d$. Then

$$\min_{n \in N^\perp} \| y - n \| = \sup_{x \in BN} | < x, y > |$$

and the minimum on the left is attained for some $n_o \in N^{\perp}$. If the supremum is attained, then there is a maximizing x_o which is aligned with $y - n_o$.

Note that in the first theorem, the minimization problem appears in the primal space X, while in the second theorem the minimization is formulated in the dual space X^d, and vice versa for the maximization problem. Existence is guaranteed in the dual space, but not in the primal space. In Theorem 6, any non-zero $y \in M^{\perp}$ yields a lower bound on the minimal attainable norm, and any $m \in M$ yields an upper bound, and likewise for Theorem 7.

We turn next to the construction of (Cartesian) product space. Given a vector space X, it is clear that $X \times X$ is a vector space, with the obvious definitions of scalar multiplication and vector addition. We will need the following observation concerning norms on product spaces. There are two notions of norm involved, so the following is not entirely trivial.

Theorem 8 If X^d is the dual of X, then $X^d \times X^d$ is the dual of $X \times X$, whenever the norms on $X \times X$ and on X obey the condition:

$$M_1^{-1}(\| a \| + \| b \|) \leq \| (a,b) \| \leq M_2(\| a \| + \| b \|) \tag{4}$$

for some real constants $M_1, M_2 > 0$ for every $(a,b) \in X \times X$ (uniformly).

Note that the condition in the theorem is enough to imply that $X \times X$ is a Banach space whenever X is one. Note also that it is not always true that $(X \times X)^d = X^d \times X^d$; some extra condition, such as that in eq. (4) above, is needed.

Proof By $(x,y) \in X^d \times X^d$ we mean of course the linear functional on $X \times X$ defined by

$$< (a,b),(x,y) >=< a,x > + < b,y >$$

Given any $x, y \in X^d$, it is trivial that (x,y) is a linear functional on $X \times X$. It needs to be shown that the boundedness of both x and y as functionals over X implies the boundedness of (x,y) as a functional over $X \times X$. Since x, y are bounded linear functionals on X, we find using the condition in eq. (4) that for all $a, b \in X$

$$\begin{aligned} | < (a,b),(x,y) > | &= | < (a,x) > + < (b,y) > | \\ &\leq | < (a,x) > | + | < (b,y) > | \\ &\leq \| a \| \| x \| + \| b \| \| y \| \\ &\leq \max\{\| x \|, \| y \|\}(\| a \| + \| b \|) \\ &\leq \max\{\| x \|, \| y \|\} M_1 \| (a,b) \| \end{aligned}$$

so (x, y) is indeed a *bounded* linear functional on $X \times X$. Hence, the vector space $X^d \times X^d$ is contained in the dual of $X \times X$, i.e.

$$X^d \times X^d \subset (X \times X)^d$$

Consider now an arbitrary bounded linear functional f over $X \times X$. The restriction of f to $X \times \{0\}$ must be a linear functional, and it can be identified with a functional x over X via the obvious (algebraic) isomorphism $X \approx X \times \{0\}$. Likewise, f yields a y, by restriction to $\{0\} \times X$. It remains only to show that the boundedness of $f = (x, y)$ as a linear functional over $X \times X$ implies that of x and y as linear functionals over X. Now

$$\begin{aligned} | < (a, 0), f > | &= | < a, x > | \\ &\leq \| (a, 0) \| \| f \| \\ &\leq M_2 \| a \| \| f \| \end{aligned}$$

and so x (and similarly y) is bounded when viewed as a functional over X, and hence, is an element of X^d. This shows that

$$X^d \times X^d \supset (X \times X)^d$$

completing the proof. □

VII. SOME VECTOR SPACES AND THEIR DUALS

We now define certain normed vector spaces. The elements of these spaces are functions in a single complex variable s (to be thought of as the usual Laplace transform variable), or of $j\omega$. When considering discrete time systems, we consider s to lie in the unit disc, and consider the unit circle in place of the imaginary axis. The dependence on s or $j\omega$ will be suppressed in our notation. See [35,36,37,38] for a treatment of these spaces.

$\mathcal{L}^\infty$ – Measurable functions of $j\omega$ with finite infinity norm, where by the infinity norm of a function f is meant

$$\| f \|_\infty := \text{ess-sup}_{\omega \in \mathcal{R}} |f(j\omega)|$$

$\mathcal{L}^1$ – Measurable functions of $j\omega$ with finite 1-norm, where by the 1-norm of a function f is meant

$$\| f \|_1 := \int_{-\infty}^{\infty} |f| d\omega$$

$\mathcal{H}^\infty$ – Functions of s that are analytic in the open right-half plane (ORHP) and are bounded (i.e. have a finite infinity-norm) there. This space can be viewed as a subspace of $\mathcal{L}^\infty$.

$\mathcal{H}^1$ – Functions of s that are analytic and have a finite 1-norm in the ORHP. This space can be viewed as a subspace of $\mathcal{L}^1$.

The following facts will be needed.

Theorem 9 [32,35,36,37,38]

1. The dual of $\mathcal{L}^1$ is $\mathcal{L}^\infty$
2. The dual of $\mathcal{L}^1/\mathcal{H}^1$ is $\mathcal{H}^\infty$
3. For $\mathcal{H}^1 \subset \mathcal{L}^1$, $(\mathcal{H}^1)^\perp = \mathcal{H}^\infty$

The first statement above, for example, says that every bounded linear functional f on $\mathcal{L}^1$ can be represented in the form

$$f(x) =< x, f >= \int_{-\infty}^{\infty} x l d\omega \quad \forall x \in \mathcal{L}^1$$

for some $l \in \mathcal{L}^\infty$ and, conversely, every $l \in \mathcal{L}^\infty$ yields a bounded linear functional f over $\mathcal{L}^1$ in this way. Moreover, the dual norm of f is just the infinity-norm of l, i.e.

$$\| f \| = \| l \|_\infty = \sup_{x \in B\mathcal{L}^1} | < x, l > |$$

In the sequel, we will use the same symbol l for both the function l and the corresponding functional f, but this slight abuse of notation will not cause confusion. Similar remarks apply to the other identities above.

VIII. APPLICATION OF DUALITY THEORY

To apply duality theory to the problem at hand, we must firstly identify the norms and spaces required. By Theorem 5 above, the norm of interest on $\mathcal{L}^\infty \times \mathcal{L}^\infty$ is

$$\| (x, y) \|_{\infty \times \infty} := \| \, |x| + |y| \, \|_\infty$$

The following norm on the space $\mathcal{L}^1 \times \mathcal{L}^1$ will also be useful:

$$\| (a, b) \|_{1 \times 1} := \int \max\{|a|, |b|\} d\omega$$

The following theorem relates these two normed spaces.

Theorem 10 The space $\mathcal{L}^\infty \times \mathcal{L}^\infty$ with norm $\| (x, y) \|_{\infty \times \infty}$ is the dual of the space $\mathcal{L}^1 \times \mathcal{L}^1$ with norm $\| (a, b) \|_{1 \times 1}$

Proof Consider the space $X = \mathcal{L}^1 \times \mathcal{L}^1$ with norm $\int \max\{|a|, |b|\} d\omega$. Since this norm satisfies the condition in eq. (4), and since $\mathcal{L}^\infty$ is the dual of $\mathcal{L}^1$, Theorem 8 demonstrates that $\mathcal{L}^\infty \times \mathcal{L}^\infty$ is the dual of $\mathcal{L}^1 \times \mathcal{L}^1$. It remains to determine the corresponding norm on the dual space. By definition,

$$\| (x,y) \|_{\infty\times\infty} = \sup_{(a,b)\in BX} |< (a,b),(x,y) >|$$

where the supremum is subject to $\| (a,b) \| = \int \max\{|a|,|b|\}d\omega = 1, \quad a,b \in \mathcal{L}^1$.

$$= \sup_{(a,b)\in BX} \left| \int (ax+by)d\omega \right|$$

$$\leq \sup_{(a,b)\in BX} \int |ax| + |by| d\omega$$

$$\leq \sup_{(a,b)\in BX} \int (|x|+|y|) \max\{|a|,|b|\}d\omega$$

$$\leq \| \, |x| + |y| \, \|_\infty \sup_{(a,b)\in BX} \int \max\{|a|,|b|\}d\omega$$

Hence,

$$\| (x,y) \|_{\infty\times\infty} \leq \| \, |x| + |y| \, \|_\infty \tag{5}$$

It remains to show that this bound can be approached arbitrarily closely. Chose any point on the $j\omega$ axis (possibly ∞) where the above infinity norm is attained, say $j\alpha$. We can multiply y by a complex scalar with absolute value one, β say, so that $x(j\alpha)$ and $\beta y(j\alpha)$ have the same argument at $j\alpha$. Thus,

$$\| x + \beta y \|_\infty = \| \, |x| + |y| \, \|_\infty$$

By the $(\mathcal{L}^1)^d = \mathcal{L}^\infty$ theory, there is a sequence l_n in $\mathcal{L}^1$, with $\| l_n \|_1 = 1$, for which

$$< x + \beta y, l_n > \rightarrow \| x + \beta y \|_\infty$$

Hence

$$< x, l_n > + < \beta y, l_n > \rightarrow \| \, |x| + |y| \, \|_\infty$$

Taking $a_n = l_n$ and $b_n = \beta l_n$ yields

$$\| (a_n, b_n) \|_{1\times 1} = \int \max\{|a_n|,|b_n|\}d\omega$$

$$= \int \max\{|l_n|,|\beta l_n|\}d\omega = \int |l_n| d\omega = \| l_n \|_1 = 1$$

and the result follows, since it is now clear that the upper bound in (5) can be approached arbitrarily closely. □

To apply the above duality theory, we need to identify the subspaces to be used.

Theorem 11 Let $W, V, W^{-1}, V^{-1} \in \mathcal{H}^\infty$. Consider the subspace of $\mathcal{L}^1 \times \mathcal{L}^1$ given by

$$N = \{(a,b) \in \mathcal{L}^1 \times \mathcal{L}^1 | a = Vl + Vh, b = Wl - Wh, l \in \mathcal{L}^1, h \in \mathcal{H}^1\} \quad (6)$$

$$\text{Then } N^\perp = \{(x,y) \in \mathcal{L}^\infty \times \mathcal{L}^\infty | (x,y) = (Wq, -Vq), q \in \mathcal{H}^\infty\} \quad (7)$$

Proof We have, by definition,

$$(x,y) \in N^\perp \Leftrightarrow < (a,b),(x,y) >= 0 \qquad \forall (a,b) \in N$$

$$\Leftrightarrow < (Vl + Vh, Wl - Wh),(x,y) >= 0 \quad \forall l \in \mathcal{L}^1, h \in \mathcal{H}^1$$

$$\Leftrightarrow \int (Vx + Wy)l + (Vx - Wy)h d\omega = 0 \quad \forall l \in \mathcal{L}^1, h \in \mathcal{H}^1$$

$$\Leftrightarrow \left\{ \begin{array}{ll} & \int (Vx + Wy)l d\omega = 0 \quad \forall l \in \mathcal{L}^1 \\ \text{and} & \int (Vx - Wy)h d\omega = 0 \quad \forall h \in \mathcal{H}^1 \end{array} \right\}$$

Viewing $(Vx + Wy)$ as a functional over $\mathcal{L}^1$, it is orthogonal to every primal vector $l \in \mathcal{L}^1$, so it must be the zero functional. Since W, V are invertible in $\mathcal{H}^\infty$, multiplication by either maps $\mathcal{H}^1$ onto itself and $\mathcal{L}^1$ onto itself. Hence, without loss of generality, we can let $x = Wq, y = Vp$, so that

$$Vx + Wy \equiv 0 \Leftrightarrow WV(q + p) \equiv 0$$

$$\Leftrightarrow p = -q$$

Also, viewing $Wx - Vy$ as a functional over $\mathcal{L}^1$, $Vx - Wy = 2WVq$ annihilates every vector h in $\mathcal{H}^1$, (since $WV\mathcal{H}^1 = \mathcal{H}^1$), and so the vector q must be an element of $(\mathcal{H}^1)^\perp = \mathcal{H}^\infty$ (See Theorem 9). Hence,

$$N^\perp = \{(x,y) \in \mathcal{L}^\infty \times \mathcal{L}^\infty | (x,y) = (Wq, -Vq), q \in \mathcal{H}^\infty\}$$

□

In Section V, the Youla Lemma was used to express our problem in the form: find the minimum distance from a point to a subspace. Section VI outlined two theorems that deal with just this situation. Combined with the above results, we are now in a position to apply duality theory to the problem at hand.

Theorem 7, with $X = \mathcal{L}^1 \times \mathcal{L}^1$, so that $X^d = \mathcal{L}^\infty \times \mathcal{L}^\infty$, with $y = (Wn^{-1}v, Vd^{-1}u)$, and with N and $N^\perp$ given by eqs. (6) and (7) above, yields

$$K_{mo}^{-1} = \min_{q \in \mathcal{H}^\infty} \| \, |Wn^{-1}v + Wq| + |Vd^{-1}u - Vq| \, \|_\infty \quad (8)$$

$$= \min_{n \in N^{\perp}} \| y - n \|_{\infty \times \infty} = \sup_{x \in BN} | < x, y > |$$

$$= \sup_{(Vl+Vh, Wl-Wh) \in BN} | < (Wn^{-1}v, Vd^{-1}u), (Vl + Vh, Wl - Wh) > |$$

$$= \sup_{(Vl+Vh, Wl-Wh) \in BN} \left| \int WV(n^{-1}vl + d^{-1}ul + n^{-1}vh - d^{-1}uh) \mathrm{d}\omega \right|$$

$$= \sup_{(Vl+Vh, Wl-Wh) \in BN} \left| \int WVn^{-1}d^{-1}(l + dvh - nuh) \mathrm{d}\omega \right| \qquad (9)$$

by the Bezout identity. The existence of a minimizing $(x, y) \in N^{\perp} \subset \mathcal{L}^{\infty} \times \mathcal{L}^{\infty}$ is assured. Using this theorem requires $y = (Wn^{-1}v, Vd^{-1}u)$ to be in $\mathcal{L}^{\infty} \times \mathcal{L}^{\infty}$, which is the case.

Bounds on the Achievable Performance:

If the supremum in eq. (9) is restricted to a subset of BN, then a lower bound on K_{mo}^{-1} results. Since the supremum is difficult to solve explicitly, we give two such bounds, obtained by taking $h = 0$ and $l = 0$ in the elements of N.

$$K_{mo}^{-1} = \sup_{(Vl+Vh, Wl-Wh) \in BN} \left| \int WVn^{-1}d^{-1}(l + dvh - nuh) \mathrm{d}\omega \right|$$

$$= \sup_{l,h} \frac{|\int WVn^{-1}d^{-1}(l + dvh - nuh)\mathrm{d}\omega|}{\int \max\{|V||l + h|, |W||l - h|\}\mathrm{d}\omega}$$

$$\geq \sup_{l \in \mathcal{L}^1} \frac{|\int WVn^{-1}d^{-1}l\mathrm{d}\omega|}{\int \max\{|W|, |V|\}|l|\mathrm{d}\omega}$$

$$= \| \frac{VW}{\max\{|W|, |V|\}} \|_{\infty}$$

$$= \| \min\{|W|, |V|\} \|_{\infty}$$

Likewise, taking $l = 0$ yields

$$K_{mo}^{-1} = \sup_{l,h} \frac{|\int WVn^{-1}d^{-1}(l + dvh - nuh)\mathrm{d}\omega|}{\int \max\{|V||l + h|, |W||l - h|\}\mathrm{d}\omega}$$

$$\geq \sup_{h \in \mathcal{H}^1} \frac{|\int WV(n^{-1}v - d^{-1}u)h\mathrm{d}\omega|}{\int \max\{|W|, |V|\}|h|\mathrm{d}\omega}$$

From the duality relations for the classical 1-block $\mathcal{H}^{\infty}$ problem [31,32,35,36], we obtain

$$= \min_{q \in \mathcal{H}_{\infty}} \| \frac{WV}{\max\{|W|, |V|\}}(n^{-1}v - d^{-1}u + q) \|_{\infty}$$

$$= \min_{q \in \mathcal{H}_\infty} \| \min\{|W|, |V|\}(n^{-1}v - d^{-1}u + q) \|_\infty$$

The problem of evaluating the right hand side above is clearly an irrational 1-block $\mathcal{H}_\infty$-problem, and so can be solved to within any $\epsilon > 0$ by a rational one-block $\mathcal{H}^\infty$ problem. We have established the following:

Theorem 12

$$K_{mo}^{-1} \geq \| \min\{|W|, |V|\} \|_\infty$$

$$K_{mo}^{-1} \geq \min_{q \in \mathcal{H}_\infty} \| \min\{|W|, |V|\}(n^{-1}v - d^{-1}u + q) \|_\infty$$

IX. THE ALL-PASS PROPERTY

In this section, the characterization of optimality by the alignment condition is investigated. Our treatment is modeled after [32]. Let

$$(x_0, y_0) = (Wq_0 + Wn^{-1}v, -Vq_0 + Vd^{-1}u) = (WS_0 n^{-1}d^{-1}, VT_0 n^{-1}d^{-1})$$

where q_0 attains the infimum in eq. (8). Suppose that $(a_0, b_0) \in \mathcal{L}^1 \times \mathcal{L}^1$ attains the supremum in eq. (9). Normalizing (a_0, b_0) so that

$$\| (a_0, b_0) \|_{1\times 1} = 1$$

allows the alignment condition to be written as

$$\begin{aligned} K_{mo}^{-1} &= \| \, |WS_0| + |VT_0| \, \|_\infty = \| (x_0, y_0) \|_{\infty\times\infty} \\ &= | < (a_0, b_0), (x_0, y_0) > | \\ &= \left| \int_{-\infty}^{\infty} (a_0 x_0 + b_0 y_0) d\omega \right| \\ &\leq \int_{-\infty}^{\infty} |a_0 x_0 + b_0 y_0| d\omega \end{aligned}$$

with equality holding when for all ω either (i) $\angle(a_0 x_0 + b_0 y_0) = 0$, or (ii) $a_0 x_0 + b_0 y_0 = 0$,

$$\leq \int_{-\infty}^{\infty} (|a_0 x_0| + |b_0 y_0|) d\omega$$

with equality holding when for all ω either (i) $\angle a_0 x_0 = \angle b_0 y_0$, or (ii) $a_0 x_0 = 0$, or (iii) $b_0 y_0 = 0$,

$$\leq \int_{-\infty}^{\infty} \max\{|a_0|, |b_0|\}(|x_0| + |y_0|) d\omega$$

with equality holding when for all ω either (i) $|a_0| = |b_0|$, or (ii) $|a_0| > |b_0|$ and $y_0 = 0$, or (iii) $|b_0| > |a_0|$ and $x_0 = 0$,

$$\leq \| \, |x_0| + |y_0| \, \|_\infty \int_{-\infty}^{\infty} \max\{|a_0|, |b_0|\} d\omega$$

with equality holding when for all ω either (i) $K_{mo}^{-1} = |x_0| + |y_0|$, or (ii) $\max\{|a_0|, |b_0|\} = 0$,

$$= \| \, |WS_0| + |VT_0| \, \|_\infty$$

since the norm of (a_0, b_0) has been normalized to one

$$= K_{mo}^{-1}$$

Since the above expressions are of the form "$K_{mo}^{-1} \leq \ldots \leq K_{mo}^{-1}$", we conclude that the alignment condition implies that each of the inequalities must, in fact, be an equality. Consequently, at every ω (strictly speaking, at almost all ω's), the following conditions must hold:

$$\angle a_0 x_0 = 0 \text{ or } a_0 x_0 = 0 \tag{10}$$

$$\angle b_0 y_0 = 0 \text{ or } b_0 y_0 = 0 \tag{11}$$

$$\text{(i)}|a_0| = |b_0| \text{ or (ii)}|a_0| > |b_0| \text{ and } y_0 = 0 \text{ or (iii)}|b_0| > |a_0| \text{ and } x_0 = 0 \tag{12}$$

$$|x_0| + |y_0| = K_{mo}^{-1} \text{ or } \max\{|a_0|, |b_0|\} = 0 \tag{13}$$

Moreover, we must have

$$(a_0, b_0) \in N, \quad \text{and} \tag{14}$$

$$(x_0, y_0) \in N^{\perp} \tag{15}$$

The conditions in equations (10)-(15) are also sufficient for optimality, [30, Corollary 1, p120]. One consequence of the above is as follows.

Theorem 13 The optimal k_m-synthesis problem of equation (8) either has an all-pass solution for some $q_0 \in \mathcal{H}^\infty$, or $K_{mo}^{-1} = \| \min\{|W|, |V|\} \|_\infty$

Proof Suppose that (a_0, b_0) attains the supremum in eq. (9). From eq. (13) above, we have that for almost every ω either

$$|x_0| + |y_0| = K_{mo}^{-1} \quad \text{or} \quad \max\{|a_0|, |b_0|\} = 0$$

Suppose that $\max\{|a_0|, |b_0|\}$ is zero on a set of positive measure, $\Omega \subset j\mathcal{R}$ say. Then

$$\max\{|a_0|, |b_0|\} = 0 \text{ on } \Omega$$

$$\Rightarrow a_0 = V(l + h) = 0 \text{ and } b_0 = W(l - h) = 0 \text{ on } \Omega$$

$$\Rightarrow V^{-1}a_0 = W^{-1}b_0 = l + h = l - h = 0 \text{ on } \Omega$$
$$\Rightarrow h = 0 \text{ on } \Omega$$

Now an element of $\mathcal{H}^1$ that is zero on a set of positive measure is identically zero [32]. Hence $a_0 = Wl$ and $b_0 = Vl$ for all ω. In this situation, eq. (9) becomes

$$K_{mo}^{-1} = \sup_l \frac{\left|\int WVn^{-1}d^{-1}l\mathrm{d}\omega\right|}{\int \max\{|W|,|V|\}|l|\mathrm{d}\omega}$$

and so

$$K_{mo}^{-1} \leq \sup_l \frac{\int |WVn^{-1}d^{-1}l|\mathrm{d}\omega}{\int \max\{|W|,|V|\}|l|\mathrm{d}\omega}$$
$$= \sup_l \frac{\int \max\{|W|,|V|\}\min\{|W|,|V|\}|l|\mathrm{d}\omega}{\int \max\{|W|,|V|\}|l|\mathrm{d}\omega}$$
$$\leq \| \min\{|W|,|V|\} \|_\infty \sup_l \frac{\int \max\{|W|,|V|\}|l|\mathrm{d}\omega}{\int \max\{|W|,|V|\}|l|\mathrm{d}\omega}$$
$$= \| \min\{|W|,|V|\} \|_\infty$$

This shows that if $\max\{|a_0|,|b_0|\}$ is zero on a set of positive measure, then

$$K_{mo}^{-1} \leq \| \min\{|W|,|V|\} \|_\infty$$

In view of Theorem 12, we conclude that if $h = 0$ on a set of positive measure, then

$$K_{mo}^{-1} = \| \min\{|W|,|V|\} \|_\infty$$

On the other hand, suppose that h is non-zero on any set of positive measure. As noted above, this means that $\max\{|a_0|,|b_0|\}$ is non-zero almost everywhere (a.e.). Consequently, from eq. (13) we have $|x_0| + |y_0| = K_{mo}^{-1}$ a.e. and so $|WS| + |VT|$ has the same magnitude K_{mo}^{-1} a.e., i.e. $|WS| + |VT|$ is "all-pass".

Establishing the all-pass property in the case where the supremum is not attained is considerably more subtle, and requires a good deal more functional analysis than that of Section VI. Theorem 1 in [39], however, is applicable to the present problem, and yields the desired result. □

The interested reader should see [39,40,41] for more abstract studies of the all-pass property. The results in [40] are not directly applicable to the present problem because of the non-differentiability of our norm.

Note that if the optimal solution is not $S = 0$ or $S = 1$, then both S and T are non-zero almost everywhere. (Otherwise, either S or T would be zero on a set of positive measure, and hence would be identically zero). If, in addition, $K_{mo}^{-1} \neq \| \min\{|W|,|V|\} \|_\infty$ then the conditions in equations (10)-(13) simplify to:

$$\angle a_0 x_0 = 0 \tag{16}$$

$$\angle b_0 y_0 = 0 \tag{17}$$

$$|a_0| = |b_0| \tag{18}$$

$$|x_0| + |y_0| = K_{mo}^{-1} \tag{19}$$

X. THE STABLE MINIMUM PHASE CASE

In this section, we consider the case where the plant G_0 is both stable and minimum phase. Our objective is to determine the optimal value of K_{mo} explicitly.

In this case all the plant's poles and zeros can be cancelled by the controller, so the Blaschke products n and d become $n = d = 1$ and we may take $u = 1/2 = v$. Thus $1 = nu + dv$, and

$$S = d(n\mathfrak{q} + v) = \mathfrak{q} + 1/2$$

$$T = n(-d\mathfrak{q} + u) = -\mathfrak{q} + 1/2$$

where $\mathfrak{q}$ is the Youla parameter. For convenience, let $q = 2\mathfrak{q}$. So the problem at hand is now

$$2K_{mo}^{-1} = 2 \inf_{q \in \mathcal{H}^\infty} \|\, |WS| + |VT| \,\|_\infty$$

$$= \inf_{q \in \mathcal{H}^\infty} \|\, |W(q+1)| + |V(-q+1)| \,\|_\infty$$

We will work with the space $X^d = \mathcal{H}^\infty \times \mathcal{H}^\infty$ equipped with the norm

$$\|\, (x, y) \,\|_{\infty \times \infty} = \|\, |x| + |y| \,\|_\infty$$

Our first task is to identify its predual.

Theorem 14 The space $\mathcal{H}^\infty \times \mathcal{H}^\infty$ with norm $\|\, |x| + |y| \,\|_\infty$ is the dual of the space $\mathcal{L}^1/\mathcal{H}^1 \times \mathcal{L}^1/\mathcal{H}^1$ with norm

$$\|\, (a, b) \,\| = \inf_{h_1, h_2 \in \mathcal{H}^1} \int \max\{|a + h_1|, |b + h_2|\} d\omega$$

Proof This is just a restatement of Theorem 10 in term of cosets. □

First of all, let us dispose with a trivial situation. If $|W| \geq |V| \quad \forall \omega$ then clearly

$$\|\, |V| \,\|_\infty = \|\, \min\{|W|, |V|\} \,\|_\infty$$

Now, the solution $T = 1, S = 0$ is realizable, (the plant being stable and minimum phase) and so

$$K_{mo}^{-1} \leq \|\, (W.0, V.1) \,\|_{\infty \times \infty} = \|\, |V| \,\|_\infty$$

Since this solution achieves the lower bound of Theorem 12, it must be optimal. Likewise, if $|W| \leq |V| \quad \forall\omega$, then $T = 0, S = 1$ is realizable, so the optimum is again

$$K_{mo}^{-1} = \| \min\{|W|, |V|\} \|_\infty$$

Consequently, we assume that there is at least one value of $j\omega$ at which W and V have the same absolute value; otherwise, we are done.

Theorem 15 Let

$$N = \{(a,b) \in \mathcal{L}^1/\mathcal{H}^1 \times \mathcal{L}^1/\mathcal{H}^1 | (a,b) = (Vl, Wl), l \in \mathcal{L}^1/\mathcal{H}^1\},$$

$$\text{Then } N^\perp = \{(x,y) \in \mathcal{H}^\infty \times \mathcal{H}^\infty | (x,y) = (Wq, -Vq), q \in \mathcal{H}^\infty\}$$

Proof By definition of $N^\perp$,

$$(x,y) \in N^\perp \Leftrightarrow < (Vl, Wl), (x,y) >= 0 \quad \forall l \in \mathcal{L}^1/\mathcal{H}^1$$

$$\Leftrightarrow \int (Vx + Wy) l d\omega = 0 \quad \forall l \in \mathcal{L}^1/\mathcal{H}^1$$

$$\Leftrightarrow Vx + Wy = 0$$

since this vector annihilates every primal vector. Hence,

$$x = Wq, \qquad y = -Vq \qquad \text{for some } q \in \mathcal{H}^\infty$$

as required. □

Actually, the above theorem is just a restatement of Theorem 11 in terms of cosets. As in the previous section, we will use Theorem 7. This yields

$$2K_{mo}^{-1} = \inf_{q \in \mathcal{H}^\infty} \| \, |W(1+q)| + |V(1-q)| \, \|_\infty \tag{20}$$

$$= \inf_{\mathfrak{n} \in N^\perp} \| \mathfrak{n} + \mathfrak{y} \|_{\infty \times \infty} = \sup_{n \in BN} < \mathfrak{y}, n >$$

$$= \min_{\mathfrak{n} \in N^\perp} \| \mathfrak{y} + \mathfrak{n} \|_{\infty \times \infty}$$

where N and $N^\perp$ are given in Theorem 15 and $\mathfrak{y} = (W, V)$. Note that the infimum in eq. (8) is achieved by Theorem 7 (since the infimum is in the dual space). We will need the following lemma:

Lemma 1 The subspace M is dense in X if and only if the zero functional is the only bounded linear functional in X^d which annihilates all of M.

Proof This well known result is an easy consequence of Theorem 6 above; see, for example, III.6.14 in [42]. □

We can now tackle the main theorem of this section.

Theorem 16 When the plant G_0 is SISO, stable and minimum phase,

$$K_{mo}^{-1} = \min_{q \in \mathcal{H}^\infty} \| \, |WS| + |VT| \, \|_\infty = \| \min\{|W|, |V|\} \, \|_\infty$$

Proof Given any element of $\mathcal{H}^\infty$, the operation of evaluating it at a given point, α say, in the open right half plane (or the complement of the closed unit disc for discrete time problems) is easily seen to be a bounded linear functional. This functional will be denoted by $\mathcal{E}^{(\alpha)}$. Indeed, it is clear that its norm is unity, since

$$\| \mathcal{E}^{(\alpha)} \| = \sup_{a \in B\mathcal{H}^\infty} | < a, \mathcal{E}^{(\alpha)} > | = \sup_{a \in B\mathcal{H}^\infty} |a(\alpha)| = 1$$

Moreover, it has an obvious representation via Cauchy's integral formula:

$$< f, \mathcal{E}^{(\alpha)} > = \frac{1}{2\pi j} \int \frac{f(z)}{z - \alpha} \mathrm{d}z$$

Consequently, it is an element of the space $\mathcal{L}^1 / \mathcal{H}^1$.

We will use couples of such evaluate functionals acting on $\mathcal{H}^\infty \times \mathcal{H}^\infty$ where the α's lie on the vertical line through $\epsilon + j0$ for some fixed $\epsilon > 0$. Define the subspace M_1 as follows:

$$M_1 = \{ f \in X | f = \sum_{i=1}^{n} \beta_i (V_i \mathcal{E}^{(\alpha_i)}, W_i \mathcal{E}^{(\alpha_i)}), \beta_i \in \mathcal{C}, \mathrm{Re}(\alpha_i) = \epsilon \}$$

where, for notational convenience, we have let

$$W_i := W(\alpha_i)$$

$$V_i := V(\alpha_i)$$

Clearly, $M_1 \subset N$. We claim now that M_1 is, in fact, dense in N. To use Lemma 1 to establish this, we must show that the zero functional is the only functional in N^* that annihilates all of M_1. Now, N^* can be identified with $X^d / N^\perp$[30,36,42], each of whose elements can be written as (Wq, Vq) for some $q \in \mathcal{H}^\infty$. Hence,

$$(Wq, Vq) \in M_1^\perp$$

$$\Leftrightarrow < (a, b), (Wq, Vq) > = 0 \quad \forall (a, b) \in M_1$$

$$\Leftrightarrow < (V(\alpha)\mathcal{E}^{(\alpha)}, W(\alpha)\mathcal{E}^{(\alpha)}), (Wq, Vq) > = 0 \quad \forall \alpha$$

$$\Leftrightarrow 2V(\alpha)W(\alpha)q(\alpha) = 0 \quad \forall \alpha$$

$$\Leftrightarrow q = 0$$

Thus, the only functional in N^* that annihilates every vector in M_1 is the zero functional, so that, by Lemma 1, M_1 is dense in N as claimed.

The value of a supremum is not altered by limiting it to a dense set. Hence, using eq. (20), we obtain

$$2K_{mo}^{-1} = \sup_{(a,b) \in BN} < (a,b), (W,V) >$$

$$= \sup_{n, Re(\alpha_i), \beta_i} \frac{| \sum_{i=1}^n \beta_i < (W,V), (V_i \mathcal{E}^{(\alpha_i)}, W_i \mathcal{E}^{(\alpha_i)}) > |}{\| \sum_{i=1}^n \beta_i (V_i \mathcal{E}^{(\alpha_i)}, W_i \mathcal{E}^{(\alpha_i)}) \|}$$

$$= 2 \sup_{n, Re(\alpha_i), \beta_i} \frac{| \sum_{i=1}^n \beta_i V_i W_i |}{\| \sum_{i=1}^n \beta_i (V_i \mathcal{E}^{(\alpha_i)}, W_i \mathcal{E}^{(\alpha_i)}) \|}$$

We need to consider the quantity in the denominator above.

$$\| \sum_{i=1}^n \beta_i (V_i \mathcal{E}^{(\alpha_i)}, W_i \mathcal{E}^{(\alpha_i)}) \|$$

$$= \sup_{(x,y) \in \mathcal{H}^\infty} \frac{| < (x,y), \sum_{i=1}^n \beta_i (V_i \mathcal{E}^{(\alpha_i)}, W_i \mathcal{E}^{(\alpha_i)}) > |}{\| \, |x| + |y| \, \|_\infty}$$

$$= \sup_{(x,y) \in \mathcal{H}^\infty} \frac{| \sum_{i=1}^n \beta_i (xV + yW)(\alpha_i) |}{\| \, |x| + |y| \, \|_\infty} \tag{21}$$

$$\leq \sum_{i=1}^n |\beta_i| \max\{|V_i|, |W_i|\} \qquad \text{since } |x| + |y| \leq 1 \tag{22}$$

If one takes the supremum of the above, the inequality becomes an equality. To see this, consider

$$a(s) := \frac{\epsilon}{s - \epsilon - j\omega}$$

so that

$$|a(j\omega)|^2 = \frac{\epsilon^2}{\epsilon^2 + (\omega - \omega)^2}$$

Clearly $|a(j\omega)| \leq 1$ at each ω, and

$$\lim_{\epsilon \to 0} |a(j\omega)| = 0 \quad \forall \omega \neq \omega \qquad \text{and}$$

$$|a(j\omega)| = 1$$

So the function $|a(j\omega)|$ has a "spike" at $\omega = \omega$, and is very small away from ω. So by chosing a and b in the supremum in eq. (21) to be linear combinations of such functions, the equality in eq. (22) can be approached arbitrarily closely.

Thus

$$K_{mo}^{-1} = \sup_{\alpha_i,\beta_i} \frac{|\sum_{i=1}^n \beta_i V_i W_i|}{\| \sum_{i=1}^n \beta_i (V_i \mathcal{E}^{(\alpha_i)}, W_i \mathcal{E}^{(\alpha_i)}) \|}$$

$$= \sup_{\alpha_i,\beta_i} \frac{|\sum_{i=1}^n \beta_i V_i W_i|}{\sum_{i=1}^n |\beta_i| \max\{|V_i|, |W_i|\}}$$

$$\leq \sup_{\alpha_i,\beta_i} \frac{\sum_{i=1}^n |\beta_i||V_i W_i|}{\sum_{i=1}^n |\beta_i| \max\{|V_i|, |W_i|\}}$$

Now, $|V_i W_i| = \max\{|V_i|, |W_i|\} \min\{|V_i|, |W_i|\}$, so

$$K_{mo}^{-1} \leq \sup_{\alpha_i,\beta_i} \frac{\sum_{i=1}^n |\beta_i| \max\{|V_i|, |W_i|\} \min\{|V_i|, |W_i|\}}{\sum_{i=1}^n |\beta_i| \max\{|V_i|, |W_i|\}}$$

$$\leq \| \min\{|V|, |W|\} \|_\infty \sup_{\alpha_i,\beta_i} \frac{\sum_{i=1}^n |\beta_i| \max\{|V_i|, |W_i|\}}{\sum_{i=1}^n |\beta_i| \max\{|V_i|, |W_i|\}}$$

$$= \| \min\{|W|, |V|\} \|_\infty$$

Combining the above inequality with Theorem 12 concludes the proof. □

It is easy to see that the above theorem does not apply if W or V are discontinuous; essential use was made of our assumptions on these functions.

XI. SUMMARY

We have considered a class of k_m-synthesis problems, namely those with a rank 1 matrix M when rearranged into the canonical form. The resulting analysis problem is easy and the resulting synthesis problem is convex. This class covers the case where the plant has one input or one output. A special case, essentially corresponding to classical "Bode style" SISO compensator design, illustrated the approach. Duality theory established an all-pass property. The optimal value of K_m was explicitly determined for stable and minimum phase plants.

Note that the methods described here clearly extend to the general rank-1 problem for both continuous time and discrete time systems. It is not difficult to see that problems involving a plant with one output, with the uncertainty of each of the elements of the plant's frequency response matrix described by a single scalar complex-valued variable, with all disturbances reflected to the output, and with the specifications of robust disturbance attenuation and robust stability lead to a rank one matrix.

An interesting and important open problem is to require the controller to be open loop stable, and to include a robust specification limiting the size of the complementary sensitivity function. Solving this problem would bring us even closer to a full and complete formal mathematical solution of the classical compensator design problem in the frequency domain. It is hoped that the results of this paper are a step forward in this direction.

XII. ACKNOWLEDGEMENT

This work was supported in part by AFOSR Grant 89-0398.

XIII. REFERENCES

[1] M. G. Safonov, A. J. Laub and G. L. Hartmann, "Feedback Properties of Multivariable Systems: The Role and Use of the Return Difference Matrix", *IEEE Transactions on Automatic Control*, Vol. 26, pp. 47-65, Feb. 1981.

[2] M. J. Chen and C. A. Desoer, "The Problem of Guaranteeing Robust Disturbance Rejection in Linear Multivariable Feedback Systems", *International Journal of Control*, Vol. 37, pp. 305-313, 1983.

[3] J. S. Bird and B. A. Francis, "On the Robust Disturbance Attenuation Problem", *Proc. IEEE Conf. on Decision and Control*, Athens, Greece, pp. 1804-1809, Dec. 1986.

[4] S. Hara and H. Katori, "On Constrained $\mathcal{H}^\infty$ Optimization Problem for SISO Systems", *IEEE Transactions on Automatic Control*, Vol. 31, pp. 856-858, Sept. 1986.

[5] P. Dorato and Y. Li, "U-Parameter Design of Robust Single-Input-Single-Output Systems", *Proc. IEEE Conf. on Decision and Control*, Tampa FL, pp. 2304-2307, Dec. 1989.

[6] M. G. Safonov and M. Athans, "A Multiloop Generalization of the Circle Criterion for Stability Margin Analysis", *IEEE Transactions on Automatic Control*, Vol. 26, pp. 415-422, April 1981.

[7] M. G. Safonov, "Stability Margins of Diagonally Perturbed Multivariable Feedback Systems", *Proc. IEE, Part D*, Vol. 129, pp. 251-256, Nov. 1982.

[8] J. C. Doyle, "Analysis of Feedback Systems with Structured Uncertainties", *Proc. IEE, Part D*, Vol. 129, pp. 242-250, Nov. 1982.

[9] M. K. H. Fan and A. L. Tits, "Characterization and Efficient Computation of the Structured Singular Value", *IEEE Transactions on Automatic Control*, Vol. 31, pp. 734-743, Aug. 1986.

[10] M. K. H. Fan, A. L. Tits and J. C. Doyle, "Robustness in the Presence of Mixed Parametric Uncertainty and Unmodeled Dynamics", *IEEE Transactions on Automatic Control*, Jan. 1991. Vol. 36, pp. 25-38, Jan. 1991.

[11] M. K. H. Fan and A. L. Tits, "m-Form Numerical Range and the Computation of the Structured Stability Margin", *IEEE Transactions on Automatic Control*, Vol. 33, pp. 284-289, Mar. 1988.

[12] M. G. Safonov and M. S. Verma, "L^∞ Optimization and Hankel Approximation", *IEEE Transactions on Automatic Control*, Vol. 30, pp. 279-280, March 1985.

[13] J. C. Doyle, "Synthesis of Robust Controllers and Filters", *Proc. IEEE Conf. on Decision and Control*, pp. 109-114, Dec. 1983.

[14] M. G. Safonov, "Optimal Diagonal Scaling for Infinity Norm Optimization", *Proc. Amer. Control Conf.*, pp. 125-128, Boston MA, June 1985.

[15] M. G. Safonov, "Optimal Diagonal Scaling for Infinity Norm Optimization", *Systems and Control Letters*, Vol. 7, pp. 257-260, July 1986.

[16] M. G. Safonov, "Optimal $\mathcal{H}^\infty$ Synthesis of Robust Controllers for Systems with Structured Uncertainty", *Proc. IEEE Conf. on Decision and Control*, pp. 1822-1825, Athens, Greece, 1986.

[17] M. G. Safonov and V. X. Le, "An Alternative Solution to the $\mathcal{H}_\infty$-Optimal Control Problem", *Systems and Control Letters*, Vol. 10, pp. 155-158, 1988.

[18] A. Tannenbaum, "Feedback Stabilization of Linear Dynamical Plants with Uncertainty in the Gain Factor", *International Journal of Control*, Vol. 32, pp. 1-16, 1980.

[19] A. Tannenbaum, "Modified Nevanlinna-Pick Interpolation and Feedback Stabilization of Linear Plants with Uncertainty in the Gain Factor", *International Journal of Control*, Vol. 36, pp. 331-336, 1982.

[20] P. P. Khargonekar and A. Tannenbaum, "Non-Euclidian Metrics and the Robust Stabilization of Systems with Parameter Uncertainty", *IEEE Transactions on Automatic Control*, Vol. 30, pp. 1005-1013, Oct. 1985.

[21] M. G. Safonov and M. Athans, "Gain and Phase Margins for Multiloop LQG Regulators", *IEEE Transactions on Automatic Control*, Vol. 22, pp. 173-179, Apr. 1977.

[22] M. G. Safonov, "*Robustness and Stability Aspects of Stochastic Multivariable Feedback System Design*", Doctoral Dissertation, M.I.T., Cambridge MA, August 1977.

[23] M. G. Safonov, "*Stability and Robustness of Multivariable Feedback Systems*", M.I.T. Press, Cambridge MA, 1980.

[24] J. C. Doyle, J. E. Wall, and G. Stein, "Performance and Robustness Analysis for Structured Uncertainty", *Proc. Conf. on Decision and Control*, pp. 629-636, Orlando FL, Dec. 1982.

[25] B. R. Barmish, P. P. Khargonekar, Z. C. Shi and R. Tempo, "Robustness Margin need not be a Continuous Function of the Problem Data", *Systems and Control Letters*, Vol. 15, pp. 91-98, 1990.

[26] C. A. Desoer, R.-W. Liu, J. Murray and R. Saeks, "Feedback System Design: The Fractional Representation Approach to Analysis and Synthesis", *IEEE Transactions on Automatic Control*, Vol. 25, pp. 399-412, June 1980.

[27] G. P. Akilov and A. M. Rubinov, "The Method of Successive Approximations for Determining the Polynomial of Best Approximation", *Soviet Math. Dokl.*, Vol. 5, pp. 951-953, 1964.

[28] M. Verma and E. Jonckheere, "L^∞-Compensation with Mixed Sensitivity as a Broadband Matching Problem", *Systems and Control Letters*, Vol. 4, pp. 125-129, May 1984.

[29] H. Kwakernaak, "Minimax Frequency Domain Performance and Robustness Optimization of Linear Feedback Systems", *IEEE Transactions on Automatic Control*, Vol. 30, pp. 994-1004, Oct. 1985.

[30] D. G. Luenberger, "*Optimization by Vector Space Methods*", Wiley, 1969.

[31] G. Zames and B. A. Francis, "Feedback, Minimax Sensitivity, and Optimal Robustness", *IEEE Transactions on Automatic Control*, Vol. 28, pp. 585-601, May 1983.

[32] W. W. Rogosinski and H. S. Shapiro, "On Certain Extremum Problems for Analytic Functions", *Acta Math.*, Vol. 90, pp. 287-318, 1953.

[33] M. A. Dahleh and J. B. Pearson, "l^1 Optimal Feedback Controllers for MIMO Discrete-Time Systems", *IEEE Transactions on Automatic Control*, Vol. 32, pp. 314-322, April 1987.

[34] M. A. Dahleh and J. B. Pearson, "L^1 Optimal Compensators for Continuous Time Systems", *IEEE Transactions on Automatic Control*, Vol. 32, pp. 889-895, Oct. 1987.

[35] P. Koosis, "*Introduction to $\mathcal{H}_p$ Spaces*", London Mathematical Society Lecture Note Series, Vol. 40, Cambridge University Press, 1980.

[36] P. Duren, "*Theory of $\mathcal{H}_p$ Spaces*", Academic Press, New York, 1970.

[37] K. Hoffman, "*Banach Spaces of Analytic Functions*", Prentice-Hall 1962, reprinted by Dover in 1988.

[38] J. B. Garnett, "*Bounded Analytic Functions*", Academic Press, 1981.

[39] J. W. Helton and R. E. Howe, "A Bang-Bang Theorem for Optimization over Spaces of Analytic Functions", *Journal of Approximation Theory*, Vol. 47, pp. 101-121, 1986.

[40] E. Wegert, "Boundary Value Problems and Best Approximation by Holomorphic Functions", *Journal of Approximation Theory*, Vol. 61, pp. 322-334, 1990.

[41] J. W. Helton, "Worst Case Analysis in the Frequency Domain: The $\mathcal{H}^\infty$ Approach to Control", *IEEE Transactions on Automatic Control*, Vol. 30, pp. 1154-1170, Dec. 1985.

[42] J. B. Conway, "*A Course in Functional Analysis*", Graduate Texts in Mathematics, Vol. 96, 2nd Edition, Springer-Verlag 1990.

A Generalized Eigenproblem Solution for Singular H^2 and H^∞ Problems

B. R. Copeland

Real Time Systems Group
Dept. of Electrical and Computer Engineering
University of the West Indies
Trinidad and Tobago, West Indies

M. G. Safonov

Dept. of Electrical Engineering-Systems
University of Southern California
Los Angeles, CA 90089-2563

I Introduction

This paper addresses the need for a theory that will permit the reliable computation of control laws for singular H^2 and H^∞ control problems involving plants with $j\omega$-axis or infinite zeros (i.e., strictly proper plants). Such problems characteristically lead to controllers having poles on the $j\omega$-axis or at infinity (i.e., improper controllers). In such cases traditional solution approaches fail. As will be shown, the difficulties presented by these singular problems can be overcome via a generalized eigenproblem approach.

In what follows we consider H^2 and H^∞problems for the plant $G(s) = \begin{bmatrix} G_{11}(s) & G_{12}(s) \\ G_{21}(s) & G_{22}(s) \end{bmatrix}$ with state space representation

$$\begin{pmatrix} \dot{x} \\ \hline y_1 \\ y_2 \end{pmatrix} = \left[\begin{array}{c|cc} A & B_1 & B_2 \\ \hline C_1 & D_{11} & D_{12} \\ C_2 & D_{21} & D_{22} \end{array}\right] \begin{pmatrix} x \\ \hline u_1 \\ u_2 \end{pmatrix} \tag{1}$$

where $x \in \Re^n$, $u_2 \in \Re^{r_2}$, $u_1 \in \Re^{r_1}$, $y_1 \in \Re^{m_1}$, $y_2 \in \Re^{m_2}$. As usual, the vectors u, y are the control and measurement vectors, respectively while u_1, y_1 are the disturbance and controlled vectors. For brevity we will often refer to these problems as H^p problems; the reader should therefore note the restriction of our arguments to the cases $p \in \{2, \infty\}$.

We now describe the particular versions of the H^p problems which will be considered. First recall that for any specified controller

$$u_2(s) = K(s)y_2(s) \tag{2}$$

CONTROL AND DYNAMIC SYSTEMS, VOL. 50

the closed loop transfer function from u_1 to y_1 is given by the lower linear fractional transformation

$$T_{y_1u_1}(s) = \mathcal{F}_l(G(s), K(s)) := G_{11}(s) + G_{12}K(s)(I - G_{22}K(s))^{-1}G_{21}(s) \quad (3)$$

The H^p problem is stated as follows: *Determine a stabilizing controller $K(s)$ for which $\|T_{y_1u_1}\|_p$ is minimized.*
We shall also make reference to the suboptimal H^p problem: *Given $\gamma > 0$ determine a stabilizing controller $K(s)$ for which $\|T_{y_1u_1}\|_p < \gamma$.*

Note that we do not require the controller $K(s)$ to be proper. We consider the class of H^p problems which are singular:

Definition 1 *An H^2 or H^∞problem will be called* **singular** *if the cross coupling transfer functions $G_{12}(s)$ and $G_{21}(s)$ have zeros on the extended imaginary axis $\mathcal{C}^{0e} := \{s = j\omega,\ w \in \Re\} \cup \{\infty\}$. Otherwise the relevant problem will be said to be* **regular**.

We shall make the following assumptions:

A1. $r_2 \leq m_1, m_2 \leq r_1$

A2. (A, B_2) is stabilizable and (C_2, A) is detectable.

A3. $D_{11} = 0, D_{22} = 0$

Of these assumptions, A2 is absolutely essential to guarantee the existence of a stable solution. The other assumptions are usually incorporated to expedite the necessary analysis. In particular, with regards to assumption A3, it should be noted that if $D_{22} \neq 0$ then the loop-shifting procedures outlined in [1] can be used to zero this term. Likewise, the matrix D_{11} can also be zeroed, provided a solution to the H^p problem exists. Note, however, that if there exists no feedback to zero D_{11} then the corresponding H^2 cost must be infinite. Similarly, in the H^∞ case, if there exists no feedback to make $\bar{\sigma}(D_{11}) < \gamma$, then no controller exists to satisfy the specified H^∞constraint (see Lemma 2 of [1]); and once $D_{11} < \gamma$ then the methods of [1, 40] enable one to zero D_{11}.

As is well known, the solutions for regular H^p problems have been completely worked out [2, 3]. In these references it has been shown that the required controller, if it exists, may be specified as a lower linear fractional transformation

$$K(s) = \mathcal{F}_l(M(s), Q(s)) \quad (4)$$

where the "central controller" $M(s)$ is obtained by solving a pair of Riccati equations and the "free parameter" $Q(s)$ is chosen from a set $S_Q(p, \gamma)$ of stable transfer functions which satisfy certain norm constraints (see Theorems 2, 4 of [2]). Given $\gamma > 0$, the set $S_Q(p, \gamma)$ gives a parameterization of all controllers for which $\|T_{y_1u_1}\|_p \leq \gamma$, $T_{y_1u_1} \in RH^\infty$. For H^2 problems, $Q(s) = 0$ yields the optimal solution; this is not the case for H^∞ problems where the optimal solution must be found by iteration on the bound γ.

The difficulty with singular problems is directly related to the unsolvability of the pertinent Riccati equations. In fact, when the problem is singular at infinity, at least one of these equations will be ill-defined by virtue of the fact that $D_{12}^T D_{12}$ and/or $D_{21} D_{21}^T$ are singular. When the only $\mathcal{C}^{0e}$ zeros are finite, the Riccati equations, though now well-defined, have no asymptotically stabilizing solutions. In either case, it becomes impossible to specify the central controller $M(s)$ and, as a result, any $K(s)$ for which $||T_{y_1 u_1}||_p < \gamma$.

In what follows, we approach the singular H^p problem by first replacing the Riccati equations by equivalent generalized eigenvalue problems. The relevant equations can then be formulated in an inverse-free setting and are therefore well-defined even when the pertinent problem is singular at infinity. Using this framework, the solution to regular problems can be determined by obtaining a minimal set of basis vectors for the n-dimensional stable eigenspaces of two specific matrix pencils (see [1] for a similar approach). The next step is to perturb the singular problem into a regular one by way of a scalar parameter ϵ. In the context of the LQ/LQG interpretation of the H^2 problem [2], ϵ is seen to enter as a weight on those directions of the control input space which are not otherwise penalized or those directions of the disturbance input space which are otherwise noise-free. A similar explanation of the role of ϵ can be forwarded for the H^∞ case using the game-theory interpretation of the problem [4]. Finally, we derive the limiting solution for the resulting family of regular problems as $\epsilon \longrightarrow 0$.

This procedure is somewhat in the vein of the "multivariable root locus" analysis carried out by Kwakernaak [5] and other authors; in fact, those who are familiar with LQ theory, will realize that we are essentially reformulating singular H^p problems as "cheap control " problems. However, the generalized eigenproblem formulation of the Riccati equations leads to a much more complete description of the solution to the problem. For example, the solution derived for the state feedback H^2 problem clearly distinguishes the singular arcs [6] and impulsive components in the control when the pertinent system has zeros at infinity. Moreover, in the measurement feedback case, use of the descriptor representation of linear systems allows for the representation of those solutions which may be improper.

Many studies have been carried out on the subject of cheap and singular H^2 problems. However, the analysis has focused on the case where the only $\mathcal{C}^{0e}$ zeros are at infinity. This is due to the fact that when there are finite magnitude imaginary axis zeros, analysis of the asymptotic root loci [5, 7, 8, 9] suggests that pole/zero cancellation of these $\mathcal{C}^{0e}$ zeros must occur. This is, in general, undesirable since the cancellation takes place on the $j\omega$-axis. Moreover, for ϵ small enough those closed-loop poles which approach the troublesome zeros are lightly damped; this is also undesirable. On the other hand, when the only zeros are at infinity, the asymptotic properties are generally much more attractive since the root loci form

Butterworth patterns of various orders and magnitudes as $\epsilon \rightarrow 0$ in the left half plane (LHP) [5]. Despite the limitation of having finite $\mathcal{C}^{0e}$ zeros, the analysis in this work is believed to be useful as a starting point for the characterization of all suboptimal solutions to the relevant H^2 and H^∞problems.

The more successful attempts at solving cheap and singular H^2 problems have employed either singular perturbation techniques [10, 11] or geometric control theory [12, 13, 14]; a neat characterization of the problem and its solution has been phrased in terms of the dissipation inequality [15, 16, 17]. In [18] and [19] the problem is analyzed using a spectral factorization approach. Frequency domain techniques are employed in these works and attention is restricted to the case of infinite $\mathcal{C}^{0e}$ zeros. By contrast, our results rely upon a simple generalization of the commonly used procedures based on the solution to two Riccati equations and directly yields a realization of the optimal controller in descriptor form. Our analysis allows for all $\mathcal{C}^{0e}$ zeros.

All of the methods mentioned thus far have clearly identified the role of the infinite frequency structure of Eq (1) in cheap and singular control problems. One major contribution of the work here is that it displays this reliance on the infinite frequency structure in a very clear fashion by determining the eigenstructure of a particular matrix pencil. More importantly, this results in a solution which may be reliably computed using standard generalized eigenspace software routines.

Apart from this, our results indicate that in those situations where the use of observer-based controllers of reduced order, $\delta < n$, are possible (see [20], for example) the solution automatically produces a controller of order δ. This is in the vein of the result obtained by Friedland [21]. Note also that the results obtained for the H^2 problem are of great significance to the LQG/LTR design procedure [22].

Much more general approaches have been taken for the H^∞problem. In [23], for example, Safonov has discussed the underlying problem and has proposed a suboptimal solution employing a bilinear transformation which maps the $j\omega$ axis to a circle in the right half plane (RHP); this also accounts for strictly proper plants since these may be considered as having zeros at $j\omega = \infty$. O'Young, Postlethwaite and Gu [24] have proposed another suboptimal solution by incorporation of sensor/actuator noise and saturation constraints into $G(s)$. It should also be noted that these constraints may be altogether avoided by the appropriate choice of the relevant design weighting functions [25]; however this method yields a solution to a slightly different problem. Hara et al. [26] have reformulated the solution for the one-block case in a descriptor setting and have removed the restriction of $j\omega$-axis zeros. Kimura et al. [27] have employed a generalization of the theory of J-lossless systems in H^∞ synthesis and reported results for the situation where the zeros at infinity are simple, i. e. of order at most unity. Stoorvogel [28] has reported success in the singular case by use of the dissipation inequality arising out of the linear quadratic optimization problem.

The chapter is organized as follows: In Section II we discuss the zero structure of finite dimensional linear systems, with particular regard to those zeros which are infinite; this discussion lays the basis for the analysis which follows. In Section III we describe how regular problems can be solved using a generalized eigenproblem approach as opposed to the orthodox method requiring the solution of a pair of Riccati equations. In Section IV we introduce a method for solving suboptimal H^p problems by perturbing them into regular ones. We state our results for the state feedback singular problems, Theorems 1 and 5, in Section V. In Sections VI and VII we treat the measurement feedback cases. We then conclude with computational algorithms and a few examples in Sections VIII and IX respectively.

Standard notation is used throughout; note, however, that if $G(s) = C(sE - A)^{-1}B + D$ then we write $G(s) \stackrel{s}{=} P_G(s)$ where

$$P_G(s) := \left[\begin{array}{c|c} -sE + A & B \\ \hline C & D \end{array}\right] \tag{5}$$

II Preliminary Considerations

In this section we review the theory of infinite zeros of linear time-invariant systems. An important component in this discussion is the concept of eigenvalues and eigenvectors of matrix pencils. This topic is covered in detail elsewhere (see [29, 30, 31], for example) and so will only be briefly considered here.

First we define the finite eigenvalues of the regular pencil $M(s) = -sE + A$ to be the zeros of the polynomial equation det $M(s) = 0$ if this polynomial is not a constant. Note that $M(s)$ may have eigenvalues at infinity if E is singular [31]. We also note that there exist nonsingular matrices R, L which reduce $M(s)$ to the Weierstrass canonical form defined by

$$LM(s)\Xi = \text{diag}\{sI - \Lambda_f, -s\Lambda_\infty + I\} \tag{6}$$

where Λ_f is a Jordan matrix whose eigenvalues correspond to the finite eigenvalues of $M(s)$ and Λ_∞ is a nilpotent matrix in Jordan form [31]. The pencil $-s\Lambda_\infty + I$ thus contains the infinite frequency structure of $M(s)$ and the number and sizes of the Jordan blocks in Λ_∞ determine the number and orders, respectively, of the infinite eigenvalues of $M(s)$. By definition the finite and infinite eigenspaces are the spaces spanned by the columns of Ξ_f and Ξ_∞ respectively where $\Xi = \left[\begin{array}{cc} \Xi_f & \Xi_\infty \end{array}\right]$ is a partitioning of Ξ corresponding to the right-hand side of Eq (6). Moreover,

$$A\Xi_f = E\Xi_f\Lambda_f \tag{7}$$

$$A\Xi_\infty\Lambda_\infty = E\Xi_\infty \tag{8}$$

Suppose that Λ_{f_j} is a $k_j \times k_j$ Jordan matrix of Λ_f such that $\det(sI - \Lambda_f) = (s - z_j)^{k_j}$; also suppose that the corresponding columns of Ξ_f are respectively $\left[\begin{array}{ccc} \xi_j^{(1)} & \cdots & \xi_j^{(k(j)} \end{array}\right]$. Then z_j is said to be a finite zero of $M(s)$ of order k_j and that $\xi_j^{(k)}$, $k \in \underline{k_j}$, is a <u>grade k eigenvector</u> corresponding to the zero z_j.

Similarly, if Λ_{∞_j} is a $k_j \times k_j$ Jordan block of Λ_∞, and the corresponding columns of Ξ_∞ are $\xi_{\infty_j}^{(1)}, \cdots, \xi_{\infty_j}^{(k_j)}$ then we say that $z_j = \infty$ is a zero of $M(s)$ of order $k_j - 1$. Note that the order of each infinite zero is one less than the dimension of the corresponding Jordan block. We define $\xi_{\infty_j}^{(k)}$ to be the grade k eigenvector at infinity of $M(s)$ corresponding to the j-th infinite zero; moreover, the vectors $\{\xi_{\infty_j}^{(i)}, \ i \in \underline{k_j}\}$ are said to form an eigenvector chain at infinity. From Eq (8) these vectors also satisfy the relation

$$E\xi_{\infty_j}^{(1)} = 0, \ \ E\xi_{\infty_j}^{(k)} = A\xi_{\infty_j}^{(k-1)} \ \ k = 2, \ \cdots, \ k_j \tag{9}$$

We will need the following:

Definition 2 *The left eigenvectors of the regular pencil $M(s)$ are defined to be the eigenvectors of $M^T(s)$.*

When $M(s)$ is a rectangular pencil of full column rank the eigenvalues and eigenvectors can be defined in a similar fashion. We note that there exist nonsingular constant matrices X, Y which transform such an $M(s)$ into its Kronecker canonical form (KCF) [32, 29] i. e.

$$YM(s)X = \text{diag}\{sE_\eta + A_\eta, sE_f + A_f\} \tag{10}$$

where $sE_\eta + A_\eta$ is injective on $\mathcal{C} \cup \{\infty\}$ and the pencil $sE_f + A_f$ is regular. The latter pencil clearly contains the column zero structure of $M(s)$. We therefore define the eigenvalues of $M(s)$ to be the eigenvalues of $sE_f + A_f$ (these include eigenvalues at infinity) which justifies our claim.

The following definition introduces the concept of trivial infinite zeros:

Definition 3 *An infinite eigenvalue of $M(s)$ of order $k = 0$ will be said to be* **trivial**. *All other infinite eigenvalues will be called* **non-trivial**.

Before proceeding we remind the reader that the eigenvector chains are of significance to the solution of the equation

$$(-sE + A)\xi(s) = E\xi_0 \tag{11}$$

for any $\xi_0 \in \Re^n$. The following lemma summarizes the pertinent result for both the finite and infinite eigenspaces:

Lemma 1 (Verghese *et al.* [33]) *For Eq (11) the following hold:*

(i) For $\xi_0 = -\xi_j^{(l)}$ a grade l eigenvector corresponding to the finite eigenvalue λ_j of $-sE + A$, the solution $\xi(s)$ is

$$\xi(s) = \sum_{i=1}^{l} \xi_j^{(l+1-i)} \frac{1}{(s - \lambda_j)^i}. \tag{12}$$

(ii) For $\xi_0 = -\xi_{\infty j}^{(l)}$, $l = 2, 3, \cdots, k_j + 1$, *corresponding to an infinite eigenvalue of order* k_j, *the solution to Eq (11) is given by*

$$\xi(s) = \sum_{i=1}^{l-1} \xi_j^{(l-i)} s^{(i-1)}. \tag{13}$$

(iii) For $\xi_0 = -\xi_{\infty j}^{(1)}$ *(i.e., for* ξ_- *a first order infinite eigenvector), one has* $E\xi_0 = 0$ *and Eq (11) admits only the trivial solution* $\xi(s) = 0$.

Comment: Let

$$\delta_\infty := \sum_{j=1}^{r} k_j \tag{14}$$

and note that for the infinite frequency spectrum, the solutions are polynomial and evolve in the δ_∞-dimensional subspace spanned by the infinite eigenvectors of all grades but the highest. The vectors ξ_0 corresponding to the polynomial solutions lie in a δ_∞-dimensional subspace spanned by infinite frequency eigenvectors of all but the first grade. The distinct roles played by the two subspaces of the infinite frequency eigenspace of a pencil, $P(s)$, motivates the following

Definition 4 *Given any pencil* $P(s)$ *we define the* **lower infinite frequency eigensubspace** *of* $P(s)$ *to be the space spanned by all but the highest grade eigenvectors corresponding to all nontrivial infinite zeros of* $P(s)$. *Similarly, we define the* **upper infinite frequency eigensubspace** *of* $P(s)$ *to be the space spanned by all but the first grade eigenvectors corresponding to all non-trivial infinite zeros of* $P(s)$.

The next two lemmas will prove useful in the ensuing analysis.

Lemma 2 *Let* $M(s)$ *be a regular pencil with* X *being a matrix whose columns form a basis for the eigenspace corresponding to some eigenvalue* $\lambda_x \in \mathcal{C} \cup \{\infty\}$ *of* $M(s)$. *Similarly, let* Y *be a matrix whose columns form a basis for the left eigenspace of* $M(s)$ *corresponding to an eigenvalue* $\lambda_y \in \mathcal{C} \cup \{\infty\}$. *If* $\lambda_x \neq \lambda_y$, *then*

$$Y^T(-sE + A)X \equiv 0. \tag{15}$$

Proof: Without loss of generality we assume that Y spans a finite frequency eigenspace of $M(s)$. It suffices to prove that $Y^T EX = 0$; the result would then follow from the fact that

$$Y^T(-sE + A)X = (-sI + \Lambda_y^T)Y^T EX \tag{16}$$

Suppose first, that λ_x is finite. Then from Eq (7, 8) there exist matrices Λ_y and Λ_x satisfying

$$Y^T A = \Lambda_y^T Y^T E \tag{17}$$

$$AX = EX\Lambda_x \tag{18}$$

where all the eigenvalues of Λ_y and Λ_x are at λ_y and λ_x respectively. Left multiply Eq (18) by Y^T and substitute from Eq (17) to get

$$Y^T E X \Lambda_x = Y^T A X = \Lambda_y^T Y^T E X \tag{19}$$

i. e.

$$(Y^T E X)\Lambda_x - \Lambda_y^T (Y^T E X) = 0 \tag{20}$$

Now note that by assumption, the eigenvalues of Λ_x and Λ_y do not correspond $\Rightarrow$ by Lemma 1.5 of [39] that the unique solution to this last equation is $Y^T E X = 0$.

For $\lambda_x = \infty$, Eq (18) becomes

$$A X \Lambda_x = E X \tag{21}$$

where Λ_x is now a nilpotent matrix. Again left multiply by Y^T and substitute from Eq (17) to get

$$Y^T E X = \Lambda_y^T Y^T E X \Lambda_x \tag{22}$$

By repeated substitution from Eq (17) and Eq (21) we get

$$\left.\begin{array}{rcl} Y^T E X & = & \Lambda_y^T Y^T E X \Lambda_x \\ & = & \Lambda_y^T Y^T A X \Lambda_x^2 \\ & = & \Lambda_y^{2T} Y^T E X \Lambda_x^2 \\ & & \vdots \\ & = & \vdots \\ & = & \Lambda_y^{nT} Y^T E X \Lambda_x^n \ \forall \text{integers n} \end{array}\right\} \tag{23}$$

Since Λ_x is nilpotent then, by definition, there exists some integer $n \geq 1$ such that $\Lambda_x^n = 0$. The result then follows.

□

Lemma 3 *Let $\Xi^{(l)}$ be a matrix whose columns span the lower infinite frequency eigensubspace of the pencil $M(s) = -sE + A$. Similarly, let $\Xi^{(u)}$ be a matrix whose columns span the upper infinite frequency eigensubspace of $M(s)$. Then there exists an invertible matrix T such that*

$$A \Xi^{(l)} T = E \Xi^{(u)} \tag{24}$$

Proof: Without loss of generality, we may assume that the columns of $\Xi^{(l)}$ and $\Xi^{(l)}$ are actual lower and upper infinite frequency eigenvectors, respectively, of the pencil $M(s)$. Further, it suffices to take the eigenvectors associated with a single Jordan block, i. e.

$$(\Xi_\infty, \Lambda_\infty) = \text{igep}(M(s)) \tag{25}$$

where the $k \times k$ matrix Λ_∞ is given by

$$\Lambda_\infty = \begin{bmatrix} 0 & I_{k-1} \\ 0 & 0 \end{bmatrix} \tag{26}$$

We also assume that the corresponding eigenvalues are $\xi^{(i)}$, $i \in \underline{k}$ so that

$$\Xi_\infty = \begin{bmatrix} \xi^{(1)}, & \cdots, & \xi^{(k)} \end{bmatrix} \tag{27}$$

Then

$$\left.\begin{array}{ccc} \Xi_\infty & = & \left[\begin{array}{cc} \Xi^{(l)} & \xi^{(k)} \end{array}\right] \\ & \text{and} & \\ \Xi_\infty & = & \left[\begin{array}{cc} \xi^{(1)} & \Xi^{(u)} \end{array}\right] \end{array}\right\} \tag{28}$$

Substitution in Eq (8) yields

$$A\left[\begin{array}{cc} \Xi^{(l)} & \xi^{(k)} \end{array}\right]\Lambda = E\left[\begin{array}{cc} \xi^{(1)} & \Xi^{(u)} \end{array}\right]. \tag{29}$$

In view of Eq (26) the result follows by equating the last $(k-1)$ columns of Eq (29).

□

We conclude the section with a brief discussion on pencils which are Rosenbrock system matrices in state space form i. e.

$$M(s) = -sE_M + A_M = \left[\begin{array}{cc} -sI + A & B \\ C & D \end{array}\right]. \tag{30}$$

A concept which is of critical importance to our study is that of the invariant zeros of the system $\{A, B, C, D\}$. The finite invariant zeros of this system are defined to be the finite eigenvalues of $M(s)$ [34, 35]. For infinite invariant zeros, we employ the following definition:

Definition 5 (Infinite Invariant Zeros) *The infinite invariant zeros of* $\{A, B, C, D\}$ *are defined to be the non-trivial infinite eigenvalues of* $M(s)$.

The definition is motivated by the fact that trivial infinite eigenvalues are associated with constant diagonal blocks in the KCF of $M(s)$. The implication for the system $\{A, B, C, D\}$ is that even when D is injective (in which case there are no transmission zeros at infinity [36]), the pencil $M(s)$ will have r trivial infinite eigenvalues. We also recall that for finite frequencies [34], the set of transmission zeros and invariant zeros coincide. The fact that this also holds in the infinite frequency case, with the definition posed, follows from Lemma 3 in [37]. Observe that the lower and upper infinite frequency eigensubspaces of $M(s)$ are spanned by all but the highest, respectively lowest, grade eigenvectors corresponding to the infinite invariant zeros of $\{A, B, C, D\}$.

Finally, we have the following result

Lemma 4 *The lower infinite frequency eigensubspace of the pencil* $M(s)$ *in Eq (30) lies in* Ker $\left[\begin{array}{cc} C & D \end{array}\right]$. *Any vector* x *which lies in the upper infinite frequency eigensubspace of* $M(s)$ *satisfies* $E_M x \neq 0$.

Proof: Follows directly from Eq (9).

□

III A New Setting for Regular H^2 and H^∞ Problems

In this section we formulate the solutions to the regular H^2 and H^∞ problems in terms of two generalized eigenproblems. This approach differs somewhat from that of [2, 3, 38] and allows for a compact description of the solution when D_{12} and D_{21} are not parts of orthogonal matrices, as required in those works. More importantly, the formulation sets the foundation upon which the analyses of the singular cases performed in later sections are based.

A H^2 Control

We begin with the H^2 problem by defining the following matrix pencils (the result is fairly well known and has been applied mostly to optimal regulator and filtering problems; see [15], for example):

$$W_{12_2}(s) := \begin{bmatrix} -sI + A & 0 & B_2 \\ -C_1^T C_1 & -sI - A^T & -C_1^T D_{12} \\ D_{12}^T C_1 & B_2^T & D_{12}^T D_{12} \end{bmatrix} \tag{31}$$

and

$$W_{21_2}(s) := \begin{bmatrix} -sI + A^T & 0 & C_2^T \\ -B_1 B_1^T & -sI - A & -B_1 D_{21}^T \\ D_{21} B_1^T & C_2 & D_{21} D_{21}^T \end{bmatrix}. \tag{32}$$

Lemma 5 *Assume that $C_2 = I$ and $D_{12} = 0$. Let the columns of $\Xi_{12} = \begin{bmatrix} X_{12} \\ \Phi_{12} \\ U_{12} \end{bmatrix}$ form a basis for the stable eigenspace of $W_{12_2}(s)$. Then a solution exists to the optimal H^2 problem (with state feedback) if and only if* rank $X_{12} = n$. *If this condition holds, then the optimal control takes the form of the state feedback $u = Kx$ where*

$$K = U_{12} X_{12}^{-1} \tag{33}$$

Proof: The proof follows directly from the fact that a solution exists if and only if the Hamiltonian matrix

$$H^{(2)} = \begin{bmatrix} A - B_2(D_{12}^T D_{12})^{-1} D_{12}^T C_1 & -B_2(D_{12}^T D_{12})^{-1} B_2^T \\ -C_1^T C_1 + C_1^T D_{12}(D_{12}^T D_{12})^{-1} D_{12}^T C_1 & -(A - B_2(D_{12}^T D_{12})^{-1} D_{12}^T C_1)^T \end{bmatrix} \tag{34}$$

has a stable eigenspace defined for some matrices $\begin{bmatrix} X \\ \Phi \end{bmatrix} \in \Re^{2n \times n}$, $\Lambda \in \Re^{n \times n}$, where X is nonsingular and Λ is strictly Hurwitz (see [12], for example), by

$$(-sI + H^2) \begin{bmatrix} X \\ \Phi \end{bmatrix} = \begin{bmatrix} X \\ \Phi \end{bmatrix} (-sI + \Lambda) \tag{35}$$

The optimal solution is then given by

$$u = Kx, \quad K = -(D_{12}^T D_{12})^{-1}(D_{12}^T C_1 + B_2^T P) \tag{36}$$

where

$$P := \Phi X^{-1} \tag{37}$$

Define the pencil

$$\hat{W}_{12_2}(s) := \begin{bmatrix} -sI + H^{(2)} & 0 \\ \begin{bmatrix} D_{12}^T C_1 & B_2^T \end{bmatrix} & D_{12}^T D_{12} \end{bmatrix} \tag{38}$$

and note that the finite right eigenstructure of $H^{(2)}$ and $\hat{W}_{12_2}(s)$ coincide; in particular

$$\hat{W}_{12_2}(s) \begin{bmatrix} X \\ \Phi \\ U \end{bmatrix} = \begin{bmatrix} X \\ \Phi \\ 0 \end{bmatrix} (-sI + \Lambda) \tag{39}$$

where

$$U := -(D_{12}^T D_{12})^{-1}(D_{12}^T C_1 X + B_2^T \Phi) \tag{40}$$

Left-multiply Eq (39) by the elementary matrix

$$L = \begin{bmatrix} I & 0 & B_2(D_{12}^T D_{12})^{-1} \\ 0 & I & -C_1^T D_{12}(D_{12}^T D_{12})^{-1} \\ 0 & 0 & I \end{bmatrix} \tag{41}$$

to obtain

$$W_{12_2}(s) \begin{bmatrix} X \\ \Phi \\ U \end{bmatrix} = \begin{bmatrix} X \\ \Phi \\ 0 \end{bmatrix} (-sI + \Lambda) \tag{42}$$

From this we conclude that the finite eigenvalues of $W_{12_2}(s)$ and $H^{(2)}$ coincide and that the stable eigenspace of $W_{12_2}(s)$ is spanned by the columns of $\begin{bmatrix} X \\ \Phi \\ U \end{bmatrix}$. Now let $X_{12} = X$, $\Phi_{12} = \Phi$ and $U_{12} = U$ and complete the proof by substituting Eq (40) in Eq (36).

□

The result for the measurement feedback problem is as follows:

Lemma 6 *Let the columns of* $\Xi_{12} := \begin{bmatrix} X_{12} \\ \Phi_{12} \\ U_{12} \end{bmatrix}$ *and of* $\Xi_{21} := \begin{bmatrix} X_{21} \\ \Phi_{21} \\ U_{21} \end{bmatrix}$ *span the respective* n*-dimensional stable subspaces of* $W_{12_2}(s)$ *and* $W_{21_2}(s)$. *Then a solution exists to the optimal* H^2 *problem for the plant* $G(s)$ *if and only if* rank X_{12} = rank $X_{21} = n$. *In this case the following hold*

(i) The unique optimal solution is

$$K(s) \stackrel{s}{=} \left[\begin{array}{c|c} -sE_k + A_k & U_{21}^T \\ \hline -U_{12} & 0 \end{array}\right] \tag{43}$$

where

$$E_k := X_{21}^T X_{12} \tag{44}$$

$$A_k := X_{21}^T A X_{12} + X_{21}^T B_2 U_{12} + U_{21}^T C_2 X_{12} \tag{45}$$

(ii) The corresponding cost is

$$J^* := \min ||T_{y_1 u_1}||_2^2 = \mathrm{Tr}(PKD_{21}D_{21}^T K^T + C_1^T C_1 Q) \tag{46}$$

where

$$P := \Phi_{12} X_{12}^{-1} \tag{47}$$
$$Q := (X_{21}^T)^{-1} \Phi_{21}^T \tag{48}$$
$$K := (QC_2^T + B_1 D_{21}^T)(D_{21}D_{21}^T)^{-1} \tag{49}$$

$||T_{y_1 u_1}||_2 < \gamma$ *is given*

Proof: The proof follows directly from the standard separation result for the LQG problem (see [39, Theorems 4.6, 5.4] for example) and Lemma 5 which establishes the relationship between the generalized eigenproblems for $W_{12_2}(s)$ and $W_{21_2}(s)$ and the usual Riccati equations.

□

Before proceeding we give a few general properties of pencils of the form $W_{12_2}(s)$. First define

$$M_G(S) := \begin{bmatrix} -sI + A & B \\ C & D \end{bmatrix} \tag{50}$$

Lemma 7 *The following properties hold for any matrix pencil of the form*

$$W(s) = \begin{bmatrix} -sI + A & R & B \\ -C^T C & -sI - A^T & -C^T D \\ D^T C & B^T & D^T D \end{bmatrix} \tag{51}$$

where $R \geq 0$:

(i) The finite zeros of $W(s)$ are symmetrically placed about the $j\omega$-axis.

(ii) If $M_G(s)$ has a complete eigenvector chain $\{\begin{pmatrix} x^{(1)} \\ u^{(1)} \end{pmatrix}, \ldots, \begin{pmatrix} x^{(k)} \\ u^{(k)} \end{pmatrix}\}$ at $s = \lambda \in C \cup \infty$, then the set of vectors $\{\begin{pmatrix} x^{(1)} \\ 0 \\ u^{(1)} \end{pmatrix}, \ldots, \begin{pmatrix} x^{(k)} \\ 0 \\ u^{(k)} \end{pmatrix}\}$ forms a partial chain at $s = \lambda$ for the pencil $W(s)$.

Proof:

Define

$$T = \mathrm{diag}\{J, I_r\} \tag{52}$$

where J is the symplectic matrix $J = \begin{bmatrix} 0 & -I_n \\ I_n & 0 \end{bmatrix}$. Then by direct calculation, we observe that

$$TW(s)T = W(s)^*, \quad W(s)^* := W(-s)^T \tag{53}$$

Since T is nonsingular, Part (i) then follows from the fact that the set of finite zeros of $W(s)$ and $W(s)^*$ coincide.

For part (ii) assume without loss of generality that $\lambda \in \mathcal{C}$. Let $X = \begin{bmatrix} x^{(1)} & \cdots & x^{(k)} \end{bmatrix}$ and $U = \begin{bmatrix} u^{(1)} & \cdots & u^{(k)} \end{bmatrix}$. Then it is readily seen that

$$M_G(s)\begin{bmatrix} X \\ U \end{bmatrix} = \begin{bmatrix} X \\ 0 \end{bmatrix}(-sI+\Lambda) \Rightarrow W(s)\begin{bmatrix} X \\ 0 \\ U \end{bmatrix} = \begin{bmatrix} X \\ 0 \\ 0 \end{bmatrix}(-sI+\Lambda)$$

A similar argument holds for $\lambda = \infty$.

□

Comment: It should be noted that the Lemmas 5 and 6 are devoid of the usual assumptions on the stabilizability of (A, B_2) or the location of the transmission zeros of (C_1, A, B_2, D_{12}). The fact that we need not explicitly state these assumptions is supported by the following lemma which can be easily proved by considering the defining equations for the zeros of $G_{12}(s)$ (a dual result can be stated for $G_{21}(s)$):

Lemma 8 *Suppose that (A, B_2) is not stabilizable or that $G_{12}(s)$ has zeros on the $j\omega$-axis or at infinity. Then the matrix X_{12} is singular.*

Proof: If G_{12} has $j\omega$-axis zeros, then so does the pencil $M_{G_{12}}(s)$. It then follows directly from Lemma 7 that $W_{12_2}(s)$ has zeros on the $j\omega$-axis $\Rightarrow$ the stable eigenspace of $W_{12_2}(s)$ is of dimension $< n$ which in turn implies the result.

If (A, B_2) is not stabilizable $\Rightarrow$ the pencil $\begin{bmatrix} -sI + A & B_2 \end{bmatrix}$ has zeros in the closed right half plane $\Rightarrow \begin{bmatrix} -sI - A^T \\ B_2^T \end{bmatrix}$ has zeros in the closed left half plane. Since the case of $j\omega$-axis zeros has already been dealt with, we assume, without loss of generality, that the real parts of these zeros are strictly negative. Therefore, let the corresponding zero structure of this last pencil be described by

$$\begin{bmatrix} -sI - A^T \\ B_2^T \end{bmatrix} X = X(-sI+\Lambda) \tag{54}$$

Where Λ is strictly Hurwitz. Then by direct calculation

$$W_{12_2}(s)\begin{bmatrix} 0 \\ X \\ 0 \end{bmatrix} = \begin{bmatrix} 0 \\ X \\ 0 \end{bmatrix}(-sI+\Lambda) \tag{55}$$

i. e. the columns of $\begin{bmatrix} 0 \\ X \\ 0 \end{bmatrix}$ span a subspace of the stable eigenspace of $W_{12_2}(s)$.

□

These assumptions are therefore embedded in the nonsingularity requirements on X_{12} and X_{21}.

Lemma 9 *Let the columns of the matrix $\begin{bmatrix} X \\ \Phi \\ U \end{bmatrix}$ form a basis for the stable eigenspace of the pencil $W(s)$ and the columns of $\begin{bmatrix} X_\infty \\ U_\infty \end{bmatrix}$ form a basis for the infinite frequency eigenspace of $M_G(s)$ defined in Eq (50).*

Then

$$\Phi^T X_\infty = 0 \tag{56}$$

Proof: From Lemma 7 the columns of $\begin{bmatrix} X_\infty \\ 0 \\ U_\infty \end{bmatrix}$ span a subspace of the infinite frequency eigenspace of $W(s)$. Moreover, it is easy to show that the columns of $\begin{bmatrix} \Phi \\ -X \\ U \end{bmatrix}$ span the stable left eigenspace of $W(s)$. The result then follows from Lemma 2.

□

It is instructive to emphasize the link between the result of Lemma 5 and standard LQ theory:

Lemma 10 *The vector pair* $(x(s), u_2(s))$ *solves the optimal control problem*

$$\min_{u_2(t)} J(x(0), u_2(t)) := \int_0^\infty \|y_1(t)\|_2^2 \mathrm{dt} \tag{57}$$

subject to $x(t)$ *asymptotically stable*

if and only if, for some $\phi(s)$, $\begin{bmatrix} x(s) \\ \phi(s) \\ u_2(s) \end{bmatrix}$ *is an asymptotically stable solution of*

$$W_{12_2}(s) \begin{bmatrix} x(s) \\ \phi(s) \\ u_2(s) \end{bmatrix} = \begin{bmatrix} x(0) \\ \phi(0) \\ 0 \end{bmatrix} \tag{58}$$

B H^∞ Control

For the H^∞ problem we define the pencils

$$W_{12_\infty}(s) := \begin{bmatrix} -sI + A & B_1 B_1^T & B_2 \\ -C_1^T C_1 & -sI - A^T & -C_1^T D_{12} \\ D_{12}^T C_1 & B_2^T & D_{12}^T D_{12} \end{bmatrix}. \tag{59}$$

and

$$W_{21_\infty}(s) := \begin{bmatrix} -sI + A^T & C_1^T C_1 & C_2^T \\ -B_1 B_1^T & -sI - A & -B_1 D_{21}^T \\ D_{21} B_1^T & C_2 & D_{21} D_{21}^T \end{bmatrix}. \tag{60}$$

The H^∞ result analogous to Lemma 5 is as follows:

Lemma 11 *Assume that* $C_2 = I$ *and* $D_{12} = 0$. *Let the columns of* $\Xi_{12} = \begin{bmatrix} X_{12} \\ \Phi_{12} \\ U_{12} \end{bmatrix}$ *form a basis for the stable eigenspace of* $W_{12_\infty}(s)$. *Then there exists a controller which stabilizes* $G(s)$ *and makes* $\|T_{y_1 u_1}\|_\infty < 1$ *if and only if* rank $X_{12} = n$ *and* $A + B_2 K_{12}$ *is strictly Hurwitz where*

$$K_{12} = U_{12} X_{12}^{-1}. \tag{61}$$

If this condition holds, then one possible solution is the state feedback $u = K_{12} x$.

Proof: First apply the input transformation

$$u_2 = (D_{12}^T D_{12})^{-\frac{1}{2}} \hat{u}_2 \tag{62}$$

and consider the resulting system described by

$$\hat{G}(s) \overset{s}{=} \left[\begin{array}{c|cc} -sI + A & B_1 & \hat{B}_2 \\ \hline C_1 & 0 & \hat{D}_{12} \\ I_n & 0 & 0 \end{array} \right] \tag{63}$$

where

$$\hat{B}_2 := B_2(D_{12}^T D_{12})^{-\frac{1}{2}} \tag{64}$$

$$\hat{D}_{12} := D_{12}(D_{12}^T D_{12})^{-\frac{1}{2}} \tag{65}$$

Now note that $\hat{D}_{12}$ is part of an orthogonal matrix and the pertinent Riccati equation is (see [2, 40])

$$(A-\hat{B}_2\hat{D}_{12}^T C_1)^T P + P(A-\hat{B}_2\hat{D}_{12}^T C_1) + P(B_1 B_1^T - \hat{B}_2^T \hat{B}_2)P + C_1^T(I-\hat{D}_{12}\hat{D}_{12}^T)C_1 = 0 \tag{66}$$

It is known that Eq (66) has a positive semidefinite stabilizing solution, P, if and only if [40]

(i) the matrix

$$H^{(\infty)} = \begin{bmatrix} A - B_2(D_{12}^T D_{12})^{-1} D_{12}^T C_1 & B_1 B_1^T - B_2^T (D_{12}^T D_{12})^{-1} B_2 \\ -C_1^T (I - D_{12}(D_{12}^T D_{12})^{-1} D_{12}^T) C_1 & -(A - B_2(D_{12}^T D_{12})^{-1} D_{12}^T C_1)^T \end{bmatrix}. \tag{67}$$

has an n-dimensional stable eigenspace and,

(ii) The stable eigenspace of $H^{(\infty)}$ is spanned by the $2n \times n$ partitioned matrix $\begin{bmatrix} X \\ \Phi \end{bmatrix}$ where X is nonsingular.

The solution to Eq (66) is then given by $P = \Phi X^{-1}$. Now let Λ be an $n \times n$ strictly Hurwitz matrix satisfying

$$H^{(\infty)} \begin{bmatrix} X \\ \Phi \end{bmatrix} = \begin{bmatrix} X \\ \Phi \end{bmatrix} \Lambda \tag{68}$$

then it is easily verified by direct substitution that

$$W_{12_\infty}(s) \begin{bmatrix} X \\ \Phi \\ U \end{bmatrix} = \begin{bmatrix} X \\ \Phi \\ 0 \end{bmatrix} (-sI + \Lambda) \tag{69}$$

where

$$U := -(D_{12}^T D_{12})^{-1}(D_{12}^T C_1 X + B_2^T \Phi) \tag{70}$$

i. e. the columns of the matrix $\begin{bmatrix} X \\ \Phi \\ U \end{bmatrix}$ span the stable eigenspace of $W_{12_\infty}(s)$.

The stabilizing feedback for the transformed system $\hat{G}(s)$ is

$$\hat{K} = -(\hat{D}_{12}^T C_1 + \hat{B}_2^T P) \tag{71}$$

and hence the actual feedback for the original systems $G(s)$ is

$$K_{12} = -(D_{12}^T D_{12})^{-1}(D_{12}^T C_1 + B_2^T P) = UX^{-1}. \tag{72}$$

The proof is completed by letting $X_{12} = X$, $\Phi_{12} = \Phi$, $U_{12} = U$. □

Again, it is instructive to emphasize the link between the result of Lemma 11 and linear quadratic differential game (LQDG) theory [4]:

Lemma 12 *The vector pair $(x(s), u_2(s))$ solves the LQDG problem*

$$\min_{u_2(t)} \max_{u_1(t)} J(x(0), u_2(t)) := \int_0^\infty \|y_1(t)\|_2^2 - \|u_1\|_2^2 dt \tag{73}$$

subject to $x(t)$ asymptotically stable

if and only if, for some $\phi(s)$, $\begin{bmatrix} x(s) \\ \phi(s) \\ u_2(s) \end{bmatrix}$ is an asymptotically stable solution of

$$W_{12_\infty}(s) \begin{bmatrix} x(s) \\ \phi(s) \\ u_2(s) \end{bmatrix} = \begin{bmatrix} x(0) \\ \phi(0) \\ 0 \end{bmatrix} \tag{74}$$

Proof: See [41, 4]. □

The result for the regular H^∞ problem with measurement feedback is as follows:

Lemma 13 *There exists a compensator $u_2(s) = K(s)y_2(s)$ which asymptotically stabilizes $G(s)$ and makes $\|T_{y_1u_1}\|_\infty < 1$ if and only if there exist matrices $\Xi_{12} := \begin{bmatrix} X_{12} \\ \Phi_{12} \\ U_{12} \end{bmatrix} \in \Re^{(2n+r)\times n}$, $\Xi_{21} := \begin{bmatrix} X_{21} \\ \Phi_{21} \\ U_{21} \end{bmatrix} \in \Re^{(2n+m)\times n}$ whose columns form bases for the respective stable eigenspaces of $W_{12_\infty}(s)$ and $W_{21_\infty}(s)$ corresponding to the eigenvalues of the matrices Λ_{12} and Λ_{21} and the following conditions hold:*

(i) rank X_{12} = rank $X_{21} = n$.

(ii) $A + B_2K_{12}$ and $A + K_{21}C_2$ are strictly Hurwitz.

(iii) $\lambda_{max}(X_{21}^{-T}\Phi_{21}^T\Phi_{12}X_{12}^{-1}) < 1$.

where

$$K_{12} = U_{12}X_{12}^{-1} \tag{75}$$

$$K_{21} = (X_{21}^T)^{-1}U_{21}^T \tag{76}$$

Furthermore, if these conditions hold, all such compensators may be parameterized by $K(s) = \mathcal{F}_l(M(s), Q(s))$ where

$$M(s) \stackrel{s}{=} P_M(s) := \left[\begin{array}{c|cc} -sE_k + A_k & B_{k_1} & B_{k_2} \\ \hline -C_{k_1} & 0 & D_{k_1} \\ -C_{k_2} & D_{k_2} & 0 \end{array}\right] \tag{77}$$

$$\left.\begin{array}{ll} E_k = X_{21}^T X_{12} - \Phi_{21}^T \Phi_{12} & \\ B_{k_1} = -U_{21}^T & B_{k_2} = X_{21}^T B_2 + \Phi_{21}^T C_1^T D_{12} \\ C_{k_1} = -U_{12} & C_{k_2} = C_2 X_{12} + D_{21} B_1^T \Phi_{12} \\ D_{k_1} = (D_{12}^T D_{12})^{\frac{1}{2}} & D_{k_2} = (D_{21} D_{21}^T)^{\frac{1}{2}} \\ A_k = E_k \Lambda_{12} + U_{21}^T C_{k_2} = \Lambda_{21}^T E_k + B_{k_2} U_{12} & \end{array}\right\} \quad (78)$$

and

$$Q(s) = D_{k_1}^{-1} \hat{Q}(s) D_{k_2}^{-1} \quad (79)$$

where $\|\hat{Q}(s)\|_\infty \leq 1$, $\hat{Q}(s) \in RH^\infty$.

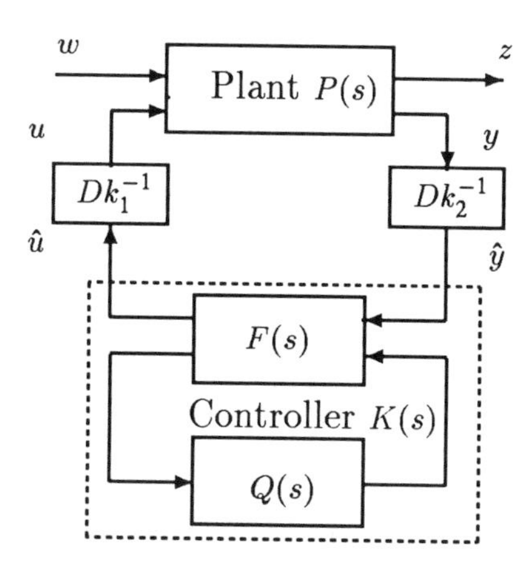

Figure 1: Plant with orthogonalizing compensators

Proof: It suffices to show that the Lemma yields the descriptor system H^∞ controller result in [40], for example. To do this, first apply the static pre- and post-compensators (see Fig. 1) $u_2 = D_{k_1}^{-1} \hat{u}_2$ and $\hat{y}_2 = D_{k_2}^{-1} y_2$. The new plant $\hat{G}(s) \stackrel{s}{=} \left[\begin{array}{c|cc} -sI + A & B_1 & B_2 D_{k_1}^{-1} \\ \hline C_1 & 0 & \hat{D}_{12} \\ D_{k_2}^{-1} C_2 & \hat{D}_{21} & 0 \end{array}\right]$, where $\hat{D}_{12}$ and $\hat{D}_{21}$ are parts of orthogonal matrices.

For this system we form the corresponding $\hat{W}_{12_\infty}(s)$, $\hat{W}_{21_\infty}(s)$ along the lines prescribed in Eqs (59, 60) and obtain the matrices $\begin{bmatrix} \hat{X}_{12} \\ \hat{\Phi}_{12} \\ \hat{U}_{12} \end{bmatrix}$ and $\begin{bmatrix} \hat{X}_{21} \\ \hat{\Phi}_{21} \\ \hat{W}_{21} \end{bmatrix}$ whose columns form bases for the respective stable eigenspaces of $\hat{W}_{12}^{(\infty)}(s)$ and $\hat{W}_{21}^{(\infty)}(s)$. By Lemma 11 there exist solutions to the usual pair of Riccati equations in [38], say, if and only if X_{12}, X_{21} are nonsingular. Hence the controller $\hat{K}(s)$ for the compensated plant may be determined from [3, Theorem 5.2].

It is easy to show that $\begin{bmatrix} \hat{X}_{12} \\ \hat{\Phi}_{12} \\ \hat{U}_{12} \end{bmatrix} = \begin{bmatrix} X_{12} \\ \Phi_{12} \\ D_{k_1}^2 U_{12} \end{bmatrix}$ and $\begin{bmatrix} \hat{X}_{21} \\ \hat{\Phi}_{21} \\ \hat{U}_{21} \end{bmatrix} = \begin{bmatrix} X_{21} \\ \Phi_{21} \\ D_{k_2}^2 U_{21} \end{bmatrix}$. The rest follows by noting that $K(s) = D_{k_1}^{-1} \hat{K}(s) D_{k_2}^{-1}$. □

Comment: If the conditions stated in Theorem 13 hold then the solutions to the usual regulator and filter Riccati equations are given by

$$P = \Phi_{12} X_{12}^{-1} \tag{80}$$

$$Q = X_{21}^{-T} \Phi_{21}^{T} \tag{81}$$

respectively.

Comment: As was proved in [40], the *strict* Hurwitz conditions in (ii) of Lemma 13 are equivalent to the numerically difficult to verify *semi*definiteness conditions $P \geq 0, Q \geq 0$.

Comment: Following is an alternative expression for A_k which will prove useful later on:

$$A_k = \Xi_{21}^T \begin{bmatrix} A & B_1 B_1^T & B_2 \\ C_1^T C_1 & A^T & C_1^T D_{12} \\ C_2 & D_{21} B_1^T & 0 \end{bmatrix} \Xi_{12} \tag{82}$$

Comment: Note that $Q(s)$ now need not be a contractive operator, although it still must be stable. In view of this fact, $Q(s)$ will be called an *auxiliary compensator*.

IV A Regular Embedding for Singular H^p Problems

We now describe what is essentially a suboptimal design method for the case where $G_{12}(s)$ and/or $G_{21}(s)$ have $\mathcal{C}^{0e}$ zeros. Specifically, we show how to solve the singular problem by determining a solution to a "nearby" regular problem. This therefore represents an extension to the work of [42, 43] in that we treat the more general case of measurement feedback. As such, we hasten to point out that the procedure is self contained and would, if conditions permit, generate a solution to the singular problem using compensators of at most n-th order.

Consider the following embedding of the system Eq (1):

$$G_\epsilon(s) \stackrel{s}{=} \left[\begin{array}{c|cc} -sI + A & \tilde{B}_1 & B_2 \\ \hline \tilde{C}_1 & 0 & \tilde{D}_{12} \\ C_2 & \tilde{D}_{21} & 0 \end{array} \right] \tag{83}$$

where $\epsilon > 0$ and

$$\left.\begin{aligned} \tilde{B}_1 &:= \begin{bmatrix} B_1 & 0 \end{bmatrix} \\ \tilde{C}_1 &:= \begin{bmatrix} C_1 \\ 0 \end{bmatrix} \\ \tilde{D}_{12} &:= \begin{bmatrix} D_{12} \\ \epsilon I_{r_2} \end{bmatrix} \\ \tilde{D}_{21} &:= \begin{bmatrix} D_{12} & \epsilon I_{m_2} \end{bmatrix} \end{aligned}\right\} \tag{84}$$

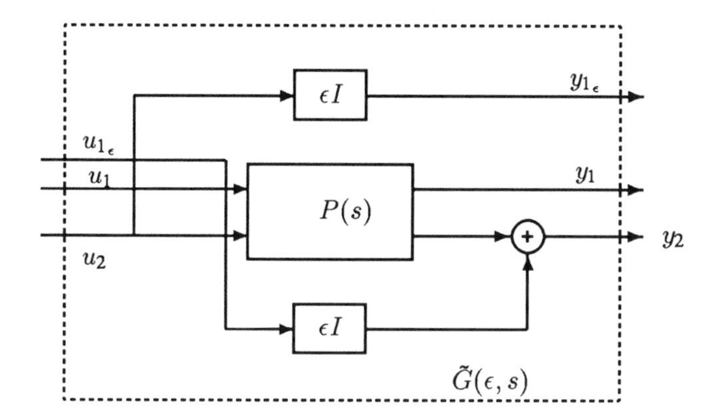

Figure 2: Modified Plant $G_\epsilon(s)$

Figure 2 gives a diagrammatic representation of this embedding.

The new exogenous signals are

$$\tilde{u_1} = \begin{bmatrix} u_1 \\ u_{1_\epsilon} \end{bmatrix} \in \Re^{r_1+m_2} \quad \text{and} \quad \tilde{y_1} = \begin{bmatrix} y_1 \\ y_{1_\epsilon} \end{bmatrix} \in \Re^{m_1+r_2} \tag{85}$$

where u_{1_ϵ} and y_{1_ϵ} are given by $y_{1_\epsilon} = \epsilon u_2$ and $y_2 = C_{22}\nu + \epsilon u_{1_\epsilon}$, ν being the state of the system Eq (83).

It is readily seen that the H^p problem for $G_\epsilon(s)$ is regular; for the H^∞ problem, for example, the relevant Riccati equations are

$$\begin{aligned} &(A - B_2(D_{12}^T D_{12} + \epsilon^2 I)^{-1} D_{12}^T C_1)^T P_\epsilon + P_\epsilon(A - B_2(D_{12}^T D_{12} + \epsilon^2 I)^{-1} D_{12}^T C_1) + \\ &P_\epsilon(B_1 B_1^T - B_2(D_{12}^T D_{12} + \epsilon^2 I)^{-1} B_2^T)P_\epsilon + C_1^T(I - D_{12}(D_{12}^T D_{12} + \epsilon^2 I)^{-1} D_{12}^T)C_1 \\ &= 0 \end{aligned} \tag{86}$$

$$\begin{aligned} &(A - B_1 D_{21}^T(D_{21} D_{21}^T + \epsilon^2 I)^{-1} C_2)Q_\epsilon + Q_\epsilon(A - B_1 D_{21}^T(D_{21} D_{21}^T + \epsilon^2 I)^{-1} C_2)^T + \\ &Q_\epsilon(C_1^T C_1 - C_2^T(D_{21}^T D_{21} + \epsilon^2 I)^{-1} C_2)Q_\epsilon + B_1(I - D_{21}^T(D_{21} D_{21}^T + \epsilon^2 I)^{-1} D_{21})B_1^T \\ &= 0 \end{aligned} \tag{87}$$

The following lemma gives necessary and sufficient conditions for the existence of a solution to the suboptimal problem for $G(s)$. We denote the closed loop transfer function from any input u to any output y for the augmented plant $G_\epsilon(s)$ by $T_{yu_\epsilon}(s)$.

Lemma 14 *There exists a controller which solves the suboptimal H^p problem for the plant $G(s)$ if and only if there exists an $\epsilon^* > 0$ for which a solution to the suboptimal problem for the plant $G_{\epsilon^*}(s)$ exists.*

Furthermore, if such an ϵ^* exists, it follows that
(i) There exists a solution to the suboptimal problem $\forall \epsilon \in [0, \epsilon^*]$.
(ii) The optimum H^p cost is monotone decreasing on $[0, \epsilon^*]$.

Proof: Suppose that there exists a solution $K^*(s)$ to the suboptimal problem for $G_{\epsilon^*}(s)$ for some $\epsilon^* > 0$. Let $K_\epsilon(s)$ denote any controller applied to the plant $G_\epsilon(s)$; furthermore, $\forall\, \epsilon \in [0, \epsilon^*]$ let $K_\epsilon(s) = K^*(s)$. Then $\forall \epsilon \in [0, \epsilon^*]$ we have

$$T_{\tilde{y}_1\tilde{u}_{1_\epsilon}}(s) = \left[\begin{array}{c|c} T_{y_1u_1}(s) & \epsilon G_{12}(s)T_{22}(s) \\ \hline \epsilon T_{22}G_{21}(s) & \epsilon^2 T_{22}(s) \end{array}\right] \tag{88}$$

where

$$T_{22}(s) := K^*(s)(I - G_{22}(s)K^*(s))^{-1} \tag{89}$$

and

$$T_{y_1u_1}(s) = G_{11}(s) + G_{12}T_{22}G_{21}(s) \tag{90}$$

We note that each partition in Eq (88) is a stable transfer function at $\epsilon = \epsilon^*$; it therefore follows that $T_{\tilde{y}_1\tilde{u}_{1_\epsilon}}(s) \in RH^\infty\ \forall \epsilon \in [0, \epsilon^*]$. Moreover, one may readily check that for $p \in \{2, \infty\}$ the quantity $||T_{\tilde{y}_1\tilde{u}_{1_{\epsilon_1}}}||_p$ is monotone increasing in ϵ $\forall \epsilon \in [0, \epsilon^*]$. Hence, $\forall \epsilon \in [0, \epsilon^*]$, $||T_{\tilde{y}_1\tilde{u}_{1_\epsilon}}||_p \leq ||T_{\tilde{y}_1\tilde{u}_{1_{\epsilon^*}}}||_p < 1$. This proves (i) while sufficiency immediately follows from the fact that $T_{y_1u_1}(s) \in RH^\infty$, and

$$\begin{aligned} ||T_{y_1u_1}||_p &= ||T_{\tilde{y}_1\tilde{u}_{1_0}}||_p && (91)\\ &\leq ||T_{\tilde{y}_1\tilde{u}_{1_{\epsilon^*}}}||_p && (92)\\ &< 1 && (93) \end{aligned}$$

For (ii), let $K_\epsilon^*(s)$ denote the optimum controller for any specified ϵ. Then for $\epsilon_2 > \epsilon_1 > 0$ we have

$$\begin{aligned} ||T_{\tilde{y}_1\tilde{u}_{1_{\epsilon_2}}}|_{K^*(\epsilon_2,s)}||_p &\geq ||T_{\tilde{y}_1\tilde{u}_{1_{\epsilon_1}}}|_{K^*(\epsilon_2,s)}||_p && (94)\\ &\geq ||T_{\tilde{y}_1\tilde{u}_{1_{\epsilon_1}}}|_{K^*(\epsilon_1,s)}||_p && (95) \end{aligned}$$

For necessity, assume that the controller $K_0(s)$ solves the problem for the plant $G(s)$. For the H^2 case no generality is lost by assuming that $K_0(s)$ is strictly proper[1]. We then have that $||T_{y_1u_1}||_p = 1 - \delta, 0 < \delta < 1$. Apply the controller $K_0(s)$ and consider the closed loop system for increasing ϵ. Since, by assumption, $K_0(s)$ internally stabilizes $\tilde{G}(0, s)$, it follows that $T_{y_1u_1}$, T_{22}, $T_{22}G_{21}$ and $G_{12}T_{22}$ are all in RH^∞. Moreover, if $T_{y_1u_1} \in RH^2$, it follows from Eq (90) and the assumption that $K_0(s)$ is strictly proper that T_{22}, $T_{22}G_{21}$ and $G_{12}T_{22}$ are all in RH^2. These matrices do not depend on ϵ, hence $T_{\tilde{y}_1\tilde{u}_1} \in H^p\ \forall\, \epsilon$. Thus, we have

[1] If this is not the case, the frequency response of the controller can be "rolled off", while maintaining the H^2 norm bound, by using a series compensator of the form $\frac{1}{(1+\eta s)}$ for some small enough η. A rigorous justification for this common approach is omitted.

that $k_{12} := \|G_{12}T_{22}\|_p < \infty$, $k_{21} := \|T_{22}G_{21}\|_p < \infty$, $k_{22} := \|T_{22}\|_p < \infty$. Moreover,

$$\|T_{\tilde{y}_1\tilde{u}_{1_\epsilon}}\|_p = \| \left[\begin{array}{c|c} T_{y_1u_1} & \epsilon G_{12}T_{22} \\ \hline \epsilon T_{22}G_{21} & \epsilon^2 T_{22} \end{array} \right] \|_p \tag{96}$$

$$\leq \|T_{y_1u_1}\|_p + \epsilon k_{12} + \epsilon k_{21} + \epsilon^2 k_{22} \tag{97}$$

(equality holds for the H^p case). Now note that for $\epsilon^* := \frac{\delta}{6\max(k_{12},k_{21},\sqrt{k_{22}})}$, $\|T_{\tilde{y}_1\tilde{u}_{1_\epsilon}}\|_p \leq 1 - \delta + \frac{\delta}{2} = 1 - \frac{\delta}{2} < 1 \ \ \forall \epsilon \in [0, \epsilon^*]$.

□

Finally, we present the following lemma which will prove to be useful when we specifically consider H^∞ problems. See Kwakernaak [39] and Bucy *et al* [15] for similar results in the H^2 case.

Lemma 15 *Suppose that there exists an $\epsilon^* > 0$ for which there is a solution to the suboptimal problem. For each $\epsilon \in (0, \epsilon^*]$, denote by P_ϵ, Q_ϵ the respective solutions of the Riccati equations (86,87). Then it holds that*
(i) $P_0 := \lim_{\epsilon\to 0} P_\epsilon$ and $Q_0 := \lim_{\epsilon \to 0} Q_\epsilon$ exist and $P_0 \geq 0$, $Q_0 \geq 0$.
(ii) $\lambda_{max}(Q_0P_0) < 1$.

To prove this lemma we again utilize the fact that the H^∞ problem can be tackled by solving the LQ differential game problem [4, 41] Eq (73).
Proof: For Eq (83) the cost in Eq (73) is

$$J(\tilde{u}_1, u_2, \epsilon) = \int_0^\infty \tilde{y}_1^T \tilde{y}_1 - \tilde{u}_1^T \tilde{u}_1 \text{ dt} \tag{98}$$

$$= \int_0^\infty y_1^T y_1 - u_1^T u_1 + \epsilon^2 u_2^T u_2 - u_{1_\epsilon}^T u_{1_\epsilon} \text{ dt} \tag{99}$$

The optimizing inputs are [4, 41]

$$u_1^*(t) = B_1^T P_\epsilon x(t) \tag{100}$$

$$u_{1_\epsilon}^*(t) = 0 \tag{101}$$

$$u_2^*(t) = -(D_{12}^T D_{12} + \epsilon^2 I)^{-1}(D_{12}^T C_1 + B_2^T P_\epsilon)x(t) \tag{102}$$

Let $y_1^*(t)$ be the output corresponding to $u_1(t) = u_1^*(t)$, $u_{1_\epsilon}^*(t) = 0$, $u_2(t) = u_2^*(t)$ and initial condition $x(0)$. Now let $\epsilon_1 > \epsilon_2 > 0$ and fix $\epsilon = \epsilon_1$ in Eq (99). Recall also that for any $x(0) = x_0 \in \Re^n$ the optimal cost in Eq (73) is

$$J^*(u_1^*, u_2^*, \epsilon) = x_0^T P_\epsilon x_0. \tag{103}$$

where P is a positive semidefinite matrix. It follows from the above that

$$x_0^T P_{\epsilon_2} x_0 \leq \int_0^\infty y_1^{*^T} y_1^* - u_1^{*^T} u_1^* + \epsilon_2^2 u_2^{*^T} u_2^* \text{ dt} \tag{104}$$

$$\leq x_0^T P_{\epsilon_1} x_0 \tag{105}$$

$\forall x_0 \in \Re^n$. Hence, for any monotone decreasing sequence $\{\epsilon_j\}_{j=0}^{\infty}$ with $\epsilon_j > 0 \forall j$, $\{P_{\epsilon_j}\}_{j=0}^{\infty}$ is a monotone decreasing sequence of non-negative definite (and therefore bounded) matrices; this implies that $P_0 := \lim_{\epsilon \to 0} P_\epsilon$ exists and is also non-negative definite. By a similar argument, $Q_0 := \lim_{\epsilon \to 0} Q_\epsilon$ exists and is non-negative definite. This proves Part (i).

Part (ii) follows from the fact that if the hypothesis of the Lemma holds then, for all $\epsilon \in (0, \epsilon^*]$, $Q_0 \leq Q_\epsilon$, $P_0 \leq P_\epsilon$ and therefore $\lambda_{\max}(Q_0 P_0) \leq \lambda_{\max}(Q_\epsilon P_\epsilon) < 1$. □

V The Singular H^p State-Feedback Problem

In the last section we perturbed the singular H^p problem into a regular one by suitable introduction of a scalar parameter $\epsilon > 0$; this allows us to obtain a suboptimal solution to the relevant problem. One setback with this approach is that, for H^∞ problems, one is never really quite sure if a solution exists to the original (unperturbed) problem unless one happens to solve a perturbed problem, for some sufficiently small value of ϵ. This indicates that an iterative process will at times be required in order to arrive at a solution. Moreover, in H^2 problems, one would generally wish to determine the optimal cost so as to ascertain the extra penalty incurred by the suboptimal design.

In this section we seek to alleviate these problems by eliminating our dependence on the perturbation parameter ϵ. This is done by analysis of the behavior of the suboptimal solutions as $\epsilon \to 0$.

A characteristic of singular problems involving finite $j\omega$=axis zeros is that the limiting state trajectories as $\epsilon \to 0$ may be only bounded, not asymptotically stable. Accordingly, in treating singular problems we do not impose the constraint that the closed-loop optimal state trajectories be asymptotically stable; instead we only require that the solutions be bounded for t >0 and that the optimal cost be finite. So, modes that are not observable through y_1 need not be asymptotically stable and impulses and their derivatives are permitted at time $t = 0$.

A The H^2 Problem

We begin with the H^2 problem. Consider the perturbed plant Eq (83) and define

$$W_{12_2}(\epsilon, s) \quad := \quad \begin{bmatrix} -sI + A & 0 & B_2 \\ -\tilde{C}_1^T \tilde{C}_1 & -sI - A^T & -\tilde{C}_1^T \tilde{D}_{12} \\ \tilde{D}_{12}^T \tilde{C}_1 & B_2^T & \tilde{D}_{12}^T \tilde{D}_{12} \end{bmatrix} \tag{106}$$

$$= \begin{bmatrix} -sI + A & 0 & B_2 \\ -C_1^T C_1 & -sI - A^T & -C_1^t D_{12} \\ D_{12}^T C_1 & B_2^T & D_{12}^T D_{12} + \epsilon^2 I \end{bmatrix} \tag{107}$$

Note that

$$W_{12_2}(0, s) = W_{12_2}(s) \tag{108}$$

Now let $\mathcal{S}_{12_2}(\epsilon)$ denote the stable eigensubspace of the pencil $W_{12_2}(\epsilon, s)$ and define

$$M_{G_{12}}(s) := \left[\begin{array}{c|c} -sI + A & B_2 \\ \hline C_1 & D_{12} \end{array}\right] \tag{109}$$

Theorem 1 *Let* $\begin{bmatrix} X_{0e}^{(l)} \\ U_{0e}^{(l)} \end{bmatrix}$ *be a matrix whose columns form a basis for the sum of the finite* C^{0e} *eigenspace and the lower infinite frequency eigensubspace of the pencil* $M_{G_{12}}(s)$. *Then it holds that*

$$\lim_{\epsilon \to 0} \mathcal{S}_{12_2}(\epsilon) = \mathcal{S}_{12_2}(0) \oplus \mathrm{Im} \left(\begin{bmatrix} X_{0e}^{(l)} \\ 0 \\ U_{0e}^{(l)} \end{bmatrix} \right) \tag{110}$$

In order to prove this theorem we will need two supporting lemmas. The first is proved in Gohberg *et al.* [44]:

Lemma 16 *Let* $A(\mu)$,$\mu \in \Omega$ *be an* $n \times n$ *complex matrix-valued function analytic in a domain* Ω *of* C. *Let* $r = \max \mathrm{rank} \ A(\mu)$. *Then there exist n-dimensional vector valued functions* $y_i(\mu)$, $i \in \underline{n}$ *such that*

(i) $y_i(\mu)$ *is analytic in* $\Omega \ \forall \ i \ \in \ \underline{n}$,

(ii) $y_1(\mu),\cdots,y_r(\mu)$ *are linearly independent* $\forall \mu \in \Omega$,

(iii) $y_{r+1}(\mu),\cdots,y_n(\mu)$ *are linearly independent* $\forall \mu \in \Omega$,

(iv) $\mathrm{Span}\{y_1(\mu),\cdots,y_r(\mu)\} = \mathrm{Im} \ A(\mu)$ *and* $\mathrm{Span}\{y_{r+1}(\mu),\cdots,y_n(\mu)\} = \mathrm{Ker} \ A(\mu)$

$\forall \ \mu \in \Omega$, *except for those isolated points,* $\mu_l \in \Omega$, *for which* $\mathrm{rank} \ A(\mu_l) < r$. *At these points the following hold*

$$\mathrm{Span}\{y_1(\mu_l),\cdots,y_r(\mu_l)\} \supset \mathrm{Im} \ A(\mu_l) \tag{111}$$

$$\mathrm{Span}\{y_{r+1}(\mu_l),\cdots,y_n(\mu_l)\} \subset \mathrm{Ker} \ A(\mu_l) \tag{112}$$

Proof: See Chapter S6 of [44]. □

The next lemma is easily proved using standard matrix theory:

Lemma 17 *Suppose* $T \in C^{n \times n}$ *has k eigenvalues at* $s = \lambda$ *of respective orders* α_i, $i \in \underline{k}$. *Define* $\alpha := \sum_{i \in \underline{k}} \alpha_i$. *Then it holds that*

$$\dim \mathrm{Ker} \ (T - \lambda I)^q \geq q \tag{113}$$

$\forall \ q \in \underline{\alpha}$.

Proof of Theorem 1: Assume that $M_{G_{12}}(s)$ has k distinct C^{0e} eigenvalues of various orders. Consider the neighborhood $\Omega_{\epsilon^*} := (-\epsilon^*, \epsilon^*)$ of $\epsilon = 0$. Let $\alpha \in \Re$ be such that, on Ω_{ϵ^*}, $W_{12_2}(\epsilon, s)$ has no eigenvalue $\lambda = -\alpha$. Now make the transformation

$$\tilde{s} = s + \alpha \tag{114}$$

and note that the eigenstructure of $W_{12_2}(\epsilon, s)$ at $s = \lambda$ is isomorphic to that of $W_{12_2}(\epsilon, \tilde{s}) = -\tilde{s}E_b + (\alpha E_b + A_b(\epsilon))$ at $\tilde{s} = \lambda + \alpha$. Also note that by construction $\alpha E_b + A_b(\epsilon)$ is nonsingular on Ω_{ϵ^*} and define

$$\Theta(\epsilon) = (\alpha E_b + A_b(\epsilon))^{-1} E_b \tag{115}$$

Now the right eigenstructure of $W_{12_2}(\epsilon, \tilde{s})$ is identical to that of

$$(\alpha E_b + A_b(\epsilon))^{-1} W_{12_2}(\epsilon, \tilde{s}) = -(\alpha E_b + A_b(\epsilon))^{-1} E_b \tilde{s} + I \tag{116}$$

It therefore follows that the eigenstructure of the pencil $W_{12_2}(\epsilon, s)$ at $s = \lambda$ is isomorphic to that of the matrix $\Theta(\epsilon)$ at $\rho = \frac{1}{\lambda+\alpha}$. Figure 3 shows how the conformal map $\rho = \frac{1}{s+\alpha}$ maps the s-plane; note that the left half s-plane is mapped onto the exterior $\mathcal{J}_0$ of the circle $\mathcal{J}$.

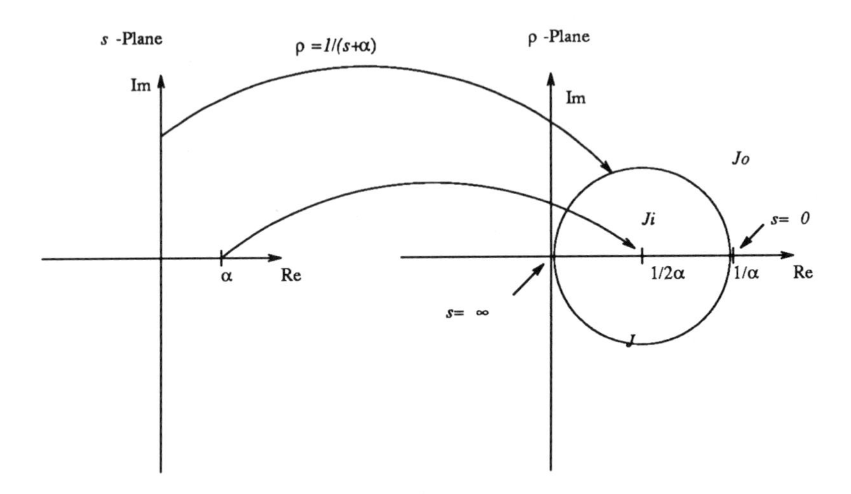

Figure 3: Mapping of the s-Plane under $\rho = \frac{1}{s+\alpha}$

From Lemma 7 we know that $\forall \epsilon\ \Theta(\epsilon)$ has a characteristic polynomial of the form

$$\pi(\epsilon, \rho) = \rho^r \pi_o(\epsilon, \rho) \pi_i(\epsilon, \rho) \tag{117}$$

where the zeros of π_o are all in $\mathcal{J}_o$ for $\epsilon \neq 0$. Without loss of generality, we assume that ϵ^* is small enough so that those eigenvalues in $\mathcal{J}_o$ which do approach $\mathcal{J}$ as $\epsilon \to 0$ are distinct from those remaining in $\mathcal{J}_o$ $\forall \epsilon \in \Omega_{\epsilon^*}$. We also assume that $\forall \epsilon \in \Omega_*$ those eigenvalues in $\mathcal{J}_o$ which approach $\rho = \rho_i$ on $\mathcal{J}$ are distinct from those approaching the value $\rho = \rho_j$ on $\mathcal{J}$, $i \neq j$. Thus, on Ω_{ϵ^*} we have the coprime factorization :

$$\pi_o = \pi_1 \pi_2, \quad \pi_2 := \pi_{2_1} \cdots \pi_{2_k}$$

where the zeros of π_1 are in $\mathcal{J}_o$ $\forall \epsilon \in \Omega_{\epsilon^*}$, and the zeros of π_{2_i}, $i \in \underline{k}$, approach the value $\rho = \rho_i$ on $\mathcal{J}$ as $\epsilon \to 0$. Note that $\Theta(\epsilon)$ is analytic in Ω_{ϵ^*} by virtue of the nonsingularity of $\alpha E_b + A_b(\epsilon)$; hence the polynomials π_1 and π_{2_i}, $i \in \underline{k}$, are continuous in ϵ on Ω^*.

Let δ be the order of π_1 and δ_i the order of π_{2_i} $i \in \underline{k}$. Now, $\forall \epsilon \in \Omega_{\epsilon^*}$ we have that

$$\begin{aligned} \mathcal{S}_{12_2}(\epsilon) &= \text{Ker } \pi_o(\epsilon, \Theta(\epsilon)) \\ &= \text{Ker } \pi_1(\epsilon, \Theta(\epsilon)) \oplus \text{Ker } \pi_{2_1}(\epsilon, \Theta(\epsilon)) \oplus \cdots \oplus \text{Ker } \pi_{2_k}(\epsilon, \Theta(\epsilon)) \end{aligned} \tag{118}$$

the latter equality following from the coprimeness of the respective polynomials. Consider, first, the subspace Ker $\pi_1(\epsilon, \Theta(\epsilon))$; since the dimension of this subspace is constant ($= \delta$) on Ω_{ϵ^*} it follows from Lemma 16 that

$$\lim_{\epsilon \to 0} \text{Ker } \pi_1(\epsilon, \Theta(\epsilon)) = \text{Ker } \pi_1(0, \Theta(0)) \tag{119}$$

Next consider the subspace Ker $\pi_{2_i}(\epsilon, \Theta(\epsilon))$. By Lemma 7 the chains corresponding to $\rho = \rho_i$ have a cumulative length $l \geq \delta_i$. It therefore follows from Lemma 17 that dim Ker $\pi_{2_i}(0, \Theta(0)) = \dim \text{Ker } (\Theta(0) - \rho_i I)^{\delta_i} \geq \delta_i$. This proves that if $\Xi_{oe}(\epsilon)$ is a matrix whose columns form a basis for Ker $\pi_2(\epsilon, \Theta(\epsilon))$, $\epsilon \neq 0$, then by Lemma 16 $\Xi_{oe} := \lim_{\epsilon \to 0} \Xi_{oe}(\epsilon)$ exists and satisfies

$$\text{Im } \Xi_{oe} \subset \text{Ker } (\Theta(0) - \rho_1 I)^{\delta_1} \oplus \cdots \oplus \text{Ker } (\Theta(0) - \rho_k I)^{\delta_k} \tag{120}$$

Summarizing, we have that

$$\lim_{\epsilon \to 0} \mathcal{S}_{12_2}(\epsilon) = \mathcal{S}_{12_2}(0) \oplus \text{Im } \Xi_{oe} \tag{121}$$

where rank $\Xi_{oe} = \sum_{i=1}^{k} \delta_i =$ rank $\Xi_{oe}(\epsilon)$ for any $\epsilon \in \Omega_*$.

The proof is now almost complete save for the characterization of Ξ_{oe}. To obtain this, we note that the eigenspace of $W_{12_2}(0, s)$ associated with the $\mathcal{C}^{0e}$ eigenvalues determines those solutions of Eq (58) with poles on the $j\omega$-axis or at infinity (polynomial solutions). It follows that the corresponding cost Eq (57) would be infinite unless all such solutions lie in Ker $\begin{bmatrix} C_1 & D_{12} \end{bmatrix}$ (in which case the cost is zero). Now note that by Lemma 7 the columns of $\begin{bmatrix} X_{0e}^{(l)} \\ 0 \\ U_{0e}^{(l)} \end{bmatrix}$ do indeed form partial chains of $W_{12_2}(\epsilon, s)$ where $\begin{bmatrix} X_{0e}^{(l)} \\ U_{0e}^{(l)} \end{bmatrix} \subset$ Ker $\begin{bmatrix} C_1 & D_{12} \end{bmatrix}$; moreover, the choice of any other vectors in the respective chains results in infinite cost in Eq (57). □

Theorem 1 verifies that the limiting optimal solution is indeed determined by the modal responses of $W_{12_2}(0, s)$ as suggested by application of the Minimum Principle at $\epsilon = 0$. The complexity in deriving the limiting subspace when

$W_{12_2}(0, s)$ has infinite eigenvalues arises from the fact that while the relevant initial conditions may depend on the highest grade eigenvector of $M_{G_{12}}(s)$, the resulting solutions are spanned by eigenvectors of lower grade. This results in a limiting feedback gain K (cf. Lemma 5) which is infinite.

Implicit in the above is the assumption that a solution exists at $\epsilon = 0$. The following lemma is helpful in determining when a solution to the problem actually exists.

Lemma 18 *Let* $\mathcal{S}_{12_2}(0) = \text{Im} \begin{bmatrix} X_{12} \\ \Phi_{12} \\ U_{12} \end{bmatrix}$ *and let* $\begin{bmatrix} X_{oe}^{(u)} \\ U_{oe}^{(u)} \end{bmatrix}$ *be a matrix whose columns form a basis for the sum of the finite* $\mathcal{C}^{0e}$ *eigenspace and the upper infinite frequency eigensubspace of* $M_{G_{12}}(s)$. *Define* $\mathcal{X} = [\ X_{12} \quad X_{oe}^{(u)}\ ]$. *Then*

(i) Im $\mathcal{X}$ *is the space of initial conditions* x_0 *for which the cost Eq (57) is finite.*

(ii) The minimal cost 57) is finite for all initial conditions $x(0)$ *if and only if* rank $\mathcal{X} = n$.

Proof: In view of Theorem 1, the minimal cost Eq (57) for the limiting solution as $\epsilon \to 0$ is finite for a given $x(0)$ if and only if there exist corresponding vectors $\phi(0)$, $x(s)$, $\phi(s)$, $u(s)$ satisfying Eq (58) such that $x(s)$, $\phi(s)$, $u(s)$ are either asymptotically stable or in Ker $[C_1 D_{12}]$. Denote by $\Xi^{(u)}$ the space spanned by all but the first order infinite eigenvectors of $W_{12_2}(s)$ and let $\Xi^{(l)}$ denote the space spanned by all but the infinite eigenvectors of highest order. Observe that the left-hand side of Eq (58)is of the form Eq (11) addressed by Lemma 1. Moreover, for any $x(0), \phi(0)$ the right-hand side of Eq (58) lies in Im E. Moreover, Im $E =$ Im $E\Xi^{(u)}$ since, by definition, $E\xi_\infty^{(1)} = 0$ for every first order infinite eigenvector $\xi_\infty^{(1)}$. Thus, for every $x(0), \phi(0) \in R^n$ there exists a vector $\xi^{(u)} \in \Xi^{(u)}$ such that

$$\begin{bmatrix} x(0) \\ \phi(0) \\ 0 \end{bmatrix} = \xi^{(u)}. \tag{122}$$

Thus the conditions of Lemma 1 are satisfied by Eq (58). It follows from Lemma 1 that the solutions $\begin{bmatrix} x(s) \\ \phi(s) \\ u(s) \end{bmatrix}$ of Eq (58) evolve in the subspace $\Xi^{(l)}$. Now, the admissible solutions for which the cost is finite are those which are either asymptotically stable (i.e., in $\mathcal{S}_{12_2}(0)$) or unobservable through y_1 (i.e., in Ker $[C_1 D_{12}]$). But by Lemmas 1, 4 and 7, these solutions evolve in the space $\mathcal{S}_{12_2}(0) \oplus \text{Im} \left(\begin{bmatrix} X_{0e}^{(l)} \\ 0 \\ U_{0e}^{(l)} \end{bmatrix} \right)$ and correspond to an $x(0) \in \mathcal{X}$, which proves part (i) of the Lemma. Part (ii) follows immediately, since $x(0)$ may assume any value in $\Re^n$. □

The following theorem gives us a formula for the limiting value of Eq (57).

Theorem 2 *Let* $\begin{bmatrix} X_{12} \\ \Phi_{12} \\ U_{12} \end{bmatrix}$ *and* $\begin{bmatrix} X_{oe}^{(u)} \\ U_{oe}^{(u)} \end{bmatrix}$ *be defined as in the previous lemma. Let* $\begin{bmatrix} X_{oe}^{(l)} \\ U_{oe}^{(l)} \end{bmatrix}$ *be the corresponding matrix whose columns span the lower infinite eigenspace of* $M_{G_{12}}$. *Assume that there exists a solution to the state feedback* H^2 *problem for* $G_{12_{\epsilon^*}}(s)$ *for some* $\epsilon^* > 0$. *Then, for* $\epsilon \to 0$, *the limiting state, costate and control trajectories have Laplace transforms:*

$$\begin{bmatrix} x(s) \\ \phi(s) \\ u(s) \end{bmatrix} = \begin{bmatrix} X_{12} & X_{oe}^{(l)} \\ \Phi_{12} & 0 \\ U_{12} & U_{oe}^{(l)} \end{bmatrix} \begin{bmatrix} H^T \\ H_{oe}^T \end{bmatrix} x_o \tag{123}$$

where H *is a partition of the nonsingular matrix* $\begin{bmatrix} H & H_{oe} \end{bmatrix}$ *which satisfies*

$$\begin{bmatrix} H^T \\ H_{oe}^T \end{bmatrix} \begin{bmatrix} X_{12} & X_{oe}^{(u)} \end{bmatrix} = I \tag{124}$$

The limiting cost Eq (57) is given by

$$J(x_0) := J(0, x_0, u_*) = x_0^T P_* x_0 \tag{125}$$

where $u_*(s)$ *is the optimal control and* P_* *is a symmetric positive semi-definite* $n \times n$ *matrix given by*

$$P_* = \Phi_{12} H^T. \tag{126}$$

Proof: By Lemma 18, the existence of a solution to the problem Eq (57) implies that $\forall\ x_0 \in \Re^n \exists$ vectors v, v_{oe} such that

$$x_0 = X_{12} v + X_{oe}^{(u)} v_{oe} \tag{127}$$

By Lemma 11 the solutions are of the form Eq (123). Moreover, the resulting cost is equivalent to that obtained for $x_0 = X_{12} v$ since the cost pertaining to the $\mathcal{C}^{0e}$ eigenspace of $W_{12_2}(0, s)$ is zero. In the time domain the optimal state and control trajectories corresponding to $x_0 = X_{12} v$ is given by $\begin{pmatrix} x(t) \\ u(t) \end{pmatrix} = \begin{bmatrix} X_{12} \\ \Phi_{12} \end{bmatrix} e^{\Lambda t} v$ for $t > 0$, where Λ is strictly Hurwitz and satisfies

$$\begin{bmatrix} A & 0 & B_2 \\ -C_1^T C_1 & -A^T & -C_1^T D_{12} \\ D_{12}^T C_1 & B_2^T & D_{12}^T D_{12} \end{bmatrix} \begin{bmatrix} X_{12} \\ \Phi_{12} \\ U_{12} \end{bmatrix} = \begin{bmatrix} X_{12} \\ \Phi_{12} \\ 0 \end{bmatrix} \Lambda \tag{128}$$

The optimal cost is thus given by

$$\begin{aligned} J(x_0, u) &= \int_0^\infty \begin{pmatrix} x \\ u \end{pmatrix}^T \begin{bmatrix} C_1^T C_1 & C_1^T D_{12} \\ D_{12}^T C_1 & D_{12}^T D_{12} \end{bmatrix} \begin{pmatrix} x \\ u \end{pmatrix} \mathrm{dt} \\ &= v^T \left(\int_0^\infty e^{\Lambda^T t} \begin{bmatrix} X_{12} \\ U_{12} \end{bmatrix}^T \begin{bmatrix} C_1^T C_1 & C_1^T D_{12} \\ D_{12}^T C_1 & D_{12}^T D_{12} \end{bmatrix} \begin{bmatrix} X_{12} \\ U_{12} \end{bmatrix} e^{\Lambda t} \mathrm{dt} \right) v \end{aligned} \tag{129}$$

Now from Eq (128) it can be shown that

$$\begin{bmatrix} X_{12} \\ U_{12} \end{bmatrix}^T \begin{bmatrix} C_1^T C_1 & C_1^T D_{12} \\ D_{12}^T C_1 & D_{12}^T D_{12} \end{bmatrix} \begin{bmatrix} X_{12} \\ U_{12} \end{bmatrix} = -(X_{12}^T \Phi_{12} \Lambda + \Lambda^T X_{12}^T \Phi_{12}) \tag{130}$$

Hence we have that

$$\begin{aligned} J(x_0,u) &= -v^T(\int_0^\infty e^{\Lambda^T t}(X_{12}^T\Phi_{12}\Lambda + \Lambda^T X_{12}^T\Phi_{12})e^{\Lambda t}dt)v \\ &= -v^T(\int_0^\infty \frac{d}{dt}(e^{\Lambda^T t}X_{12}^T\Phi_{12}e^{\Lambda t})dt)v \\ &= v^T X_{12}^T\Phi_{12}v \end{aligned} \tag{131}$$

the last equality following from the fact that Λ is Hurwitz. Since $\begin{bmatrix} X_{12} & X_{0e}^{(u)} \end{bmatrix}$ is nonsingular, there exists a matrix $\begin{bmatrix} H & H_{oe} \end{bmatrix}$ such that

$$\begin{bmatrix} H^T \\ H_{oe}^T \end{bmatrix}\begin{bmatrix} X_{12} & X_{0e}^{(u)} \end{bmatrix} = I_n. \tag{132}$$

Left multiply Eq (127) by H^T to obtain

$$v = H^T x_0 \tag{133}$$

and substitute in Eq (131) to get

$$J(x_0,u) = x_0^T H X_{12}^T\Phi_{12}H^T x_0 \tag{134}$$

Now we note that from Lemma 9 that $X_{oe}^T\Phi_{12} = 0$. So, $H_{oe}X_{oe}^T\Phi_{12} = 0$ and hence

$$(HX_{12}^T - I)\Phi_{12} = 0 \tag{135}$$

the last relationship following from Eq (132). Rewrite Eq (135) as $HX_{12}^T\Phi_{12} = \Phi_{12}$ and substitute in Eq (134) to give

$$\begin{aligned} J &= x_o^T\Phi_{12}H^T x_0 \\ &= x_0^T P_* x_0 \end{aligned} \tag{136}$$

The equations above show that P_* is indeed a matrix of a positive semi-definite form. To show that it is also symmetric define $N := \Phi_{12}^T X_{12} - X_{12}^T\Phi_{12}$ and pre-multiply both sides of Eq (128) by $\begin{bmatrix} \Phi_{12}^T & -X_{12}^T & 0 \end{bmatrix}$ to obtain

$$N\Lambda = \Phi_{12}^T AX_{12} + (\Phi_{12}^T AX_{12})^T + X_{12}^T C_1^T C_1 X_{12} - U_{12}^T D_{12}^T D_{12}U_{12}. \tag{137}$$

Since the right-hand side of Eq (137) is symmetric and N is skew symmetric, it follows that

$$(-\Lambda^T)N - N\Lambda = 0 \tag{138}$$

Since Λ is Hurwitz, and the eigenvalues of $-\Lambda^T$ and Λ are disjoint, it follows by Lemma 1.5 of [39] that the unique solution to the above is $N = 0$ and hence $\Phi_{12}^T X_{12} = X_{12}^T\Phi_{12}$.

□

Let us now consider a simple example taken from [6, pg. 248]. Here, we have $G(s) \stackrel{s}{=} M_G(s) = \left[\begin{array}{cc|c} 0 & 1 & 1 \\ 0 & 0 & -1 \\ \hline 1 & 0 & 0 \end{array}\right]$. $W_{12_2}(0,s)$ has a finite LHP zero at $s = -1$ with corresponding eigenvector $(-2\ 1\ 2\ 2\ 1)^T$. The infinite eigenvectors for $M_{G_{12}}(s)$ are the columns of $\left[\begin{array}{cc} 0 & 1 \\ 0 & -1 \\ \hline 1 & 0 \end{array}\right]$. Hence, corresponding to an initial condition $k(1 \quad -1)^T$ the optimal input is $-k\delta(t)$. For initial conditions along $(-2 \quad 1)^T$ the required input is of the form ke^{-t}; this vector is indeed the "singular arc" for the infinite horizon problem. These results may be compared with those of [6].

B The H^∞ Problem

We now consider the limiting solution to the H^∞ problem for the modified plant Eq (83) as $\epsilon \to 0$. Let $\mathcal{S}_{12_\infty}(\epsilon)$ denote the stable eigenspace of the matrix pencil

$$W_{12_\infty}(\epsilon, s) := \begin{bmatrix} -sI + A & \tilde{B}_1\tilde{B}_1^T & B_2 \\ -\tilde{C}_1^T\tilde{C}_1 & -sI - A^T & -\tilde{C}_1^T\tilde{D}_{12} \\ \tilde{D}_{12}^T\tilde{C}_1 & B_2^T & \tilde{D}_{12}^T\tilde{D}_{12} \end{bmatrix} \tag{139}$$

$$= \begin{bmatrix} -sI + A & B_1B_1^T & B_2 \\ -C_1^TC_1 & -sI - A^T & -C_1^TD_{12} \\ D_{12}^TC_1 & B_2^T & D_{12}^TD_{12} + \epsilon^2 I_{r_2} \end{bmatrix} \tag{140}$$

We now have the following theorem:

Theorem 3 *Assume that there is an $\epsilon^* > 0$ for which there exists a solution to the state-feedback H^∞ problem for $G_{\epsilon^*}(s)$. Let the columns of $\begin{bmatrix} X_{12_s} \\ \Phi_{12_s} \\ U_{12_s} \end{bmatrix}$ form a basis for the stable eigenspace of $W_{12_\infty}(s)$. Let $\begin{bmatrix} X_{12_{0e}}^{(l)} \\ U_{12_{0e}}^{(l)} \end{bmatrix}$ be a matrix whose columns forms a basis for the sum of the finite $\mathcal{C}^{0e}$ zero eigenspace and the lower infinite frequency eigensubspace of $M_{G_{12}}(s)$. Then*

$$\lim_{\epsilon \to 0} S_{12_\infty}(\epsilon) = S_{12_\infty}(0) \oplus \mathrm{Im} \begin{bmatrix} X_{12_{0e}}^{(l)} \\ 0 \\ U_{12_{0e}}^{(l)} \end{bmatrix} \tag{141}$$

Proof: Note first of all that since a solution to the H^∞ problem for $G_\epsilon(s)$ exists for some $\epsilon = \epsilon^*$, then by Lemmas 11 and 14 and $W_{12_\infty}(\epsilon, s)$ has an n-dimensional stable eigenspace on $[\epsilon^*, 0)$. The proof then follows the same argument presented in the proof of Theorem 1. The following deviations should be noted:

(1) As in the previous section the cost index used here is that which pertains to the LQ differential game Eq (73).

(2) As in [4] the "worst case" u_1 is given by $u_1^* = B_1^T\phi$ where ϕ is obtained from the solution to

$$W_{12_\infty}(0, s) \begin{pmatrix} x(s) \\ \phi(s) \\ u(s) \end{pmatrix} = \begin{pmatrix} x(0) \\ \phi(0) \\ 0 \end{pmatrix} \tag{142}$$

Note that by Lemma 7 u_1^* is identically zero on the limiting infinite eigensubspace. Moreover, using a similar argument to that given in the proof to Theorem 1, it can be shown that, for any other subspace spanned by infinite eigenvectors of higher grade, there exist initial conditions for which $J = \infty$. The pertinent trajectories lie in the subspace spanned by the infinite eigenvectors of $W_{12_\infty}(0, s)$ related to the highest grade infinite eigenvectors of $M_{G_{12}}(s)$. □

Lemma 19 *Let* $\mathcal{S}_{12_\infty}(0) = \mathrm{Im}\begin{bmatrix} X_{12_s} \\ \Phi_{12_s} \\ U_{12_s} \end{bmatrix}$ *and let* $\begin{bmatrix} X^{(u)}_{12_{oe}} \\ U^{(u)}_{12_{oe}} \end{bmatrix}$ *be a matrix whose columns form a basis for the sum of the finite* $\mathcal{C}^{0e}$ *eigenspace and the upper infinite frequency eigensubspace of* $M_{G_{12}}(s)$. *Define* $\mathcal{X} = [\; X_{12_s} \quad X^{(u)}_{12_{oe}} \;]$. *Then*

(i) Im $\mathcal{X}$ *is the space of initial conditions* x_0 *for which the cost Eq (73) is finite.*

(ii) If rank $\mathcal{X} < n$, *then no solution to the* H^∞ *problem for* $G_{12}(s)$ *exists.*

Proof: Follows the same arguments as for the proof to Lemma 18. Note that in this case, unlike the H^2 case treated by Lemma 18 the nonsingularity of $\mathcal{X}$ is not enough to guarantee that a solution to the problem exists. □

Theorem 4 *Let* $\begin{bmatrix} X_{12_s} \\ \Phi_{12_s} \\ U_{12_s} \end{bmatrix}$, $\begin{bmatrix} X^{(u)}_{12_{oe}} \\ U^{(u)}_{12_{oe}} \end{bmatrix}$ *and* $\mathcal{X}$ *be as defined in Lemma 19. Assume that* rank $\mathcal{X} = n$, *and that there exists an* $\epsilon^* > 0$ *for which there is a solution to the* H^∞ *problem for* $G_{\epsilon^*}(s)$. *Then the limiting solution* P_0 *to the Riccati equation (86) is the positive semidefinite matrix*

$$P_0 = \Phi_{12_s} H^T_{12_s} \tag{143}$$

where

$$\begin{bmatrix} H^T_{12_s} \\ H^T_{12_{0e}} \end{bmatrix} [\; X_{12_s} \quad X_{12_{0e}} \;] = I \tag{144}$$

Proof: The proof is a slight modification of the proof to Theorem 2. First note that by Lemma 19, the existence of a solution to the H^∞ problem implies that $\forall$ $x_0 \in \Re^n$ there exist vectors v, v_{oe} such that

$$x_0 = X_{12_s} v + X^{(u)}_{oe} v_{oe} \tag{145}$$

Moreover, the resulting cost is equivalent to that obtained for $x_0 = X_{12_s} v$ since the cost pertaining to the $\mathcal{C}^{0e}$ eigenspace of $W_{12_\infty}(0, s)$ is zero. In the time domain the optimal state and control trajectories corresponding to x_0 is given by $\begin{pmatrix} x(t) \\ u_2(t) \end{pmatrix} = \begin{bmatrix} X_{12_s} \\ U_{12_s} \end{bmatrix} e^{\Lambda_{12_s} t} v$ for $t > 0$, where Λ_{12_s} is strictly Hurwitz and satisfies

$$\begin{bmatrix} A & B_1 B_1^T & B_2 \\ -C_1^T C_1 & -A^T & -C_1^T D_{12} \\ D_{12}^T C_1 & B_2^T & D_{12}^T D_{12} \end{bmatrix} \begin{bmatrix} X_{12_s} \\ \Phi_{12_s} \\ U_{12_s} \end{bmatrix} = \begin{bmatrix} X_{12_s} \\ \Phi_{12_s} \\ 0 \end{bmatrix} \Lambda_{12_s} \tag{146}$$

The optimal cost is thus given by

$$\begin{aligned} J(x_0) &= \int_0^\infty \begin{pmatrix} x \\ u_2 \end{pmatrix}^T \begin{bmatrix} C_1^T C_1 & C_1^T D_{12} \\ D_{12}^T C_1 & D_{12}^T D_{12} \end{bmatrix} \begin{pmatrix} x \\ u_2 \end{pmatrix} - u_1^T u_1 \mathrm{dt} \\ &= v^T \Big(\int_0^\infty e^{\Lambda^T t} \Big(\begin{bmatrix} X_{12_s} \\ U_{12_s} \end{bmatrix}^T \begin{bmatrix} C_1^T C_1 & C_1^T D_{12} \\ D_{12}^T C_1 & D_{12}^T D_{12} \end{bmatrix} \begin{bmatrix} X_{12_s} \\ U_{12_s} \end{bmatrix} \\ &\quad - \Phi^T_{12_s} B_1 B_1^T \Phi_{12_s} \Big) e^{\Lambda t} \mathrm{dt} \Big) v \end{aligned} \tag{147}$$

Now from Eq (146) it can be shown that the bracketed term in the integrand of Eq (147) is equal to $-(X_{12_s}^T \Phi_{12_s} \Lambda_{12_s} + \Lambda_{12_s}^T X_{12_s}^T \Phi_{12_s})$. Hence we have that

$$
\begin{aligned}
J(x_0) &= -v^T\left(\int_0^\infty e^{\Lambda_{12_s}^T t}(X_{12_s}^T \Phi_{12_s}\Lambda_{12_s} + \Lambda_{12_s}^T X_{12_s}^T \Phi_{12_s})e^{\Lambda_{12_s} t}\mathrm{dt}\right)v \\
&= -v^T\left(\int_0^\infty \frac{\mathrm{d}}{\mathrm{dt}}(e^{\Lambda_{12_s}^T t}X_{12_s}^T \Phi_{12_s} e^{\Lambda_{12_s} t})\mathrm{dt}\right)v \\
&= v^T X_{12_s}^T \Phi_{12_s} v
\end{aligned} \tag{148}
$$

the last equality following from the fact that Λ_{12_s} is Hurwitz. Since $\begin{bmatrix} X_{12_s} & X_{0e}^{(u)} \end{bmatrix}$ is nonsingular, there exists a matrix $\begin{bmatrix} H_{12_s} & H_{oe} \end{bmatrix}$ such that

$$
\begin{bmatrix} H_{12_s}^T \\ H_{oe}^T \end{bmatrix} \begin{bmatrix} X_{12_s} & X_{0e}^{(u)} \end{bmatrix} = I_n. \tag{149}
$$

Left multiply Eq (145) by $H_{12_s}^T$ to obtain

$$
v = H_{12_s}^T x_0 \tag{150}
$$

and substitute in Eq (148) to get

$$
J(x_0) = x_0^T H_{12_s} X_{12_s}^T \Phi_{12_s} H_{12_s}^T x_0 \tag{151}
$$

Now we note that from Lemma 9 that $X_{oe}^T \Phi_{12_s} = 0$. So $H_{oe} X_{oe}^T \Phi_{12_s} = 0$, and hence

$$
(H_{12_s} X_{12_s}^T - I)\Phi = 0 \tag{152}
$$

the last relationship following from Eq (149). Rewrite Eq (152) as $HX^T\Phi = \Phi$ and substitute in Eq (151) to give

$$
\begin{aligned}
J &= x_o^T \Phi_{12_s} H_{12_s}^T x_0 \\
&= x_0^T P_* x_0
\end{aligned} \tag{153}
$$

Since $J(u_1^*, u_2^*, x(0)) \geq 0$ for all ϵ it follows that P_* is indeed a matrix of a positive semi-definite form. To show that it is also symmetric define $N := \Phi_{12_s}^T X_{12_s} - X_{12_s}^T \Phi_{12_s}$ and pre-multiply both sides of Eq (146) by $\begin{bmatrix} \Phi_{12_s}^T & -X_{12_s}^T & 0 \end{bmatrix}$ to obtain

$$
\begin{aligned}
N\Lambda &= \Phi_{12_s}^T A X_{12_s} + (\Phi_{12_s}^T A X_{12_s})^T + X_{12_s}^T C_1^T C_1 X_{12_s} - U_{12_s}^T D_{12}^T D_{12} U_{12_s} \\
&\quad + \Phi_{12_s}^T B_1 B_1^T \Phi_{12_s}
\end{aligned} \tag{154}
$$

Since the right-hand side of Eq (154) is symmetric and N is skew symmetric, it follows that

$$
(-\Lambda_{12_s}^T)N - N\Lambda_{12_s} = 0 \tag{155}
$$

Since Λ_{12_s} is Hurwitz, and the eigenvalues of $-\Lambda_{12_s}^T$ and Λ_{12_s} are disjoint; so by Lemma 1.5 [39] that the unique solution to the above is $N = 0$ and hence $\Phi_{12_s}^T X_{12_s} = X_{12_s}^T \Phi_{12_s}$. □

VI System Matrix Parameterizations of Well-Posed Systems

In this section, we digress momentarily to consider the following problem: Suppose we are given the polynomial matrix

$$M(\epsilon, s) = \left[\begin{array}{c|c} M_{11}(\epsilon, s) & M_{12}(\epsilon, s) \\ \hline M_{21}(\epsilon, s) & M_{22}(\epsilon, s) \end{array}\right] \tag{156}$$

which is polynomial in s and analytic in ϵ for all ϵ in some region $\mathcal{W} \subset \Re$. Define the mapping

$$G(\epsilon, s) = -M_{21}(\epsilon, s) M_{11}(\epsilon, s)^{-1} M_{12}(\epsilon, s) + M_{22}(\epsilon, s). \tag{157}$$

The problem is to determine whether $\lim_{\epsilon \to \epsilon_0} G(\epsilon, s)$ exists for some given $\epsilon_0 \in \mathcal{W}$. $M(\epsilon, s)$ can thus be considered as the Rosenbrock system matrix [35], say, of a linear system which is parameterized in terms of a scalar ϵ.

Directly related to this issue is the notion of what we term a well-posed or an i/o well-posed system matrix.

Definition 6 *The Rosenbrock system matrix*

$$P(s) = \left[\begin{array}{c|c} P_{11}(s) & P_{12}(s) \\ \hline P_{21}(s) & P_{22}(s) \end{array}\right] \tag{158}$$

is said to be **i/o well posed** *if given any $u(s)$ in the field of rational functions, there exists a unique $y(s)$ satisfying*

$$P_{11}(s)x(s) + P_{12}(s)u(s) = 0 \tag{159}$$

$$P_{21}(s)x(s) + P_{22}(s)u(s) = y(s) \tag{160}$$

for some rational vector function $x(s)$. If $x(s)$ is also unique for each $u(s)$, the system matrix will be said to be **well-posed** .

Equivalently, it is seen that $P(s)$ is i/o well-posed if and only if there exists a unique function, $T(s)$, such that $y(s) = T(s)u(s)$, where the pair $(y(s), u(s))$ satisfy Eq (159, 160), even when $P_{11}(s)$ is singular. Further thought shows that $P(s)$ is i/o well posed if and only if the following hold

(i) $\exists$ at least one solution $x(s)$ to Eq (159) for any given $u(s)$.

(ii) If $x_1(s)$ and $x_2(s)$ are solutions to Eq (159) corresponding to some $u(s)$ then

$$P_{21}(s)(x_1(s) - x_2(s)) = 0, \text{a.e.} \tag{161}$$

This is embodied in the following lemma:

Lemma 20 *The system matrix $P(s)$ is i/o well posed if and only if the following condition holds*

$$\operatorname{rank} P_{11}(s) = \operatorname{rank} \begin{bmatrix} P_{11}(s) & P_{12}(s) \end{bmatrix} \tag{162}$$

$$= \operatorname{rank} \begin{bmatrix} P_{11}(s) \\ P_{21}(s) \end{bmatrix} \tag{163}$$

for almost all $s \in \mathcal{C}$. It is well-posed if and only if Eq (162) holds and $P_{11}(s)$ is injective almost everywhere in $\mathcal{C}$.

Proof: From elementary linear algebra (see [45], for example) we know that given any $u(s)$, that Eq (162) and the injectivity of $P_{11}(s)$ are necessary and sufficient conditions for the existence of a unique solution $x(s)$ in Eq (159). Note that an equivalent expression for Eq (162) is that Im $P_{12}u(s) \subset$ Im $P_{11}(s)$ almost everywhere.

Again, we note that, by itself, Eq (162) is necessary and sufficient for the existence of at least one $x(s)$ given any $u(s)$. We also note that Eq (163) is equivalent to saying that Ker $P_{21}(s) \subset$ Ker $P_{11}(s)$ almost everywhere. We claim that this is a necessary and sufficient condition for $y(s)$ to be uniquely determined by Eq (160) for all $x(s)$ which is a solution to Eq (159) for a given $u(s)$. To see this we recall that any solution $x(s)$ to Eq (159) takes the form

$$x(s) = x_i(s) + x_k(s) \tag{164}$$

where $x_i(s) \in \text{Ker}^{\perp} P_{11}(s)$ and is therefore uniquely determined by $u(s)$ and $x_k(s) \in \text{Ker } P_{11}(s)$. The result then follows by the fact that

$$y(s) = P_{21}x_i(s) + P_{21}x_k(s) + P_{22}u(s) \tag{165}$$

and is unique, given $u(s)$ if and only if $P_{21}x_k(s) = 0$.

□

The following corollary to this Lemma provides us with a means for calculating the transfer function of an i/o well-posed system matrix.

Corollary 1 *If the conditions of Lemma 20 hold the corresponding transfer function from $u(s)$ to $y(s)$ is given by*

$$G(s) = -P_{21}P_{11}^{\dagger}P_{12}(s) + P_{22}(s) \tag{166}$$

where $P_{11}^{\dagger}(s)$ is a generalized inverse of $P_{11}(s)$ evaluated over the field of polynomial functions and defined by [44]

$$P_{11} = P_{11}P_{11}^{\dagger}P_{11}, \quad P_{11}^{\dagger} = P_{11}^{\dagger}P_{11}P_{11}^{\dagger} \tag{167}$$

Proof: Follows from Lemma 20 and elementary linear algebra considerations. □
Note that for $P_{11}(s)$ square the definition of a well-posed system matrix coincides with the orthodox definition. Note also that if $P(s)$ is well-posed then it is also i/o well-posed.

The motivation for studying this problem arises from the fact that the stochastic regulators which solve the cheap LQG problem will be characterized by system matrices which are analytic functions of the weighting parameter ϵ. In general, the exact dependence on ϵ will not be known. Since we will be trying to determine the limiting form of these regulators as $\epsilon \to 0$, it is prudent to ask, for example, if $G(\epsilon_0, s)$ exists even when the limiting system matrix $M(\epsilon_0, s)$ is not well posed.

To demonstrate this limiting behavior, consider the following example:

$$M(s) = \left[\begin{array}{cc|c} s\epsilon^2 & s\epsilon & \alpha(\epsilon) \\ s\epsilon & 1 & \beta \\ \hline \gamma & 1 & 0 \end{array}\right] \tag{168}$$

For $\epsilon \neq 0$ $M(\epsilon, s)$ is well-posed with corresponding transfer function

$$G(\epsilon, s) = -\frac{\alpha(\epsilon)(\gamma - \epsilon s) + \beta\epsilon(\epsilon - 1)s}{\epsilon^2 s(1 - s)} \tag{169}$$

For $\epsilon = 0$ the (1,1) block is singular and $M(0, s)$ may or may not be i/o well-posed. If it is well posed then the corresponding transfer function may be determined from Eq (166); in this case we have $P_{11}^{\dagger}(s) = \left[\begin{smallmatrix} 0 & 0 \\ 0 & 1 \end{smallmatrix}\right]$. Therefore, if $M(0, s)$ is i/o well-posed then the corresponding transfer function is $G(0, s) = -\beta$.

Now let us define

$$G_0(s) := \lim_{\epsilon \to 0} G(\epsilon, s) \tag{170}$$

if this limit exists. The well-posedness of this system matrix for $\epsilon \to 0$ is summarized in Table I for various values of α, β and γ. Observe, in particular, that the existence of an i/o well-posed limiting system matrix does not guarantee the existence of $G_0(s)$.

Table I: Examples of Parameterized System Matrices and their Limits

$\alpha(\epsilon)$	β	γ	i/o well posed limit?	$G(\epsilon, s)$, $\epsilon > 0$	$G(0, s)$	$G_0(s)$
ϵ	0	1	no	$\frac{\epsilon s - 1}{\epsilon s(1-s)}$	not defined	not defined
ϵ^2	0	1	no	$\frac{\epsilon s - 1}{s(1-s)}$	not defined	$\frac{-1}{s(1-s)}$
ϵ	0	0	yes	$\frac{1}{1-s}$	0	$\frac{1}{1-s}$
ϵ	1	0	yes	$\frac{s}{\epsilon s(1-s)}$	-1	not defined

We can obtain considerable insight into the problem by first considering the case where $M(\epsilon, s)$ is a polynomial matrix in ϵ. In this case the required limit exists if and only if the corresponding transfer function $G(\epsilon, s)$ has no poles at $\epsilon = \epsilon_0$. Thus the problem is seen to involve pole/zero cancellations at $\epsilon = \epsilon_0$; the required conditions must be determined from a knowledge of the relevant pole and zero directions at $\epsilon = \epsilon_0$ [34]. It is important to note that the vectors which specify these pole/zero directions will themselves be functions of ϵ.

In the case where $M(\epsilon, s)$ is analytic in ϵ, we claim that the same type of analysis can still be carried out. To see this, observe that the poles and zeros of $G(\epsilon, s)$ are still defined, at least in the scalar case, since $G(\epsilon, s)$ is a matrix valued meromorphic function[2] in ϵ and since the definition of poles and zeros for

[2] Meromorphic functions are functions which can be written as the ratio of two analytic functions. See [46][Ch. 10]for a more formal definition. Rational functions are examples of meromorphic functions.

rational scalar functions are, in fact, directly derived from the relevant definition for scalar meromorphic functions. For matrices, the zeros of $G(\epsilon, s)$ can be defined as the values of ϵ for which $M(\epsilon, s)$ loses full (column/row) rank. The poles will, as usual, be defined to be the zeros of $M_{11}(\epsilon, s)$. It is therefore seen that the definitions of poles and zeros for matrix valued meromorphic functions are natural extensions of the usual definitions for rational transfer function matrices (defined on $\mathcal{C}$) except that now, there will, in general be an infinite number of poles and zeros to be accounted for.

If we now apply the pole/zero cancellation arguments given above, it is seen that the required vectors will be analytic functions of ϵ and will be difficult to determine even when the elements of $M(\epsilon, s)$ are explicitly known. This will not be the case in the study we will soon undertake. Presumably, the required condition can also be determined by deriving a sufficient number of terms in the power series expansion of $M(\epsilon, s)$ about $\epsilon = \epsilon_0$; however, this may also require an explicit knowledge of $M(\epsilon, s)$.

In lieu of a complete description of the solution to the subproblem defined in this section, we state the following lemmas which give sufficient conditions pertaining to the existence of the required limit. Both follow from the discussions given above; the first states that the limit exists if $G(\epsilon, s)$ has no poles at ϵ_0. The second states that the limit cannot exist if $G(\epsilon, s)$ has at least one pole, but no zero at $\epsilon = \epsilon_0$.

Lemma 21 *If it holds that* $\det M_{11}(\epsilon_0, s) \neq 0$, *then* $\lim_{\epsilon \to \epsilon_0} G(\epsilon, s) = G(\epsilon_0, s)$ *exists.*

Lemma 22 *If it holds that* $M_{11}(\epsilon_0, s)$ *is singular but* $M(\epsilon, s)$ *has full column or row rank at* $\epsilon = \epsilon_0$ *then* $\lim_{\epsilon \to \epsilon_0} G(\epsilon, s)$ *does not exist.*

VII Singular Measurement Feedback Problems

We now consider the behavior of the solution to the measurement feedback H^p problem for Eq (83) as $\epsilon \to 0$. In addition to the matrix pencil $M_{G_{12}}$ defined in Eq (109), we shall utilize the system matrix

$$M_{G_{21}}(s) \quad := \quad \begin{bmatrix} -sI + A & B_1 \\ C_2 & D_{21} \end{bmatrix} \tag{171}$$

The left and right $\mathcal{C}^{0e}$ eigenspaces of $M_{G_{12}}$ and $M_{G_{21}}$ respectively play a key role in the determination of the limiting solution to the H^p problem as $\epsilon \to 0$. We therefore also define the following matrices

$\begin{bmatrix} X_{12_\infty}^{(l)} \\ U_{12_\infty}^{(l)} \end{bmatrix}$ and $\begin{bmatrix} X_{21_\infty}^{(l)} \\ U_{21_\infty}^{(l)} \end{bmatrix}$, whose columns form bases for the lower infinite zero eigensubspaces of $M_{G_{12}}(s)$ and $M_{G_{21}}^T(s)$ respectively

$\begin{bmatrix} X_{12_\infty}^{(u)} \\ U_{12_\infty}^{(u)} \end{bmatrix}$ and $\begin{bmatrix} X_{21_\infty}^{(u)} \\ U_{21_\infty}^{(u)} \end{bmatrix}$ whose columns form bases for the upper infinite zero eigensubspaces of the respective pencils $M_{G_{12}}(s)$ and $M_{G_{21}}^T(s)$

$\begin{bmatrix} X_{12_j} \\ U_{12_j} \end{bmatrix}$ and $\begin{bmatrix} X_{21_j} \\ U_{21_j} \end{bmatrix}$ whose columns form bases for the finite $\mathcal{C}^{0e}$ eigenspaces of $M_{G_{12}}(s)$ and $M_{G_{21}}^T(s)$ respectively, corresponding to the eigenvalues of some matrices Λ_{12_j} and Λ_{21_j} i. e. (see Eq (7))

$$M_{G_{12}}(s) \begin{bmatrix} X_{12_j} \\ U_{12_j} \end{bmatrix} = \begin{bmatrix} X_{12_j} \\ 0 \end{bmatrix} (-sI + \Lambda_{12_j}) \tag{172}$$

and

$$M_{G_{21}}(s) \begin{bmatrix} X_{21_j} \\ U_{21_j} \end{bmatrix} = \begin{bmatrix} X_{21_j} \\ 0 \end{bmatrix} (-sI + \Lambda_{21_j}) \tag{173}$$

As before, we begin with the H^2 problem.

A The H^2 Problem

The H^2 problem is equivalent to the LQG problem of determining the limiting form of the stochastic regulator which minimizes

$$J(\epsilon) = \mathrm{E}\,(y_1^T y_1) \tag{174}$$

as $\epsilon \to 0$. For $\epsilon > 0$ $\tilde{D}_{21}^T(\epsilon)$ and $\tilde{D}_{12}(\epsilon)$ are injective (see eqn. 84) and the standard result (see [39]) yields an n-th order compensator

$$F_\epsilon(s) = -K_{12}(\epsilon)(sI - A + B_2 K_{12}(\epsilon) + K_{21}(\epsilon)C_2)^{-1} K_{21}(\epsilon) \tag{175}$$

where

$$K_{12}(\epsilon) := (\tilde{D}_{12}(\epsilon)^T \tilde{D}_{12}(\epsilon))^{-1}(B_2^T P_{12}(\epsilon) + \tilde{D}_{12}^T \tilde{C}_1) \tag{176}$$

$$K_{21}(\epsilon) := (P_{21}(\epsilon)C_2^T + \tilde{B}_1 \tilde{D}_{21}^T(\epsilon))(\tilde{D}_{21}(\epsilon)\tilde{D}_{21}(\epsilon)^T)^{-1} \tag{177}$$

and $P_{12}(\epsilon)$, $P_{21}(\epsilon)$ are the unique positive definite stabilizing solutions to (we drop the ϵ arguments for brevity)

$$A^T P_{12} + P_{12} A + C_1^T C_1 - K_{12}^T \tilde{D}_{12}^T \tilde{D}_{12} K_{12} = 0 \tag{178}$$

$$P_{21} A^T + A P_{21} + B_1 B_1^T - K_{21} \tilde{D}_{21} \tilde{D}_{21}^T K_{21}^T = 0 \tag{179}$$

To set up the appropriate eigenproblems we define, in addition to the pencil $W_{12_2}(\epsilon, s)$ in Eq (107), the matrix pencil

$$W_{21_2}(\epsilon, s) := \begin{bmatrix} -sI + A^T & 0 & C_2^T \\ -\tilde{B}_1 \tilde{B}_1^T & -sI - A & -\tilde{B}_1 \tilde{D}_{21}^T(\epsilon) \\ \tilde{D}_{21}(\epsilon)\tilde{B}_1^T & C_2 & \tilde{D}_{21}(\epsilon)\tilde{D}_{21}^T(\epsilon) \end{bmatrix} \tag{180}$$

$$= \begin{bmatrix} -sI + A^T & 0 & C_2^T \\ -B_1 B_1^T & -sI - A & -B_1 D_{21}^T \\ D_{21} B_1^T & C_2 & D_{21} D_{21}^T + \epsilon^2 I_{m_2} \end{bmatrix} \tag{181}$$

Now let the columns of $\begin{bmatrix} X_{12_s} \\ \Phi_{12_s} \\ U_{12_s} \end{bmatrix}$ and $\begin{bmatrix} X_{21_s} \\ \Phi_{21_s} \\ U_{21_s} \end{bmatrix}$ form bases for the stable subspaces of $W_{12_2}(0, s)$ and $W_{21_2}(0, s)$ respectively, corresponding to the matrices $\Lambda_f = \Lambda_{12_s}$ and $\Lambda_f = \Lambda_{21_s}$ in Eq (7).

Finally define

$$\begin{bmatrix} X_{12} \\ \Phi_{12} \\ U_{12} \end{bmatrix} = \begin{bmatrix} X_{12_s} & X_{12_j} & X_{12_\infty}^{(l)} \\ \Phi_{12_s} & 0 & 0 \\ U_{12_s} & U_{12_j} & U_{12_\infty}^{(l)} \end{bmatrix} \tag{182}$$

$$\begin{bmatrix} X_{21} \\ \Phi_{21} \\ U_{21} \end{bmatrix} = \begin{bmatrix} X_{21_s} & X_{21_j} & X_{21_\infty}^{(l)} \\ \Phi_{21_s} & 0 & 0 \\ U_{21_s} & U_{21_j} & U_{21_\infty}^{(l)} \end{bmatrix} \tag{183}$$

and note that for all $\epsilon > 0$ the controller $F_\epsilon(s)$ has system matrix representation

$$P_{F_\epsilon}(s) \stackrel{s}{=} \left[\begin{array}{c|c} X_{21}^T(\epsilon)(-sI + A)X_{12}(\epsilon) + X_{21}^T(\epsilon)B_2U_{12}(\epsilon) + U_{21}^T(\epsilon)C_2X_{12}(\epsilon) & U_{21}^T(\epsilon) \\ \hline -U_{12}(\epsilon) & 0 \end{array}\right] \tag{184}$$

where the columns of $\Xi_{12}(\epsilon) := \begin{bmatrix} X_{12}(\epsilon) \\ \Lambda_{12}(\epsilon) \\ U_{12}(\epsilon) \end{bmatrix}$ and $\Xi_{21}(\epsilon) := \begin{bmatrix} X_{21}(\epsilon) \\ \Lambda_{21}(\epsilon) \\ U_{21}(\epsilon) \end{bmatrix}$ form bases for the respective stable eigenspaces of $W_{12_2}(\epsilon, s)$ and $W_{21_2}(\epsilon, s)$, i. e. $\Xi_{12}(\epsilon)$ and $\Xi_{21}(\epsilon)$ satisfy Eq (7) with $\Lambda_f = \Lambda_{12}(\epsilon)$ and $\Lambda_f = \Lambda_{21}(\epsilon)$ respectively where $\Lambda_{12}(\epsilon)$ and $\Lambda_{21}(\epsilon)$ are asymptotically stable matrices $\forall \epsilon > 0$.

The limiting form of the required compensator is given by the following

Theorem 5 *Suppose that* $\exists \epsilon > 0$ *for which a solution to the LQG problem Eq (174) exists.*

Then, as $\epsilon \to 0$ *the following hold:*

(i) The limiting value of the controller system matrix is

$$P_F(s) := \left[\begin{array}{c|c} X_{21}^T(-sI + A)X_{12} + X_{21}^TB_2U_{12} + U_{21}^TC_2X_{12} & U_{21}^T \\ \hline -U_{12} & 0 \end{array}\right]. \tag{185}$$

(ii) The limiting closed loop transfer function is asymptotically stable and is given by

$$T_{y_1u_1}(s) \stackrel{s}{=} \left[\begin{array}{cc|c} -sI + \Lambda_{12_s} & H_{12_s}^T(H_{21_s}\Lambda_{21_s}^T - AH_{21_s}) & H_{12_s}^T B_* \\ 0 & (-sI + \Lambda_{21_s})^T & -B_{1*} \\ \hline C_1X_{12_s} + D_{12}U_{12_s} & -C_1H_{21_s} & 0 \end{array}\right] \tag{186}$$

where

$$B_* := B_1 - H_{21_s}^T B_{1*} \tag{187}$$

$$B_{1*} := X_{21_s}^T B_1 + U_{21_s}^T D_{21} \tag{188}$$

$$H_{12}^T := \begin{bmatrix} H_{12_s}^T \\ H_{12_j}^T \\ H_{12_\infty}^T \end{bmatrix} = \begin{bmatrix} X_{12_s} & X_{12_j} & X_{12_\infty}^{(u)} \end{bmatrix}^{-1} \tag{189}$$

$$H_{21}^T := \begin{bmatrix} H_{21_s}^T \\ H_{21_j}^T \\ H_{21_\infty}^T \end{bmatrix} = \begin{bmatrix} X_{21_s} & X_{21_j} & X_{21_\infty}^{(u)} \end{bmatrix}^{-1}. \tag{190}$$

(iii) The limiting cost Eq (174) is given by

$$J_* := \lim_{\epsilon \to 0} J(\epsilon) = \text{Tr}(\Pi_*) \tag{191}$$

where

$$\Pi_* = P_{c*}B_*B_*^T + C_1^T C_1 P_{f*} \tag{192}$$

$$P_{c*} = \Phi_{12_s} H_{12_s}^T \tag{193}$$

$$P_{f*} = H_{21_s} \Phi_{21_s}^T. \tag{194}$$

(iv) If the system matrix $P_F(s)$ is well-posed, then $\lim_{\epsilon \to 0} F_\epsilon(s)$ exists and is given by

$$F(s) \stackrel{s}{=} P_F(s) \tag{195}$$

and this value of $F(s)$ achieves the limiting transfer function $Ty_1u_1(s)$ and cost J_*.

Proof: It is easy to show that for $\epsilon > 0$ the stochastic regulator is given by

$$F_\epsilon(s) \stackrel{s}{=} P_{F_\epsilon}(s) \tag{196}$$

The fact that $P_{F_\epsilon}(s) \to P_F(s)$ follows from Theorem 1 and the analyticity of these matrices in ϵ. For part (ii) assume that ϵ^* is small enough so that those stable eigenvalues of $W_{12_2}(\epsilon, s)$ which approach locations at infinity, the finite $j\omega$-axis and the open LHP are all in disjoint groups on $[0, \epsilon^*]$. For $\epsilon > 0$ we know that the closed loop transfer function has Rosenbrock system matrix representation

$$P_{T_{y_1u_{1_\epsilon}}}(s) = \left[\begin{array}{cc|c} -sI + A - B_2K_{12} & -B_2K_{12} & B_1 \\ 0 & -sI + A - K_{21}C_2 & K_{21}D_{21} - B_1 \\ \hline C_1 - D_{12}K_{12} & -D_{12}K_{12} & 0 \end{array}\right] \tag{197}$$

where $K_{12}(\epsilon)$, $K_{21}(\epsilon)$ are given in Eq (176,177). Perform the following equivalence transformations:

column 2=column 2-column 1
column 1= column 1 $\times X_{12}$
row 2= $X_{21}^T \times$ row 2.

and use the eigenvector relationships for $W_{12_2}(\epsilon, s)$ and $W_{21_2}(\epsilon, s)$ in Eq (7) to obtain the descriptor form

$$P_{T_{y_1u_{1_\epsilon}}}(s) = \left[\begin{array}{cc|c} X_{12}(\epsilon)(-sI + \Lambda_{12}(\epsilon)) & sI - A & B_1 \\ 0 & (-sI + \Lambda_{21}(\epsilon))^T X_{21}^T(\epsilon) & -B_{1\epsilon} \\ \hline C_1X_{12}(\epsilon) + D_{12}(\epsilon)U_{12}(\epsilon) & -C_1 & 0 \end{array}\right] \tag{198}$$

where

$$B_{1\epsilon} := U_{21}^T(\epsilon)D_{21}(\epsilon) + X_{21}^T(\epsilon)B_1 \tag{199}$$

For ϵ small enough we can partition $\Xi_{12}(\epsilon)$ and $\Xi_{21}(\epsilon)$ as follows:

$$\Xi_{12}(\epsilon) := \begin{bmatrix} X_{12}(\epsilon) \\ \Lambda_{12}(\epsilon) \\ U_{12}(\epsilon) \end{bmatrix} = \left[\begin{array}{c|c|c} X_{12_s}(\epsilon) & X_{12_j}(\epsilon) & X_{12_\infty}(\epsilon) \\ \Phi_{12_s}(\epsilon) & \Phi_{12_j}(\epsilon) & \Phi_{12_\infty}(\epsilon) \\ U_{12_s}(\epsilon) & U_{12_j}(\epsilon) & U_{12_\infty}(\epsilon) \end{array}\right] \tag{200}$$

$$\Xi_{21}(\epsilon) := \begin{bmatrix} X_{21}(\epsilon) \\ \Lambda_{21}(\epsilon) \\ U_{21}(\epsilon) \end{bmatrix} = \left[\begin{array}{c|c|c} X_{21_s}(\epsilon) & X_{21_j}(\epsilon) & X_{21_\infty}(\epsilon) \\ \Phi_{21_s}(\epsilon) & \Phi_{21_j}(\epsilon) & \Phi_{21_\infty}(\epsilon) \\ U_{21_s}(\epsilon) & U_{21_j}(\epsilon) & U_{21_\infty}(\epsilon) \end{array}\right] \tag{201}$$

where

$$\lim_{\epsilon\to 0}\left[\begin{array}{c|c|c} X_{12_s}(\epsilon) & X_{12_j}(\epsilon) & X_{12_\infty}(\epsilon) \\ \Phi_{12_s}(\epsilon) & \Phi_{12_j}(\epsilon) & \Phi_{12_\infty}(\epsilon) \\ U_{12_s}(\epsilon) & U_{12_j}(\epsilon) & U_{12_\infty}(\epsilon) \end{array}\right] = \left[\begin{array}{c|c|c} X_{12_s} & X_{12_j} & X_{12_\infty}^{(l)} \\ \Phi_{12_s} & 0 & 0 \\ U_{12_s} & U_{12_j} & U_{12_\infty}^{(l)} \end{array}\right] \tag{202}$$

and

$$\lim_{\epsilon\to 0}\left[\begin{array}{c|c|c} X_{21_s}(\epsilon) & X_{21_j}(\epsilon) & X_{21_\infty}(\epsilon) \\ \Phi_{21_s}(\epsilon) & \Phi_{21_j}(\epsilon) & \Phi_{21_\infty}(\epsilon) \\ U_{21_s}(\epsilon) & U_{21_j}(\epsilon) & U_{21_\infty}(\epsilon) \end{array}\right] = \left[\begin{array}{c|c|c} X_{21_s} & X_{21_j} & X_{21_\infty}^{(l)} \\ \Phi_{21_s} & 0 & 0 \\ U_{21_s} & U_{21_j} & U_{21_\infty}^{(l)} \end{array}\right] \tag{203}$$

Associated with those zeros of interest which remain finite as $\epsilon \to 0$ are the matrices $\Lambda_{12_s}(\epsilon)$, $\Lambda_{12_j}(\epsilon)$ whose eigenvalues correspond to the eigenvalues of $W(\epsilon, s)$ on each eigensubspace in the closed left half plane; similarly, $\Lambda_{12_s}(\epsilon)$, $\Lambda_{12_j}(\epsilon)$ are matrices whose eigenvalues correspond to those of $W_{21_2}(\epsilon, s)$ on the pertinent eigensubspaces. Moreover, we have that

$$\lim_{\epsilon\to 0}\Lambda_{12_s}(\epsilon) = \Lambda_{12_s}, \ \lim_{\epsilon\to 0}\Lambda_{12_j}(\epsilon) = \Lambda_{12_j} \tag{204}$$

and

$$\lim_{\epsilon\to 0}\Lambda_{21_s}(\epsilon) = \Lambda_{21_s}, \ \lim_{\epsilon\to 0}\Lambda_{21_j}(\epsilon) = \Lambda_{21_j} \tag{205}$$

Substituting Eq (200–201) into Eq (197) and using the equations which define each eigensubspace, we obtain $P_{T_{y_1 u_{1\epsilon}}}(s)$ which is listed in Table II.a. Now from Lemmas 3 and 4 we have that there exist invertible matrices T_{12} and T_{21} such that

$$\begin{bmatrix} A & B_2 \\ C_1 & D_{12} \end{bmatrix}\begin{bmatrix} X_{12_\infty}^{(l)} \\ U_{12_\infty}^{(l)} \end{bmatrix} T_{12} = \begin{bmatrix} X_{12_\infty}^{(u)} \\ 0 \end{bmatrix} \tag{206}$$

$$\begin{bmatrix} A^T & C_{21}^T \\ B_1^T & D_{21} \end{bmatrix}\begin{bmatrix} X_{21_\infty}^{(l)} \\ U_{21_\infty}^{(l)} \end{bmatrix} T_{21} = \begin{bmatrix} X_{21_\infty}^{(u)} \\ 0 \end{bmatrix} \tag{207}$$

Now multiply column 3 on the right by T_{12}, row 4 on the left by T_{21}^T, row 1 on the left by H_{12}^T and column 4 on the right by H_{21} defined in the theorem and take the limit as $\epsilon \to 0$ to obtain $P_{T_{y_1 u_1}}(s)$ in Table II.b.

Here we have used Eq (189, 190). Clearly the (1,1) block in $P_{T_{y_1 u_1}}(s)$ is regular, and therefore by Lemma 21 that $T_{y_1 u_1}(s) := \lim_{\epsilon\to 0} T_{y_1 u_{1\epsilon}}(s)$ exists. Observe that this also follows from the fact that $\forall \epsilon > 0$

Table II: System Matrices for $T_{y_1u_1}$

(a)

$$P_{T_{y_1u_{1_\epsilon}}}(s) =$$

$$\left[\begin{array}{cccc|c} X_{12_s}(\epsilon)(-sI+\Lambda_{12_s}(\epsilon)) & X_{12_j}(\epsilon)(-sI+\Lambda_{12_j}(\epsilon)) & (-sI+A)X_{12_\infty}(\epsilon)+B_2U_{12_\infty}(\epsilon) & sI-A & B_1 \\ 0 & 0 & 0 & (-sI+\Lambda_{21_s}(\epsilon))^T X_{21_s}^T(\epsilon) & -(U_{21_s}^T(\epsilon)D_{21}+X_{21_s}^T(\epsilon)B_1) \\ 0 & 0 & 0 & (-sI+\Lambda_{21_j}(\epsilon))^T X_{21_j}^T(\epsilon) & -(U_{21_j}^T(\epsilon)D_{21}+X_{21_j}^T(\epsilon)B_1) \\ 0 & 0 & 0 & X_{21_\infty}^T(\epsilon)(-sI+A)+U_{21_\infty}^T C_{21}^T & -(U_{21_\infty}^T(\epsilon)D_{21}+X_{21_\infty}^T(\epsilon)B_1) \\ \hline C_1X_{12_s}(\epsilon)+D_{12}U_{12_s}(\epsilon) & C_1X_{12_j}(\epsilon)+D_{12}U_{12_j}(\epsilon) & (C_1X_{12_\infty}(\epsilon)+D_{12}U_{12_\infty}(\epsilon)) & -C_1 & 0 \end{array}\right]$$

(b)

$$P_{T_{y_1u_1}}(s) := \lim_{\epsilon\to 0} P_{T_{y_1u_{1_\epsilon}}}(s) =$$

$$\left[\begin{array}{cccccc|c} -sI+\Lambda_{12_s} & 0 & 0 & * & * & * & H_{12_s}^T B_1 \\ 0 & -sI+\Lambda_{12_j} & 0 & * & * & * & H_{12_j}^T B_1 \\ 0 & 0 & -sH_{12_\infty}^T X_{12_\infty}+I & * & * & * & H_{12_\infty}^T B_1 \\ 0 & 0 & 0 & (-sI+\Lambda_{21_s})^T & 0 & 0 & -(U_{21_s}^T D_{21}+X_{21_s}^T B_1) \\ 0 & 0 & 0 & 0 & (-sI+\Lambda_{21_j})^T & 0 & 0 \\ 0 & 0 & 0 & 0 & 0 & -sX_{21_\infty}^T H_{21_\infty}+I & 0 \\ \hline C_1X_{12_s}+D_{12}U_{12_s} & 0 & 0 & -C_1H_{21_s}^T & -C_1H_{21_j}^T & -C_1H_{21_\infty}^T & 0 \end{array}\right]$$

By Lemma 14 $T_{y_1 u_{1_\epsilon}}(s)$ is an RH^2 function with monotonically decreasing norm. It is readily ascertained that the $\mathcal{C}^{0e}$ blocks in $P_{T_{y_1 u_1}}(s)$ are either unobservable or uncontrollable. This gives

$$T_{y_1 u_1}(s) \stackrel{s}{=} \left[\begin{array}{cc|c} -sI + \Lambda_{12_s} & H_{12_s}^T (sI - A) H_{21_s} & H_{12_s}^T B_1 \\ 0 & (-sI + \Lambda_{21_s})^T & -(U_{21_s}^T D_{21} + X_{21_s}^T B_1) \\ \hline C_1 X_{12_s} & -C_1 H_{21_s} & 0 \end{array}\right] \tag{208}$$

Now subtract $(H_{12_s} H_{21_s}^T)$ times row 2 from row 1 to obtain the result. To prove (iii) it suffices to note that the limit exists since, by Lemma 14, $\|T_{\tilde{y}_1 \tilde{u}_{1_\epsilon}}\|_2$ is a positive semidefinite, monotone decreasing function of ϵ. The cost may be obtained directly from Eq (186); the reader is referred to Subsection C for the details.

Finally for (iv) note that the existence of the limit in Eq (195) is guaranteed by Lemma 21. The rest follows from the fact that

$$T_{y_1 u_1}(s) = \mathcal{F}_l(G(s), F(s)) \tag{209}$$

The details are worked out in Subsection D

□

Comment: Observe that, as expected, the $j\omega$-axis zeros are cancelled in the transfer function $T_{y_1 u_1}$; hence, the limiting feedback control law is not internally stabilizing when there are (finite) $j\omega$-axis zeros.

Comment: (i) The structure in Eq (185) automatically compensates for rank deficiencies in D_{21} <u>and</u> D_{12}. The controller Eq (185) therefore incorporates the usual minimal order observer and/or dual observer structures [21, 20]. This issue will shortly be discussed in more detail.

(ii) Since $\text{rank}\ (X_{21}^T X_{12}) \leq n$ it follows that the compensator $F(s)$ may be improper; since the rank deficiency in X_{21}, X_{12} is due solely to the existence of infinite zeros in G_{12}, G_{21} this confirms the fact that $F(s)$ can only be improper if D_{12} and/or D_{21}^T are column rank deficient (see example in Section IX). The compensator order δ satisfies

$$\delta \leq \min\{\text{rank}\ [X_{21_s}\ X_{21_\infty}],\ \text{rank}\ [X_{12_s}\ X_{12_\infty}]\} \leq n$$

We complete this subsection by showing that Eq (185) embodies a minimal order observer structure [39, 20] under the following conditions [21]:

A4 $D_{21}(0) = 0$.

A5 $D_{12}(0)$ is injective.

A6 $C_2 B_1$ is surjective.

We state the result as a corollary to Theorem 5.

Corollary 2 *Suppose that assumptions A4-A6 hold. Then the limiting controller exists and has $(n - m_2)$-th order state-space realization*

$$F(s) \stackrel{s}{=} \left[\begin{array}{c|c} -sI + X_{21_s}^{\dagger} A_k (I - X_{21_s}^{\perp} (C_2 X_{21_s}^{\perp})^{-1} C_2) X_{21_s} & -X_{21_s}^{\dagger} A_k X_{21_s}^{\perp} (C_2 X_{21_s}^{\perp})^{-1}) \\ \hline K_{12} (I - X_{21_s}^{\perp} (C_2 X_{21_s}^{\perp})^{-1} C_2) X_{21_s} & -K_{12} X_{21_s}^{\perp} (C_2 X_{21_s}^{\perp})^{-1} \end{array}\right] \tag{210}$$

where the columns of $X_{21_s}^{\perp}$ *form a basis for* $\mathrm{Im}\,^{\perp} X_{21_s}$ *and*

$$\begin{aligned} X_{21_s}^{\dagger} &:= (X_{21_s}^T X_{21_s})^{-1} X_{21_s}^T && (211)\\ A_k &:= A - B_2 K_{12} && (212)\\ K_{12} &= -U_{12} X_{12}^{-1} && (213) \end{aligned}$$

Proof: Assumption A6 implies that all the infinite zeros of $C_2(sI - A)^{-1}B_1 + D_{21}(0)$ are of order 1. By the usual recursions, we see that $M_{G_{21}}^T(s)$ has m_2 chains at infinity of length 2; in fact the columns of the matrix $\begin{bmatrix} 0 & C_2^T \\ I_{m_2} & 0 \end{bmatrix}$ form a minimal basis for the lower and upper the infinite frequency eigenspaces, respectively, of $M_{G_{21}}^T(s)$. Thus, $X_{21_\infty}^{(l)} = 0$ and $U_{21_\infty}^{(l)} = I_{m_2}$. It can be shown that, as a consequence of Assumption A5, that $W_{12_2}(0, s)$ has an n-dimensional stable eigenspace and so we obtain from Theorem 5

$$F(s) = \left[\begin{array}{c|c} X_{21_s}^T(-sI + A_k)X_{12_s} + U_{21_s}^T C_2 X_{12_s} & U_{21_s}^T \\ C_2 X_{12_s} & I_{m_2} \\ \hline -U_{12} & 0 \end{array}\right] \tag{214}$$

If X_{12_s} is of rank n then there exists an $n \times n$ matrix T such that $[X_{21_s}\ X_{21_s}^{\perp}] = X_{12_s}T$, where $X_{21_s}^{\perp} \in \Re^{n \times m_2}$, $\mathrm{Im}\ X_{21_s}^{\perp} = \mathrm{Im}\,^{\perp} X_{21_s}$. Multiply the first column of Eq (214) by T and partition to get

$$\left[\begin{array}{cc|c} X_{21_s}^T(-sI + A_k)X_{21_s} + U_{21_s}^T C_2 X_{21_s} & X_{21_s}^T A X_{21_s}^{\perp} + U_{21_s}^T C_2 X_{21_s}^{\perp} & U_{21_s}^T \\ C_2 X_{21_s} & C_2 X_{21_s}^{\perp} & I_{m_2} \\ \hline K_{12} X_{21_s} & K_{12} X_{21_s}^{\perp} & 0 \end{array}\right] \tag{215}$$

Subtract $U_{21_s}^T \times$ row 2 from row 1 to get

$$\left[\begin{array}{cc|c} X_{21_s}^T(-sI + A_k)X_{21_s} & X_{21_s}^T A_k X_{21_s}^{\perp} & 0 \\ C_2 X_{21_s} & C_2 X_{21_s}^{\perp} & I_{m_2} \\ \hline K_{12} X_{21_s} & K_{12} X_{21_s}^{\perp} & 0 \end{array}\right] \tag{216}$$

Now note first of all, that X_{21_s} is of column dimension $n - m_2$. Moreover, by Lemma 18, the existence of a solution $\Rightarrow [X_{21_s}\ C_2^T]$ is nonsingular $\Rightarrow C_2 X_{21_s}^{\perp}$ is also a nonsingular matrix. Thus Eq (214) has an $(n - m_2)$-th order non-strictly proper state-space realization Eq (210). □

Comment: By a similar argument, it can be shown that $F(s)$ has a reduced order dual-observer structure [20] for $D_{21}(0)$ surjective, $D_{12}(0)$ column rank deficient and $C_1 B_2$ injective.

Comment: The assumption on $D_{12}(0)$ is simplified for ease of exposition; however, the above arguments lead to a similar result when $D_{12}(0)$ does not satisfy A4 but is row rank deficient with all other assumptions remaining unchanged.

B The H^∞ Problem

The analysis for the H^∞ problem is similar to that performed for the H^2 problem; the main difference is in the increased complexity of the resulting equations. For the H^∞ problem, we set up the appropriate eigenproblem, by defining the matrix pencil $W_{12_\infty}(\epsilon, s)$ as before Eq (140) and

$$W_{21}(\epsilon, s) := \begin{bmatrix} -sI + A^T & \tilde{C}_1^T \tilde{C}_1 & C_2^T \\ -\tilde{B}_1 \tilde{B}_1^T & -sI - A & -\tilde{B}_1 \tilde{D}_{21}^T \\ \tilde{D}_{21} \tilde{B}_1^T & C_2 & \tilde{D}_{21} \tilde{D}_{21}^T \end{bmatrix} \tag{217}$$

$$= \begin{bmatrix} -sI + A^T & C_1 C_1^T & C_2^T \\ -B_1 B_1^T & -sI - A & -B_1 D_{21}^T \\ D_{21} B_1^T & C_2 & D_{21} D_{21}^T + \epsilon^2 I_{m_2} \end{bmatrix} \tag{218}$$

Now let the columns of $\begin{bmatrix} X_{12_s} \\ \Phi_{12_s} \\ U_{12_s} \end{bmatrix}$ and $\begin{bmatrix} X_{21_s} \\ \Phi_{21_s} \\ U_{21_s} \end{bmatrix}$ form bases for the stable subspaces of $W_{12_\infty}(0, s)$ and $W_{21}(0, s)$ respectively, corresponding to the eigenvalues of the matrices Λ_{12_s} and Λ_{21_s} i. e.

$$W_{12_\infty}(0, s) \begin{bmatrix} X_{12_s} \\ \Phi_{12_s} \\ U_{12_s} \end{bmatrix} = \begin{bmatrix} X_{12_s} \\ \Phi_{12_s} \\ 0 \end{bmatrix} (-sI + \Lambda_{12_s}) \tag{219}$$

and

$$W_{21}(0, s) \begin{bmatrix} X_{21_s} \\ \Phi_{21_s} \\ U_{21_s} \end{bmatrix} = \begin{bmatrix} X_{21_s} \\ \Phi_{21_s} \\ 0 \end{bmatrix} (-sI + \Lambda_{21_s}) \tag{220}$$

Finally define

$$\Xi_{12} := \begin{bmatrix} X_{12} \\ \Phi_{12} \\ U_{12} \end{bmatrix} = \begin{bmatrix} X_{12_s} & X_{12_j} & X_{12_\infty}^{(l)} \\ \Phi_{12_s} & 0 & 0 \\ U_{12_s} & U_{12_j} & U_{12_\infty}^{(l)} \end{bmatrix} \tag{221}$$

$$\Xi_{21} := \begin{bmatrix} X_{21} \\ \Phi_{21} \\ U_{21} \end{bmatrix} = \begin{bmatrix} X_{21_s} & X_{21_j} & X_{21_\infty}^{(l)} \\ \Phi_{21_s} & 0 & 0 \\ U_{21_s} & U_{21_j} & U_{21_\infty}^{(l)} \end{bmatrix} \tag{222}$$

and note that for all $\epsilon > 0$ the central controller $K_0(\epsilon, s)$ has system matrix representation

$$M_{K_0}(\epsilon, s) = \left[\begin{array}{c|c} \Xi_{21}^T(\epsilon) \begin{bmatrix} -sI + A & B_1 B_1^T & B_2 \\ C_1^T C_1 & sI + A^T & \tilde{C}_1^T \tilde{D}_{12}(\epsilon) \\ C_2 & \tilde{D}_{21}(\epsilon) \tilde{B}_1^T & 0 \end{bmatrix} \Xi_{12}(\epsilon) & U_{21}^T(\epsilon) \\ \hline -U_{12}(\epsilon) & 0 \end{array} \right] \tag{223}$$

where the columns of $\Xi_{12}(\epsilon) := \begin{bmatrix} X_{21}(\epsilon) \\ \Lambda_{21}(\epsilon) \\ U_{21}(\epsilon) \end{bmatrix}$ and $\Xi_{21}(\epsilon) := \begin{bmatrix} X_{12}(\epsilon) \\ \Lambda_{12}(\epsilon) \\ U_{12}(\epsilon) \end{bmatrix}$ form bases for the respective stable eigenspaces of $W_{21}(\epsilon, s)$ and $W_{12_\infty}(\epsilon, s)$, i. e.

$$W_{12_\infty}(\epsilon, s) \begin{bmatrix} X_{12}(\epsilon) \\ \Phi_{12}(\epsilon) \\ U_{12}(\epsilon) \end{bmatrix} = \begin{bmatrix} X_{12}(\epsilon) \\ \Phi_{12}(\epsilon) \\ 0 \end{bmatrix} (-sI + \Lambda_{12}(\epsilon)) \tag{224}$$

and

$$W_{21}(\epsilon, s) \begin{bmatrix} X_{21}(\epsilon) \\ \Phi_{21}(\epsilon) \\ U_{21}(\epsilon) \end{bmatrix} = \begin{bmatrix} X_{21}(\epsilon) \\ \Phi_{21}(\epsilon) \\ 0 \end{bmatrix} (-sI + \Lambda_{21}(\epsilon)) \tag{225}$$

where $\Lambda_{12}(\epsilon)$ and $\Lambda_{21}(\epsilon)$ are asymptotically stable matrices $\forall \epsilon > 0$.

The limiting form of the required compensator is given by the following

Theorem 6 *Suppose that $\exists \epsilon = \epsilon^* > 0$ for which a solution to the H^∞ problem for Eq (83) exists.*

Then, for $\epsilon \to 0$ the following hold:

(i) The limiting value of the controller system matrix is given by

$$M_{K_0}(s) := \left[\begin{array}{c|c} \Xi_{21}^T \begin{bmatrix} -sI + A & B_1 B_1^T & B_2 \\ C_1^T C_1 & sI + A^T & C_1^T D_{12} \\ C_2 & D_{21} B_1^T & 0 \end{bmatrix} \Xi_{12} & U_{21}^T \\ \hline -U_{12} & 0 \end{array}\right]. \tag{226}$$

(ii) The limiting closed loop transfer function, $T_{y_1 u_1}(s)$, is asymptotically stable with $\|T_{y_1 u_1}\| < 1$ and is given by

$$T_{y_1 u_1}(s) = \mathcal{F}_l(G(s), K_0(s)) \stackrel{s}{=} \left[\begin{array}{cc|c} -sI + \Lambda_{12_s} - H_{12_s}^T B_1 B_1^T \Phi_{12_s} & H_{12_s}^T(-sI + A) H_{21_s} & H_{12_s}^T B_1 \\ \Theta^* & -sI + \Lambda_{21_s}^T - \Phi_{21_s}^T C_1^T C_1 H_{21_s} & X_{21_s}^T B_1 + U_{21_s}^T D_{21} \\ \hline C_1 X_{12_s} + D_{12} U_{12_s} & C_1 H_{21_s} & 0 \end{array}\right] \tag{227}$$

where

$$\Theta^* := \Phi_{21_s}^T(-sI + A)\Phi_{12_s} + \Phi_{21_s}^T \Phi_{12_s} \Lambda_{12_s} + \Lambda_{21_s}^T \Phi_{21_s}^T \Phi_{12_s} \tag{228}$$

and where H_{12_s}, H_{21_s} satisfy

$$H_{12}^T := \begin{bmatrix} H_{12_s}^T \\ H_{12_j}^T \\ H_{12_\infty}^T \end{bmatrix} = \begin{bmatrix} X_{12_s} & X_{12_j} & X_{12_\infty}^{(u)} \end{bmatrix}^{-1} \tag{229}$$

$$H_{21}^T := \begin{bmatrix} H_{21_s}^T \\ H_{21_j}^T \\ H_{21_\infty}^T \end{bmatrix} = \begin{bmatrix} X_{21_s} & X_{21_j} & X_{21_\infty}^{(u)} \end{bmatrix}^{-1}. \tag{230}$$

(iii) If the system matrix $M_{K_0}(s)$ is well-posed, then $\lim_{\epsilon \to 0} K_0(\epsilon, s)$ exists and is given by

$$K_0(s) \stackrel{s}{=} M_{K_0}(s) \tag{231}$$

and this value of $K_0(s)$ achieves the limiting transfer function $T_{y_1 u_1}(s)$.

Proof: As noted before, for $\epsilon > 0$ the central controller has system matrix given by Eq (223). The fact that $M_{K_0}(\epsilon, s) \to M_{K_0}(s)$ follows from Theorem 3 and the analyticity of $\Xi_{12}(\epsilon)$, $\Xi_{21}(\epsilon)$.

For part (ii) assume that ϵ^* is small enough so that those stable eigenvalues of $W_{12_\infty}(\epsilon, s)$ which approach locations at infinity, the finite $j\omega$-axis and the open

LHP are all in disjoint groups on $[0, \epsilon^*]$. For $\epsilon > 0$ we know that the closed loop transfer function from u_1 to y_1 has Rosenbrock system matrix

$$M_{T_{y_1 u_{1\epsilon}}}(s) = \left[\begin{array}{cc|c} -sI + A & -B_2 U_{12}(\epsilon) & B_1 \\ U_{21}^T(\epsilon) C_2 & -sE_k + A_k & -U_{21}^T(\epsilon) D_{21} \\ \hline C_1 & -D_{12} U_{12}(\epsilon) & 0 \end{array}\right] \tag{232}$$

where E_k, A_k are given in Eq (78) and Eq (82), respectively. For $\epsilon < \epsilon^*$ we can partition $\Xi_{12}(\epsilon)$ and $\Xi_{21}(\epsilon)$ as follows:

$$\Xi_{12}(\epsilon) = \left[\begin{array}{c|c|c} X_{12_s}(\epsilon) & X_{12_j}(\epsilon) & X_{12_\infty}(\epsilon) \\ \Phi_{12_s}(\epsilon) & \Phi_{12_j}(\epsilon) & \Phi_{12_\infty}(\epsilon) \\ U_{12_s}(\epsilon) & U_{12_j}(\epsilon) & U_{12_\infty}(\epsilon) \end{array}\right] \tag{233}$$

$$\Xi_{21}(\epsilon) = \left[\begin{array}{c|c|c} X_{21_s}(\epsilon) & X_{21_j}(\epsilon) & X_{21_\infty}(\epsilon) \\ \Phi_{21_s}(\epsilon) & \Phi_{21_j}(\epsilon) & \Phi_{21_\infty}(\epsilon) \\ U_{21_s}(\epsilon) & U_{21_j}(\epsilon) & U_{21_\infty}(\epsilon) \end{array}\right] \tag{234}$$

where

$$\lim_{\epsilon \to 0} \left[\begin{array}{c|c|c} X_{12_s}(\epsilon) & X_{12_j}(\epsilon) & X_{12_\infty}(\epsilon) \\ \Phi_{12_s}(\epsilon) & \Phi_{12_j}(\epsilon) & \Phi_{12_\infty}(\epsilon) \\ U_{12_s}(\epsilon) & U_{12_j}(\epsilon) & U_{12_\infty}(\epsilon) \end{array}\right] = \left[\begin{array}{c|c|c} X_{12_s} & X_{12_j} & X_{12_\infty}^{(l)} \\ \Phi_{12_s} & 0 & 0 \\ U_{12_s} & U_{12_j} & U_{12_\infty}^{(l)} \end{array}\right] \tag{235}$$

and

$$\lim_{\epsilon \to 0} \left[\begin{array}{c|c|c} X_{21_s}(\epsilon) & X_{21_j}(\epsilon) & X_{21_\infty}(\epsilon) \\ \Phi_{21_s}(\epsilon) & \Phi_{21_j}(\epsilon) & \Phi_{21_\infty}(\epsilon) \\ U_{21_s}(\epsilon) & U_{21_j}(\epsilon) & U_{21_\infty}(\epsilon) \end{array}\right] = \left[\begin{array}{c|c|c} X_{21_s} & X_{21_j} & X_{21_\infty}^{(l)} \\ \Phi_{21_s} & 0 & 0 \\ U_{21_s} & U_{21_j} & U_{21_\infty}^{(l)} \end{array}\right] \tag{236}$$

Associated with those zeros of interest which remain finite as $\epsilon \to 0$ are the matrices $\Lambda_{12_s}(\epsilon)$, $\Lambda_{12_j}(\epsilon)$ whose eigenvalues correspond to the eigenvalues of $W(\epsilon, s)$ on each subspace pertaining to the closed right half plane; similarly, $\Lambda_{12_s}(\epsilon)$, $\Lambda_{12_j}(\epsilon)$ are matrices whose eigenvalues correspond to those of $W_{21}(\epsilon, s)$ on the pertinent subspaces. Moreover, we have that

$$\lim_{\epsilon \to 0} \Lambda_{12_s}(\epsilon) = \Lambda_{12_s}, \qquad \lim_{\epsilon \to 0} \Lambda_{12_j}(\epsilon) = \Lambda_{12_j} \tag{237}$$

$$\lim_{\epsilon \to 0} \Lambda_{21_s}(\epsilon) = \Lambda_{21_s}, \qquad \lim_{\epsilon \to 0} \Lambda_{21_j}(\epsilon) = \Lambda_{21_j} \tag{238}$$

Substituting and partitioning in Eq (232) we obtain the system matrix in Table III.

Now take the limit as $\epsilon \to 0$ to get

$$M_{T_{y_1 u_1}}(s) = \lim_{\epsilon \to 0} M_{T_{y_1 u_{1\epsilon}}}(s) \tag{239}$$

Perform the following transformations

column 2 = column 2 - column 1 $\times X_{12_s}$
column 3 = column 3 - column 1 $\times X_{12_j}$
column 4 = column 4 - column 1 $\times X_{12_\infty}$
row 2 = row 2 + $X_{21_s}^T \times$ row 1
row 3 = row 3 + $X_{21_j}^T \times$ row 1
row 4 = row 4 + $X_{21_\infty}^T \times$ row 1
row 1 = $H_{12}^T \times$ row 1
column 1 = column 1 $\times H_{21}$

Table III: System Matrix for $M_{T_{y_1 u_{1_e}}}(s)$

$$
M_{T_{y_1 u_{1_e}}}(s) =
\left[\begin{array}{cccc|c}
-sI + A & -B_2 U_{12_s} & -B_2 U_{12_j} & -B_2 U_{12_\infty} & B_1 \\
U_{21_s}^T C_2 & \begin{array}{c} X_{21_s}^T(-sI + A)X_{12_s} + \Phi_{21_s}^T(sI + A^T)\Phi_{12_s} \\ +X_{21_s}^T B_1 B_1^T \Phi_{12_s} + \Phi_{21_s}^T C_1^T C_1 X_{12_s} \\ +X_{21_s}^T B_2 U_{12_s} + U_{21_s}^T C_2 X_{12_s} \\ +\Phi_{21_s}^T C_1^T D_{12} U_{12_s} + U_{21_s}^T D_{21} B_1^T \Phi_{12_s} \end{array} & \begin{array}{c} X_{21_s}^T[(-sI + A)X_{12_j} + B_2 U_{12_j}] \\ \Phi_{21_s}^T C_1^T (C_1 X_{12_j} + D_{12} U_{12_j}) \\ U_{21_s}^T C_2 X_{12_j} \end{array} & \begin{array}{c} X_{21_s}^T[(-sI + A)X_{12_\infty} + B_2 U_{12_\infty}] \\ \Phi_{21_s}^T C_1^T (C_1 X_{12_\infty} + D_{12} U_{12_\infty}) \\ U_{21_s}^T C_2 X_{12_\infty} \end{array} & -U_{21_s}^T D_{21} \\
U_{21_j}^T C_2 & \begin{array}{c} X_{21_j}^T[(-sI + A)X_{12_s} + B_2 U_{12_s}] \\ +[X_{21_j}^T B_1 + U_{21_j}^T D_{21}]B_1^T \Phi_{12_s} \\ +U_{21_j}^T C_2 X_{12_s} \end{array} & \begin{array}{c} X_{21_j}^T[(-sI + A)X_{12_j} + B_2 U_{12_j}] \\ +U_{21_j}^T C_2 X_{12_j} \end{array} & \begin{array}{c} X_{21_j}^T[(-sI + A)X_{12_\infty} + B_2 U_{12_\infty}] \\ +U_{21_j}^T C_2 X_{12_\infty} \end{array} & -U_{21_j}^T D_{21} \\
U_{21_\infty} C_2 & \begin{array}{c} X_{21_\infty}^T[(-sI + A)X_{12_s} + B_2 U_{12_s}] \\ +[X_{21_\infty}^T B_1 + U_{21_\infty}^T D_{21}]B_1^T \Phi_{12_s} \\ +U_{21_\infty}^T C_2 X_{12_s} \end{array} & \begin{array}{c} X_{21_\infty}^T[(-sI + A)X_{12_j} + B_2 U_{12_j}] \\ +U_{21_\infty}^T C_2 X_{12_j} \end{array} & \begin{array}{c} X_{21_\infty}^T[(-sI + A)X_{12_\infty} + B_2 U_{12_\infty}] \\ +U_{21_\infty}^T C_2 X_{12_\infty} \end{array} & -U_{21_\infty}^T D_{21} \\
\hline
C_1 & -D_{12} U_{12_s} & -D_{12} U_{12_j} & -D_{12} U_{12_\infty} & 0
\end{array}\right]
$$

where H_{12} and H_{21} are as given in the theorem. Partition the first column and row in conformity with the partitioning of H_{21} and H_{12}^T and rearrange columns and rows to get the system matrix in Table IV. We can now eliminate all but the first two rows and columns in the (1,1) block since the remaining pertain to diagonal blocks which are either completely unobservable or completely uncontrollable. This gives the result in Eq (227). Note that the transfer functions $T_{y_1 u_{1_e}}(s)$ are all RH^∞ functions with less than unit norm. It therefore follows that $T_{y_1 u_1} \in RH^\infty$ with $||T_{y_1 u_1}|| < 1$.

For part (iii) simply note that $\mathcal{F}_l(G(s), K_0(s)) \stackrel{s}{=} M_{T_{y_1 u_1}}(s)$ in Eq (239).

□

Comment: As in the H^2 case the limiting controller may not exist in cases in which the limiting controller system matrix (226) is not well posed. This is shown by example in the following section. This situation may be shown to arise from the fact that there is an "almost disturbance decoupling problem" embedded in the singular H^∞ problem [28, 47].

Comment: For singular problems the matrix E_k will be rank deficient. This follows from the fact that the rank deficiency of $\begin{bmatrix} X_{12} \\ \Phi_{12} \end{bmatrix}$ is r_2 since the r_2 grade 1 infinite eigenvectors of $M_{G_{12}}(s)$ are of the form $\begin{bmatrix} 0 \\ u \end{bmatrix}$. A similar result holds for $M_{G_{21}}(s)$. This indicates that, in general, the central controller $K_0(s)$ may be improper.

Comment: By Theorem 3, the limiting solution to Eq (87) is given by

$$Q_0 = H_{21_s} \Phi_{21_s}^T. \tag{240}$$

Theorem 6 gives the limiting form for the H^∞ controller, if indeed the limit does exist. If this is the case, then it may be improper; this is seen in the examples described in the following section. There are cases, however, when the limiting compensator is proper and is also of reduced order, i. e. of order $< n$. This has been noted in [27] and is the subject of the following corollary to Theorem 6:

Corollary 3 (Reduced Order Proper Controller) *Let* $r_{2_1} := \text{rank } D_{12}$ *and* $r_{2_2} := r_2 - r_{2_1}$. *Assume, without loss of generality that*

$$D_{12} = \begin{bmatrix} D_{12_1} & 0 \end{bmatrix} \tag{241}$$

where $D_{12_1} \in \Re^{r_2 \times r_{2_1}}$ *is injective and partition* B_2 *in conformity with the right hand side of Eq (241), i. e.*

$$B_2 = \begin{bmatrix} B_{2_1} & B_{2_2} \end{bmatrix} \tag{242}$$

Let $D_{11} = 0$, $D_{22} = 0$, D_{21} *be surjective and* $C_1 B_{2_2}$ *be injective. Suppose that* $W_{12_\infty}(s)$ *and* $W_{21}(s)$ *have no finite* $\mathcal{C}^{0e}$ *zeros and that there exists an* $\epsilon^* > 0$ *for which there is a solution to the* H^∞ *problem for* $G_{\epsilon^*}(s)$ *with* $\lambda_{\max}(Q_0 P_0) < 1$. *Then there exists a proper controller* $K(s)$ *which solves the* H^∞ *problem for* $G(s)$*; the order of this controller is at most* $n - r_{2_2}$.

Table IV: $M_{T_{y_1 u_1}}(s)$ System Matrix

$$M_{T_{y_1 u_1}}(s) = \left[\begin{array}{cccccc|c} \Theta_{11} & \Theta_{12} & 0 & 0 & * & * & H_{12_s}^T B_1 \\ \Theta_{21} & \Theta_{22} & 0 & 0 & * & * & X_{21_s}^T B_1 + U_{21_s}^T D_{21} \\ * & * & H_{12_\infty}^T X_{12_\infty} s - I & 0 & * & * & H_{12_\infty}^T B_1 \\ * & * & 0 & -sI + \Lambda_{12_j} & * & * & H_{12_j}^T B_1 \\ 0 & 0 & 0 & 0 & -sI + \Lambda_{21_j}^T & 0 & 0 \\ 0 & 0 & 0 & 0 & 0 & X_{21_\infty}^T H_{21_\infty} s - I & 0 \\ \hline C_1 H_{21_s} & -(C_1 X_{12_s} + D_{12} U_{12_s}) & 0 & 0 & C_1 H_{21_j} & C_1 H_{21_\infty} & 0 \end{array}\right]$$

$$\begin{aligned} \Theta_{11} &= H_{12_s}^T(-sI + A)H_{21_s} \\ \Theta_{12} &= -(-sI + \Lambda_{12_s}) + H_{12_s}^T B_1 B_1^T \Phi_{12_s} \\ \Theta_{21} &= (-sI + \Lambda_{21_s}^T) - \Phi_{21_s}^T C_1^T C_1 H_{21_s} \\ \Theta_{22} &= \Phi_{21_s}^T(-sI + A)\Phi_{12_s} + \Phi_{21_s}^T \Phi_{12_s} \Lambda_{12_s} + \Lambda_{21_s}^T \Phi_{21_s}^T \Phi_{12_s} \end{aligned}$$

Proof: The condition $C_1 B_{2_2}$ injective implies that $M_{G_{12}}$ has a simple infinite zero structure, i. e. all infinite zeros of $M_{G_{12}}(s)$ are of order unity.

The infinite zero eigenvectors are then the columns of $\begin{bmatrix} 0 & B_{2_2} \\ \hat{I}_{r_{2_2}} & 0 \end{bmatrix}$ and $\Xi_{12} = \begin{bmatrix} X_{12_s} & 0 \\ \Phi_{12_s} & 0 \\ U_{12_s} & \hat{I}_{r_{2_2}} \end{bmatrix}$ where,

$$\hat{I}_{r_{2_2}} := \begin{bmatrix} 0_{r_{2_1} \times r_{2_2}} \\ I_{r_{2_2}} \end{bmatrix} \tag{243}$$

Suppose that $\Xi_{21} = \begin{bmatrix} X_{21} \\ \Phi_{21} \\ U_{21} \end{bmatrix}$, $\Xi_{21} \in \Re^{(2n+m)\times n}$. Then if a solution exists it follows by Theorem 3 that $\begin{bmatrix} X_{12_s} & B_{2_2} \end{bmatrix}$ and X_{21} are nonsingular and

$$P_0 = \begin{bmatrix} \Phi_{12_s} & 0 \end{bmatrix} \begin{bmatrix} X_{12_s} & B_{2_2} \end{bmatrix}^{-1} \geq 0 \tag{244}$$

$$Q_0 = X_{21}^{-1^T} \Phi_{21}^T \geq 0 \tag{245}$$

We are given that $\lambda_{\max}(Q_0 P_0) < 1$. Hence,

$$\lambda_{max}(-s X_{21}^T \begin{bmatrix} X_{12_s} & B_{2_2} \end{bmatrix} + \begin{bmatrix} \Phi_{21}^T \Phi_{12_s} & 0 \end{bmatrix}) < 1 \tag{246}$$

and therefore, by taking $s = 1$, we have

$$\text{rank} \begin{bmatrix} X_{21}^T X_{12_s} - \Phi_{21}^T \Phi_{12_s} \mid X_{21}^T B_{2_2} \end{bmatrix} = n \tag{247}$$

Substitution in Eq (226) gives

$$K_0(s) \stackrel{s}{=} \left[\begin{array}{cc|c} (-sI + \Lambda_{21}^T)\begin{bmatrix} X_{21}^T X_{12_s} - \Phi_{21}^T \Phi_{12_s} \end{bmatrix} + X_{21}^T B_2 U_{12_s} & X_{21}^T B_{2_2} & U_{21}^T \\ \hline U_{12_s} & \hat{I}_{r_{2_2}} & 0 \end{array}\right] \tag{248}$$

Let $T = \begin{bmatrix} T_1 \\ T_2 \end{bmatrix}$ be a row compression on $X_{21}^T X_{12_s} - \Phi_{21}^T \Phi_{12_s}$.
Then from Eq (247)

$$T \begin{bmatrix} -(X_{21}^T X_{12_s} - \Phi_{21}^T \Phi_{12_s}) & X_{21}^T B_{2_2} \end{bmatrix} = \begin{bmatrix} E_{11} & A_{12} \\ 0 & A_{22} \end{bmatrix} \tag{249}$$

where $E_{11} \in \Re^{(n-r)\times(n-r)}$ and $A_{22} \in \Re^{r\times r}$ are nonsingular. By simple equivalence operations, it can be shown that

$$K_0(s) \stackrel{s}{=} \left[\begin{array}{cc|c} -sE_{11} + A_{11} & A_{12} & T_1 U_{21}^T \\ A_{21} & A_{22} & T_2 U_{21}^T \\ \hline U_{12_s} & \hat{I}_{r_{2_2}} & 0 \end{array}\right] \tag{250}$$

$$\cong \left[\begin{array}{c|c} -sI + E_{11}^{-1}(A_{11} - A_{12}A_{22}^{-1}A_{21}) & E_{11}^{-1}(T_1 - A_{12}A_{22}^{-1}T_2)U_{21}^T \\ \hline U_{12_s} - \hat{I}_{r_{2_2}} A_{22}^{-1} A_{21} & -\hat{I}_{r_{2_2}} A_{22}^{-1} T_2 U_{21}^T \end{array}\right]. \tag{251}$$

where

$$A_{11} := T_1(\Lambda_{21}^T[X_{21}^T X_{12_s} - \Phi_{21}^T \Phi_{12_s}] + X_{21}^T B_2 U_{12_s}) \tag{252}$$

and

$$A_{21} := T_2(\Lambda_{21}^T[X_{21}^T X_{12_s} - \Phi_{21}^T \Phi_{12_s}] + X_{21}^T B_2 U_{12_s}) \tag{253}$$

which is a realization of $K_0(s)$ of order at most $n - r_{2_2}$. □

Comment: It should be noted that Lemma 3 cannot be readily extended to the case where $M_{G_{12}}(s)$ has a more complex zero structure at infinity. If for example all of the infinite zeros were of order 2, then a similar analysis to that carried out above reveals that $A_{22} = 0$ in Eq (249) would be singular. This indicates that if the pencil $\begin{bmatrix} -sE_{11} + A_{11} & A_{12} \\ A_{21} & A_{22} \end{bmatrix}$ were regular then, in general, $K_0(s)$ in Eq (250)would be improper. It can also be shown that for this controller, the modes at infinity are uncontrollable (see [48] for explanation of this terminology) from the input channel corresponding to B_{k_2} in Eq (78) ; hence $K(s)$ cannot be made proper by any auxiliary feedback $Q(s)$ in Eq (79). Indications are, however, that the only viable $Q(s)$ (viable in the sense that it produces a well-posed system matrix for $K(s)$ at least) is one which may be parameterized in terms of ϵ with a system matrix which is itself not well-posed in the limit as $\epsilon \to 0$. It is not clear how this can be constructed in general.

There is, however, a simpler alternative which we will discuss, somewhat briefly, in the next section.

C Optimal H^2 Cost

In what follows, we derive the cost associated with the transfer function $T_{y_1U_1}(s)$ in Eq (186).

First let

$$B_T := \begin{bmatrix} H_{12_s}^T B_* \\ -(X_{21_s}^T B_1 + U_{21_s}^T D_{21}) \end{bmatrix} \tag{254}$$

$$C_T := \begin{bmatrix} C_1 X_{12_s} + D_{12} U_{12_s} & -C_1 H_{21_s} \end{bmatrix} \tag{255}$$

$$A_T := \begin{bmatrix} \Lambda_{12} & A_{T_{12}} \\ 0 & \Lambda_{21_s}^T \end{bmatrix} \tag{256}$$

where

$$A_{T_{12}} = H_{12_s}^T (H_{21_s} \Lambda_{21_s}^T - A H_{21_s}) \tag{257}$$

Then note that from Theorem 1.53 of [39] the limiting variance matrix $Q := \begin{bmatrix} Q_{11} & Q_{12} \\ Q_{12}^T & Q_{22} \end{bmatrix}$ in response to unit intensity white noise is the unique solution to

$$A_T Q + Q A_T^T + B_T B_T^T = 0 \tag{258}$$

Moreover, the optimal cost J_* is given by

$$J_* = Tr(C_T^T C_T Q) \tag{259}$$

Expanding Eq (258) we get

$$\Lambda_{12}Q_{11} + Q_{11}\Lambda_{12}^T + A_{T_{12}}Q_{12}^T + Q_{12}A_{T_{12}}^T + H_{12_s}^T B_* B_*^T H_{12_s} = 0 \quad (260)$$

$$\Lambda_{21_s}^T Q_{22} + Q_{22}\Lambda_{21_s} + (X_{21_s}^T B_1 + U_{21_s}^T D_{21})(X_{21_s}^T B_1 + U_{21_s}^T D_{21})^T = 0 \quad (261)$$

$$\Lambda_{12_s} Q_{12} + Q_{12}\Lambda_{21_s} + A_{T_{12}}Q_2 - H_{12_s}^T B_*(X_{21_s}^T B_1 + U_{21_s}^T D_{21})^T = 0 \quad (262)$$

We prove the result by a series of assertions:

Assertion 1 *The unique solution to Q_{22} in Eq (261) is*

$$Q_{22} = X_{21_s}\Phi_{21_s} \quad (263)$$

Proof: From

$$W_{21_2}(0,s)\begin{bmatrix} X_{21_s} \\ \Phi_{21_s} \\ U_{21_s} \end{bmatrix} = \begin{bmatrix} X_{21_s} \\ \Phi_{21_s} \\ 0 \end{bmatrix}(-sI + \Lambda_{21}) \quad (264)$$

one obtains

$$\begin{aligned}
&(X_{21_s}^T B_1 + U_{21_s}^T D_{21})(X_{21_s}^T B_1 + U_{21_s}^T D_{21})^T && (265)\\
&= X_{21_s}^T (X_{21_s}^T B_1 + U_{21_s}^T D_{21})^T + U_{21_s}^T (X_{21_s}^T B_1 + U_{21_s}^T D_{21})^T \\
&= -X_{21_s}^T (A\Phi_{21_s} + \Phi_{21_s}\Lambda_{21}) - U_{21_s}^T C_2 \Phi_{21_s} \\
&= -(X_{21_s}^T A + U_{21_s}^T C_2)\Phi_{21_s} - X_{21_s}^T \Phi_{21_s}\Lambda_{21_s} \\
&= -(\Lambda_{21_s}^T X_{21_s}^T \Phi_{21_s} + X_{21_s}^T \Phi_{21_s}\Lambda_{21_s}) && (266)
\end{aligned}$$

Hence Eq (261) is satisfied by Eq (263). By Lemma 1.5 of [39] the fact that Λ_{21} is strictly Hurwitz implies that this solution is unique.

□

Assertion 2 *The unique solution Q_{12} in Eq (262) is given by*

$$Q_{12} = 0 \quad (267)$$

Proof: By direct substitution for $A_{T_{12}}$ we have

$$\begin{aligned}
&A_{T_{12}}Q_{12} - H_{12_s}^T B_*(X_{21_s}^T B_1 + U_{21_s}^T D_{21})^T = \\
&\quad -H_{12_s}^T (AH_{21_s}X_{21_s}^T \Phi_{21_s} + B_1(X_{21_s}^T B_1 + U_{21_s}^T D_{21})^T \\
&\quad + H_{21_s}X_{21_s}^T \Phi_{21_s}\Lambda_{21_s}) && (268)\\
&= -H_{12_s}^T (A(H_{21_s}X_{21_s}^T - I)\Phi_{21_s} \\
&\quad + (H_{21_s}X_{21_s}^T - I)\Phi_{21_s}\Lambda_{21}) && (269)
\end{aligned}$$

Note that in the last equation we have used the following relationship:

$$B_1(B_1^T X_{21_s} + D_{21}^T U_{21_s}) = -(A\Phi_{21_s} + \Phi_{21_s}\Lambda_{21}) \quad (270)$$

By Eq (135) $(H_{21_s}X_{21_s}^T - I)\Phi_{21_s} = 0$. Hence the left hand side expression in Eq (268) is zero. Substitution in Eq (262) gives

$$\Lambda_{12_s}Q_{12} + Q_{12}\Lambda_{21_s} = 0 \quad (271)$$

The result follows by application of Lemma 1.5 in [39] and the fact that Λ_{12_s} and Λ_{21_s} are both strictly Hurwitz.

□

Now expanding Eq (259) we get

$$J_* = \mathrm{Tr}[(C_1 X_{12_s} + D_{12} U_{12_s})^T (C_1 X_{12_s} + D_{12} U_{12_s}) Q_{11}] + \mathrm{Tr}[H_{21_s}^T C_1^T C_1 H_{21_s} Q_{22}] \quad (272)$$

The last term is

$$\begin{aligned}\mathrm{Tr}[H_{21_s}^T C_1^T C_1 H_{21_s} Q_{22}] &= \mathrm{Tr}[C_1^T C_1 H_{21_s} Q_{22} H_{21_s}^T] \\ &= \mathrm{Tr}[C_1^T C_1 H_{21_s} X_{21_s}^T \Phi_{21_s} H_{21_s}^T] \\ &= \mathrm{Tr}[C_1^T C_1 \Phi_{21_s} H_{21_s}^T] \\ &= \mathrm{Tr}[C_1^T C_1 P_{21_*}] \end{aligned} \quad (273)$$

Use

$$\begin{aligned}&(C_1 X_{12_s} + D_{12} U_{12_s})^T (C_1 X_{12_s} + D_{12} U_{12_s}) \\ &= -(X_{12_s}^T (A\Phi_{12_s} + \Phi_{12_s} \Lambda_{12_s}) + U_{12_s}^T B_2^T \Phi_{12_s}) \\ &= -(\Lambda_{12_s}^T X_{12_s}^T \Phi_{12_s} + X_{12_s}^T \Phi_{12_s} \Lambda_{12_s}) \end{aligned} \quad (274)$$

to get

$$\begin{aligned}&\mathrm{Tr}[(C_1 X_{12_s} + D_{12} U_{12_s})^T (C_1 X_{12_s} + D_{12} U_{12_s}) Q_{11}] \\ &= -\mathrm{Tr}[\Lambda_{12_s}^T X_{12_s}^T \Phi_{12_s} Q_{11} + X_{12_s}^T \Phi_{12_s} \Lambda_{12_s} Q_{11}] \\ &= -\mathrm{Tr}[(X_{12_s} \Phi_{12_s})(\Lambda_{12_s} Q_{11} + Q_{11} \Lambda_{12_s}^T)] \end{aligned} \quad (275)$$

$$(276)$$

From Eq (262) and Assertion 2 the first term on the right hand side of Eq (272) is

$$\mathrm{Tr}(X_{12_x}^T \Phi_{12_s} H_{12_s}^T B_* B_*^T H_{12_s}) = \mathrm{Tr} P_{12_*} B_* B_*^T \quad (277)$$

giving

$$J_* = \mathrm{Tr}(P_{12_*} B_* B_*^T + C_1^T C_1 P_{21_*}). \quad (278)$$

D Limiting Transfer Function

We now complete the details of the proof to Part (iv) of Theorem 5. We begin with the following

Assertion 3 *The limit transfer function $T_{y_1 u_1}(s)$ is given by*

$$T_{y_1 u_1}(s) = \mathcal{F}_l(G(s), F(s)) \quad (279)$$

Proof: By direct substitution

$$\mathcal{F}_l(G(s), F(s)) = \left[\begin{array}{cc|c} -sI + A & -B_2 U_{12} & B_1 \\ U_{21}^T C_2 & \Theta & U_{21}^T D_{21} \\ \hline C_1 & -D_{12} U_{12} & 0 \end{array}\right] \tag{280}$$

where

$$\Theta := X_{21}^T(-sI + A)X_{12} + X_{21}^T B_2 U_{12} + U_{21}^T C_2 X_{12} \tag{281}$$

Now add $X_{21}^T \times$ row 1 to row 2, subtract column 1 $\times$ X_{12} from column 2, multiply column 2 by -I and interchange columns 1 and 2 to get:

$$\mathcal{F}_l(G(s), F(s)) =$$
$$\left[\begin{array}{cc|c} (-sI + A)X_{12} + B_2 U_{12} & -sI + A & B_1 \\ 0 & X_{21}^T(-sI + A) + U_{21}^T C_2 & U_{21}^T D_{21} + X_{21}^T B_1 \\ \hline C_1 X_{12} + D_{12} U_{12} & C_1 & 0 \end{array}\right] \tag{282}$$

Finally multiply row 1 by H_{12}^T and column 2 by $-H_{21}$ to obtain the system matrix in Table 4.2(b) and apply the same arguments as in the main body of the proof of the theorem to get the result. □

VIII A New Embedding for the Singular H^∞ Problem

We saw in the last section that when the infinite $\mathcal{C}^{0e}$ zeros are of order unity, then it may be possible to obtain a reduced order proper solution to H^p problems. In this section we propose a new embedding of such problems which exploits this property; this embedding allows us to obtain reduced order proper solutions for problems with infinite zeros of all orders. We demonstrate the method for H^∞ problems but stress that the method is applicable to the H^2 case as well.

Suppose first that $D_{11} = 0$, $D_{22} = 0$ and that D_{21} is surjective. We have already discussed the alleviation of the first two restrictions; we will shortly describe how the final constraint can be accommodated. Finally, we assume that G_{12} and G_{21} have no finite $\mathcal{C}^{0e}$ zeros.

If $C_1 B_{2_2}$ is injective then Corollary 3 provides us with a solution to the problem. If, however, $C_1 B_{2_2}$ is column rank deficient (in which case $M_{G_{12}}$ will have infinite zeros of minimum order 2) then we can proceed as follows: Choose a matrix C of the same column dimension as C_1 such that $\begin{bmatrix} C_1 \\ C \end{bmatrix} B_{2_2}$ is injective (this can be done by using the singular value decomposition, for example [49]). Then by construction $\hat{C}_\epsilon B_{2_2}$ is also injective, where

$$\hat{C}_\epsilon := \begin{bmatrix} C_1 \\ \epsilon C \end{bmatrix} \tag{283}$$

Now consider the new embedding (see Fig. 4: Note that, for simplicity, the D_{21} block has been omitted)

$$\tilde{G}_\epsilon(s) = \begin{bmatrix} \tilde{G}_{11_\epsilon}(s) & \tilde{G}_{12_\epsilon}(s) \\ \tilde{G}_{21_\epsilon}(s) & \tilde{G}_{22_\epsilon}(s) \end{bmatrix}$$

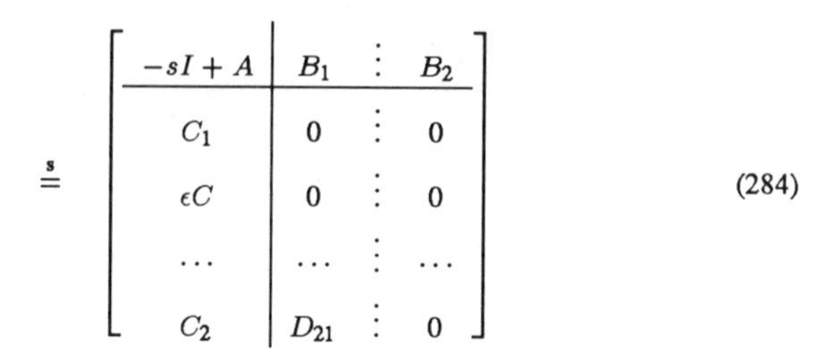

$$\stackrel{s}{=} \left[\begin{array}{c|c:c} -sI + A & B_1 & B_2 \\ \hline C_1 & 0 & 0 \\ \epsilon C & 0 & 0 \\ \cdots & \cdots & \cdots \\ C_2 & D_{21} & 0 \end{array}\right] \tag{284}$$

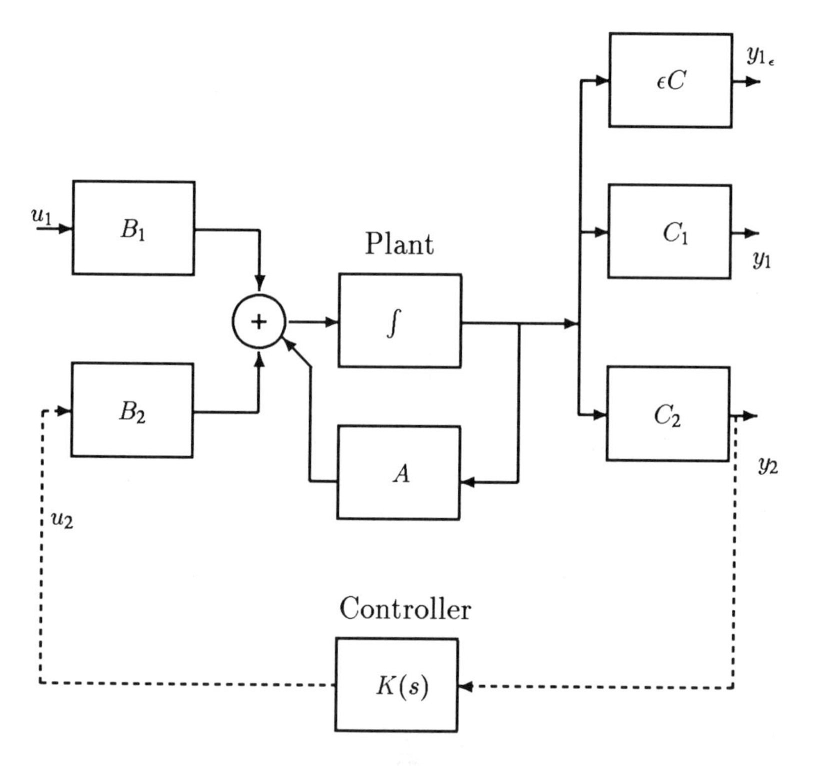

Figure 4: Another Embedding of the plant $G(s)$

Note that the injectivity of CB_{2_2} guarantees that $\tilde{G}_{12_\epsilon}(s)$ has zeros at infinity of order no greater than 1.

Lemma 23 *There exists a controller which solves the suboptimal H^p problem for the plant $G(s)$ if and only if there exists an $\epsilon^* > 0$ for which a solution to the suboptimal problem for the plant $\tilde{G}_{\epsilon^*}(s)$ exists.*

Furthermore, if such an ϵ^* exists, it follows that

(i) There exists a solution to the suboptimal problem $\forall \epsilon \in [0, \epsilon^*]$.

(ii) The optimum H^∞ cost is monotone decreasing on $[0, \epsilon^*]$.

Proof: The proof follows the same line of argument given in the proof to Lemma 14. Suppose that there exists a compensator $K^*(s)$ which solves the suboptimal problem for $\tilde{G}_{\epsilon^*}(s)$ for some $\epsilon^* > 0$. Let $K_\epsilon(s)$ denote any controller applied to the plant $\tilde{G}_\epsilon(s)$; furthermore, $\forall\, \epsilon \in [0, \epsilon^*]$ let $K_\epsilon(s) = K^*(s)$. Then $\forall \epsilon$ we have

$$T_{\tilde{y}_1 u_{1_\epsilon}}(s) = \begin{bmatrix} T_{y_1 u_1}(s) \\ \epsilon(\hat{G}_{11}(s) + \hat{G}_{12}T_{22}(s)G_{21}(s)) \end{bmatrix} \tag{285}$$

where

$$\hat{G}_{11}(s) = C(sI - A)^{-1}B_1 \tag{286}$$

$$\hat{G}_{12}(s) = C(sI - A)^{-1}B_2 \tag{287}$$

$$T_{y_1 u_1}(s) = \mathcal{F}_l(G(s), K(s)) \tag{288}$$

and, as before,

$$\tilde{y}_1 := \begin{bmatrix} y_1 \\ y_{1_\epsilon} \end{bmatrix} \tag{289}$$

$$T_{22}(s) := K(s)(I - G_{22}(s)K(s))^{-1} \tag{290}$$

Since each partition in Eq (285) is a stable transfer function at $\epsilon = \epsilon^*$ it follows that $T_{\tilde{y}_1 u_{1_\epsilon}}(s) \in RH^\infty$ $\forall \epsilon$ and, in particular, $\forall \epsilon \in [0, \epsilon^*]$. By simple matrix manipulations, it follows that $\forall \epsilon \in [0, \epsilon^*]$ $||T_{\tilde{y}_1 u_{1_\epsilon}}||_\infty \leq ||T_{\tilde{y}_1 u_{1_{\epsilon^*}}}||_\infty < 1$.This proves (i) and (ii). Sufficiency follows from the fact that $T_{y_1 u_1}(s) \in RH^\infty$, $||T_{y_1 u_1}||_\infty \leq ||T_{\tilde{y}_1 u_{1_{\epsilon^*}}}||_\infty < 1$.

For necessity, assume that the controller $K_0(s)$ solves the problem for the plant $G(s)$. Then we have that $||T_{y_1 u_1}||_\infty = 1 - \delta$, $1 > \delta > 0$. Apply the controller $K_0(s)$ and consider the closed loop system for increasing ϵ. Since $K_0(s)$ internally stabilizes $G_{22}(s) \Rightarrow \hat{G}_{11}(s) + \hat{G}_{12}T_{22}(s)G_{21}(s)$ is also in RH^∞; moreover this matrix does not depend on $\epsilon \Rightarrow T_{\tilde{y}_1 u_1} \in RH^\infty$ $\forall\, \epsilon$. Hence we have that $k := ||\hat{G}_{11} + \hat{G}_{12}T_{22}G_{21}||_\infty < \infty$. Moreover,

$$||T_{\tilde{y}_1 u_{1_\epsilon}}||_\infty \leq ||T_{y_1 u_1}||_\infty + \epsilon k \tag{291}$$

Now note that for $\epsilon^* := \frac{\delta}{2k}$ $||T_{\tilde{y}_1 u_{1_\epsilon}}||_\infty \leq 1 - \delta + \frac{\delta}{2} = 1 - \frac{\delta}{2} < 1$ $\forall \epsilon \in [0, \epsilon^*]$. □

Thus, as for our original embedding, this method provides us with a method for solving singular H^∞ problems. There is, however, an added advantage to using this new embedding as described in the following lemma:

Lemma 24 *Assume that there is ϵ^* for which there exists a solution to the H^∞ problem for $\tilde{G}_{\epsilon^*}(s)$ with $\lambda_{\max}(Q_0 P_0) < 1$. Then there exists a controller of order at most $n - r_{2_2}$ which solves the H^∞ problem for $G(s)$.*

Proof: Follows from the fact that the infinite zeros of $\tilde{G}_{12_\epsilon}(s)$ are, by construction, of order unity, at most, and the application of Corollary 3 and Lemma 23. □

Comment: Lemma 24 and Corollary 3 provide us a means of obtaining reduced order solutions to all singular H^∞ problems.

Comment: It should be noted that the $\mathcal{C}^{0e}$ structure of singular problems will generically be simple, i. e. the $\mathcal{C}^{0e}$ zeros which pertain to such problems will almost always be of order unity. This easily follows from the known sensitivity of Jordan or Jordan-like structures (for example, Weierstrass and Kronecker Canonical structures) associated with zeros of order 2 or higher [29, 50]. The embedding procedure above clarifies this for the case of infinite $\mathcal{C}^{0e}$ zeros: for any $\epsilon > 0$ $\tilde{G}_1(\epsilon, s)$ always has a simple $\mathcal{C}^{0e}$ structure. In view of this, the result stated in Corollary 3 is therefore made much more significant.

Comment: Finally, we note that the constraint on D_{21} may be eliminated by employing a combination of both embeddings discussed in this chapter, i. e.

$$\tilde{G}_2(\epsilon, s) \stackrel{s}{=} \left[\begin{array}{c|cc} -sI + A & \tilde{B}_1 & B_2 \\ \hline C_1 & 0 & 0 \\ \epsilon C & 0 & 0 \\ \cdots & \cdots & \cdots \\ C_2 & \tilde{D}_{21} & 0 \end{array}\right] \tag{292}$$

where $\tilde{B}_1$ and $\tilde{D}_{21}$ are as defined in Eq (84).

Since the results described above can be dualized to the case where D_{21}, say, is rank deficient it is seen that the embedding procedures described here allow us to obtain solutions to singular problems of order $n - \min(\eta_{12}, \eta_{21})$ at most, where η_{12} and η_{21} are the respective nullities of D_{12} and D_{21}.

IX Computational Aspects

We will now discuss the issue of the computation of the subspaces described in Theorems 1, 5. The finite subspaces of $W_{21}(\epsilon, s)$, $W_{12}(\epsilon, s)$ may, of course, be determined by application of the QZ algorithm [49]. Van Dooren [32] , however, has pointed out that this approach may be rendered numerically unreliable by the infinite frequency structure of these pencils. Our computational procedures are therefore based on the "pencil algorithms" discussed in [32].

Algorithm 1: Finite subspace calculation

Input: State space parameters of $G(s) \stackrel{s}{=} \left[\begin{array}{c|c} -sI + A & B \\ \hline C & D \end{array}\right]$.

Step 1 *Form the pencil* $W(s) = \left[\begin{array}{cc|c} -sI + A & 0 & B \\ -C^TC & -sI - A^T & -C^TD \\ \hline D^TC & B^T & D^TD \end{array}\right]$.

Step 2 *Use the dual form of Algorithm 3.6 [32] to obtain unitary matrices S_1, T_1 such that*

$$S_1 W(s) T_1 = \left[\begin{array}{c|c} -sE_\infty + A_\infty & 0 \\ \hline * & -sE_f + A_f \end{array}\right] \tag{293}$$

where the (2,2) block is regular and contains the finite frequency structure of $W(s)$.

Step 3 *Using the QZ algorithm or block Schur procedures [51] determine unitary matrices S_2, T_2 such that*

$$S_2(-sE_f + A_f)T_2 = \left[\begin{array}{c|c} -sE_+ + A_+ & 0 \\ \hline * & -sE_- + A_- \end{array}\right] \tag{294}$$

where the (2,2) block contains all the LHP zeros of $-sE_f + A_f$ and hence $W(s)$.

Step 4 *Let* $S = \left[\begin{array}{c|c} I & 0 \\ \hline 0 & S_2 \end{array}\right] S_1$, $T = T_1 \left[\begin{array}{c|c} I & 0 \\ \hline 0 & T_2 \end{array}\right]$. *Then, with the obvious partitioning,*

$$\left[\begin{array}{c|c} S_{11} & S_{12} \\ \hline S_{21} & S_{22} \end{array}\right] W(s) \left[\begin{array}{c|c} T_{11} & T_{12} \\ \hline T_{21} & T_{22} \end{array}\right] = \left[\begin{array}{c|c} * & 0 \\ \hline * & -sE_- + A_- \end{array}\right] \tag{295}$$

where the (2,2) on the right hand side again contains all the LHP zeros of $W(s)$. The required basis vectors are then the columns of $\begin{bmatrix} T_{12} \\ T_{22} \end{bmatrix}$.

Output: $\begin{bmatrix} T_{12} \\ T_{22} \end{bmatrix}$ whose columns form a minimal basis for the stable eigenspace of $W(s)$.

Algorithm 2: Infinite frequency subspace

Input: State space parameters of $G(s) \stackrel{s}{=} M(s)$ where $M(s) = \left[\begin{array}{c|c} -sI + A & B \\ \hline C & D \end{array}\right]$.

Step 1 *Use algorithm 4.1 of [32] to obtain unitary matrices S, T such that (partitioning S, T accordingly)*

$$\left[\begin{array}{c|c} S_{11} & S_{12} \\ \hline S_{21} & S_{22} \end{array}\right] M(s) \left[\begin{array}{c|c} T_{11} & T_{12} \\ \hline T_{21} & T_{22} \end{array}\right] = \left[\begin{array}{c|c} * & 0 \\ \hline * & -sE_\infty + A_\infty \end{array}\right] \tag{296}$$

where the (2,2) block contains the infinite frequency structure of $M(s)$.

Step 2 *The columns of* $\Xi_\infty = \begin{bmatrix} T_{12} \\ T_{22} \end{bmatrix}$ *form a minimal basis for the required eigenspace. Apply the SVD [49] to obtain an orthogonal matrix V_1 which spans the right null-space of* $\begin{bmatrix} C & D \end{bmatrix} \begin{bmatrix} T_{12} \\ T_{22} \end{bmatrix}$.

Then the columns of $\hat{\Xi}_\infty^{(l)} = \begin{bmatrix} T_{12} \\ T_{22} \end{bmatrix} V_1$ *form a minimal basis for the lower infinite eigenspace of $M(s)$.*

Step 3 *Again use the SVD to determine a unitary matrix V_2 whose columns form a minimal basis for* Ker $^{\perp}T_{12}$. *Then the columns of the matrix* $\Xi_{\infty}^{(u)} \begin{bmatrix} T_{12} \\ T_{22} \end{bmatrix} V_2$ *form a minimal basis for the upper infinite frequency eigensubspace of $M(s)$.*

Output: $\begin{bmatrix} T_{12} \\ T_{22} \end{bmatrix}$, $\Xi^{(l)}\infty$ and $\Xi_{\infty}^{(u)}$.

Comment: Justification for Steps 2 and 3 above follows directly from Lemma 4.

X Some Examples

Let us now consider a few simple examples to demonstrate the theory developed. The first pertains to the singular H^2 problem.
Example 1:
Consider the system

$$\begin{pmatrix} \dot{x}_1 \\ x_2 \\ x_3 \end{pmatrix} = \begin{bmatrix} -b & 1 & 0 \\ 0 & -a & 1 \\ 0 & 0 & 0 \end{bmatrix} \begin{pmatrix} x_1 \\ x_2 \\ x_3 \end{pmatrix} + \begin{bmatrix} 0 & 0 & 0 \\ 0 & 1 & 0 \\ 1 & 0 & 0 \end{bmatrix} u_1 + \begin{bmatrix} 0 \\ 0 \\ 1 \end{bmatrix} u_2 \quad (297)$$

$$y_1 = \begin{bmatrix} 0 & 0 & 1 \\ 0 & 0 & 0 \end{bmatrix} \begin{pmatrix} x_1 \\ x_2 \\ x_3 \end{pmatrix} + \begin{bmatrix} 0 \\ 1 \end{bmatrix} u_2 \quad (298)$$

$$y_2 = \begin{bmatrix} 1 & 0 & 0 \end{bmatrix} \begin{pmatrix} x_1 \\ x_2 \\ x_3 \end{pmatrix} + \begin{bmatrix} 0 & 0 & \epsilon \end{bmatrix} u_1 \quad (299)$$

where $a > 0$, $b > 0$ and $u_1 = \begin{pmatrix} \omega_1 \\ \omega_2 \\ \omega_3 \end{pmatrix}$ is unit intensity white noise.

Forming the appropriate pencils we find that $W_{12_2}(0, s)$ has LHP zeros at $s = -1, -b, -a$ with

$$X_{12_s} = \begin{bmatrix} -1 & 1 & 1 \\ 1-b & 0 & b-a \\ (a-1)(1-b) & 0 & 0 \end{bmatrix}, \Phi_{12_s} = \begin{bmatrix} 0 & 0 & 0 \\ 0 & 0 & 0 \\ (a-1)(1-b) & 0 & 0 \end{bmatrix}$$

and $U_{12_s} = [(1-a)(1-b)\ 0\ 0]$. Also $W_{21_2}(0, s)$ has only one LHP eigenvalue (at $s = -1$). The required matrices here are

$$X_{21_s} = \begin{bmatrix} a-1 \\ 1 \\ -1 \end{bmatrix}, \Phi_{21_s} = \begin{bmatrix} 0 \\ 0 \\ -1 \end{bmatrix}, U_{21_s} = [(1-a)(1-b)]$$

for the infinite frequency eigenspace

$$X_{21_\infty} = \begin{bmatrix} 0 & 1 \\ 0 & 0 \\ 0 & 0 \end{bmatrix}, U_{21_\infty} = \begin{bmatrix} 1 & 0 \end{bmatrix}.$$

Substitution in Eq (185) yields

$$F(s) \stackrel{s}{=} \left[\begin{array}{ccc|c} (s+1)(2(a+b)-ab-3) & (s+b)(1-a) & (s+a)(1-b) & 0 \\ -1 & 1 & 1 & 1 \\ s+1 & -(s+b) & -(s+a) & 0 \\ \hline (a-1)(1-b) & 0 & 0 & 0 \end{array}\right] \tag{300}$$

which is well posed for $a \neq 1, b \neq 1, a \neq b$ with corresponding transfer function

$$F(s) = -\frac{(s+a)(s+b)}{s+2}. \tag{301}$$

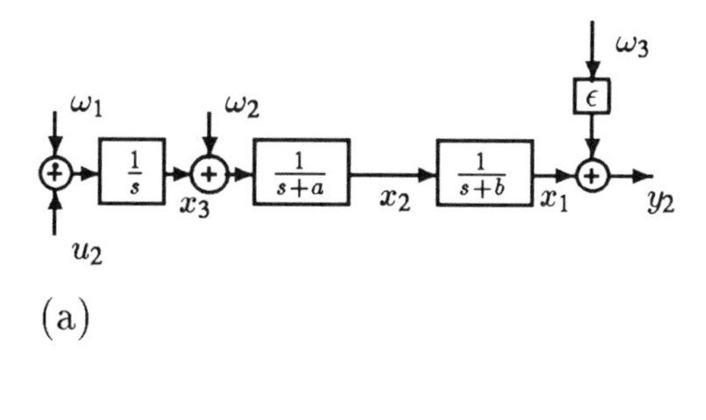

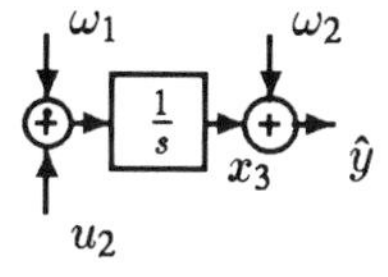

(b)

Figure 5: Block Diagrams for LQG Example

There is a nice interpretation of the structure derived. Note first that the system in question has the cascade structure shown in Fig 5a; for $\epsilon = 0$ the output is noise free and can therefore be (theoretically) differentiated via the transfer function $(s+a)(s+b)$ to obtain the precise value of $\hat{y} = \begin{pmatrix} x_3 \\ \omega_2 \end{pmatrix}$. Since the noise intensities for the remaining subsystem (see Fig 5b) are nonsingular we can now consider the reduced (regular) LQG problem of determining $u_2(t)$ to minimize $\mathrm{E}(x_3^2 + u_2^2)$. The dynamic constraint equations are $\dot{x}_3 = u_2 + \omega_1$, with measured variable $\hat{y}$. The stabilizing solution to this problem is $\hat{F}(s) = -\frac{1}{s+2}$. Now note that $F(s) = \hat{F}(s)(s+a)(s+b)$, implying that the procedure generated by Theorems 1, 5 automatically assumes this particular cancellation structure.

Finally, observe that while the transfer function Eq (301) exists $\forall a,\ b$, the system matrix Eq (300) is not i/o well-defined for $a = b = 1$. In this case the problem arises due to the fact that $W_{12}(0, s)$ assumes a third order pole at $s = -1$;

recalculation of the appropriate eigenspace for these parameter values yields, as expected, $F(s) = \frac{-(s+1)^2}{s+2}$.

Corollary 3 indicates that when u has dimension $r = n$ the H^∞ compensator will be a constant gain. This is shown more clearly in the next example.

Example 2:
If in Corollary 3 we have rank $B_2 = n$, all other assumptions remaining unchanged then we obtain $\Xi_{12} = \begin{bmatrix} 0 \\ 0 \\ I_n \end{bmatrix}$ and

$$K_0(s) \stackrel{s}{=} \left[\begin{array}{c|c} X_{21}^T B_2 & U_{21}^T \\ \hline I & 0 \end{array}\right] \tag{302}$$

which is a static compensator

$$\left.\begin{array}{rcl} K_0(s) = K & = & -B_2^{-1} X_{21}^{-T} U_{21}^T \\ & = & B_2^{-1}(B_1 D_{21}^T + Q_0 C_2^T)(D_{21} D_{21}^T)^{-1} \end{array}\right\} \tag{303}$$

Where Q_0 is as given in Eq (240). This solution is implementable by output injection and may be compared to the result in [42].

The next example illustrates the fact that, in some cases, the limiting H^∞ compensator can also be improper.

Example 3 (Francis [12, page 64]:
Here we have

$$\mathrm{G(s)} = \left[\begin{array}{c|c} \frac{1}{s+1} & \frac{1}{(s+1)^2} \\ \hline 1 & 0 \end{array}\right] \stackrel{s}{=} \left[\begin{array}{cc|cc} -s-1 & 1 & 1 & 0 \\ 0 & -s-1 & 0 & 1 \\ \hline 1 & 0 & 0 & 0 \\ 0 & 0 & 1 & 0 \end{array}\right] \tag{304}$$

Here $D_{12} = 0$ so we require the infinite zero structure of $C_1(sI - A)^{-1}B_2$; this transfer function has all of its zeros at infinity. We therefore obtain $\Xi_{12} = \begin{bmatrix} 0 & 0 \\ 0 & 1 \\ 0 & 0 \\ 0 & 0 \\ 1 & 1 \end{bmatrix}$. $W_{21}(0, s)$ has $2n$ finite zeros with $\Lambda_{21} = \begin{bmatrix} -1 & 1 \\ 0 & -1 \end{bmatrix}$ and $\Xi_{21} = \begin{bmatrix} 0 & 1 \\ 1 & 0 \\ 0 & 0 \\ 0 & 0 \\ 0 & -1 \end{bmatrix}$. It then follows that the central controller is given by $K_0(s) = -(s+1)$ which is easily shown to achieve the H^∞ optimum (see also [12]) $T_{y_1 u_1} \equiv 0$.

Example 4:
Take

$$G(s) = \begin{bmatrix} \frac{1}{s+1} & \frac{1}{(s+1)^2} \\ \frac{s+2}{s+1} & \frac{s+2}{(s+1)^2} \end{bmatrix} \stackrel{s}{=} \left[\begin{array}{cc|cc} -s-1 & 1 & 1 & 0 \\ 0 & -s-1 & 0 & 1 \\ \hline 1 & 0 & 0 & 0 \\ 1 & 1 & 1 & 0 \end{array}\right]. \tag{305}$$

Here $W_{12}(0, s)$ has no finite zeros and we obtain Ξ_{12} and Ξ_{21} as in Example 5.2. In this case, however, $\Lambda_{21} = \begin{bmatrix} -1 & 0 \\ 0 & -2 \end{bmatrix}$ and the central controller is

$$K_0(s) \stackrel{s}{=} \left[\begin{array}{cc|c} 1 & -s & 0 \\ 0 & 0 & -1 \\ \hline -1 & -1 & 0 \end{array}\right] \tag{306}$$

By Lemma 22 the controller in Eq (306) is not well-posed.

Let us now attempt to solve the problem for this plant using the embedding technique illustrated in Fig. 4. We take $\epsilon = 1$ and $C = \begin{bmatrix} 0 & 1 \end{bmatrix}$ to get

$$\tilde{G}_1(1, s) \stackrel{s}{=} \left[\begin{array}{cc|cc} -s-1 & 1 & 1 & 0 \\ 0 & -s-1 & 0 & 1 \\ \hline 1 & 0 & 0 & 0 \\ 0 & 1 & 0 & 0 \\ \cdots & \cdots & \cdots & \cdots \\ 1 & 1 & 1 & 0 \end{array}\right] \tag{307}$$

Solving the pertinent eigenproblems, we see that $W_{12}(s)$ now has a finite eigenvalue at $s = -\sqrt{3}$ and that Λ_{21} and Ξ_{21} are as before; moreover

$$\Xi_{12} = \begin{bmatrix} -(\sqrt{3}+1) & 0 \\ 1 & 0 \\ 1 & 0 \\ 0 & 0 \\ 1-\sqrt{3} & 1 \end{bmatrix}, \quad H_{12} = \begin{bmatrix} -\sqrt{3}-1 & 0 \\ 1 & 1 \end{bmatrix} \tag{308}$$

Note that H_{12} is nonsingular and that $\lambda_{\max}(P_0 Q_0) = 0 < 1$. Substituting for $K_0(s)$ we obtain

$$K_0(s) \stackrel{s}{=} \left[\begin{array}{cc|c} -s-1 & -1 & 0 \\ (\sqrt{3}+1)(s+2) & -1 & 1 \\ \hline \sqrt{3}-1 & -1 & 0 \end{array}\right] \tag{309}$$

which has transfer function $K_0(s) = K_0 := -\frac{1}{\sqrt{3}+2}$. For this compensator, $T_{y_1u_1}(s) = \frac{s+1}{s^2+(2-K_0)s+1-2K_0}$ which is stable (the poles are at $s = -1.13 \pm 0.5j$) and has infinity norm $\frac{1}{1-2K_0} < 1$.

XI Summary and Conclusion

We have studied a general form of the H^2 and H^∞ control problems. The main results are stated in Theorems 5 and 6 which describe the system matrices for the controllers, closed loop transfer functions, $T_{y_1u_1}(s)$, and the associated costs. The results are in a form which permits singular H^2 and H^∞ problems involving, for example, rank deficient D_{12} or D_{21} matrices to be handled with nearly the same ease as conventional nonsingular problems. The analysis of the singular case has been carried out by examining a sequence of non-singular perturbed problems. In particular, it has been shown that the limiting controller system matrices are as given in Eq (185) and Eq (226), and that the limiting controllers may be computed provided these system matrices are well posed, as defined in Section VI.

As discussed in Section IX, the results lead to reliable algorithms for the computation of H^2 and H^∞ controllers which apply equally well to both singular and nonsingular cases. Moreover, for singular cases involving simple zeros at infinity, we have shown the resultant control laws enjoy an order reduction at least as great as the rank deficiency.

XII Acknowledgment

This research was supported in part by the U. S. Air Force Office of Scientific Research under Grant 89-0398.

References

[1] M. G. Safonov and D. J. N. Limebeer. Simplifying the H^∞ theory via loop shifting. In *Proceedings of the IEEE Conference on Decision and Control*, December 1988. Austin, TX.

[2] J. Doyle, K. Glover, P. Khargonekar, and B. Francis. State space solutions to standard H_2 and H_∞ control problems. *IEEE Transactions on Automatic Control*, AC-24(8):731–747, August 1988.

[3] K. Glover, D. J. Limebeer, J. C. Doyle, E. M. Kasenally, and M. G. Safonov. A characterization of all solutions to the four block general distance problem. *SIAM J. Control and Optimization*, 29:283–324, March 1991.

[4] D. J. N. Limebeer, B. D. O. Anderson, P. Khargonekar, and M. Green. A game theoretic approach to H^∞ control for time varying systems. Preprint.

[5] H. Kwakernaak. Asymptotic root loci of multivariable linear optimal regulators. *IEEE Transactions on Automatic Control*, AC-17:378–382, 1976.

[6] A. E. Bryson, Jr. and Y. Ho. *Applied Optimal Control*. John Wiley & Sons, 1975.

[7] B. Kouvaritakis. The optimal root loci of linear multivariable systems. *International Journal of Control*, 28(1):33–62, 1978.

[8] U. Shaked. The asymptotic behaviour of the root-loci of multivariable optimal regulators. *IEEE Transactions on Automatic Control*, AC-23(3):425–430, June 1978.

[9] W. M. Wonham. *Linear Multivariable Control: A Geometric Approach*. Springer, New York, 2nd edition, 1979.

[10] Jr. R. E. O'Malley. A more direct solution of the nearly singular linear regulator problem. *SIAM J. Contr. Optimiz.*, 14:1063–1077, 1976.

[11] P. Sannuti and H. Wason. Multiple time-scale decomposition in cheap control problems - singular control. *IEEE Transactions on Automatic Control*, AC-30(7):633–644, July 1985.

[12] B. A. Francis. *A Course in H_∞ Control Theory*. Springer-Verlag: Heidelberg, 1987.

[13] M. L. J. Hautus and L. M. Silverman. System structure and singular control. *Linear Algebra anad its Applications*, 50:369–402, 1983.

[14] Harry L. Trentelman. Families of linear-quadratic problems: Continuity properties. *IEEE Transactions on Automatic Control*, AC-32(4):323–329, April 1987.

[15] R. S. Bucy and E. Jonckheere. Singular filtering problems. *Syst. and Contr. Lett.*, 13:339–344, 1989.

[16] E. Jonckheere. On the existence of a negative semidefinite, antistabilising solution to the discrete-time algebraic riccati equation. *IEEE Transactions on Automatic Control*, AC-26(3):707–712, 1981.

[17] J. M. Schumacher. The role of the dissipation matrix in singular optimal control. *Syst. Contr. Lett.*, 2:262–266, 1983.

[18] V. Kucera. Stationary LQG control of singular systems. *IEEE Transactions on Automatic Control*, AC-31:31–39, January 1986.

[19] E. Soroka and U. Shaked. The LQG optimal regulation problem for systems with perfect measurements: Explicit solution, properties and application to practical designs. *IEEE Transactions on Automatic Control*, AC-33(10):941–944, October 1988.

[20] J. O'Reilly. *Observers for Linear Systems*. Academic Press, 1983. Math. in Science and Engin., Vol 170.

[21] B. Friedland. Limiting forms of optimal stochastic linear regulators. *Trans. ASME (J. Dynam. Syst. Meas. Contr.)*, 93:134–141, 1971. Ser. G.

[22] G. Stein and M. Athans. The LQG/LTR procedure for multivariable feedback control design. *IEEE Transactions on Automatic Control*, AC-32(2):105–114, February 1987.

[23] M. G. Safonov. Imaginary-axis zeros in multivariable H^∞ optimal control. In R. F. Curtain, editor, *Modelling, Robustness and Sensitivity Reduction*. Springer-Verlag, New York, 1987.

[24] S. D. O'Young, I. Postlethwaite, and D. W. Gu. A treatment of $j\omega$-axis model-matching transformation zeros in the optimal H^2 and H^∞ control designs. *IEEE Transactions on Automatic Control*, AC-34(5):551–553, May 1989.

[25] M. G. Safonov and R. Y. Chiang. Cacsd using the state-space L^∞ theory – a design example. In *Proceedings of the IEEE Conference on Decision and Control*, 1988.

[26] S. Hara, T. Sugie, and R. Kondo. Descriptor form of solution for H^∞ control problem with $j\omega$-axis zeros. Submitted to Automatica, May 1989.

[27] H. Kimura, Y. Lu, and R. Kawatani. On the structure of H^∞ control systems and related extensions. *IEEE Transactions on Automatic Control*, AC-36(6):653–667, June 1991.

[28] A.A. Stoorvogel and H. L. Trentelman. The quadratic matrix inequality in singular H^∞ control with state feedback. *SIAM J. Opt. and Contr.*, 28(5):1190–1208, September 1990.

[29] F. R. Gantmacher. *Matrix Theory*. New York: Chelsea, 1959.

[30] P. Lancaster and M. Tismenetsky. *The Theory of Matrices*. Academic Press, Inc., second edition, 1985.

[31] F. L. Lewis. A survey of linear singular systems. *Circuits, Systems, Signal Processes*, 5(1):3–35, 1986.

[32] P. Van Dooren. The computation of Kronecker's canonical form of a singular pencil. *Linear Algebra and its Applications*, 27:103–140, 1979.

[33] G. C. Verghese and T. Kailath. Eigenvector chains for finite and infinite zeros of rational matrices. In *Proceedings of the 18th Conference on Decision and Control*, pages 31–32, Ft. Lauderdale, FL, December 1979.

[34] A. G. J. MacFarlane and N. Karcanias. Poles and zeros of linear multivariable systems: A survey of the algebraic, geometric and complex-variable theory. *International Journal of Control*, 24(1):33–74, 1976.

[35] H. H. Rosenbrock. *State Space and Multivariable Theory*. NY:Wiley, 1970.

[36] G. Verghese and T. Kailath. Impulsive behaviour in dynamical systems: Structure and significance. In *Proceedings of the 4th International Symposium on Mathematical Theory of Networks and Systems*, pages 162–168, Delft, The Netherlands, July 1979.

[37] B. R. Copeland and M. G. Safonov. Zero cancelling compensators for singular control problems and their application to the inner-outer factorization problem. Submitted to *Int J. Robust and Nonlinear Control*, July 1991.

[38] D. J. Limebeer, E. M. Kasenally, I. Jaimouka, and M. G. Safonov. All solutions to the four block general distance problem. In *Proceedings of the IEEE Conference on Decision and Control*, December 1988. Austin, TX.

[39] H. Kwakernaak and R. Sivan. *Linear Optimal Control Systems*. New York: Wiley, 1972.

[40] M. G. Safonov, D. J. N. Limebeer, and R. Y. Chiang. Simplifying the H^{∞} theory via loop-shifting, matrix-pencil and descriptor concepts. *International Journal of Control*, 50(6):2467–2488, 1989.

[41] E. F. Mageirou and Y. C. Ho. Decentralized stabilization via game theoretic methods. *Automatica*, 13:393–399, 1977.

[42] K. Zhou and P. Khargonekar. An algebraic Riccati equation approach to H^{∞} optimization. *Systems and Control Letters*, 11:85–91, 1988.

[43] I. R. Petersen. Disturbance attenuation and H^{∞} optimization: A design method based on the algebraic riccati equation. *IEEE Transactions on Automatic Control*, AC-32(5):427–429, 1987.

[44] I. Gohberg, P. Lancaster, and L. Rodman. *Matrix Polynomials*. Academic Press, Inc., 1982.

[45] G. Strang. *Linear Algebra and its Applications*. Academic Press, Inc., second edition, 1980.

[46] W. Rudin. *Real and Complex Analysis*. McGraw-Hill, Inc., 1987.

[47] A.A. Stoorvogel. The singular H^{∞} control problem with dynamic measurement feedback. *SIAM J. Opt. and Contr.*, 29(1):160–184, January 1991.

[48] D. J. Bender and A. J. Laub. The linear-quadratic regulator for descriptor systems. *IEEE Transactions on Automatic Control*, AC-32(8):672–688, August 1987.

[49] J. J. Dongarra, C. B. Moler, J. R. Bunch, and G. W. Stewart. *LINPACK User's Guide*. SIAM, 1979.

[50] P. Van Dooren. The generalized eigenstructure problem in linear system theory. *IEEE Transactions on Automatic Control*, AC-26(1):111–129, February 1981.

[51] R. Y. Chiang and M. G. Safonov. *Robust Control Toolbox User's Guide*. Mathworks, South Natick, MA, 1988.

Techniques in Stability Robustness Bounds for Linear Discrete-Time Systems

James B. Farison
Department of Electrical Engineering
The University of Toledo
Toledo, Ohio 43606

Sri R. Kolla
Department of General Engineering
Pennsylvania State University - Shenango Campus
Sharon, Pennsylvania 16146

I. INTRODUCTION

One of the important steps in designing controllers for physical systems is the development of mathematical models. These models are often inaccurate. Controller designs based on these models may not perform adequately when applied to the actual physical systems. Sometimes, even if the original model is correct, the physical system it represents might change during operation. In this case also, the controller designed for the model may not perform well with the changed plant. It is, therefore, desirable to design controllers that perform adequately when the model imperfectly represents the system. A robust control design is a

CONTROL AND DYNAMIC SYSTEMS, VOL. 50

design which behaves in an acceptable fashion even in the presence of model errors [1,2].

Physical systems are usually modeled as transfer function matrices in the frequency domain or as state-space matrices in the time domain [3]. In the frequency domain, the robustness is measured either as gain and phase margins or the tolerance of plant perturbations [4]. Some results on frequency-domain robustness can be found in [4-6]. In the time domain, robustness is usually measured by the tolerance of state-space matrix perturbations [7]. This chapter deals with the robustness of linear discrete-time systems in the time domain.

There are two types of system specifications: stability and performance. The eigenvalues and/or a Lyapunov function of the system model determine the stability of discrete-time systems [3]. The important performance measures of the system are regulation and time response. In terms of these specifications, we can conceive two corresponding types of robustness: stability robustness and performance robustness. Stability robustness means that closed-loop system stability is maintained in the presence of parameter variations. Similarly, performance robustness means that satisfactory levels of performance are maintained in the presence of parameter variations. This chapter deals with the stability robustness of linear discrete-time systems.

For continuous-time systems, there is considerable literature on these robustness topics [1,2,7-12]. Recently, robustness issues for discrete-time systems have also begun to receive attention. This chapter summarizes existing stability robustness techniques for discrete-time systems. Section II explains different types of system uncertainties, and gives a mathematical representation for each of these uncertainty models. A detailed account of some of the existing techniques on stability robustness bounds is given in Section III. Section IV gives controller design methods for robust stability.

Some of these stability robustness results are applied to practical examples in Section V. Concluding observations are given in Section VI.

II. PERTURBATION MODELS

In this section, different types of perturbations (uncertainties, errors) are explained. Mathematical models that describe these perturbations are given. Robust stability bounds on these perturbations are given in Section III.

Consider the discrete-time system described by the state-space equation

$$x(k+1) = Ax(k) + f(k, x(k)), \tag{1}$$

where A is an n x n real, time-invariant, asymptotically stable matrix, x is the n-dimensional state vector and f is the uncertainty. Depending upon the type of uncertainty, f can be classified as nonlinear or linear and as stochastic or deterministic. A brief description of these classifications follows.

A. Nonlinear vs. Linear Perturbations

In the nonlinear case, the perturbation f is a nonlinear function of x(k) and is norm bounded as

$$\| f(k, x(k)) \| \leq \gamma \| x(k) \| . \tag{2}$$

The problem is to determine the maximum allowable value (bound) on γ so that the system (1) is stable. This problem is usually solved using the Lyapunov approach. References [13-15] give sufficient conditions for robust stability. A brief account of these bounds is given in Section III.

In the linear case, the perturbation f is modeled as

$$f(k, x(k)) = E\,x(k) , \tag{3}$$

where E is an n x n matrix. These linear perturbations can be of two types. If E is a function of time k, represented as E(k), then the perturbations are linear and time varying. Otherwise, they are linear and time invariant. The Lyapunov approach is generally used for time-varying perturbations, and eigenvalue and M-matrix methods are generally used for time-invariant perturbations.

Linear perturbations (time-varying or time-invariant) can be either of unstructured or structured type. In the case of unstructured perturbations, bounds are given on the norm $\| E \|$ of the E matrix. In the structured perturbations case, bounds are given on the magnitudes $|e_{ij}|$ of the elements e_{ij} of E. The structured perturbations are independent if the elements of E vary independently. On the other hand, if the elements vary dependently they can be modeled as

$$E = \sum_{i=1}^{m} k_i E_i, \tag{4}$$

where k_i are uncertain parameters (time-varying or time-invariant) and E_i are n x n constant matrices. In this case, bounds are obtained on k_i. There is considerable literature on robust stability bounds for different types of these linear perturbations [16]. Section III gives a brief account of some of these results.

It may be noted that the results for nonlinear perturbations can be applied to linear perturbations (but not vice versa). However, these results tend to be conservative compared to the results derived for the linear case. Similarly, time-varying perturbation results can be applied to time-invariant perturbations (but not vice versa). Again, however, the results tend to be conservative.

B. Stochastic vs. Deterministic Perturbations

The perturbation f in (1) can be either a deterministic function, as assumed in Section II,A, or it can be a stochastic function. References [17-20] consider the perturbations as stochastic and obtain robust stability bounds. For nonlinear perturbations, [20] assumes f to be bounded on its second moment as

$$\mathcal{E}\left\{\|f(k,\ x(k),\ \omega)\|^2\right\} \le \gamma^2 \mathcal{E}\left\{\|x(k)\|^2\right\}, \tag{5}$$

where $\mathcal{E}$ is the expectation operator, $\omega \in \Omega$, and $(\Omega, \mathcal{F}, \mathcal{P})$ is the probability space. For this model, bounds are derived on γ which can be tolerated by the system without causing instability [20].

For the linear stochastic perturbation case [17-19], f is modeled as

$$f(k,\ x(k),\ \omega) = \left[\sum_{i=1}^{m} k_i\ (k,\omega)\ E_i\right] x(k)\ , \tag{6}$$

where k_i are zero-mean, stationary, mutually uncorrelated scalar random sequences and E_i are known constant matrices. Robust stability bounds are derived [17-19] on the vector ν of perturbation variances ν_i,

$$\nu = \left[\nu_1\ \ \nu_2\ \ \ldots\ \ \nu_m\right]^T, \tag{7}$$

where $v_i = \mathcal{E}\left\{k_i^2\right\}$.

The Lyapunov stability method is used to obtain bounds on these stochastic linear and nonlinear perturbation cases. These results parallel the results for the deterministic perturbation case. In Section III, deterministic perturbation bound results are given. The results for the stochastic case can be found in [17-20].

The classification of different types of perturbations explained so far can be easily visualized with Fig. 1.

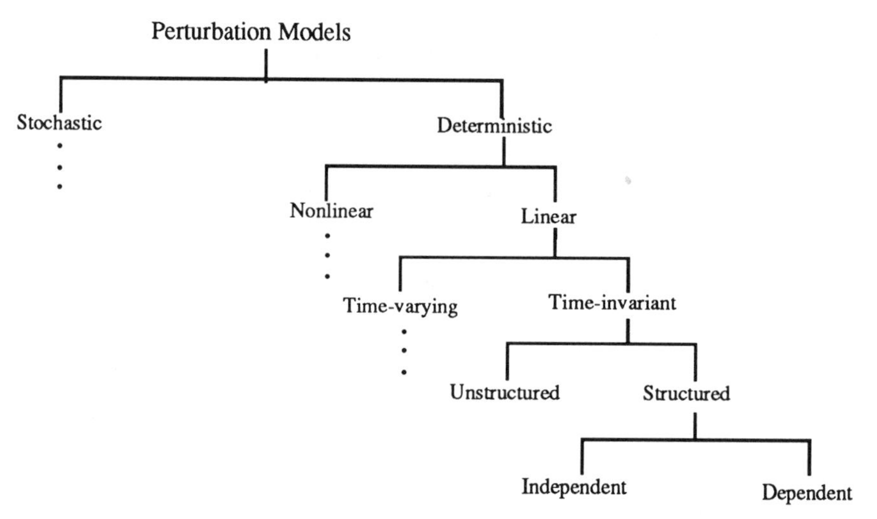

Fig. 1. Perturbation models.

C. Interval Systems: A Related Model

In the system model (1), A is a constant and stable matrix, and f is the perturbation. If f is completely known, one can check the stability robustness conditions and see whether the system (1) is stable in the presence of f. On the other hand, if f is not known, one can find the stability robustness bounds and get an idea of the type of f that can be tolerated. A related model to (1) is the interval system [21] described by

$$x(k+1) = [A]\, x(k) \quad , \tag{8}$$

where the interval matrix $[A] = [\underline{A}, \bar{A}]$, $\underline{A} \le \bar{A}$, $\underline{A} = \{\underline{a}_{ij}\}$ and $\bar{A} = \{\bar{a}_{ij}\}$ are n x n matrices. The notation $A_1 \le A_2$ represents two matrices whose elements satisfy $a_{1ij} \le a_{2ij}$ for all i and j. Any matrix of the family [A] is thus a matrix of the form $A = \{a_{ij}\}$ where $a_{ij} \in [\underline{a}_{ij}, \bar{a}_{ij}]$ for all i and j. Knowing [A] completely, it is of interest to see whether the system (8) is stable. There are two types of results available for this problem. In the first type, system (8) is transformed into system (1) and stability robustness conditions are checked [22]. The second type of results considers system (8) directly and gives stability conditions based on the corner matrices $\underline{A}$ and $\bar{A}$ [21]. A brief review of these two types of results is given in Section III.

In addition to the matrix-based model (8) for interval systems, we can also have interval polynomial models [21]. These polynomials might come directly from the transfer function representation of the system or from the characteristic polynomial of matrix [A] in (8). An interval polynomial can be represented as

$$\phi(z) = \sum_{i=0}^{n} a_{n-i}\, z^{i}, \tag{9}$$

where $a_i \in [\underline{a}_i, \bar{a}_i]$, $\underline{a}_i \le \bar{a}_i$. In this case we are interested in determining whether the polynomial (9) is a Schur polynomial when the coefficients a_i vary in the interval [21]. Based on the Kharitonov theorem for continuous-time system interval polynomials [23], some results are obtained for the discrete-time system polynomial (9). Some of these results are presented in Section III.

III. STABILITY ROBUSTNESS BOUNDS

In the preceding section, different types of parameter perturbations are explained. In this section, stability robustness bounds are given for these perturbations so that the nominally stable system remains stable in the presence of these perturbations. Different approaches used to obtain these bounds can be roughly classified as Lyapunov function methods, eigenvalue methods, and M-matrix methods. Before giving the bounds on the perturbations, these approaches are briefly explained.

A. Methodology

As noted, there are several methods used to obtain robust stability bounds. Each of these methods is used for a particular type of perturbation model. A brief review of these methods follows.

1. Lyapunov stability theory

The direct method of Lyapunov is a powerful tool to study the stability of linear and nonlinear systems, whether time invariant or time varying [24]. This method is based on the fact that if the system has an asymptotically stable equilibrium state, then the stored energy of the system displaced within a domain of attraction decays with increasing time until it finally assumes its minimum value at the equilibrium state. The Lyapunov stability theory as applied to a linear time-invariant discrete-time system can be summarized as the following theorem.

Theorem 1: Consider the linear discrete-time system

$$x(k+1) = Ax(k) , \tag{10}$$

where x is the n-dimensional state vector and A is an $n \times n$ matrix. A necessary and sufficient condition for the equilibrium state $x = 0$ to be asymptotically stable is that, given any symmetric positive definitive matrix Q, there exists a unique symmetric positive definite matrix P such that

$$A^T PA - P + Q = 0 . \tag{11}$$

The Lyapunov energy function is given by $V = x^T Px$, and the forward difference is given by $\Delta V = -x^T Qx$.

This theorem gives a necessary and sufficient condition for the stability of a linear time-invariant system. Lyapunov stability theory is extensively used for the robust stability analysis of both linear and nonlinear systems. However, it gives only a sufficient condition for nonlinear and linear time-varying systems.

2. Eigenvalue approach

The stability of linear time-invariant systems is usually studied using root-locus or eigenvalue approaches [3]. These approaches are generally used to obtain robustness bounds for linear time-invariant perturbations [25-27]. The main result used in these studies is motivated by the Gastinel-Kahan theorem and can be summarized by the following theorem [25].

Theorem 2: Let A and E be n x n matrices. If $\rho(A) < 1$, and $\rho(A+E) \geq 1$, then there exists $\delta \in (0,1)$ and $\theta \in [0, \pi]$ such that the matrix $e^{j\theta}I - A - \delta E$ is singular. $\rho(\cdot)$ is the spectral radius of $(\cdot)$.

3. M-matrix method

Another method generally used for the stability analysis of discrete-time systems is the M-matrix approach [28]. As applied to linear time-invariant systems, the main result can be stated as the following theorem [28].

Theorem 3: The origin x = 0 of the system (10) is asymptotically stable if $I - |A|$ is an M-matrix. The notation $|A|$ represents the matrix whose elements are the magnitudes of the elements a_{ij} of A.

In the remainder of this section, stability robustness bounds for discrete-time systems based on the above methods are presented. Both nonlinear and linear deterministic perturbations are considered.

B. Nonlinear Perturbations

Consider the system (1) with the nonlinear perturbation f described by (2). To obtain the bound on γ for robust stability, generally the Lyapunov stability theory is used. References [13-15,29-31] give these bounds. The initial bound by Lee and Lee [29] is later corrected by Tsay, et al. [30]. This result can be stated as the following theorem.

Theorem 4: The discrete-time system (1) with the perturbation (2) is stable if

$$\gamma < \gamma_1 = \frac{\sqrt{||PA||^2 + ||P|| \, \lambda_{min}(Q)} - ||PA||}{||P||} , \tag{12}$$

where P and Q are the matrices in the Lyapunov equation (11) and $\lambda(\cdot)$ are the eigenvalues of $(\cdot)$.

A similar result is also given by Shen and Kung [31]. Sezer and Siljak [13] give two bounds which can be stated as the following theorem.

Theorem 5: The discrete-time system (1) with the perturbation (2) is stable if

$$\gamma < \gamma_2 = \frac{\sqrt{\left[\sigma_{min}(Q) + \sigma_{max}(P-Q)\right]} - \sqrt{\sigma_{max}(P-Q)}}{\sqrt{\sigma_{max}(P)}} , \tag{13}$$

or

$$\gamma < \gamma_3 = \frac{\sigma_{min}(Q)}{\sigma_{max}(P) + \sqrt{\sigma_{max}(P)}\sqrt{\sigma_{max}(P-Q)}} , \tag{14}$$

where $\sigma(\cdot)$ are the singular values of $(\cdot)$, and P and Q are the matrices in the Lyapunov equation (11).

These two bounds (13) and (14) are obtained using two different types of Lyapunov functions. Some comments on the selection of the matrix Q in the Lyapunov equation (11) to get maximum bounds, and other methods to improve these bounds, are also presented in [13].

For nonlinear perturbations of the type described by (2), Yaz and Niu [14] also give a stability robustness bound. Their bound can be stated as the following theorem.

Theorem 6: The discrete-time system (1) with the perturbation (2) is stable if

$$\gamma < \gamma_4 = -\sigma_{max}(A) + \sqrt{\left[\sigma_{max}(A)\right]^2 + \lambda_{min}(Q)\, \lambda_{max}^{-1}(P)} \quad , \qquad (15)$$

where $\sigma(\cdot)$, $\lambda(\cdot)$, P and Q are as defined in Theorems 4 and 5.

Reference [14] also offers some comments on the selection of the Q matrix in (11) for maximum bound.

Apart from the bounds (12) - (15) that apply to a general asymptotically stable A matrix (eigenvalues within the unit circle), improved results can be obtained if one knows the location of the eigenvalues of A. References [14,29-31] also give results that incorporate the eigenvalue location information. Bourles, et al., [15] give bounds for ρ-stability of the system. A linear time-invariant system whose eigenvalues λ are such that $|\lambda| < 1/\rho$ is ρ-stable.

In conclusion, there are no theoretically established results that show which of the bounds (12) - (15) is better. The superiority may depend on the properties of the system matrix A. In all these results, the symmetric positive definite Q matrix in (11) is arbitrary. In general, Q = I gives the largest bound for these nonlinear perturbations [13,14]. It is also possible to improve these bounds using state transformation [13,14]. The concept of state transformation is explained in the linear perturbations case. It is also possible to show that the bound cannot exceed a certain value for a given A. For (13) - (15), this value is $1 - \rho(A)$ [13,14].

C. Linear Perturbations

Consider the perturbation f described by the linear function (3). As explained in Section II, E can be time varying or time invariant. In the following, existing bounds for these two perturbation classes are explained.

1. Time-varying perturbations

To obtain robust stability bounds, Lyapunov stability theory is usually applied for this class of perturbations. Results for unstructured perturbations are given in [14,22,32-38]. Results for structured perturbations can be found in [17,22,33-44].

a. Unstructured perturbations: The nonlinear perturbation results of Section III,B can be directly applied to linear time-varying unstructured perturbations. In this case, γ is replaced by $\| E \|$ in (2). Ishihara [32], Kolla, et al. [22], and Yaz and Niu [14] independently derive the following result which is similar to Theorem 6.

Theorem 7: The discrete-time system (1) with the perturbations (3) is stable if

$$\sigma_{max}(E) < \mu_{u1} = -\sigma_{max}(A) + \sqrt{\left[\sigma_{max}(A)\right]^2 + \frac{\sigma_{min}(Q)}{\sigma_{max}(P)}}, \quad (16)$$

where P and Q are the matrices in the Lyapunov equation (11).

The Q matrix in (11) is any symmetric positive definite matrix. The following remark suggests the selection of Q that gives the maximum bound [14,22].

Remark 1: The bound (16) is maximum when $Q = I$, for which $\sigma_{min}(Q) = 1$.

As shown by Yaz and Niu [14] and Farison and Kolla [33], it is possible to get an indication of the size of μ_{u1} for a given A matrix as summarized in the following remark.

Remark 2: The bound μ_{u1} for any A satisfies

$$\mu_{u1} \leq 1 - \rho(A) \; . \tag{17}$$

The equality in (17) holds when A is a normal matrix ($A^T A = AA^T$).

The bound (16) gives a sufficient condition for stability. The conservatism of the bound can be reduced by transforming the system to a different coordinate frame using a similarity transformation. The concept is explained in [13,14,34,35] and summarized below.

Theorem 8: For a given nonsingular n x n matrix T, the discrete-time system (1) with perturbations (3) is stable if

$$\sigma_{max}(E) < \mu^*_{u1} = \frac{\hat{\mu}_{u1}}{\sigma_{max}(T^{-1})\, \sigma_{max}(T)} \; , \tag{18}$$

where

$$\hat{\mu}_{u1} = -\sigma_{max}(\hat{A}) + \sqrt{[\sigma_{max}(\hat{A})]^2 + \frac{\sigma_{min}(Q)}{\sigma_{max}(\hat{P})}} \; . \tag{19}$$

In (19), $\hat{A} = T^{-1}AT$, and $\hat{P}$ is the solution of the Lyapunov equation

$$\hat{A}^T \hat{P}\hat{A} - \hat{P} + Q = 0 \ . \tag{20}$$

The state transformation is an elegant method to reduce the conservatism of the bounds. However, there are no established methods to determine the best transformation matrix T.

Shen and Kung [35] give several robustness bounds for unstructured perturbations by extending the nonlinear perturbation results in [13,30]. The following is one of the bounds derived for the linear case [35].

Theorem 9: The discrete-time system (1) with the perturbations (3) is stable if

$$\| E \| < \mu_{u2} = 1 - \sqrt{1 - \frac{1}{\| P \|}} , \tag{21}$$

where P is the solution of the Lyapunov equation (11) with Q = I.

Zhu and Skelton [36], and Hsieh and Skelton [37], give bounds similar to (21) for discrete-time system covariance controllers. Eslami [38] also uses the Lyapunov method to obtain bounds for both unstructured and structured perturbations. Reference [38] uses iterative methods to get improved bounds. As explained in [14,35], if one knows the locations of the eigenvalues of A, improved results can be obtained.

b. Structured perturbations: The structured perturbations could be independent or dependent. Stability robustness bounds for the independent case are given in [22,34,39,40]. On the other hand, [17,18,22,36,38,41-44] consider dependent perturbations.

To get bounds on independent perturbations, define constants ε_{ij} and ε such that the elements $e_{ij}(k)$ of E(k) satisfy

$$e_{ij}(k) \leq \max_{k} | e_{ij}(k) | = \varepsilon_{ij} \text{ and } \varepsilon = \max_{i,j} \varepsilon_{ij} . \tag{22}$$

The following result is due to Kolla, et al. [22].

Theorem 10: The discrete-time system (1) with the perturbations (3) is stable if

$$\varepsilon < \mu_{i1} = - \frac{\sigma_{max}(U^T \mid PA \mid)_s}{\sigma_{max}(U^T \mid P \mid U)} + \sqrt{\left[\frac{\sigma_{max}(U^T \mid PA \mid)_s}{\sigma_{max}(U^T \mid P \mid U)}\right]^2 + \frac{\sigma_{min}(Q)}{\sigma_{max}(U^T \mid P \mid U)}} , \tag{23}$$

where P and Q are the matrices in the Lyapunov equation (11). The notation $(\cdot)_s$ means the symmetric part of the square matrix $(\cdot)$: $(\cdot)_s = \left[(\cdot) + (\cdot)^T\right]/2$, and $U = \left[u_{ij}\right]$, $u_{ij} = \varepsilon_{ij}/\varepsilon$.

Note that $0 \leq u_{ij} \leq 1$. If the perturbation e_{ij} of some a_{ij} is not explicitly known, one can take u_{ij} as some positive number, for instance 1, to get an idea of the stable perturbation range by computing the RHS of (23). If the perturbation e_{ij} of a_{ij} is known to be zero, $u_{ij} = 0$. To get a bound on the relative variation of the elements of A, one can take $u_{ij} = | a_{ij} | / \max | a_{ij} |$. Fong [39] also gives similar results for the relative variation case.

As in the unstructured perturbation case, it is possible to get better bounds by coordinate transformation for the structured perturbation case also. For this case, we present an improved bound using a diagonal transformation as applied by Kolla and Farison [34].

Theorem 11: Given $T = \text{diag}[t_1, t_2, \ldots, t_n]$, the discrete-time system (1) with the perturbations (3) is stable if

$$\varepsilon < \mu_{i1}^* = \frac{\hat{\mu}_{i1}}{\max\limits_{i,j} \left| \frac{t_j}{t_i} \right| u_{ij}}, \tag{24}$$

where

$$\hat{\mu}_{i1} = -\frac{\sigma_{max}(\hat{U}^T \mid \hat{P}\hat{A} \mid)_s}{\sigma_{max}(\hat{U}^T \mid \hat{P} \mid \hat{U})} + \sqrt{\left[\frac{\sigma_{max}(\hat{U}^T \mid \hat{P}\hat{A} \mid)_s}{\sigma_{max}(\hat{U}^T \mid \hat{P} \mid \hat{U})}\right]^2 + \frac{\sigma_{min}(Q)}{\sigma_{max}(\hat{U}^T \mid \hat{P} \mid \hat{U})}}, \tag{25}$$

$$\hat{u}_{ij} = \frac{\hat{\varepsilon}_{ij}}{\hat{\varepsilon}}, \quad \hat{\varepsilon}_{ij} = \left| \frac{t_j}{t_i} \right| \varepsilon_{ij} \text{ and } \hat{\varepsilon} = \max_{i,j} \hat{\varepsilon}_{ij} . \tag{26}$$

As in the unstructured case, state transformation gives improved bounds. However, there are no general guidelines to find the best transformation matrix T.

Han and Wu [40] give the following alternative result.

Theorem 12: The discrete-time system (1) with the perturbations (3) is stable if

$$\varepsilon < \frac{1}{\sigma_{max} (\mid A \mid^T \mid P \mid U)_s} \tag{27}$$

and

$$\varepsilon^2 < \frac{1}{\sigma_{max}\,(U^T \mid P \mid U)}\,, \tag{28}$$

where U is as defined in Theorem 10 and P is the solution of the Lyapunov equation (11) with Q = 3I.

The structured dependent perturbations represented by (4) have received considerable interest, as the model is general and includes independent perturbations as a special case. When specialized to the independent case, $m = n^2$ in (4) and the k_i represent the variations in each element [12]. Yaz [17] gives the following result.

Theorem 13: The discrete-time system (1) with the perturbations (3) and (4) is stable if any one of the following conditions holds:

$$\| \tilde{k} \|_1 < \left[\max_i \sigma_{max}\,(E_i) \right]^{-1} \mu_{u1} \tag{29}$$

$$\| \tilde{k} \|_2 < \sigma_{max}^{-1}\,(E_e)\, \mu_{u1} \tag{30}$$

$$\| \tilde{k} \|_\infty < \sigma_{max}^{-1} \left[\sum_{i=1}^{m} \mid E_i \mid \right] \mu_{u1}\,, \tag{31}$$

where

$$\tilde{k} = [k_1 \;\; k_2 \; \ldots \; k_m]^T\,, \tag{32}$$

$$E_e = [E_1 \;\; E_2 \; \ldots \; E_m]\,\cdot \tag{33}$$

The factor μ_{u1} is as defined in (16), and P is the solution of the Lyapunov equation (11) with Q = I.

Similar types of bounds are given by Niu, et al. [18]. Yaz [17] gives other types of bounds which are similar to Theorem 15 below.

Kolla, et al., [22] also give bounds for dependent perturbations. Their result can be summarized as the following theorem.

Theorem 14: The discrete-time system (1) with the perturbations (3) and (4) is stable if

$$\sum_{i=1}^{m} |k_i|^2 \sigma_{max}(P_{ee}) + 2 \sum_{i=1}^{m} |k_i| \sigma_{max}(P_{aei}) < \sigma_{min}(Q) \tag{34}$$

or

$$|k_j| < -\frac{\sigma_{max}\left(\sum_{i=1}^{m} |P_{aei}|\right)}{m\,\sigma_{max}(|P_{ee}|)} + \sqrt{\left[\frac{\sigma_{max}\left(\sum_{i=1}^{m} |P_{aei}|\right)}{m\,\sigma_{max}(|P_{ee}|)}\right]^2 + \frac{\sigma_{min}(Q)}{m\,\sigma_{max}(|P_{ee}|)}} \tag{35}$$

for $j = 1, 2, \ldots, m$,

where
$$P_{ee} = \begin{bmatrix} (E_1^T P E_1) & (E_1^T P E_2)_s & \cdots & (E_1^T P E_m)_s \\ (E_1^T P E_2)_s & (E_2^T P E_2) & \cdots & (E_2^T P E_m)_s \\ \vdots & & & \\ (E_1^T P E_m)_s & (E_2^T P E_m)_s & \cdots & (E_m^T P E_m) \end{bmatrix}, \tag{36}$$

$$P_{aei} = \left(A^T P E_i\right)_s , \tag{37}$$

and P and Q are the matrices in the Lyapunov equation (11).

Niu and Abreu-Garcia [41] combine the concepts of [17] and [22] and give the following results. These results require knowledge of the location of the eigenvalues of A in (1), and assume that the eigenvalues are inside a disk of radius $(1 + \alpha)^{-1/2}$, $\alpha > 0$.

Theorem 15: The discrete-time system (1) with perturbations (3) and (4) is stable if any of the following conditions hold:

$$\| \tilde{k} \|_1 < \frac{1}{\sqrt{\left(1+\alpha^{-1}\right) \max\limits_{i,j} \{\sigma_{max}(P_{ij})\}}} \tag{38}$$

$$\| \tilde{k} \|_2 < \frac{1}{\sqrt{\left(1+\alpha^{-1}\right) \sigma_{max}(P_e)}} \tag{39}$$

$$\| \tilde{k} \|_\infty < \frac{1}{\sqrt{\left(1+\alpha^{-1}\right) \sigma_{max}\left(\sum\limits_{i,j=1}^{m} | P_{ij} |\right)}} , \tag{40}$$

where

$$P_{ij} = E_i^T PE_j \ , \tag{41}$$

$$P_e = \begin{bmatrix} E_1^T PE_1 & \ldots & E_i^T PE_j & \ldots & E_m^T PE_m \end{bmatrix} , \tag{42}$$

$$i, j = 1, 2, \ldots, m,$$

and P is the solution of the Lyapunov equation

$$(1+\alpha)\ A^T PA - P + I = 0 \ . \tag{43}$$

As in the independent perturbation case, it is possible to improve dependent perturbation results using state transformation. Some of these results can be found in [41].

The results in Theorems 13-15 provide bounds on the absolute values of the uncertain parameters; that is, the stability region in the parameter space is always symmetric with respect to the origin. This may introduce conservatism in the results. Gao and Antsaklis [42] give results that remove the symmetry restriction. Their results can be stated as the following theorem.

Theorem 16: The discrete-time system (1) with perturbations (3) and (4) is stable if

$$\sum_i k_i \lambda_i + \sum_{i,j} k_i k_j f_{ij} < 1 \ , \tag{44}$$

where

$$\lambda_i = \begin{Bmatrix} \lambda_{max}(P_{aei}) & \text{for } k_i \ \geq \ 0 \\ \lambda_{min}(P_{aei}) & \text{for } k_i \ < \ 0 \end{Bmatrix} \ i = 1, \ldots m \tag{45}$$

$$f_{ij} = \begin{cases} \lambda_{max}\left(\frac{P_{ij}}{2}\right) & \text{for } k_i k_j \geq 0 \\ \lambda_{min}\left(\frac{P_{ij}}{2}\right) & \text{for } k_i k_j < 0 \end{cases} \quad i, j = 1, \ldots, m \quad . (46)$$

P_{aei} is as defined in (37), P_{ij} is as defined in (41) and P is the solution of the Lyapunov equation (11) with Q = 2I.

There are several other results for the dependent variations case. Most of these results follow concepts similar to Theorems 13-16. Gu and Chen [43] and Eslami [38] suggest iterative techniques to obtain improved bounds. Shen and Kung [35] extend the unstructured perturbation results to structured dependent perturbations. Zhu and Skelton [36] give stability robustness bounds for covariance controllers. Gu [44] suggests optimization methods for discrete-time systems to obtain robust stability bounds for quadratic stability. In general, all of the methods for time-varying perturbations use the Lyapunov stability concept.

2. Time-invariant perturbations

The eigenvalue approach is generally used to obtain bounds on time-invariant perturbations. Some results also use the M-matrix approach. Unlike the time-varying perturbation case, these are necessary and sufficient conditions for certain classes of time-invariant systems. All of the sufficient-condition results for time-varying perturbations in Section III,C,1 can be directly applied to time-invariant perturbations. However, these results tend to give conservative bounds. In the following, the results obtained directly for time-invariant perturbations are explained. As in the time-varying case, unstructured [25-27,45-48] and structured [26,49,50] perturbations are considered separately.

a. Unstructured perturbations: Using the eigenvalue approach, Martin [25], Juang, et al., [26] and Qu and Dorsey [27] independently derive the following sufficient condition.

Theorem 17: The discrete-time system (1) with the time-invariant perturbations (3) is stable if

$$\| E \| < \mu_{u3} = \frac{1}{\| (e^{j\theta} I - A)^{-1} \|} \tag{47}$$

for all $\theta \; \varepsilon \; [0,\pi]$.

Martin [25] also gives the following necessary and sufficient condition.

Theorem 18: The discrete-time system (1) with the time-invariant perturbations (3) and $E \; \varepsilon \; \beta(\delta)$ is stable if and only if

$$\delta < \mu_{u3} = \frac{1}{\| \left(e^{j\theta} I - A \right)^{-1} \|} \tag{48}$$

for all $\theta \; \varepsilon \; [0, \pi]$, where $\beta(\delta) \equiv \left\{ E \; \varepsilon \; \mathcal{C}^{n x n} : \| E \| \leq \delta \right\}$, $\mathcal{C}$ is the field of complex numbers and $\delta > 0$.

Mori [45] establishes a result similar to Remark 2 for the bound μ_{u3}. Hinrichsen and Pritchard [46] also give bounds using the eigenvalue approach.

Qiu and Davison [47] use the properties of a Kronecker product and give several bounds for unstructured perturbations. One of these results can be stated as the following theorem.

Theorem 19: The discrete-time system (1) with the time-invariant perturbations (3) is stable if

$$\sigma_{max}(E) < \mu_{u4} = \min \Big\{ \sigma_{min}(A-I) \ , \ \sigma_{min}(A+I) \ ,$$

$$-\sigma_{max}(A) + \sqrt{\sigma^2_{max}(A) + \sigma_{n^2-1}(A \otimes A - I)} \ \Big\} \ , \quad (49)$$

where $\otimes$ is the Kronecker product and $\sigma_{n^2-1}(\cdot)$ is the n^2-1 singular value, with order $\sigma_1(\cdot) \geq \sigma_2(\cdot) \geq \cdots$.

Reference [47] also establishes a result similar to Remark 2 for the bound μ_{u4}. Hyland and Collins [48] use Kronecker product properties to obtain results for both robust stability and robust performance.

b. Structured perturbations: For structured independent perturbations, Juang, et al., [26] extend Theorem 17 and give the following bound on ε of (22) considering e_{ij} are time-invariant.

Theorem 20: The discrete-time system (1) with the time-invariant perturbations (3) is stable if

$$\varepsilon < \mu_{i2} = \frac{1}{\rho\left[\left|\left(e^{j\theta}I - A\right)^{-1}\right| U\right]} \quad (50)$$

for all $\theta \ \varepsilon \ [0,\pi]$, where U is as defined in Theorem 10.

Rachid [49] uses the M-matrix concept [28] and gives the following sufficient condition for independent perturbations.

Theorem 21: The discrete-time system (1) with the perturbations (3) and $|E| \leq \varepsilon |A|$ is stable if

$$\varepsilon < \mu_{i3} = \frac{1-\rho(|A|)}{\rho(|A|)} . \tag{51}$$

Rachid [49] also gives the following necessary and sufficient condition if A is a positive matrix (a matrix with all positive elements).

Theorem 22: If A is a positive matrix, the discrete-time system (1) with the perturbations (3) and $|E| \leq \varepsilon |A|$ ($\varepsilon < 1$) is stable if and only if

$$\varepsilon < \mu_{i3} = \frac{1-\rho(|A|)}{\rho(|A|)} . \tag{52}$$

This theorem is also valid if A is a negative matrix (a matrix with all negative elements).

For structured dependent perturbations, Kolla and Das [50] extend Theorem 20 and give the following result.

Theorem 23: The discrete-time system (1) with the time-invariant perturbations (3) and (4) is stable if

$$|k_i| < \mu_{d1} = \frac{1}{\rho\left[\sum_{i=1}^{m} \left|(e^{j\theta}I - A)^{-1} E_i\right|\right]} , \quad i=1, 2, \ldots, m, \tag{53}$$

for all $\theta \; \varepsilon \; [0,\pi]$. For m=1 (single parameter dependent uncertainty), the condition is

$$| k_1 | < \frac{1}{\rho\left[(e^{j\theta}I - A)^{-1} E_1\right]} \tag{54}$$

for all $\theta \varepsilon [0,\pi]$.

Remark 3: Bound (50) is a special case of bound (53) if we define

$$E_{n(i-1)+j} = u_{ij}\, e_i\, e_j^T \text{ , for i, j = 1, 2, . . . , n,} \tag{55}$$

where the elements u_{ij} are as defined in Theorem 10 and e_i is an n-dimensional column vector with unity in the ith entry and zero elsewhere. With this definition, $| k_i |$ in (53) corresponds exactly to ε in (50) and μ_{d1} in (53) is equal to μ_{i2} in (50) [50].

Apart from these bounds, some of the results presented below for interval systems can also be used to obtain robust stability bounds. Similarly, the bound results of Sections III,C,1 and III,C,2 can be used for interval matrix stability analysis.

D. Stability Analysis of Interval Systems

As explained in Section II,C, we can conceive two types of models for interval systems: interval matrices and interval polynomials [21,51]. For interval matrices, there are two types of results. In the first type, system (8) is transformed into system (1) and stability robustness conditions are checked. The second type of results directly consider system (8) and give stability results. Similarly, interval polynomials also have two types of results. In the following, some of these results are summarized.

1. Interval matrices

Consider the system (8), and define the n x n average matrix

$$A_0 = \tfrac{1}{2}\,(\bar{A} + \underline{A}) \tag{56}$$

and the deviation matrix

$$D = \tfrac{1}{2}\,(\bar{A} - \underline{A}) \ , \tag{57}$$

where $d_{ij} \geq 0$. In terms of these matrices, system (8) is similar in form to that of (1). References [22,49,52] extend stability robustness bound results and give sufficient conditions for the stability of interval matrices. The result by Kolla, et al., [22] can be stated as the following theorem.

Theorem 24: The interval system (8) is stable if the average matrix A_0 is stable and

$$d < -\frac{\sigma_{max}\,(U^T \mid PA_0 \mid)_s}{\sigma_{max}\,(U^T \mid P \mid U)} + \sqrt{\left[\frac{\sigma_{max}(U^T \mid PA_0 \mid)_s}{\sigma_{max}(U^T \mid P \mid U)}\right]^2 + \frac{\sigma_{min}(Q)}{\sigma_{max}(U^T \mid P \mid U)}} \ , \tag{58}$$

where $d = \max(d_{ij})$, $U = [u_{ij}]$ with $u_{ij} = d_{ij}/d$, and P is the solution of the Lyapunov equation

$$A_0^T P A_0 - P + Q = 0 \ . \tag{59}$$

It should be mentioned that Theorem 24 is valid even if the elements of [A] in (8) are time-varying. For time-invariant systems, Theorem 20 or Theorem 21 can be used [49]. Juang, et al., [52] give the following result for time-invariant systems.

Let T be a similarity transformation matrix such that A_0 is transformed into

$$A_{0T} = \left[\bar{a}_{0ij}\right] = T^{-1} A_0 T \ . \tag{60}$$

Let Λ be a diagonal n x n matrix,

$$\Lambda = \text{diag}\left(\bar{a}_{011}, \ldots, \bar{a}_{0nn}\right) , \tag{61}$$

and define the non-negative matrix $\hat{A}_0$ by

$$\hat{A}_0 = | A_{0T} - \Lambda | + | T^{-1} | \ D | T | \ . \tag{62}$$

Theorem 25: The interval system (8) is stable if

$$| \bar{a}_{0ii} |_{max} + \rho(\hat{A}_0) < 1 \ , \tag{63}$$

where $| \bar{a}_{0ii} |_{max}$ is the maximum magnitude of $\bar{a}_{0ii}$, $i = 1, 2, \ldots, n$.

Results similar to Theorem 25 can also be obtained using Gershgorin's diagonal dominance theorem, as shown by Argoun [53].

There are several results that directly consider the matrices of corner elements [21,51-59]. A recent review paper by Jury [21] summarizes several of these results. Juang, et al., [52] and Zhou and Deng [57] give the following simple condition.

Theorem 26: The interval system (8) is stable if

$$\rho\,(A_m) < 1\ , \tag{64}$$

where $A_m = [a_{mij}]$, $a_{mij} = \max(\,|\,\underline{a}_{ij}\,|,\ |\,\bar{a}_{ij}\,|\,)$.

An important observation concerning this theorem is that it gives a necessary and sufficient condition for the cases $A_m = -\underline{A}$ and $A_m = \bar{A}$ [52]. This includes positive and negative interval matrices.

Lin, et. al., [58] give the following sufficient condition.

Theorem 27: The interval system (8) is stable if

$$\max_{1 \le i \le n} \left\{ \sum_{j=1}^{n} \max\Big(\,|\,\underline{a}_{ij}\,|\ ,\ \ |\,\bar{a}_{ij}\,|\,\Big) \right\} < 1 \tag{65}$$

or

$$\max_{1 \le i \le n} \left\{ \sum_{j=1}^{n} \max\Big(\,|\,\underline{a}_{ji}\,|\ ,\ \ |\,\bar{a}_{ji}\,|\,\Big) \right\} < 1 \tag{66}$$

or

$$\sum_{i=1}^{n} \sum_{j=1}^{n} \max\left(\underline{a}_{ij}^{2}\ ,\ \bar{a}_{ij}^{2}\right) < 1\ . \tag{67}$$

A non-negative interval matrix is an interval matrix for which $\underline{a}_{ij} \ge 0$ in (8). Shafai, et al., [59] give the following necessary and sufficient condition for non-negative interval matrices.

Theorem 28: The non-negative interval system (8) is stable if and only if all the principle minors of the matrix $I - \bar{A}$ are positive.

Reference [59] also gives results for a non-negative interval matrix using interval arithmetic. Jiang [60] proposes a necessary and sufficient condition for the stability of interval matrices based on the stability of the corner matrices, with a subsequent counterexample by Kolla and Farison [61].

2. Interval polynomials

One of the important areas in the polynomial approach is the extension of Kharitonov's Hurwitz polynomial theory [23] for continuous-time systems to discrete-time system Schur polynomials. A recent review paper by Jury [21] gives a brief account of several of these results. The Kharitonov theorem states that the family of polynomials of type (9) is strictly Hurwitz (all roots are in the left half of the complex plane) if and only if the 2^n corner polynomials are strictly Hurwitz. As a further simplification of this theorem, Kharitonov proved that the family of polynomials is strictly Hurwitz if and only if four specially formulated polynomials are strictly Hurwitz.

Several researchers have tried to obtain similar conditions for discrete-time systems assuming that the family of polynomials are Schur polynomials (all roots are inside the unit circle) [21,62-65]. It has been shown that the answer is positive for $n \leq 3$ but is negative for $n > 3$. For n=3, it is not sufficient that only four polynomials have their roots inside the unit circle; one must require that all eight polynomials have their roots inside the unit circle. Counterexamples have been given for systems of order four [62,65].

However, Bose and Zeheb [65] extend Kharitonov's result to Schur polynomials by using a bilinear transformation. The varying coefficients of the Schur polynomial are transferred into coefficients of a Hurwitz polynomial. The method is shown by Bose and Zeheb [65] and Lin, et al. [66]. The Hurwitz property of the transformed polynomial is tested by checking the four specially formulated polynomials. However, these results give only sufficient conditions for robust stability of Schur polynomials. This is in contrast to the necessary and sufficient conditions of Kharitonov for robust stability of Hurwitz polynomials.

Using a system theoretic approach as in Bose [67], several results have been given by Jury [21], Kraus, et al. [68], and Benidir and Picinbono [69] for discrete-time systems analogous to the Kharitonov result for continuous-time systems. Consider the polynomial (9), and define the symmetric and anti-symmetric parts

$$\phi_e(z) = \frac{1}{2}\left[\phi(z) + z^n \phi(z^{-1})\right] \tag{68}$$

and

$$\phi_o(z) = \frac{1}{2}\left[\phi(z) - z^n \phi(z^{-1})\right] . \tag{69}$$

Note that $\phi(z) = \phi_e(z) + \phi_0(z)$.

Theorem 29: All of the roots of the polynomial (9) are less than one in their absolute value if and only if the roots of $\phi_e(z)$ and $\phi_0(z)$ are simple and are located on the circle $|z| = 1$, alternately, and $|a_n/a_0| < 1$.

Jury [21] reports several variations of this theorem. These theorems require checking the 2^{n+1} polynomials. As explained in [21,70] this number can be further reduced. Another important result for polynomials is the edge theorem by Bartlett, et al. [71], applicable for both continuous-time and discrete-time systems. Several extensions of these results can be found in [21,72-76].

Apart from the results similar to the Kharitonov theorem described so far, there are results that give conditions on the coefficients a_i in (9) for robust stability [21,77-80]. Consider the discrete-time system characteristic equation

$$\phi_1(z) = z^n + a_{n-1} z^{n-1} + \ldots + a_1 z + a_0 = 0 , \tag{70}$$

where a_i are real numbers for $i = 0, 1, \ldots, n-1$. Let $a_i \geq 0$ and consider the following equation associated with (70),

$$\phi_2(z) = z^n + b_{n-1} z^{n-1} + \ldots + b_1 z + b_0 = 0 , \tag{71}$$

where b_i are real numbers and

$$| b_i | \leq a_i , \quad \text{for } i = 0, 1, \ldots, n-1 . \tag{72}$$

Mori and Kokame [79] give the following necessary and sufficient condition.

Theorem 30: The class of systems described by (71) is stable if and only if

$$\sum_{i=0}^{n-1} a_i < 1. \tag{73}$$

A detailed discussion of several other results for the polynomial case can be found in [21]. Soh and Evans [81] use the polynomial results for interval matrices.

E. Examples

Some of the perturbation bound results presented so far are illustrated with simple examples in this section.

Example 1: Consider the system matrix

$$A = \begin{bmatrix} 0.2 & 0.3 \\ 0.1 & -0.15 \end{bmatrix}.$$

The eigenvalues of the matrix are 0.2712 and -0.2212. The system is, therefore, stable. The solution of the Lyapunov equation (11) with $Q = I$ is

$$P = \begin{bmatrix} 1.0552 & 0.0466 \\ 0.0466 & 1.1159 \end{bmatrix},$$

which is positive definite. For the nonlinear perturbation described by (2), (15) gives

$$\gamma < \gamma_4 = 0.6373 .$$

For linear unstructured time-varying perturbations (3), (16) gives

$$\sigma_{max}(E) < \mu_{u1} = 0.6373 .$$

For linear structured independent time-varying perturbations represented by the following U matrices, (23) gives

U	$\begin{bmatrix} 1 & 0 \\ 0 & 0 \end{bmatrix}$	$\begin{bmatrix} 1 & 0 \\ 1 & 0 \end{bmatrix}$	$\begin{bmatrix} 1 & 1 \\ 1 & 0 \end{bmatrix}$	$\begin{bmatrix} 1 & 1 \\ 1 & 1 \end{bmatrix}$
μ_{i1}	0.7322	0.4935	0.4009	0.3247 .

For dependent variations represented by (4), consider the case

$$E = \begin{bmatrix} k_1 & 0 \\ -k_1 & 0 \end{bmatrix}.$$

With $E_1 = \begin{bmatrix} 1 & 0 \\ -1 & 0 \end{bmatrix}$, solution of (35) gives

$$|k_1| < 0.5705 .$$

<u>Example 2</u>: This example illustrates the use of state transformation to get improved bounds. Consider the system matrix

$$A = \begin{bmatrix} 0.0 & 1.0 \\ -0.2 & -0.9 \end{bmatrix}.$$

With Q = I, (16) gives

$$\mu_{u1} = 0.0735.$$

With $M = \begin{bmatrix} 0.99 & -0.30 \\ 0.02 & 0.96 \end{bmatrix}$, (19) and (18) give the bounds

$$\hat{\mu}_{u1} = 0.1195 \quad \text{and} \quad \mu^*_{u1} = 0.0897$$

in the transformed and original coordinate frames, respectively. As $\mu^*_{u1} > \mu_{u1}$, there is an improvement in the bound.

Time-varying perturbation bounds, when applied to the time-invariant case, tend to give conservative results. The following example is intended to emphasize the fact, on the other hand, that a system may become unstable with smaller time-varying perturbations than for time-invariant perturbations [22].

Example 3: Consider the system matrix

$$A = \begin{bmatrix} 0.0 & 1.0 \\ -0.5 & 0.5 \end{bmatrix}$$

and the perturbation matrix

$$E = \begin{bmatrix} 0 & 0 \\ 0 & e \end{bmatrix}.$$

Let e(k) be a periodic time-varying perturbation with period 3 and values e(0) = e(1) = 0.58 and e(2) = -0.58. With initial condition $[1.0\ 1.0]^T$, the initial condition response of this system grows indefinitely (system is unstable) for this time-varying perturbation with maximum modulus variation $| e(k) | = 0.58$. There could be other types of time-varying perturbations for which instability occurs with even smaller modulus variation.

For time-invariant perturbations, this system remains stable for $-2 < e < 1$. The maximum modulus variation (0.58) that causes instability for time-varying perturbation is much less than the allowable time-invariant perturbation.

Example 4: This example illustrates some of the linear time-invariant perturbation bound results. Consider the system matrix

$$A = \begin{bmatrix} 0 & 1 \\ -0.5 & -1.0 \end{bmatrix}.$$

For unstructured perturbations, (47) gives

$$\sigma_{max}(E) < \mu_{u3} = 0.2141.$$

For structured independent perturbations, with $U = \begin{bmatrix} 1 & 1 \\ 1 & 1 \end{bmatrix}$, (50) gives

$$\varepsilon < \mu_{i2} = 0.1096.$$

For dependent variations, consider the case

$$E = \begin{bmatrix} k_1 & k_1 \\ k_1 & k_1 \end{bmatrix}.$$

With $E_1 = \begin{bmatrix} 1 & 1 \\ 1 & 1 \end{bmatrix}$, the solution of (54) gives

$$| k_1 | < 0.2483.$$

<u>Example 5</u>: The application of the interval matrix results is illustrated with this example. Consider the interval matrix [A] with

$$\underline{A} = \begin{bmatrix} -0.50 & 0.0 \\ -0.25 & 0.0 \end{bmatrix} \text{ and } \bar{A} = \begin{bmatrix} 0.50 & 0.60 \\ 0.75 & 0.00 \end{bmatrix}.$$

Then, d = 0.5 and

$$A_0 = \begin{bmatrix} 0.00 & 0.3 \\ 0.25 & 0.0 \end{bmatrix}, \quad D = \begin{bmatrix} 0.5 & 0.3 \\ 0.5 & 0.0 \end{bmatrix}, \quad U = \begin{bmatrix} 1.0 & 0.6 \\ 1.0 & 0.0 \end{bmatrix}.$$

Since (58) gives $d < 0.5027$, the interval matrix [A] is stable.

The application of these stability robustness results to some practical systems is given in Section V.

IV. CONTROL DESIGN FOR ROBUST STABILITY

In Section III, several stability robustness bounds are given on the perturbations so that the nominally stable system remains stable under perturbations. In this section, these results are used in the design of controllers to stabilize uncertain discrete-time systems. There are two principal approaches for this control design problem. In the first approach, existing control design techniques, such as linear quadratic (LQ) regulator theory or pole placement, are used to stabilize the nominal system. Then stability robustness bounds are applied to the resulting system to check its robustness. In the second approach, controllers are directly designed to stabilize the nominal system and maximize the stability robustness bounds. In the following, these two approaches are briefly explained.

A. Stability Robustness of Well-Known Control Design Methods

Consider the linear discrete-time system described by

$$x(k+1) = Ax(k) + Bu(k), \quad x(0) = x_0 \tag{74}$$

$$y(k) = Cx(k), \tag{75}$$

where x is the n-dimensional state vector, u is the m-dimensional control vector and y is the ℓ-dimensional output vector. A, B and C are constant matrices of appropriate dimensions. If the system (74), (75) satisfies the controllability condition, it is possible to design a full-state feedback controller

$$u(k) = Gx(k) \tag{76}$$

so that the closed-loop system matrix

$$A_c = A + BG \tag{77}$$

is stable [3]. The controller matrix G can be obtained using well-known techniques like LQ regulator theory or pole-placement design [3].

Let ΔA, ΔB and ΔG be the perturbation matrices of matrices A, B and G, respectively. Then the perturbed closed-loop system matrix is

$$\begin{aligned} A_{cp} &= (A + \Delta A) + (B + \Delta B)(G + \Delta G) \\ &= (A + BG) + \left[\Delta A + (\Delta B)G + (B + \Delta B)\Delta G\right] \\ &= A_c + E, \end{aligned} \tag{78}$$

where

$$E = \Delta A + (\Delta B)G + B(\Delta G) + (\Delta B)(\Delta G). \tag{79}$$

Let $\overline{\Delta A} = |\,\Delta A\,|_{max}$, etc., and define

$$\tilde{E} = \overline{\Delta A} + (\overline{\Delta B})\,|\,G\,| + |\,B\,|\,(\overline{\Delta G}) + (\overline{\Delta B})(\overline{\Delta G}) \tag{80}$$

and

$$\tilde{\varepsilon} = \max_{i,j} \tilde{e}_{ij} . \tag{81}$$

The structured perturbation bound results of Section III can be directly applied to the perturbed closed-loop system matrix in (78). If the perturbations ΔA, ΔB and/or ΔG are time-varying, Theorems 10, 11 and 12 can be applied. On the other hand, if the perturbations are time-invariant, Theorems 20 and 21 can be used. The following result by Kolla and Farison [82] is based on Theorem 10.

Design Observation 1: The perturbed closed-loop system (78) is stable for all perturbations in A, B and G, in the sense of (23), if

$$\tilde{\varepsilon} < \mu_{i1} \ , \tag{82}$$

where μ_{i1} is given by

$$\mu_{i1} = -\frac{\sigma_{max}\left(U^T \mid PA_c \mid\right)_s}{\sigma_{max}\left(U^T \mid P \mid U\right)} + \sqrt{\left[\frac{\sigma_{max}\left(U^T \mid PA_c \mid\right)_s}{\sigma_{max}\left(U^T \mid P \mid U\right)}\right]^2 + \frac{\sigma_{min}(Q)}{\sigma_{max}\left(U^T \mid P \mid U\right)}} \tag{83}$$

and P satisfies the Lyapunov equation

$$A_c^T PA_c - P + Q = 0 \ . \tag{84}$$

It is clear from (80) and (83) that both $\tilde{\varepsilon}$ and μ_{i1} are functions of the control gain G. The freedom in selecting this gain matrix can be used to help satisfy the stability robustness condition (82). References [15,30-32,58,82-84] use LQ regulator theory, and references [26,35,40,47,52,58,85,86] use the pole-placement design method to obtain robust controllers that satisfy different stability robustness criteria.

Using LQ regulator theory, the gain matrix G in (76) is given by

$$G = -(R + B^T KB)^{-1} B^T KA \; . \tag{85}$$

The symmetric matrix K is the positive definite solution of the algebraic discrete-time Riccati equation

$$K = \tilde{Q} + A^T KA - A^T KB (R + B^T KB)^{-1} B^T KA \; . \tag{86}$$

The matrices $\tilde{Q}$ and R are the weighting matrices in the performance index

$$J = \sum_{k=0}^{\infty} \left[x^T(k)\tilde{Q}\, x(k) + u^T(k)R\, u(k) \right] \tag{87}$$

and can be used as design parameters to get different stabilizing controllers. The following example illustrates the use of this method [82].

<u>Example 6</u>: Consider the linear uncertain discrete-time system

$$x(k+1) = \begin{bmatrix} 0 & 1 \\ a_1 & a_2 \end{bmatrix} x(k) + \begin{bmatrix} 0 \\ 1 \end{bmatrix} u(k) \; ,$$

where

$$-1.5 < a_1(k) < -0.5$$

$$-2.4 < a_2(k) < -1.6 \; .$$

The system can be represented as

$$x(k+1) = (A + \Delta A)\, x(k) + (B + \Delta B)\, u(k) \; ,$$

where $A = \begin{bmatrix} 0 & 1 \\ -1 & -2 \end{bmatrix}$, $B = \begin{bmatrix} 0 \\ 1 \end{bmatrix}$, $\overline{\Delta A} = \begin{bmatrix} 0.0 & 0.0 \\ 0.5 & 0.4 \end{bmatrix}$ and $\overline{\Delta B} = \begin{bmatrix} 0 \\ 0 \end{bmatrix}$.

Assuming $\Delta G = 0$,

$$\tilde{E} = \begin{bmatrix} 0.0 & 0.0 \\ 0.5 & 0.4 \end{bmatrix}.$$

With $\tilde{\varepsilon} = 0.5$,

$$U = \begin{bmatrix} 0.0 & 0.0 \\ 1.0 & 0.8 \end{bmatrix}.$$

For different $\tilde{Q}$ and R matrices in the LQ regulator design, different control gains can be obtained that stabilize the nominal system. For $\tilde{Q} = I$ and $R = 0.01$, (85) gives

$$G = [0.9951 \quad 1.9807] .$$

With this gain matrix, and $Q = I$, (83) gives

$$\mu_{i1} = 0.5383 .$$

As $\tilde{\varepsilon}$ (= 0.5) < μ_{i1} (= 0.5383), the designed gain matrix G stabilizes the uncertain system for all given perturbations.

B. Control Design to Improve Robust Stability

In the following, a design method is presented such that controller determination includes a stability robustness component. The method uses a robustness component based on Lyapunov stability theory [87].

As explained in Section III, Lyapunov stability theory is generally used to obtain robust stability bounds for nonlinear and/or time-varying perturbations. To obtain the largest stability robustness bound, a controller could be designed that maximizes γ_1, γ_2, γ_3 or γ_4 in (12)-(15) for nonlinear perturbations, or μ_{u1} or μ_{u2} in (16) and (21) for linear time-varying perturbations. However, these performance indices are computationally and analytically complex and maximizing them is a formidable task. Hence we modify the performance index such that it becomes more tractable. The bounds γ_4, μ_{u1} and μ_{u2} are maximized by maximizing

$$\frac{1}{\sigma_{max}(P)} . \tag{88}$$

As explained in [87], this can be accomplished by minimizing the Frobenius norm of P ,

$$\| P \|_F^2 = \mathrm{Tr}\,(P^T P) . \tag{89}$$

To get good regulation in addition to stability robustness, (89) can be added to (87). Unlike the full-state feedback used in (76), a reduced-order dynamic compensator is used to illustrate this method. The control design problem can be stated as follows.

For the system (74) and (75), design the controller

$$u(k) = G_{11}\, y(k) + G_{12}\, \eta(k) \tag{90}$$

$$\eta(k+1) = G_{21}\, y(k) + G_{22}\, \eta(k) , \quad \eta(0) = \eta_0 , \tag{91}$$

so that the performance index

$$J = \frac{1}{2}\,\mathrm{Tr}(P^T P) + \sum_{k=0}^{\infty} \left[x^T(k)\, \tilde{Q}\, x(k) + u^T(k)\, R u(k) \right] \tag{92}$$

is minimized.

A solution to this problem can be obtained using the parameter optimization technique as shown by Kolla and Farison [87]. The controller gain matrix

$$G = \begin{bmatrix} G_{11} & G_{12} \\ G_{21} & G_{22} \end{bmatrix} \tag{93}$$

can be obtained by solving the following five coupled, nonlinear, algebraic matrix equations:

$$A_c^T P_1 A_c - P_1 + I = 0 \tag{94}$$

$$A_c^T P_2 A_c - P_2 + Q_c = 0 \tag{95}$$

$$A_c L_1 A_c^T - L_1 + P_1 = 0 \tag{96}$$

$$A_c L_2 A_c^T - L_2 + X_{a0}^T = 0 \tag{97}$$

$$B_a^T (P_1 A_a L_1 + P_2 A_a L_2) C_a^T + B_a^T P_1 B_a G C_a L_1 C_a^T + (B_a^T P_2 B_a + R_a) G C_a L_2 C_a^T = 0 , \tag{98}$$

where

$$A_a = \begin{bmatrix} A & 0 \\ 0 & 0 \end{bmatrix}, \quad B_a = \begin{bmatrix} B & 0 \\ 0 & I \end{bmatrix}, \quad C_a = \begin{bmatrix} C & 0 \\ 0 & I \end{bmatrix}$$

$$Q_a = \begin{bmatrix} \tilde{Q} & 0 \\ 0 & 0 \end{bmatrix}, \quad R_a = \begin{bmatrix} R & 0 \\ 0 & 0 \end{bmatrix} \tag{99}$$

$$A_c = A_a + B_a G C_a, \quad Q_c = Q_a + C_a^T G^T R_a G C_a \tag{100}$$

and X_{a0} can be taken as zI to reflect a random initial state with zero mean and uniformly distributed over a sphere of radius z. The controller design can be specialized to output feedback or state feedback cases by appropriately defining the elements of the G matrix in (93), and A_c and Q_c in (100) [87]. This control design method is illustrated by the following example.

Example 7: Consider the simple scalar (n=1), discrete-time system

$$x(k+1) = 0.5\, x(k) + u(k), \qquad x(0) = 1.0.$$

For this system, we investigate the stability robustness properties of two control designs: 1) the standard LQ regulator that minimizes

$$J_1 = \sum_{k=0}^{\infty} \left[x^2(k) + r\, u^2(k) \right],$$

and 2) the robust state feedback design that minimizes

$$J_2 = \frac{1}{2} p_1^2 + \sum_{k=0}^{\infty} \left[x^2(k) + r\, u^2(k) \right],$$

where p_1 satisfies

$$(0.5 + g)^2 p_1 - p_1 + 1 = 0$$

from (94) and g is the control gain.

For different r values, the bounds obtained from the standard and robust state feedback designs are shown in Fig. 2. It is clear that, for a given r, the robust state feedback design from J_2 gives a larger bound μ_{u1} in (16) than the standard LQ regulator design from J_1.

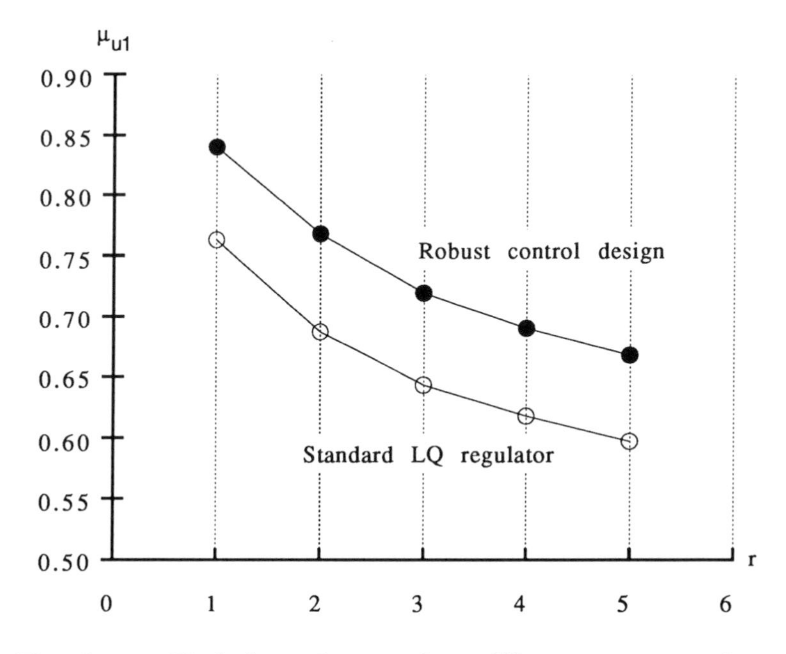

Fig. 2. Variation of μ_{u1} for different values of r.

Apart from the controller design based on stability robustness bounds described in this section, there are several other control design methods for robust stability of discrete-time systems. These include designs based on the parametric space by Ackermann [88] and Hu and Loh [89]; "matching conditions" by Corless and Manela [90], Hollot and Arabacioglu [91], Magana and Zak [92], Bahnasawi, et al. [93], and Farison and Kolla [94]; optimal projection method by Bernstein [95]; and Kharitonov polynomial approach by Soh, et al. [96].

V. PRACTICAL APPLICATIONS

In this section, applications of the preceding stability robustness bound results are given for three different examples of practical systems. The application areas are digital filters [33], economic systems [34] and process control systems [47]. An application to ARMAX models is given by Farison and Kolla [97]. The robust stability analysis and control design techniques of Sections III and IV are applied to interconnected systems by Shen and Kung [35] and Kolla and Farison [98].

The results presented in this Chapter are applicable to naturally discrete-time systems with uncertainties in the discrete-time system model matrices. However, Kolla and Farison [99] indicate that care is required when discretization is used to apply these results to study the stability of systems with uncertainties in the continuous-time model. Such a study should use results similar to those presented by Bernstein and Hollot [100], Ackermann and Hu [101], and Abreu-Garcia and Niu [102].

A. Application to Recursive Digital Filter

The structured perturbation bound (23) is applied to the coefficient truncation of recursive digital filter realizations, for both direct form and normal form realizations, by Farison and Kolla [33].

Consider the second-order Butterworth lowpass filter with cutoff at 1 rad/s, for which

$$H(z) = \frac{0.0373z}{z^2 - 1.7z + 0.745} .$$

In the direct form (canonical) realization, the A matrix is

$$A = \begin{bmatrix} 0 & 1 \\ -0.745 & 1.7 \end{bmatrix} .$$

With Q = I, the structured perturbation bound (23) for different parameter perturbation cases represented by the following U matrices gives

$$\begin{array}{cccc} U & \begin{bmatrix} 0 & 0 \\ 1 & 0 \end{bmatrix} & \begin{bmatrix} 0 & 0 \\ 0 & 1 \end{bmatrix} & \begin{bmatrix} 0 & 0 \\ 1 & 1 \end{bmatrix} \\ & & & \\ \mu_{i1} & 0.0057 & 0.0051 & 0.0033 . \end{array}$$

The third U above consisting of identity elements in the second row could be viewed as a model for coefficient truncation where $\varepsilon = 0.5 \times 10^{-r}$ for truncation to r decimals. Thus, for this example, truncation of the direct form coefficients to three decimals is still guaranteed stable.

In the normal form realization, the A matrix is

$$A = \begin{bmatrix} 0.8500 & 0.1500 \\ -0.1500 & 0.8500 \end{bmatrix} .$$

Again, with Q = I, the structured perturbation bound (23) for different parameter perturbation cases represented by the following U matrices gives

$$U \quad \begin{bmatrix} 0 & 1 \\ 0 & 0 \end{bmatrix} \quad \begin{bmatrix} 1 & 0 \\ 0 & 1 \end{bmatrix} \quad \begin{bmatrix} 1 & 1 \\ 1 & 0 \end{bmatrix} \quad \begin{bmatrix} 1 & 1 \\ 1 & 1 \end{bmatrix}$$

$$\mu_{i1} \quad 0.2087 \quad 0.1203 \quad 0.0754 \quad 0.0601 \; .$$

The fourth U above consisting of all identity elements could be viewed as a model for coefficient truncation for the normal form. Thus, for this example, truncation of the normal form coefficients to one decimal is still guaranteed stable. This is a stronger result than for the direct form.

B. Application to Macroeconomic System

The structured perturbation bounds (23), (24) are applied by Kolla and Farison [34] to a fourth-order macroeconomic system model by Runyan [103] and Perkins, et al. [104]. This linear discrete-time model combines an inventory sector and a capital goods sector. The closed-loop system matrix with state feedback is

$$A = \begin{bmatrix} 0.3299 & 0.7354 & -2.7947 & -2.9267 \\ 0.4255 & 0.5508 & 1.0163 & 1.0643 \\ 0 & 0 & 1 & 1.0000 \\ 0.1915 & 0.1498 & -0.4638 & -0.4797 \end{bmatrix} .$$

To investigate the stability robustness properties of this system under elemental variations, consider the matrix

$$U = \begin{bmatrix} 1 & 1 & 1 & 1 \\ 1 & 1 & 1 & 1 \\ 0 & 0 & 0 & 1 \\ 1 & 1 & 1 & 1 \end{bmatrix} .$$

With Q = I, (23) gives

$$\mu_{i1} = 0.0085 \; .$$

With the diagonal transformation matrix

$$T = \text{diag}\,[\ 1.1,\ \ 1.0,\ \ 0.5,\ \ 0.4\ ]\ ,$$

(24) gives the improved bound

$$\mu_{i1}^{*} = 0.0167\ .$$

This means the closed-loop system remains stable for all elemental variations that satisfy the perturbation matrix condition

$$|\,E(k)\,| \ < \ \mu_{i1}^{*}\, U\ .$$

C. Application to Process Control System

The unstructured perturbation bound (49) is applied by Qiu and Davison [47] to a seventh-order process control system of a heated rod modeled by Leden [105]. The closed-loop system matrix with dead-beat controller [47] is

$$A = \begin{bmatrix} -0.1373 & 0.2139 & -0.2831 & 0.2792 & -0.2177 & -0.1298 & 0.0666 \\ 0.0002 & -0.0163 & -0.0438 & -0.0657 & -0.0669 & -0.0473 & -0.0275 \\ 0.0469 & 0.0718 & 0.0896 & 0.0782 & 0.0493 & 0.0224 & 0.0074 \\ 0.0373 & 0.0712 & 0.1124 & 0.1292 & 0.1124 & 0.0712 & 0.0373 \\ 0.0074 & 0.0224 & 0.0498 & 0.0782 & 0.0896 & 0.0718 & 0.0469 \\ -0.0275 & -0.0473 & -0.0669 & 0.0657 & -0.0438 & -0.0163 & 0.0002 \\ -0.0666 & -0.1298 & -0.2177 & -0.2792 & -0.2831 & -0.2139 & -0.1373 \end{bmatrix} .$$

For this system matrix, the unstructured perturbation bound μ_{u4} in (49) is

$$\mu_{u4} = 0.4451\ .$$

This means that the closed-loop system remains stable under any perturbation matrix E that satisfies the condition

$$\sigma_{max}(E) < \mu_{u4} .$$

VI. CONCLUSION

A. Review

This chapter describes stability robustness bound techniques for discrete-time systems. Several of the existing methods are combined into a unified treatment and a brief summary of these results is given. The perturbations are classified into different categories, and stability robustness bounds for each type are described. Controller design methods based on these bounds are explained. These analysis and design results are then applied to some practical systems.

B. Areas for Future Research

Several of the bounds presented here give sufficient conditions for robust stability. In the present literature, the superiority of one bound over another is shown through examples. A future research direction could be to establish theoretical results showing which of the bounds is better for what types of systems.

There are necessary and sufficient conditions for certain classes of systems. It is not an easy task to obtain such conditions for general systems. However, future research could be directed to relaxing the restrictions on the type of systems for which these conditions apply.

Khammash and Pearson [106] have applied frequency-domain stability robustness results to performance robustness. Future research could explore the possibility of a similar application of the time-domain results of this chapter to performance robustness.

There seems to be no control design methods that can stabilize a discrete-time system for arbitrarily large perturbations using a linear controller, unlike the continuous-time case, as noted by Farison and Kolla [94]. Future research could look into the type of systems for which a linear controller can be designed. Jury [21] suggests several other new research directions in discrete-time system robust control.

Though discrete-time robust stability results have been applied to some practical systems, attention could be given to identifying other types of physical systems where these discrete-time system results can be applied. Finally, future research could also be directed to the sampled-data systems area to study robust stability for uncertainties in a continuous-time plant with a discrete-time controller [100-102].

VII. REFERENCES

1. P. Dorato (ed.), "Robust Control," IEEE Press, New York, 1987.

2. P. Dorato and R. K. Yedavalli (eds.), "Recent Advances in Robust Control," IEEE Press, New York, 1990.

3. K. Ogata, "Discrete-Time Control Systems," Prentice Hall, Englewood Cliffs, NJ, 1987.

4. C. Barratt and S. Boyd, "Examples of Exact Trade-Offs in Linear Controller Design," *IEEE Control Systems Magazine* **9**, 46-52 (1989).

5. U. Shaked, "Guaranteed Stability Margins for the Discrete-Time Linear Quadratic Regulators," *IEEE Transactions on Automatic Control* **AC-31**, 161-165 (1986).

6. T. Ishihara and H. Takeda, "Loop Transfer Recovery Techniques for Discrete-Time Optimal Regulators Using Prediction Estimators," *IEEE Transactions on Automatic Control* **AC-31**, 1149-1151 (1986).

7. S. P. Bhattacharyya, "Robust Stabilization Against Structured Perturbations," Springer-Verlag, Berlin, Heidelberg, 1987.

8. D. D. Siljak, "Parametric Space Methods for Robust Control Design," *IEEE Transactions on Automatic Control* **34**, 674-689 (1989).

9. M. P. Polis, A. W. Olbrot, and M. Fu, "An Overview of Recent Results on the Parametric Approach to Robust Stability," *Proceedings of 28th Conference on Decision and Control*, Tampa, FL, 23-29 (1989).

10. R. V. Patel and M. Toda, "Quantitative Measures of Robustness for Multivariable Systems," *Proceedings of Joint Automatic Control Conference*, San Francisco, CA, TP8-A (1980).

11. R. K. Yedavalli, "Perturbation Bounds for Robust Stability in Linear State Space Models," *International Journal of Control* **42**, 1507-1517 (1985).

12. K. Zhou and P. P. Khargonekar, "Stability Robustness Bounds for Linear State Space Models with Structured Uncertainty," *IEEE Transactions on Automatic Control* **AC-32**, 621-623 (1987).

13. M. E. Sezer and D. D. Siljak, "Robust Stability of Discrete Systems," *International Journal of Control* **48**, 2055-2063 (1988).

14. E. Yaz and X. Niu, "Stability Robustness of Linear Discrete-Time Systems in the Presence of Uncertainty," *International Journal of Control* **50**, 173-182 (1989).

15. H. Bourles, Y. Joannic, and O. Mercier, "ρ-Stability and Robustness: Discrete-Time Case," *International Journal of Control* **52**, 1217-1239 (1990).

16. S. R. Kolla, "Analysis and Design of Controllers for Robust Stability of Linear Multivariable Discrete-Time Systems," Ph.D. Dissertation, The University of Toledo, OH (1989).

17. E. Yaz, "Deterministic and Stochastic Robustness Measures for Discrete Systems," *IEEE Transactions on Automatic Control* **33**, 952-955 (1988).

18. X. Niu, J. A. De Abreu-Garcia, T. T. Hartley, and E. Yaz, "Robustness Measures for Discrete-Time Systems with Deterministic and Stochastic Perturbations," *Proceedings of 33rd Midwest Symposium on Circuits and Systems*, Calgary, Canada, 1168-1170 (1990).

19. E. Yaz and X. Niu, "New Results on the Robustness of Discrete-Time Systems with Stochastic Perturbations," *Proceedings of American Control Conference*, Boston, MA, 2698-2699 (1991).

20. E. Yaz, "Robustness of Discrete-Time Systems for Unstructured Stochastic Perturbations," *IEEE Transactions on Automatic Control* **36**, 867-869 (1991).

21. E. I. Jury, "Robustness of a Discrete System," *Automation and Remote Control* **51**, 571-592 (1990).

22. S. R. Kolla, R. K. Yedavalli, and J. B. Farison, "Robust Stability Bounds on Time-Varying Perturbations for State-Space Models of Linear Discrete-Time Systems," *International Journal of Control* **50**, 151-159 (1989).

23. V. L. Kharitonov, "Asymptotic Stability for an Equilibrium Position of a Family of Linear Differential Equations," *Differential'ne Uravneniya* **14**, 2086-2088 (1978).

24. R. E. Kalman and J. E. Bertram, "Control System Analysis and Design via the 'Second Method' of Lyapunov: II Discrete-Time Systems," *Journal of Basic Engineering, Transactions ASME* **82**, 394-400 (1960).

25. J. M. Martin, "State-Space Measures for Stability Robustness," *IEEE Transactions on Automatic Control* **AC-32**, 509-512 (1987).

26. Y.-T. Juang, T.-S. Kuo, and C.-F. Hsu, "Stability Robustness Analysis of Digital Control Systems in State Space Models," *International Journal of Control* **46**, 1547-1556 (1987).

27. Z. Qu and J. Dorsey, "Stability Robustness of Discrete Systems with Perturbations in State Equation," *Proceedings of American Control Conference*, San Diego, CA, 3054-3057 (1990).

28. J. Chegancas and C. Burgat, "Polyhedral Cones Associated to M-Matrices and Stability of Time-Varying Discrete-Time Systems,"

Journal of Mathematical Analysis and Applications **118**, 88-96 (1986).

29. T.-T. Lee and S.-H. Lee, "Discrete Optimal Control with Eigenvalues Assigned Inside a Circular Region," *IEEE Transactions on Automatic Control* **AC-31**, 958-962 (1986).

30. S.-C. Tsay, I.-K. Fong, T.-S. Kuo, and C.-F. Hsu, "Comments on 'Discrete Optimal Control with Eigenvalues Assigned Inside a Circular Region'," *IEEE Transactions on Automatic Control* **34**, 479-480 (1989).

31. J.-C. Shen and F.-C. Kung, "Robust Stability Bounds for Discrete Optimal Quadratic Control Design," *Control Theory and Advanced Technology* **6**, 273-282 (1990).

32. T. Ishihara, "Robust Stability Bounds for a Class of Discrete-Time Regulators with Computation Delays," *Automatica* **24**, 697-700 (1988).

33. J. B. Farison and S. R. Kolla, "Relationship of Singular Value Stability Robustness Bounds to Spectral Radius for Discrete Systems with Application to Digital Filters," *IEE Proceedings-G* **138**, 5-8 (1991).

34. S. R. Kolla and J. B. Farison, "Improved Stability Robustness Bounds Using State Transformation for Linear Discrete Systems," *Automatica* **26**, 933-935 (1990).

35. J.-C. Shen and F.-C. Kung, "Designing Robust Discrete-Time Systems by Pole Assignment," *Control Theory and Advanced Technology* **6**, 669-681 (1990).

36. G. Zhu and R. E. Skelton, "Robustness of Covariance Controllers for Discrete Systems," *Proceedings of IEEE International Conference on Systems Engineering*, Dayton, OH, 585-587 (1989).

37. C. Hsieh and R. E. Skelton, "All Covariance Controllers for Linear Discrete-Time Systems," *IEEE Transactions on Automatic Control* **35**, 908-915 (1990).

38. M. Eslami, "Computer-Aided Determination of Stability Robustness Measures of Linear Discrete-Time Systems, *Automatica* **26**, 623-627 (1990).

39. I.-K. Fong, "Analysis and Synthesis of Linear Uncertain Control Systems," Ph.D. Dissertation, National Taiwan University, Taipei (1985).

40. Q.-L. Han and Q.-B. Wu, "Stability Robustness Analysis of Discrete Linear Systems," *International Journal of Systems Science* **22**, 165-172 (1991).

41. X. Niu and J. A. De Abreu-Garcia, "Some Discrete-Time Counterparts to Continuous-Time Stability Robustness Bounds," *Proceedings of American Control Conference*, Boston, MA, 1947-1948 (1991).

42. Z. Gao and P. Antsaklis, "New Bounds on Parameter Uncertainties for Robust Stability," *Proceedings of American Control Conference*, Boston, MA, 879-880 (1991).

43. X. Gu and W. Chen, "Robust Stability Analysis for State Space Models of Discrete-Time Systems," *Proceedings of American Control Conference*, Boston, MA, 890-891 (1991).

44. K. Gu, "Quadratic Stability Bounds of Discrete-Time Uncertain Systems," *Proceedings of American Control Conference*, Boston, MA 1951-1955 (1991).

45. T. Mori, "On the Relationship Between the Spectral Radius and Stability Radius for Discrete Systems," *IEEE Transactions on Automatic Control* **35**, 835 (1990).

46. D. Hinrichsen and A. J. Pritchard, "New Robustness Results for Linear Systems Under Real Perturbations," *Proceedings of 27th Conference on Decision and Control*, Austin, TX, 1375-1379 (1988).

47. L. Qiu and E. J. Davison, "A New Method for the Stability Robustness Determination of State Space Models with Real Perturbations," *Proceedings of 27th Conference on Decision and Control*, Austin, TX, 538-543 (1988).

48. D. C. Hyland and E. G. Collins, Jr., "Some Majorant Robustness Results for Discrete-Time Systems," *Automatica* **27**, 167-172 (1991).

49. A. Rachid, "Robustness of Discrete Systems Under Structured Uncertainties," *International Journal of Control* **50**, 1563-1566 (1989).

50. S. R. Kolla and I. S. Das, "Stability Robustness Bounds for Linear Discrete Systems with Dependent Uncertainty," *Proceedings of 22nd Annual Pittsburgh Conference on Modeling and Simulation*, Pittsburgh, PA, (1991).

51. B. R. Barmish, M. Fu, and S. Saleh, "Stability of a Polytope of Matrices: Counterexamples," *IEEE Transactions on Automatic Control* **33**, 569-572 (1988).

52. Y.-T. Juang, T.-S. Kuo, and S.-L. Tung, "Stability Analysis of Continuous and Discrete Interval Systems," *Control Theory and Advanced Technology* **6**, 221-235 (1990).

53. M. B. Argoun, "On Sufficient Conditions for the Stability of Interval Matrices," *International Journal of Control* **44**, 1245-1250 (1986).

54. T. Mori and H. Kokame, "Convergence Property of Interval Matrices and Interval Polynomials," *International Journal of Control* **45**, 481-484 (1987).

55. M. Mansour, "Sufficient Conditions for the Asymptotic Stability of Interval Matrices," *International Journal of Control* **47**, 1973-1974 (1988).

56. C. V. Hollot and A. C. Bartlett, "On the Eigenvalues of Interval Matrices," *Proceedings of 26th Conference on Decision and Control*, Los Angeles, CA, 794-799 (1987).

57. C.-S. Zhou and J.-L. Deng, "Stability Analysis of Gray Discrete-Time Systems," *IEEE Transactions on Automatic Control* **34**, 173-175 (1989).

58. S.-H. Lin, Y.-T. Juang, I.-K. Fong, C.-F. Hsu, and T.-S. Kuo, "Dynamic Interval Systems Analysis and Design, *International Journal of Control* **48**, 1807-1818 (1988).

59. B. Shafai, K. Perev, D. Cowley, and Y. Chehab, "A Necessary and Sufficient Condition for the Stability of Nonnegative Interval Discrete Systems," *IEEE Transactions on Automatic Control* **36**, 742-746 (1991).

60. C.-L. Jiang, "Sufficient and Necessary Condition for the Asymptotic Stability of Discrete Linear Interval Systems," *International Journal of Control* **47**, 1563-1565 (1988).

61. S. R. Kolla and J. B. Farison, "Counter-Examples to 'Sufficient and Necessary Condition for the Asymptotic Stability of Discrete Linear Interval Systems'," *International Journal of Control* **48**, 1751-1752 (1988).

62. J. Cieslik, "On the Possibility of the Extension of Kharitonov's Stability Criterion for Interval Polynomials to the Discrete-

Time Case," *IEEE Transactions on Automatic Control* **AC-32**, 237-239 (1987).

63. C. V. Hollot and A. C. Bartlett, "Some Discrete-Time Counterparts to Kharitonov's Stability Criterion for Uncertain Systems," *IEEE Transactions on Automatic Control* **AC-31**, 355-357 (1986).

64. K. S. Yeung and S. S. Wang, "Linear Third-Order Discrete System Stability Under Parameter Variations," *Electronics Letters* **23**, 266-267 (1987).

65. N. K. Bose and E. Zeheb, "Kharitonov's Theorem and Stability Test of Multidimensional Digital Filters," *IEE Proceedings-G* **133**, 187-190 (1986).

66. H. Lin, C. V. Hollot, and A. C. Bartlett, "Stability of Families of Polynomials: Geometric Considerations in Coefficient Space," *International Journal of Control* **45**, 649-660 (1987).

67. N. K. Bose, "A System Theoretic Approach to Stability of Sets of Polynomials," *Contemporary Mathematics* **47**, 25-34 (1985).

68. F. J. Kraus, B. D. O. Anderson, and M. Mansour, "Robust Schur Polynomial Stability and Kharitonov's Theorem," *International Journal of Control* **47**, 1213-1225 (1988).

69. M. Benidir and B. Picinbono, "Comparison Between Some Stability Criteria of Discrete-Time Filters," *IEEE Transactions on Acoustics, Speech and Signal Processing* **ASSP-36**, 993-1001 (1988).

70. M. Mansour, F. J. Kraus, and B. D. O. Anderson, "Strong Kharitonov Theorem for Discrete Systems," *in* Robustness in Identification and Control (M. Milanse, R. Tempo and A. Vicino, eds.), Plenum Press, New York, 1989.

71. A. C. Bartlett, C. V. Hollot, and H. Lin, "Root Location of an Entire Polytope of Polynomials: It Suffices to Check the Edges," *Mathematics of Control, Signals and Systems* **1**, 61-71 (1987).

72. B. R. Barmish and C. L. DeMarco, "Criteria for Robust Stability of Systems with Structured Uncertainty: A Perspective," *Proceedings of American Control Conference*, Minneapolis, MN, 476-481 (1987).

73. J. E. Ackermann and B. R. Barmish, "Robust Schur Stability of a Polytope of Polynomials," *IEEE Transactions on Automatic Control* **33**, 984-986 (1988).

74. A. Katbab and E. I. Jury, "Robust Schur-Stability of Control Systems with Interval Plants," *International Journal of Control* **51**, 1343-1352 (1990).

75. H. Chapellat, S. P. Bhattacharyya, and M. Dahleb, "Robust Stability of a Family of Disk Polynomials," *International Journal of Control* **51**, 1353-1362 (1990).

76. J. Peterson and L. R. Pujara, "Some Robust Stability Theorems for Polygons of Discrete Polynomials," *Proceedings of IEEE International Conference on Systems Engineering*, Dayton, OH, 288-290 (1991).

77. K. S. Yeung, "Linear Discrete-Time System Stability Under Parameter Variations," *International Journal of Control* **40**, 855-862 (1984).

78. C. B. Soh, C. S. Berger, and K. P. Dabke, "On the Stability Properties of Polynomials with Perturbed Coefficients," *IEEE Transactions on Automatic Control* **AC-30**, 1033-1036 (1985).

79. T. Mori and H. Kokame, "A Necessary and Sufficient Condition for Stability of Linear Discrete Systems with Parameter Variations," *Journal of Franklin Institute* **321**, 135-138 (1986).

80. A. Vicino, "Some Results on Robust Stability of Discrete-Time Systems," *IEEE Transactions on Automatic Control* **33**, 844-847 (1988).

81. Y. C. Soh and R. J. Evans, "Stability Analysis of Interval Matrices - Continuous and Discrete Systems," *International Journal of Control* **47**, 25-32 (1988).

82. S. R. Kolla and J. B. Farison, "Time-Domain Control Design for Robust Stability of Discrete-Time Linear Regulators," *Proceedings of 26th Annual Allerton Conference on Communication, Control and Computing*, Monticello, IL, 893-902 (1988).

83. J. B. Farison and S. R. Kolla, "Robust Stability Bounds for Discrete System Reduced-Order Dynamic Compensator Designs," *Proceedings of IEEE International Conference on Systems Engineering*, Dayton, OH, 99-102 (1989).

84. S. R. Kolla and J. B. Farison, "Improved Robust Stability Bounds for Discrete-Time Linear Regulators with Computational Delays," *Automatica* **26**, 619-621 (1990).

85. J.-H. Chou, "Pole-Assignment Robustness in a Specified Disk," *Systems and Control Letters* **16**, 41-44 (1991).

86. Y.-T. Juang, "A Fundamental Multivariable Robustness Theorem for Robust Eigenvalue Assignment," *IEEE Transactions on Automatic Control* **33**, 940-941 (1988).

87. S. R. Kolla and J. B. Farison, "Reduced-Order Dynamic Compensator Design for Stability Robustness of Linear Discrete-Time Systems," *IEEE Transactions on Automatic Control* **36** 1077-1081 (1991).

88. J. Ackermann, "Parameter Space Design of Robust Control Systems," *IEEE Transactions on Automatic Control* **AC-25**, 1058-1072 (1980).

89. H. X. Hu and N. K. Loh, "Design of Robust Parametric LQ Control for Systems with Different Operating Conditions," *Proceedings of IEEE International Conference on Systems Engineering*, Dayton, OH, 115-118 (1989).

90. M. Corless and J. Manela, "Control of Uncertain Discrete-Time Systems," *Proceedings of American Control Conference*, Seattle, WA, 515-520 (1986).

91. C. V. Hollot and M. Arabacioglu, "ℓ^{th}-Step Lyapunov Min-Max Controllers: Stabilizing Discrete-Time Systems Under Real Parameter Variations," *Proceedings of American Control Conference*, Minneapolis, MN, 496-501 (1987).

92. M. E. Magana and S. H. Zak, "Robust State Feedback Stabilization of Discrete-Time Uncertain Dynamical Systems," *IEEE Transactions on Automatic Control* **33**, 887-891 (1988).

93. A. A. Bahnasawi, A. S. Al-Fuhaid, and M. S. Mahmoud, "Linear Feedback Approach to the Stabilization of Uncertain Discrete Systems," *IEE Proceedings-D* **136**, 47-52 (1989).

94. J. B. Farison and S. R. Kolla, "Comments on 'Asymptotic Stability for a Class of Linear Discrete Systems with Bounded Uncertainties'," *IEEE Transactions on Automatic Control* **35**, 382-384 (1990).

95. D. S. Bernstein, L. D. Davis, and D. C. Hyland, "The Optimal Projection Equations for Reduced-Order Discrete-Time Modeling, Estimation and Control," *AIAA Journal of Guidance Control and Dynamics* **9**, 288-293 (1986).

96. Y. C. Soh, R. J. Evans, I. R. Peterson, and R. E. Betz, "Robust Pole Assignment," *Automatica* **23**, 601-610 (1987).

97. J. B. Farison and S. R. Kolla, "Robust LQG Control for ARMAX Models," *Proceedings of 20th Annual Pittsburgh Conference on Modeling and Simulation*, Pittsburgh, PA, 2229-2233 (1989).

98. S. R. Kolla and J. B. Farison, "Analysis and Design of Controllers for Robust Stability of Interconnected Discrete Systems, *Proceedings of American Control Conference*, Boston, MA, 881-885 (1991).

99. S. R. Kolla and J. B. Farison, "Comments on 'Uncertain Discrete Systems: Uniform Ultimate Bounded Stabilizations'," *International Journal of Systems Science* **22**, 621-623 (1991).

100. D. S. Bernstein and C. V. Hollot, "Robust Stability for Sampled-Data Control Systems," *Systems and Control Letters* **13**, 217-226 (1989).

101. J. E. Ackermann and H. Z. Hu, "Robustness of Sampled-Data Control Systems with Uncertain Physical Plant Parameters," *Automatica* **27**, 705-710 (1991).

102. J. A. De Abreu-Garcia and X. Niu, "Stability Robustness of p-Step Matrix Integrators with Uncertainty in the System Model," *Proceedings of American Control Conference*, Boston, MA, 1959-1960 (1991).

103. H. M. Runyan, "Cybernetics of Economic Systems," *IEEE Transactions on Systems, Man, and Cybernetics* **SMC-1**, 8-18 (1971).

104. W. R. Perkins, J. B. Cruz, and N. Sundararajan, "Feedback Control of a Macroeconomic System Using an Observer," *IEEE Transactions on Systems, Man and Cybernetics* **SMC-2**, 275-278 (1972).

105. B. Leden, "Multivariable Dead-Beat Control," *Automatica* **13**, 185-188 (1977).

106. M. Khammash and J. B. Pearson, Jr., "Performance Robustness of Discrete-Time Systems with Structured Uncertainty," *IEEE Transactions on Automatic Control* **36**, 398-412 (1991).

INDEX

D

G

H

I

J

K

L

M

T

W

Y